W9-BEA-106

STRUCTURAL ANALYSIS

STRUCTURAL ANALYSIS

Second Edition

Harold I. Laursen

Professor of Civil Engineering
Oregon State University

McGraw-Hill Book Company

New York St. Louis San Francisco Auckland Bogotá Düsseldorf
Johannesburg London Madrid Mexico Montreal New Delhi Panama
Paris São Paulo Singapore Sydney Tokyo Toronto

STRUCTURAL ANALYSIS

Copyright © 1978, 1969 by McGraw-Hill, Inc. All rights reserved.
Printed in the United States of America. No part of this publication
may be reproduced, stored in a retrieval system, or transmitted, in any
form or by any means, electronic, mechanical, photocopying, recording, or
otherwise, without the prior written permission of the publisher.

1 2 3 4 5 6 7 8 9 0 DODO 7 8 3 2 1 0 9 8 7

This book was set in Modern 8A by Bi-Comp, Incorporated.
The editors were B. J. Clark and Madelaine Eichberg;
the cover was done by Ruth Reis;
the production supervisor was Dominick Petrellese.
New drawings were done by J & R Services, Inc.
R. R. Donnelley & Sons Company was printer and binder.

Library of Congress Cataloging in Publication Data

Laursen, Harold I.
Structural analysis.

Includes index.
1. Structures, Theory—Matrix methods.
I. Title.
TA642.L33 1978 624'.171 77-21575
ISBN 0-07-036643-8

To Karla, Tod, and Krista

CONTENTS

PREFACE xi

1 INTRODUCTION 1

1-1 Types of Structures 1
1-2 Analysis and Design 4
1-3 Loads 4
1-4 Structural Components 15
1-5 Structure Representation 17
1-6 Supports and Connections 19

2 BASIC STATICS 22

2-1 Free-Body Diagrams 22
2-2 Equations of Equilibrium 26
2-3 Sign Conventions 27
2-4 Shear and Moment Diagrams 30
2-5 Relationships of Load, Shear, and Bending Moment 36
2-6 Principle of Superposition 38
2-7 Graphic Statics 39

3 PLANE TRUSSES 52

3-1 Introduction 52
3-2 Truss Notation 54
3-3 Method of Joints 54
3-4 Method of Sections 58
3-5 Geometric Stability and Static Determinacy of Trusses 61
3-6 Graphical Analysis of Trusses 65

4 SPACE TRUSSES 77

4-1 Equilibrium Equations and Static Determinacy 77
4-2 Analysis of Space Trusses 80

5 ARCHES AND CABLE STRUCTURES 92

5-1 Arch Characteristics 92
5-2 Analysis of Three-hinged Arches 94
5-3 Graphical Analysis of Three-hinged Arches 96
5-4 Cable Characteristics 100

6 APPROXIMATE METHODS OF ANALYSIS 111

6-1 Frame Characteristics 111
6-2 Approximate Analyses 113
6-3 Portal Method 119
6-4 Cantilever Method 123

7 DEFLECTIONS: GEOMETRIC METHODS 132

7-1 Deflected Shapes 132
7-2 Moment-Area Method 134
7-3 Conjugate-Beam Method 147
7-4 Newmark's Method 152
7-5 Williot-Mohr Diagrams 161

8 DEFLECTIONS: ENERGY METHODS 173

8-1 Introduction 173
8-2 Real Work 174
8-3 Forms of Internal Work 177
8-4 Deflections by Real Work 186
8-5 Virtual Work 189
8-6 Forms of Internal Virtual Work 192
8-7 Deflections by Virtual Work 197
8-8 Castigliano's Theorem 210

9 INFLUENCE COEFFICIENTS 223

9-1 Introduction 223
9-2 Flexibility Influence Coefficients 224
9-3 Stiffness Influence Coefficients 231
9-4 Member-Stiffness Matrices 235

10 INFLUENCE LINES 249

10-1 Development of Influence Lines 249
10-2 Influence Lines from Deflected Shapes 259

10-3 Uses of Influence Lines 266
10-4 Absolute Maximum Shear and Moment 276

11 METHOD OF CONSISTENT DISPLACEMENTS 284
11-1 Basic Concepts of the Method 284
11-2 Applications of the Method 290
11-3 Three-Moment Equation 298
11-4 Evaluation of Fixed-End Moments 304

12 SLOPE-DEFLECTION METHOD 315
12-1 Basic Concepts of the Method 315
12-2 Analysis of Continuous Beams 317
12-3 Analysis of Frames 319

13 MOMENT-DISTRIBUTION METHOD 330
13-1 Terminology 330
13-2 Development of the Method 334
13-3 Analysis of Frames 340
13-4 Members with Variable Moment of Inertia 351
13-5 Matrix Formulation of the Moment-Distribution Method 353

14 MATRIX DISPLACEMENT (STIFFNESS) METHOD 362
14-1 Introduction 362
14-2 Element-Stiffness Matrices 364
14-3 The Concept of Assembling Element Stiffnesses 365
14-4 Transformation of Element-Stiffness Matrices 368
14-5 General Procedures for Developing the **K** Matrix 376
14-6 Distributed and Intermediate Loads 383
14-7 Support (Boundary) Conditions 386
14-8 Banding of the **K** Matrix 387
14-9 Development of the Displacement Method Using Virtual Work 389
14-10 Temperature and Lack-of-Fit Analyses 391
14-11 Evaluation of Influence Lines 392
14-12 An Analogous Approach—the Matrix Force Method 394

15 AN INTRODUCTION TO THE FINITE-ELEMENT METHOD 407
15-1 The Basic Concept 407
15-2 Development of a Plane-Stress Element 409
15-3 Application of the Method 413

A MATRIX ALGEBRA 416

A-1 Matrix Definition 416
A-2 Types of Matrices 417
A-3 Matrix Algebra 418
A-4 Determinants 424
A-5 Solution of Linear Equations 426
A-6 Matrix Inversion 429
A-7 Matrix Integration and Differentiation 432

B COMPUTER PROGRAMS 436

B-1 Solution of Simultaneous Equations and Matrix Inversion 436
B-2 Moment-Distribution Method 440
B-3 Frame Analysis Using the Code-Number Technique 444

C CONVERSIONS BETWEEN SI AND U.S. CUSTOMARY UNITS 449

ANSWERS TO EVEN-NUMBERED PROBLEMS 454

INDEX 463

PREFACE

This second edition presents the principles and applications of structural analysis as a blend of classical concepts and current computer techniques. The text begins with the basic principles of structural analysis and builds up to the treatment of complex statically indeterminate structures. The text culminates with a detailed presentation of the matrix-displacement (stiffness) method and a brief introduction to the finite-element method. The classical concepts of analysis that are believed to be valuable for a broad understanding of structural analysis have been retained from the first edition, and in some cases reinforced.

In recognition of the trend to the use of the International System of units (SI units) and the necessity of having to work with design manuals based on the U.S. Customary System of units, examples and problems are presented in both systems. Guidelines for conversion of units between the two systems are presented in Appendix C.

Prerequisites for the text material are courses commonly required in a lower-division engineering curriculum: elementary mechanics, strength of materials, and calculus. The text is intended to be used primarily at the undergraduate level, but the more advanced topics are also applicable to graduate courses.

There is strong emphasis throughout the text on such fundamentals of structural analysis as the use of free-body diagrams, equilibrium equations, shear and moment diagrams, and appropriate sign conventions for displacement and force. With regard to energy methods of analysis, the concepts of real and virtual work are fully developed prior to their use. Matrix notation is used throughout the text. For those requiring an introduction to matrix algebra, the fundamentals of matrix algebra are presented in Appendix A. Fortran IV computer programs for the solution of simultaneous equations, the moment-distribution method, and the matrix displacement (stiffness) method for frame analysis are contained in Appendix B.

Illustrative examples are used throughout, and a variety of problems is included at the ends of the chapters. As mentioned above, SI and U.S. Customary units are used in the examples and problems. The answers to even-numbered problems are given at the end of the text.

The author is grateful to Dr. James M. Gere at Stanford University and Professor James D. Kriegh at The University of Arizona, for their helpful comments upon reading the manuscript.

Harold I. Laursen

CHAPTER

ONE

INTRODUCTION

1-1 TYPES OF STRUCTURES

As a prelude to the analysis of structures we shall discuss the various types of structures encountered in structural work. It should be kept in mind that the use of a particular type of structure is contingent on such factors as the intended purpose of the structure, esthetic considerations, material available, and cost. The purpose of this text is to provide a thorough background in the methods that are available to perform the analyses of the various structures.

One of the simplest of all structures is the simply supported beam, a one-span beam supported on a roller at one end and a pin at the other. These beams have been discussed in earlier courses on statics and strength of materials and need no introduction. It will be recalled from strength-of-materials courses that such a beam supports its load by shear and moment in the beam. A resistive moment is generated in the beam, causing a compressive force in the top fibers of the beam and a tensile force in the lower fibers for a load directed downward. More complicated forms of beams from the standpoint of analysis are those with fixed ends and those that are continuous over supports. Both require additional con-

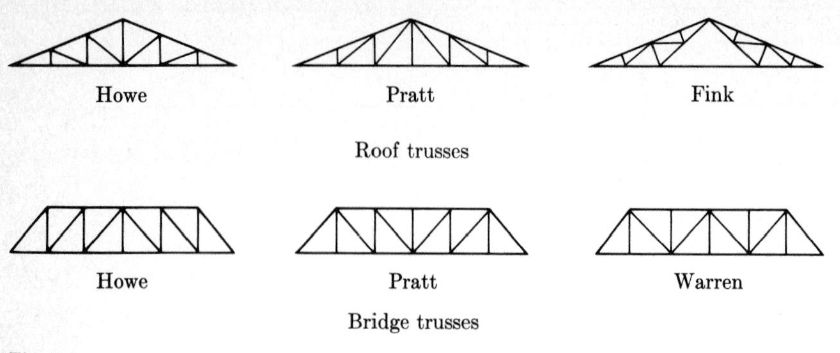

Fig. 1.1

siderations for analysis because they are *statically indeterminate;* that is, they cannot be analyzed solely by equations of statics.

For long spans truss structures are often used. Instead of resisting the loading by shear and moment, as beams do, trusses support the loads primarily by axial forces in their members. The supporting action of a truss can be likened to that of a beam. For vertical loading on a simply supported truss the upper members are subjected to a concentrated compressive force and the lower members subjected to a concentrated tensile force. Under the same conditions for a beam the upper and lower fibers are subjected, respectively, to compressive and tensile stresses. Various triangular arrangements of members are used in trusses. Some of the more conventional roof and bridge trusses are shown in Fig. 1-1. The trusses shown are referred to as *planar trusses;* that is, they lie in one plane. Three-dimensional trusses are sometimes used; these are referred to as *space trusses.*

For long spans another type of structure is used; this is an *arch,* as shown in Fig. 1-2*a*. From a structural standpoint an arch is characterized by relatively low bending moment and high thrust (compressive force), which result from a combination of the shape of the structure and supports capable of developing horizontal resistance. The action of an arch can be described as nearly opposite to the action of a cable, as shown in Fig. 1-2*b*.

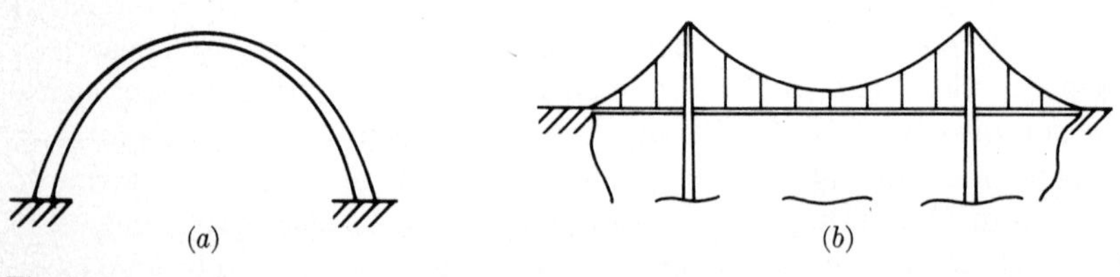

Fig. 1.2

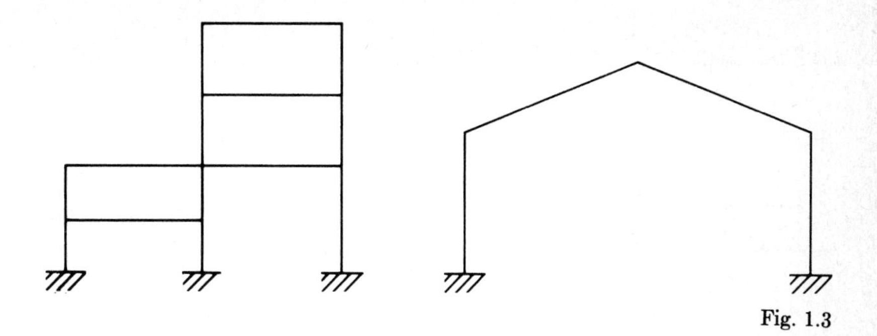

Fig. 1.3

A common type of structure used in buildings is the *frame*. Typical frames are shown in Fig. 1-3. Frames are characterized by the fact that they are assembled from individual moment-resisting members, with moment-resisting connections between some or all of the members. The resulting structure is rigid and, from an analytic standpoint, often highly statically indeterminate. As in the case of trusses, frames can be three dimensional. Owing to the complications of three-dimensional analysis, however, frames are generally treated by two planar analyses.

These structures have been described in terms of their primary members. In addition to the primary members, a structure usually contains secondary members which intermediately support floors, walls, roofs, and other surfaces.

In addition to being assembled from discrete straight elements, structures known as *shells* can be made up of continuous surfaces. Examples of such shell structures are shown in Fig. 1-4. Like arches, shells derive the major portion of their strength from their shape. The analysis of shells is generally quite complicated because of the geometry of the surface and the three-dimensional interaction of the material. Entire

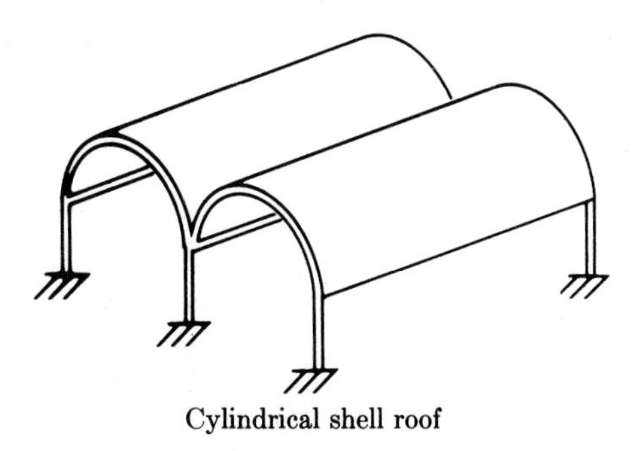

Cylindrical shell roof

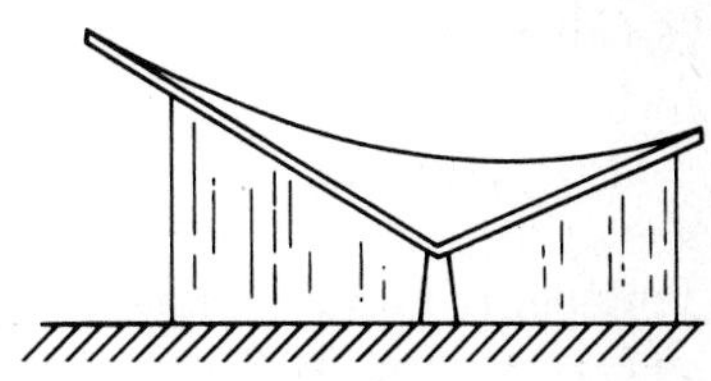

Hyperbolic paraboloid shell

Fig. 1.4

texts are devoted to shells, and their analysis is therefore not included in this text.

1-2 ANALYSIS AND DESIGN

In the following pages the student will often be asked to analyze a structure for which the member lengths and section properties are given. A logical question is: How were these member properties obtained? The answer lies in the overall concept of analysis and design.

Analysis should properly be thought of as an integral part of design. The prime objective of the structural engineer is to produce a designed structure. To design a structure he must first know how it will respond to a given loading. This, then, is the intended purpose of analysis, and the purpose of this text. The member properties that are given throughout are intended to help the beginning student of analysis to develop proficiency in analyzing the response of a given structure. Conversely, in design the member properties are chosen to result in a specific structure, and the designer may often have to readjust his initial selection of properties in order to get the desired response from the structure. Were it not for the enormous size of the resulting text, there would be some advantage in integrating the operations of analysis and design into one text. However, if the student is aware of the above-stated purpose of analysis, he will find that treating analysis separately and before design is an expedient approach.

1-3 LOADS

The loads acting on a structure can be classified into two general types, dead loads and live loads. Dead loads are loads due to the weight of the components making up the structure. For example, the weights of the roofing, the floor slabs, the individual beams, and so on, are all considered to be *dead loads* in a building. Any other loads applied to the existing structure, such as wind, contents of a building, or vehicles, are considered to be *live loads*.

Obviously, the magnitude of the dead loads can be computed if the size and type of material of the structural components are known. The problem is to have some knowledge of the sizes before the analysis and design are begun. The dead load can often be estimated from weights in similar structures. For some structures, such as trusses, the dead-weight estimates can be expressed by general formulas. Such formulas are derived from the known weights of previously built structures.

Another approach is to perform an analysis of the structure neglecting dead loads. From the results of such an analysis the dead loads can be estimated, and the structure can then be analyzed for the dead loads. Dead loads are analyzed by the same methods as other loads, and for simplicity they will not, unless otherwise stated, be considered in the examples and problems of this text.

The live load to be used in the analysis and design of structures might be difficult to determine were it not for all the information available from past experience and study. In all but exceptional cases the engineer can obtain appropriate live-load values from such sources as codes, reports, and design specifications. For the exceptional cases the engineer must resort to a separate investigation or a literature survey for information on possible investigations carried out in the past.

In the paragraphs that follow some of the most frequently encountered types of live loads are discussed. To acquaint the student with the sources of information on live loads, reference is made to typical codes and specifications. The discussion is not intended to provide an extensive source of live-load information, but rather to remove the mystery of the source of live-load values.

Building live loads

An example of the manner in which live loads for buildings are presented is shown in Table 1-1. The values given are the minimum values to be used. There are other local, state, federal, and organization-developed codes which also provide such live-load values. Most of these are similar to, and sometimes derived from, those given in Table 1-1. In later chapters we shall discuss the manner in which the loads must be placed to generate maximum stresses and deflections in a structure.

Bridge live loads

Another type of live load is that of vehicles on a highway bridge. As with buildings, there are minimum suggested values to be used in designing a bridge. A standard vehicle to be placed on the bridge, according to the standards of AASHTO (American Association of State Highway and Transportation Officials), is shown in Fig. 1-5.[1] The vehicle shown is the so-called *H-S loading*, which consists of a combination tractor and trailer. The specifications state that the live load on the bridge shall consist of such a vehicle or a uniform lane load, the magnitude of which is also specified. Similar information is available for railroad-bridge loadings.

[1] Standard Specifications for Highway Bridges, 1973, American Association of State Highway and Transportation Officials, sec. 1.2.5.

TABLE 1-1

Use or occupancy		Uniform load[1]*	Concentrated load†
Category	Description		
1. Armories		150	0
2. Assembly areas[4] and auditoriums and balconies therewith	Fixed seating areas	50	0
	Moveable seating and other areas	100	0
	Stage areas and enclosed platforms	125	0
3. Cornices, marquees and residential balconies		60	0.
4. Exit facilities, public[5]		100	0
5. Garages	General storage and/or repair	100	[3]
	Private pleasure-car storage	50	[3]
6. Hospitals	Wards and rooms	40	1000[2]
7. Libraries	Reading rooms	60	1000[2]
	Stack rooms	125	1500[2]
Manufacturing	Light	75	2000[2]
	Heavy	125	3000[2]
8. Offices		50	2000[2]
9. Printing plants	Press rooms	150	2500[2]
	Composing and linotype rooms	100	2000[2]
10. Residential[6]		40	0
11. Rest rooms[7]			
12. Reviewing stands, grandstands and bleachers		100	0
13. Schools	Classrooms	40	1000[2]
14. Sidewalks and driveways	Public access	250	[3]
15. Storage	Light	125	
	Heavy	250	
16. Stores	Retail	75	2000[2]
	Wholesale	100	3000[2]

[1] See sec. 2306 for live-load reductions.

[2] See sec. 2304(c), first paragraph, for area of load application.

[3] See sec. 2304 (c), second paragraph, for concentrated loads.

[4] Assembly areas include such occupancies as dance halls, drill rooms, gymnasiums, playgrounds, plazas, terraces, and other occupancies which are generally accessible to the public.

[5] Exit facilities include such uses as corridors and exterior exit balconies, stairways, fire escapes, and similar uses.

[6] Residential occupancies include private dwellings, apartments, and hotel guest rooms.

[7] Rest room loads shall be not less than the load for the occupancy with which they are associated but need not exceed 50 lb/ft^2.

* Pounds per square foot of horizontal projection.

† Pounds.

Reproduced from the 1976 edition of the *Uniform Building Code*, copyright 1976, with permission of the publisher, International Conference of Building Officials.

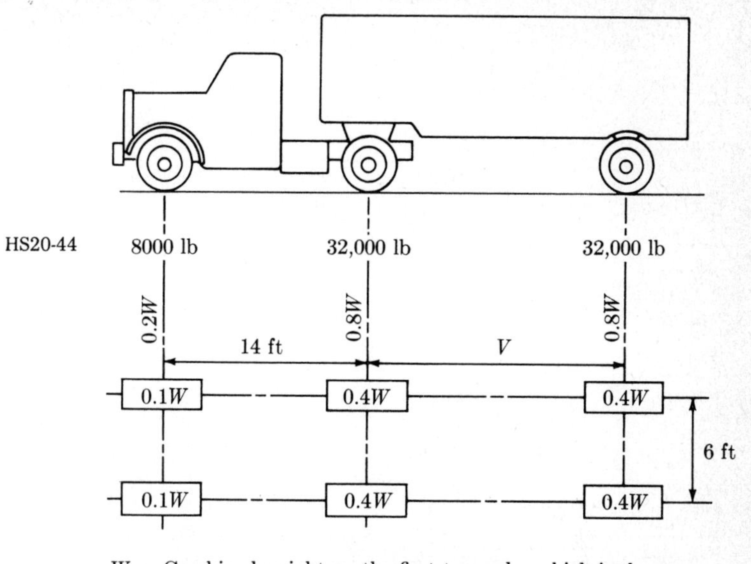

W = Combined weight on the first two axles which is the same as for the corresponding H truck.

V = Variable spacing—14 to 30 ft inclusive. Spacing to be used is that which produces maximum stresses.

Fig. 1.5

In moving live loads such as these there is a question of where the vehicle should be placed to produce maximum stresses in the bridge structure. Such problems are considered in later chapters of this text.

Wind loads

The determination of wind loads acting on a structure is basically a problem in aerodynamics. The values of static wind pressure on a structure can be obtained by first considering the general expression for dynamic pressure as a function of the wind velocity. The dynamic pressure of the wind, q, is equal to one-half the product of the mass density of the air and the square of the wind velocity,

$$q = \frac{\rho V^2}{2} = \frac{1}{2}\frac{w}{g}V^2 \tag{1-1}$$

For standard air with a unit weight of 0.0765 lb/ft³ Eq. (1-1) reduces to

$$q = \left(\frac{1}{2}\right)\left(\frac{0.0765}{32.2}\right)\left(\frac{5280}{3600}\right)^2 V^2 = 0.00256V^2 \tag{1-2}$$

where the dynamic pressure q is in pounds per square foot and the

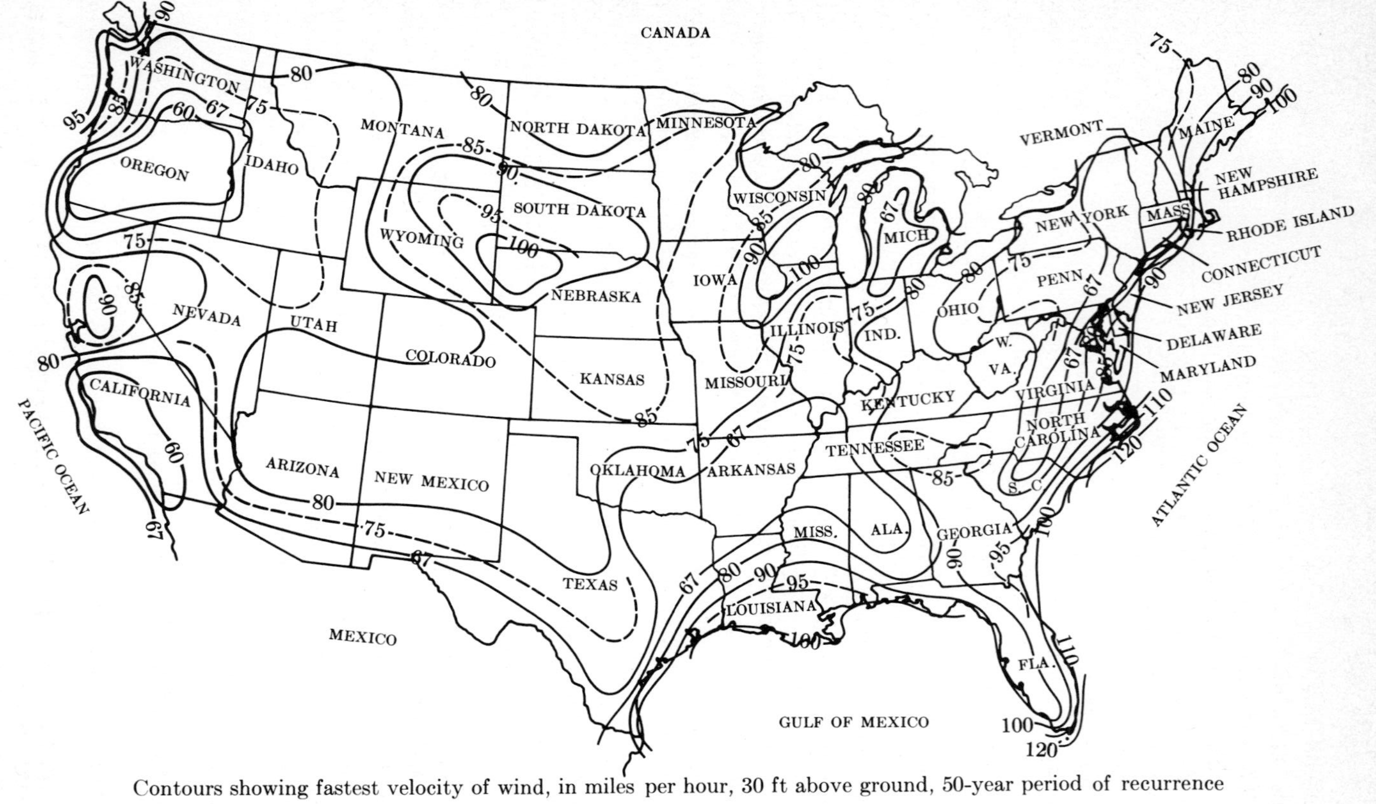

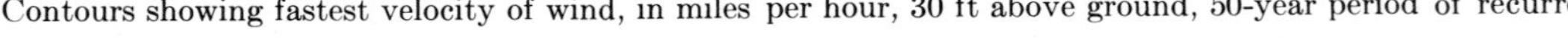
Contours showing fastest velocity of wind, in miles per hour, 30 ft above ground, 50-year period of recurrence

Fig. 1.6

velocity of the wind is in miles per hour. The dynamic pressure of the wind can thus be obtained if the velocity of the wind is known.

The extreme wind velocity to be expected in a particular location can be obtained from past wind records. Such records are generally kept by weather bureaus. The records of the United States Weather Bureau have been used to construct the contour map of extreme wind velocities in the United States shown in Fig. 1-6.[1] The wind velocities are given in miles per hour measured 30 ft from the ground. Because wind velocities increase with the distance above the ground, values such as those in Fig. 1-6 should be increased somewhat for high structures. The factor by which the velocities are increased can be found in the report from which Fig. 1-6 was taken.

The static wind pressures on a structure are also dependent on the shape of the structure and the manner in which the structure resists the wind. A convenient way of accounting for this is to apply a shape or drag factor to the value of dynamic pressure obtained from Eq. (1-2). Such factors are obtained from detailed wind studies and tests. For some of the more common structures the shape factors are available from graphs as shown in Fig. 1-7.[2] The graphs in Fig. 1-7 represent the values of the shape factor for a gabled-roof structure as recommended by the ASCE Subcommittee 31, the Swiss building code, and the Danish building code. The graphs are not intended to convey exact values, as they are not a function of the dimension of the structure, but they do serve as general guidelines for such structures.

Graphs similar to those in Fig. 1-7 have also been developed for other types of structures. In many codes the static wind pressure to be used is simply expressed in tabular form, with recommendations for adjusting the tabulated values to the type of structure under consideration. For unusual structures and where more extensive knowledge of the wind forces is required, models of the structure can be tested in wind tunnels to determine the distribution of wind forces. Much additional information on wind loads can be found in the final report of the wind study conducted by the American Society of Civil Engineers.[3]

Snow loads

Recommended values of snow loading can be found in codes and other forms of design information. Those values are based on past weather records. In many cases they are tabulated according to sections of a

[1] Wind Forces on Structures (Final Report), *Trans. ASCE*, vol. 126, part II, p. 1132, 1961.

[2] *Ibid.*, p. 1152.

[3] *Ibid.*, pp. 1124–1198.

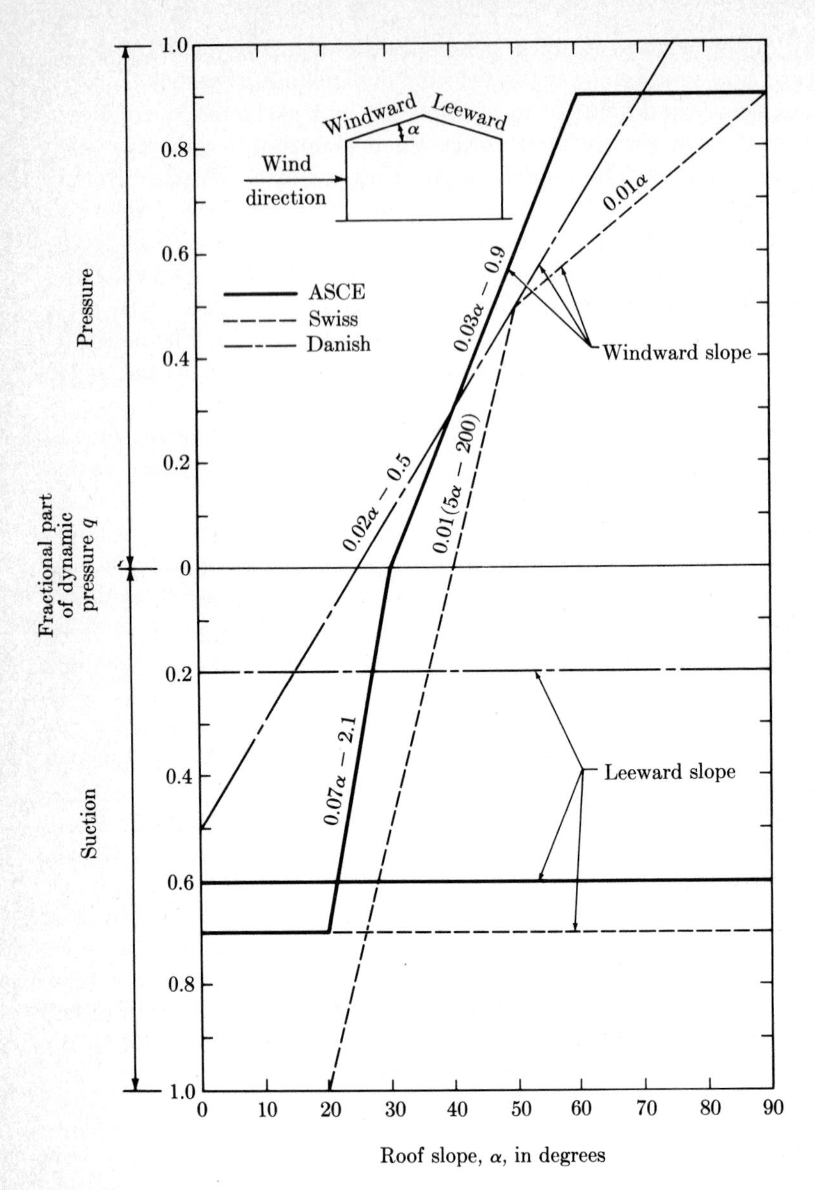
Windward
Leeward
α
Wind
direction
ASCE
Swiss
Danish
Pressure
Suction
Fractional part of dynamic pressure q
0.01α
0.03α − 0.9
0.02α − 0.5
0.01(5α − 200)
0.07α − 2.1
Windward slope
Leeward slope
1.0
0.8
0.6
0.4
0.2
0
0.2
0.4
0.6
0.8
1.0
0
10
20
30
40
50
60
70
80
90
Roof slope, α, in degrees

Fig. 1.7

country. Snow loads can also be given in the form of contour maps such as that shown in Fig. 1-8.[1] In addition, contour maps of local areas are often developed by building committees for use in their areas.

Impact loads

As evidence of impact loads, let us consider the elementary case of a weight being dropped from some height onto a beam. The resulting maximum deflection and forces in the beams are larger than if the weight were gradually applied to the beam. The dynamic force acting on the beam when the weight is dropped can conveniently be expressed in terms of a factored increase of the static weight. The factor by which the static weight is increased by dynamic application is referred to as the *impact factor*. A weight that is dropped is referred to as an *impact load*.

Impact loads can be generated in structures by various types of moving loads. For example, a truck traveling across a bridge produces a certain amount of impact. Obviously, the wheels of the truck are not dropped onto the bridge deck from considerable heights, but in moving across the bridge they generate a certain amount of impact. For bridges and other large structures it would be difficult, if not impossible, to determine the impact factor by an exact rational analysis. However, from the results of tests and past performances, general expressions have been developed to indicate the magnitude of impact factors. The specifications of the American Association of State Highway and Transportation Officials state that an impact factor must be applied to the standard truck loading shown in Fig. 1-5. The impact factor I is given by

$$I = \frac{50}{L + 125}$$

but is not to exceed 0.3. In this expression L is the loaded length of the span, defined more fully in the specifications. Similar expressions for impact factors can be found in other specifications and codes.

Earthquake loads

Another form of dynamic loading is that due to earthquakes. The effects of an earthquake on a building can be visualized from the model of a building attached to the top of a table which is then suddenly shaken.

[1] This chart is based on records furnished by the U.S. Weather Bureau, entitled "Greatest Snow Depth on Ground at Any One Time," covering the period from 1871 to 1944. Snow loads within the hatched areas are to be determined by local investigations and are in no case to be less than 45 lb/ft^2.

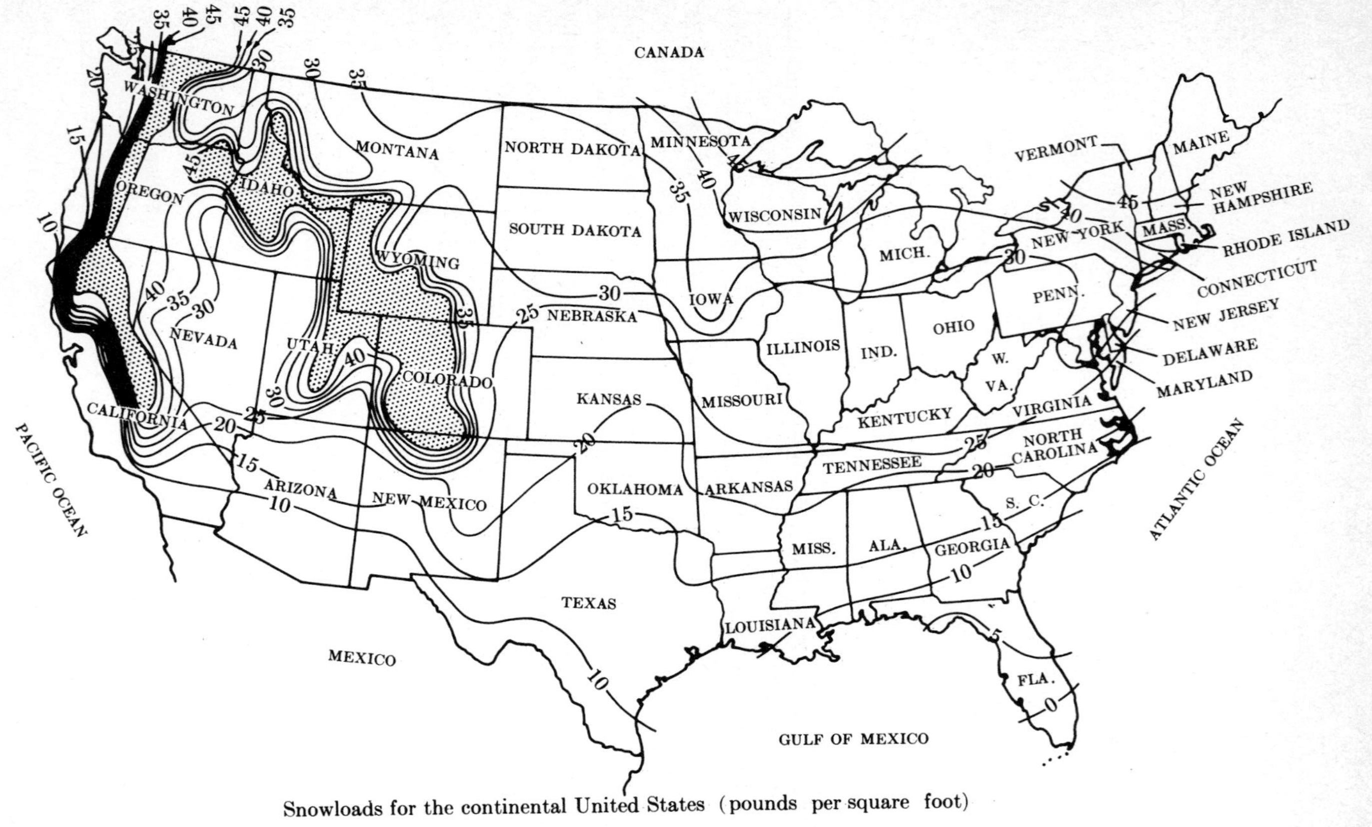

Snowloads for the continental United States (pounds per square foot)

Fig. 1.8

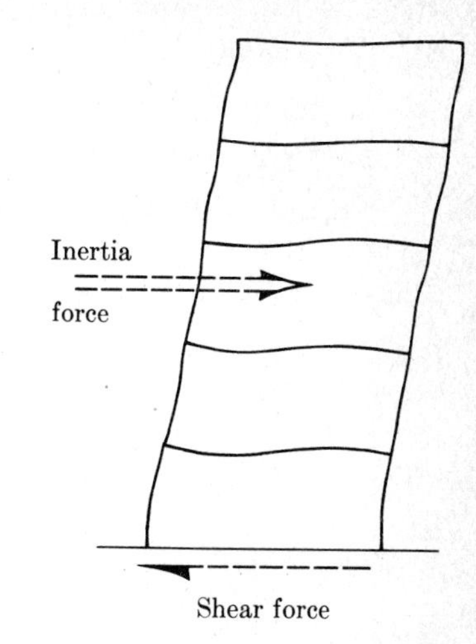

Fig. 1.9

The inertia of the building resists the sudden motion of its support and causes distortion of the building, as shown in Fig. 1-9. This distortion and the ensuing vibrations generate horizontal shear forces in the structure and between the structure and the ground. Within limits a structure can be designed to absorb the effects of the earthquake motion.

For practical purposes, the effects of an earthquake on a structure can be expressed in terms of lateral forces. The magnitude and distribution of these lateral forces are presented in various building codes. The recommendations of these codes represent the results of extensive earthquake studies and are stated in a form that is readily applicable to various types of structures. For example, the Uniform Building Code states that the minimum earthquake force for which a building is to be designed is given by the formula[1]

$$V = ZIKCSW$$

where V = lateral load or shear at the base

Z = a numerical coefficient dependent on the geographical location of the structure

I = an occupancy importance factor

K = a numerical coefficient reflecting the type of building frame

C = a numerical coefficient reflecting the natural period of vibration of the structure

S = a numerical coefficient for site-structure resonance

W = the dead load of the structure

[1]Reproduced from the 1976 edition of the *Uniform Building Code*, copyright 1976, with permission of the publisher, International Conference of Building Officials.

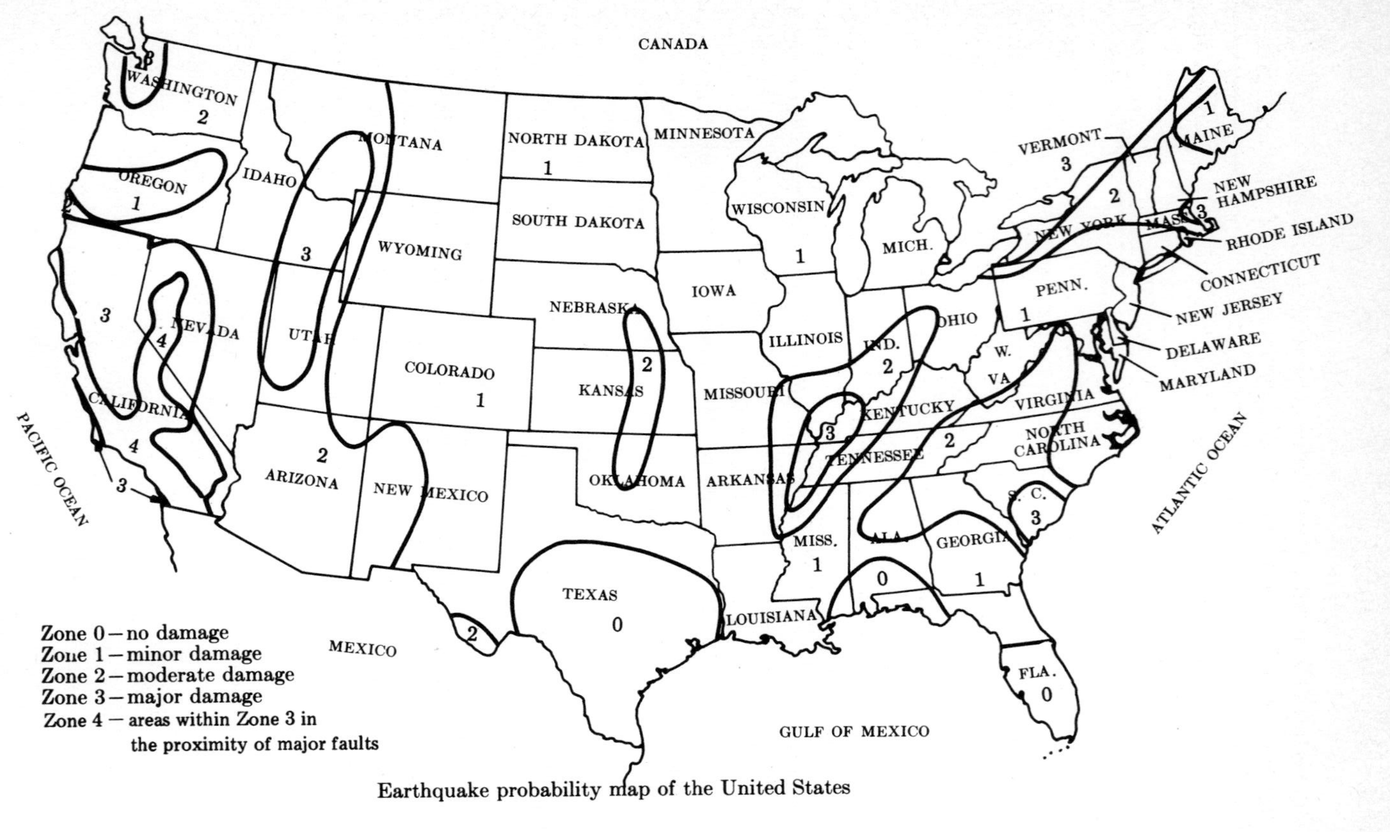

Earthquake probability map of the United States

Fig. 1.10

The values of Z for the continental United States are obtained from the map shown in Fig. 1-10. For structures in zone 1, $Z = \frac{3}{16}$; in zone 2, $Z = \frac{3}{8}$; in zone 3, $Z = \frac{3}{4}$; and in zone 4, $Z = 1$. Values of I and K are tabulated in the Uniform Building Code. The value of C is determined from the expression

$$C = \frac{1}{15\sqrt{T}}$$

where T is the fundamental period of vibration of the structure. T can be determined by dynamic analysis of the structure, or its value can be approximated from general formulas. Formulas for determining the value of S are given in the Uniform Building Code. The distribution of the lateral earthquake load to the different floor levels and to various parts of the structure is also included in the code.

In the discussion above we have not considered sources of loading, such as erection loads, differential settlement of footings, and temperature changes. Nor have we discussed combination loads on a structure, such as simultaneous wind and snow. A more complete discussion of these topics properly belongs in a design course. The intent here is simply to introduce loading so that the methods of analysis in later chapters will be more meaningful.

1-4 STRUCTURAL COMPONENTS

Some knowledge of the names of the structural components and of the manner in which they function is necessary for a better understanding of the analysis of such systems. Figures 1-11 to 1-13 show typical structures encountered in analysis, and later in design. The bridge structure of Fig. 1-11 has as its primary components two trusses on the left span and plate girders on the right span. The student should attempt to form a mental picture of how loads are transmitted through such structures

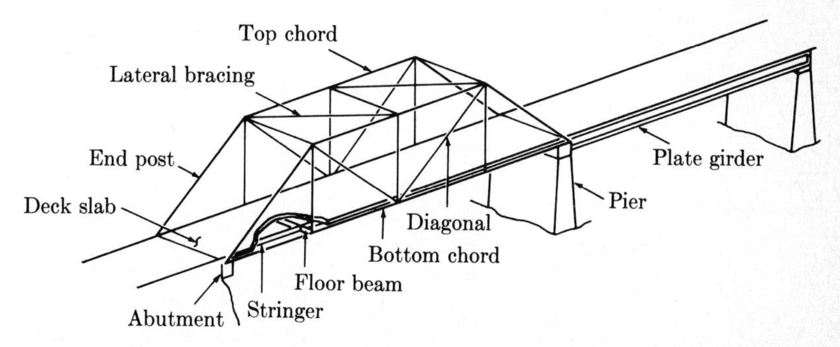

Fig. 1.11

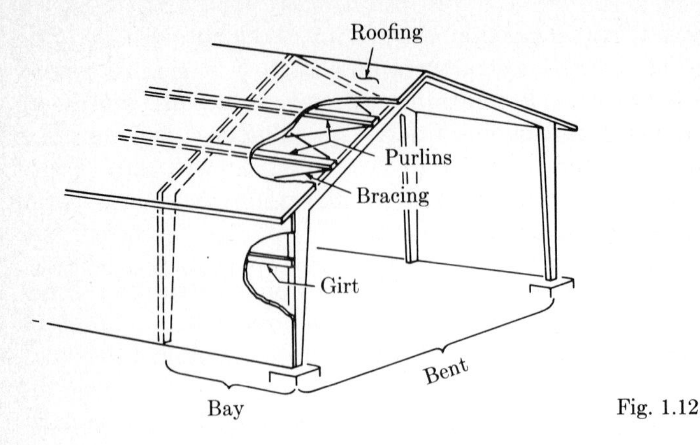

Fig. 1.12

into their supports. For example, note that on the truss span the load of a vehicle goes into the deck slab, then to the stringers, the floor beams, and the trusses, and then into the supporting abutment and pier. The individual trusses are thus subjected to concentrated loads from the floor beams. In later discussions we shall determine the magnitude and type of forces in the components resulting from such a load. In working with these structures we commonly refer to the components by the names shown.

The structure of Fig. 1-12 has as its primary components a series of frames, referred to as *bents*. Loads are brought into the bents at concentrated points by the purlins in the roof section and by the girts in the wall section. Thus, if we were to consider the loading on such a frame, the loads would be concentrated. By comparison, the loading on the individual purlins from the roofing would be distributed as illustrated.

A typical building frame is shown by the schematic diagram of Fig. 1-13. The primary components of such a structure are the roof- and

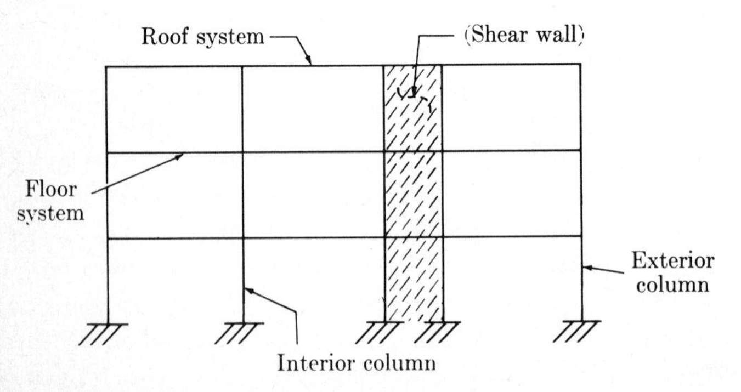

Fig. 1.13

floor-framing system and the columns. The floor loads are brought into the columns by slab, stringer, and beam arrangements, similar to the manner in which the deck loads are brought into the bridge trusses. Often a very important component known as a *shear wall* is included in the design of such structures. The purpose of the shear wall is to absorb lateral forces, such as earthquake and wind, rather than have them absorbed by the floor and column framing. If properly designed, the stairwells and elevator shafts can serve as shear walls in such structures.

1-5 STRUCTURE REPRESENTATION

Structures are idealized to a certain extent for purposes of analysis. Seldom in the overall analysis of structures are the dimensions of joints and depths of members considered. These dimensions are included only when their magnitudes in relation to the lengths of the members are sufficiently large to influence the results, or when the forces are applied in such a manner that these dimensions become significant. The members of the structure can therefore be represented by single lines. The single lines often represent the centroidal axis or an edge of the member. Two lines are sometimes used to show the member in depth, but unless the depth is specifically given, it is disregarded in the analysis. The overpass structure of Fig. 1-14*a* can be represented in a simplified, or idealized, form by the solid-line diagram in Fig. 1-14*b*. The deflected shape of the structure is represented by the dashed lines. Joints and supports are similarly represented in abbreviated form. The symbols used for joints and supports are discussed in the following section.

The use of appropriate coordinate systems is essential in the analysis of structures, particularly in the analysis of complex structures. A coordinate system commonly used in describing the forces and deflections of an individual member is the $xy\theta$ coordinate system as shown next to member AB in Fig. 1-14*b*. The x axis is commonly chosen to be parallel to the axis of the member. The student should be familiar with such a coordinate system from previous studies in strength of materials where it was used to analyze simple beam problems. Because the $xy\theta$ coordinate system is oriented with respect to the individual member, it is often referred to as a *local coordinate system*. The axes of all members in a structure are not necessarily oriented in the same direction, and to analyze the entire structure we often use a *framework coordinate system* or *global coordinate system* such as the $XY\Theta$ system shown in Fig. 1-14*b*. In the analysis of complete structures, the displacements of a point and the loading on the structure can conveniently be expressed in components referenced to the framework coordinate system. The truck wheel loads of Fig. 1-14*a*, for

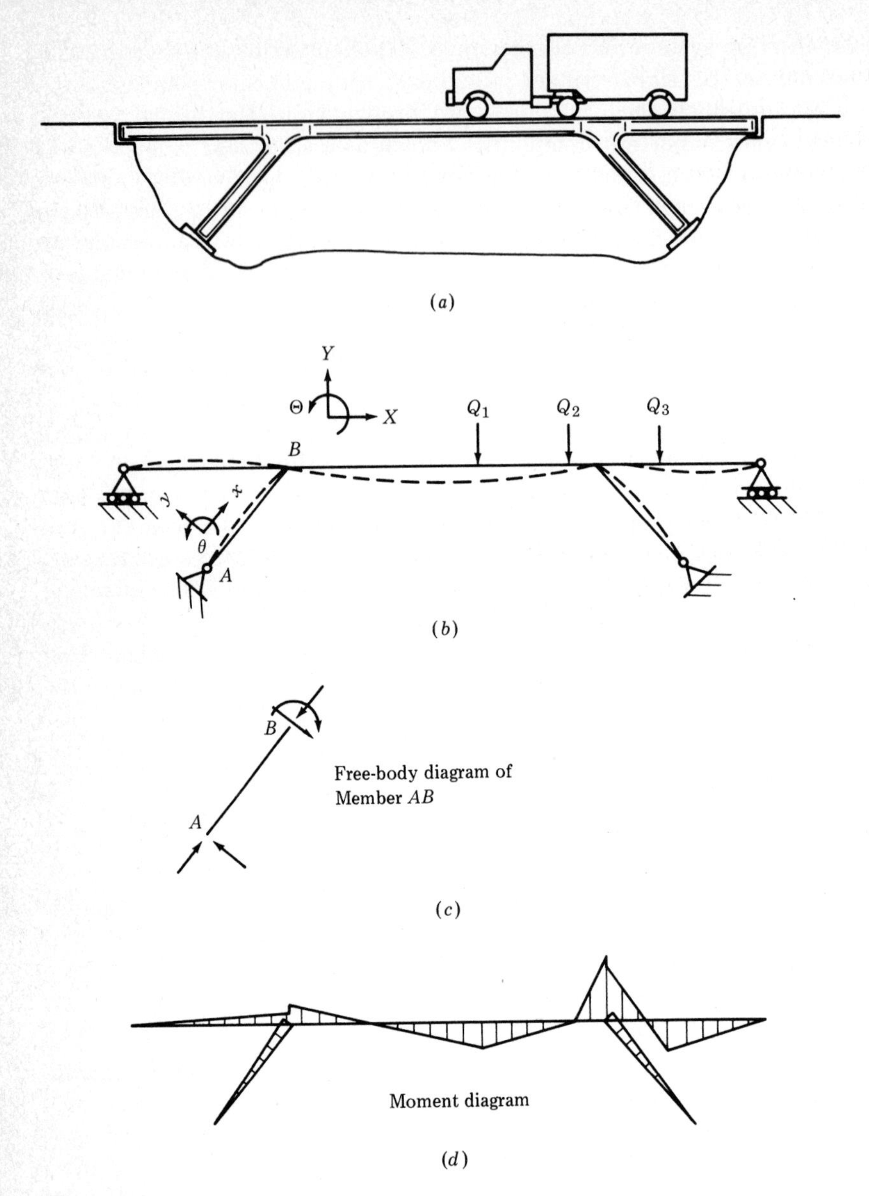

Fig. 1.14

example, can be represented by force vectors Q_1, Q_2, and Q_3, as shown in Fig. 1-14*b*, which act in the negative Y direction of the $XY\Theta$ coordinate system.

The $xy\theta$ and $XY\Theta$ coordinate systems referred to in Fig. 1-14*b* are appropriate for two-dimensional analysis of the structure. For more involved, three-dimensional analyses of structures, these coordinate systems can be extended to three dimensions.

Two additional aspects of a complete analysis of a structure such as that of Fig. 1-14*a* are shown in Fig. 1-14*c* and *d*. These aspects are the use of free-body diagrams and the construction of bending-moment diagrams. The ability to construct free-body diagrams is an invaluable aid to analysis, and the construction of the resulting moment diagram is an essential prerequisite for the design of such a structure. These two subjects are discussed in detail in the next chapter.

1-6 SUPPORTS AND CONNECTIONS

In serving their purpose of supporting loads structures are prevented from moving freely in space by *restraints* or *supports*. An essential part of structural analysis is thus to determine the manner in which the supports react. The reactions of the supports to the structure are dependent on the type of support condition used. As a first step in determining all reactions on a structure it is essential to understand the interaction between that part of the structure at the support and the supporting device.

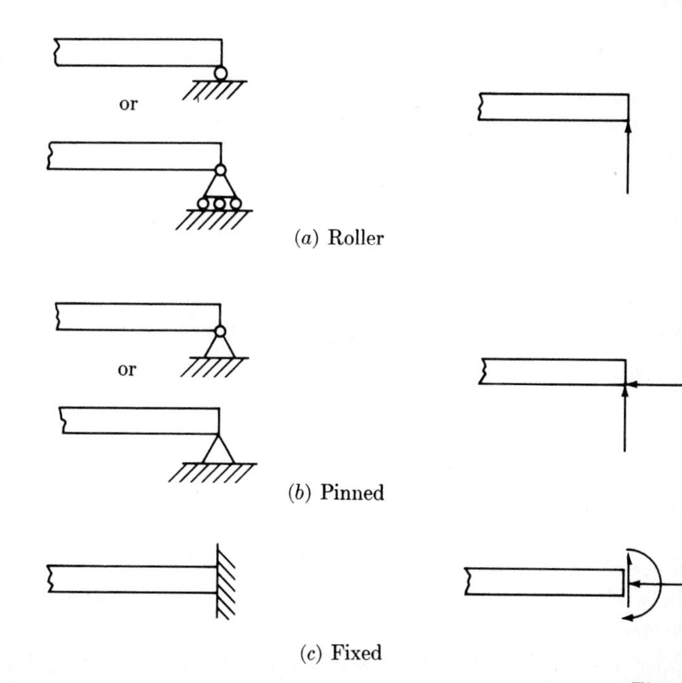

(*a*) Roller

(*b*) Pinned

(*c*) Fixed

Fig. 1.15

Various types of supports are used in structures. Figure 1-15 shows the types most often encountered. At the left of each figure is the symbol or symbols used to represent the given type of support. At the right of each figure are the corresponding components of reaction that can be transmitted to the structure by such a support. In addition to knowing the components of force that each type of support can transmit, the student should also be able to recognize the type of motion that is permitted by each. Thus for the roller of Fig. 1-15*a* rotation and horizontal translation are permitted; for the fixed support of Fig. 1-15*c* no rotation or translation is permitted.

The manner in which members of a structure are connected to each other is generally described by either a *rigid connection* or a *pinned (hinged) connection*. Joint B of the structure in Fig. 1-14*b* is considered to be a rigid connection, i.e., the rotation of each member end at the joint is the same, and moment is transmitted from one member to another. Joints that do not transmit moment from one to another are considered pinned. Although true frictionless pins are seldom encountered in reality, the assumption that joints are pinned in structures such as trusses yields acceptable results.

PROBLEMS

1-1. Estimate the live load, in pounds per square foot, on the floor of one of your lecture rooms in an area of full occupancy. Make the same estimate in a heavily occupied hallway between classes. Compare your estimated values with the recommended values of Table 1-1.

1-2. For the following, determine the value of F for conversion between the SI and U.S. Customary systems. As indicated in Appendix C, specifically show how you account for the units of each term.

(*a*) 1 ksi = F Pa (Pascals)
(*b*) 1 kip/ft = F kN/m
(*c*) 1 in.^{4} = F mm^4
(*d*) 1 m^4 = F in.^{4}
(*e*) 1 in.^{2} = F mm^2
(*f*) 1 ft-kip = F kN-m
(*g*) 1 psf = F Pa

1-3. Express the recommended uniform building live loads of Table 1-1 in SI units of Pascals for the following types of occupancy: (*a*) residential, (*b*) offices, and (*c*) heavy storage.

1-4. Convert the dimensions and weights of a standard vehicle as shown in Fig. 1-5 to SI units, and show the results on a similar sketch.

1-5. Express the following wind velocities in SI units of kilometers per hour: (*a*) 60 mph, (*b*) 100 mph.

1-6. Determine the expression for wind pressure, in SI units, equivalent to Eq. (1-2). Assume V is expressed in kilometers per hour, and q is to be expressed in Pascals.

1-7. The residential structure shown in Fig. P1-7 is subjected to a wind blowing from left to right. (*a*) Determine the recommended design pressure in pounds per square foot on each surface of the structure for your locality, using the wind velocities from Fig. 1-6 and the ASCE recommendations of Fig. 1-7. Use a 100-mph wind if you are not located in the United States. (*b*) Express the dimensions of the structure in Fig. P1-7 and the

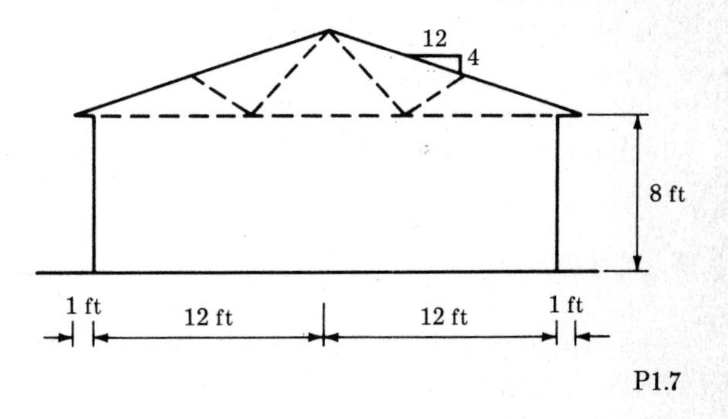

P1.7

wind pressures determined in (*a*) in SI units. (*c*) For a 2-ft depth of the structure shown in Fig. P1-7, determine the resultant horizontal wind force, in kips, acting on the structure. (*d*) Determine the resultant vertical forces, in kips, acting on the roof section.

1-8. A snow load of 40 psf is equivalent to how many kilonewtons per square meter?

CHAPTER

TWO

BASIC STATICS

2-1 FREE-BODY DIAGRAMS

The usefulness of free-body diagrams in studying structural analysis cannot be overemphasized. Free-body diagrams can be constructed for various parts of a structure, and also for the entire structure. The basic steps in constructing a free-body diagram can be stated as follows:

1. Remove the body under consideration from its original state. To do this we must hypothetically cut the structure or disengage some connections and supports. A drawing of the free body is then made.
2. On the drawing of the free body denote all possible forces acting in the given structure at the cuts and disengagements by appropriate force vectors. All external forces acting on the body in its original state are also included on the diagram. All forces acting on the free body should be clearly labeled. Proper labeling of the forces will greatly facilitate the writing of the equilibrium equations.
3. For a structure that is being broken down into a number of free-body diagrams the procedure for each diagram is the same as above. However, in dealing with forces acting on the free bodies, internal forces common to two free bodies are denoted as equal but oppositely

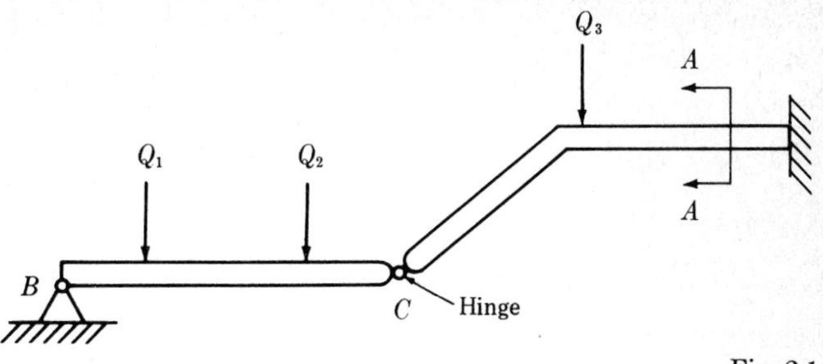

Fig. 2.1

directed force vectors. As we shall see later, this avoids confusion in writing the equilibrium equations.

The application of this procedure is illustrated in the following examples.

EXAMPLE 2-1 The free-body diagram of that part of the structure to the left of section A-A in Fig. 2-1 is to be constructed. The resulting free-body diagram is shown in Fig. 2-2. The pinned support at B is capable of transmitting to the body two orthogonal components of reaction, as shown. The assumed directions of the components are arbitrary, but it is convenient to choose them either in the vertical and horizontal directions or in directions corresponding to some orthogonal coordinate axes. These components, chosen in the vertical and horizontal directions, are denoted as B_V and B_H. At section A-A, where the cut is made, the member is capable of transmitting shear, moment, and axial forces. Thus the three force vectors V, M, and P are shown on the free body.

If the free-body diagrams of the individual elements comprising the free body of Fig. 2-2 were to be constructed, they would appear as shown in Fig. 2-3. Note the equal but oppositely directed representation of the connecting forces at C.

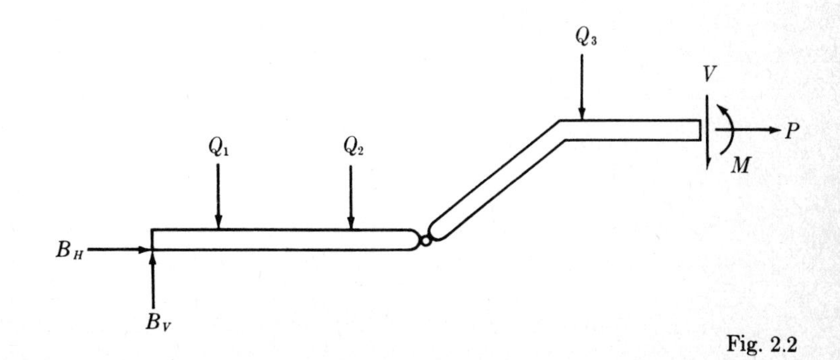

Fig. 2.2

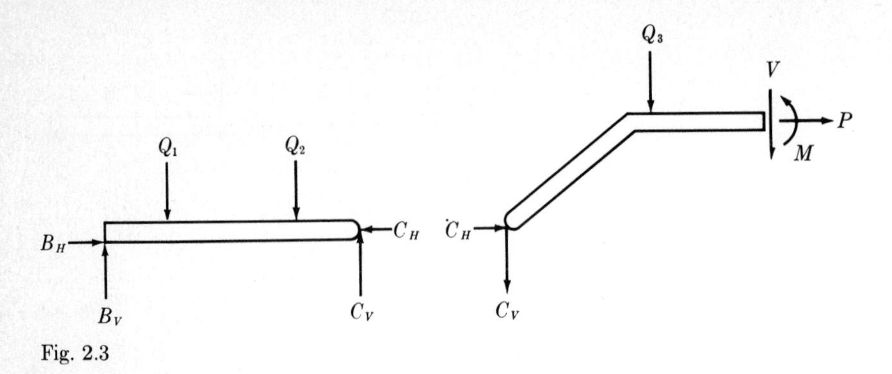

Fig. 2.3

EXAMPLE 2-2 The free-body diagrams of the elements comprising the system of Fig. 2-4 are to be constructed. To help clarify an often troublesome problem, the system is broken down into the basic parts, and separate diagrams are constructed for member AB, the pulley, and the pin, as shown in Fig. 2-5. Once the function of the pin with regard to free bodies is clearly understood, it is not essential to construct a free body of the pin. The additional subscripts P-P and P-M in Fig. 2-5

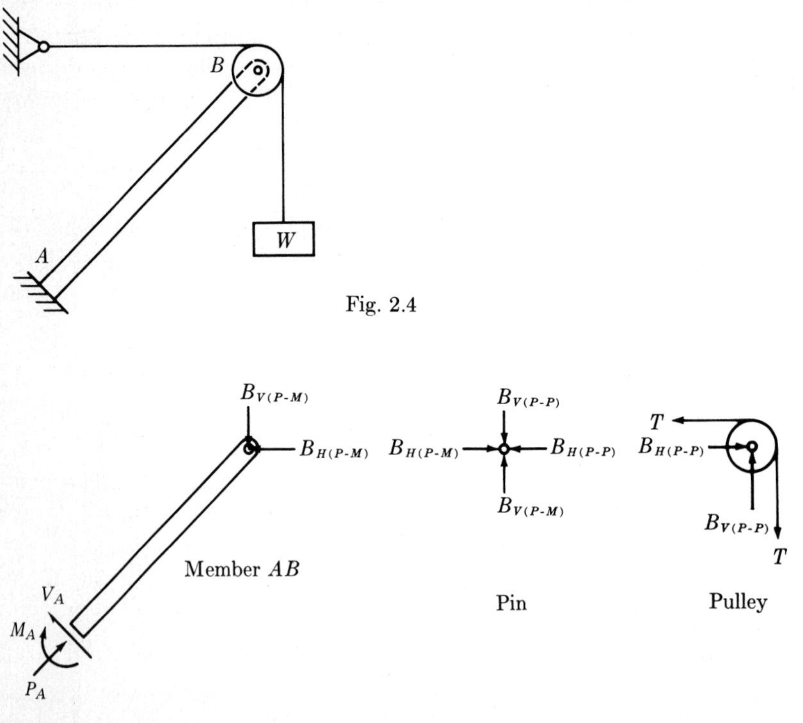

Fig. 2.4

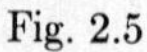

Fig. 2.5

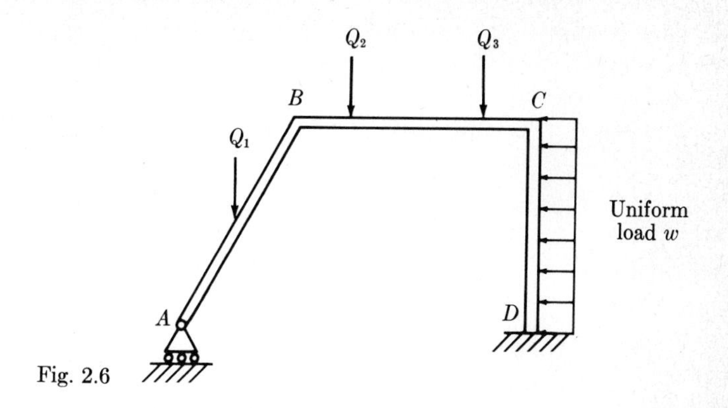

Fig. 2.6

refer to forces existing between the pulley and the pin and between the pin and the member, respectively.

EXAMPLE 2-3 Free-body diagrams of each of the three members of the frame in Fig. 2-6 are to be constructed. The left support of the frame is a roller support and the right support is a fixed support. Recall from the course in strength of materials that in analyzing individual members we generally wish the forces acting at the ends of a member to be in the form of shear normal to the axis of the member, axial force coincident with the axis of the member, and moment. Thus the internal member forces are constructed on the free-body diagrams as shown in Fig. 2-7. In the later process of ascertaining the magnitude of these forces we

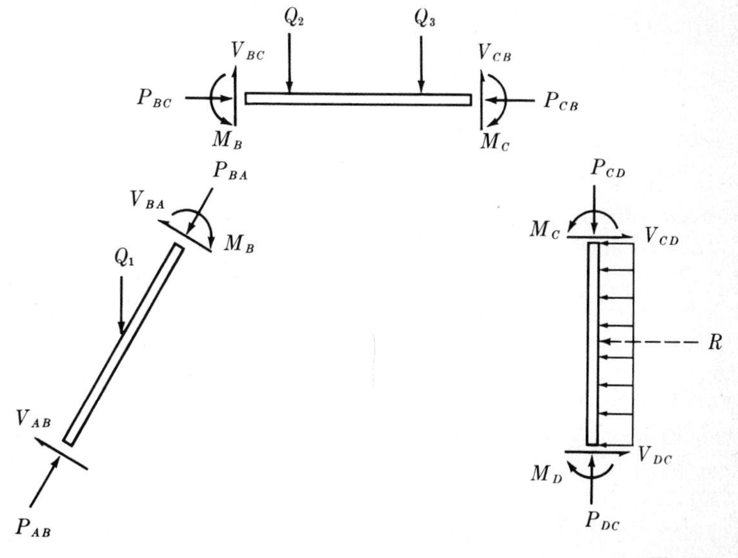

Fig. 2.7

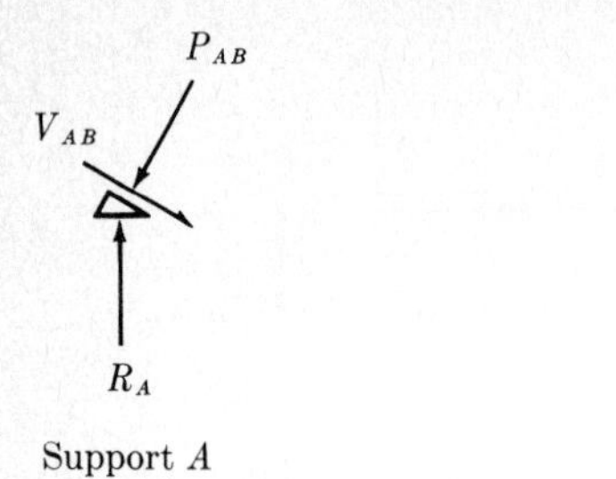

Support A

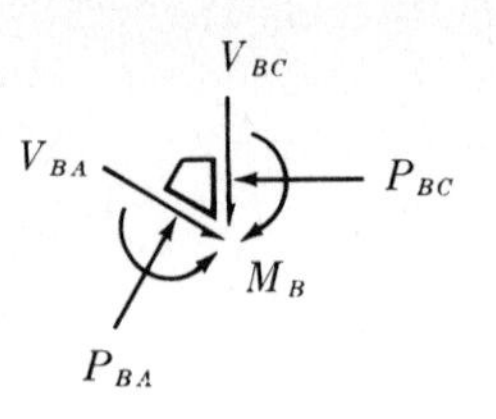

Joint B

Fig. 2.8

shall find that the relationships of forces such as those at A and B can be found from free bodies of a differential element at these points, as shown in Fig. 2-8. Recall also that the uniform load on member CD can be replaced by its resultant value R in determining the external forces acting on CD. However, the uniform load is shown on the free body of Fig. 2-7 because such a diagram must be used in determining the internal stresses of CD. It is not correct to consider the resultant R when the internal stresses of CD are being determined.

2-2 EQUATIONS OF EQUILIBRIUM

A body that is initially at rest and remains at rest when acted upon by a system of forces is said to be in a state of *static equilibrium*. The analytic expressions for equilibrium can be written in various forms. Regardless of the form used, the free-body diagrams of the previous section are an invaluable aid in realizing this condition.

From statics, recall that the scalar equations of equilibrium for a planar system can be written as

$$\Sigma F_H = 0 \qquad \Sigma F_V = 0 \qquad \Sigma M = 0$$

In this form the condition for equilibrium is stated in terms of the algebraic sum of the horizontal and vertical components of force being equal to zero and the algebraic sum of moments about any point being zero.

Instead of describing the forces in terms of horizontal and vertical components, we can write the equations of equilibrium in terms of X and Y coordinate components. The equations of equilibrium thus become

$$\Sigma F_X = 0 \qquad \Sigma F_Y = 0 \qquad \Sigma M = 0$$

Vector expressions are another form that is often desirable for equilibrium:

$$\mathbf{F}_R = \mathbf{0} \qquad \mathbf{M}_R = \mathbf{0}$$

These expressions state that the resultant vector of all forces $\mathbf{F}_R$ is equal to zero and the resultant moment vector $\mathbf{M}_R$ is equal to zero. The equivalence of this vector form and the previous scalar expressions can be seen in the following example.

EXAMPLE 2-4 A planar body in equilibrium is acted upon by three force vectors, $\mathbf{F}_1$, $\mathbf{F}_2$, and $\mathbf{F}_3$. A partial requirement for equilibrium is

$$\mathbf{F}_R = \mathbf{F}_1 + \mathbf{F}_2 + \mathbf{F}_3 = \mathbf{0}$$

Recall that a force vector can be written in terms of its scalar components and the unit base vectors. Thus

$$\mathbf{F}_1 = F_{1X}\mathbf{i} + F_{1Y}\mathbf{j}$$
$$\mathbf{F}_2 = F_{2X}\mathbf{i} + F_{2Y}\mathbf{j}$$
$$\mathbf{F}_3 = F_{3X}\mathbf{i} + F_{3Y}\mathbf{j}$$

The equilibrium equation can therefore be written as

$$\mathbf{F}_R = F_{1X}\mathbf{i} + F_{1Y}\mathbf{j} + F_{2X}\mathbf{i} + F_{2Y}\mathbf{j} + F_{3X}\mathbf{i} + F_{3Y}\mathbf{j} = \mathbf{0}$$

Equating coefficients of like base vectors, we have

$$F_{1X} + F_{2X} + F_{3X} = 0 = \Sigma F_X$$
$$F_{1Y} + F_{2Y} + F_{3Y} = 0 = \Sigma F_Y$$

The result is the first two of the previously stated scalar equations.

For three-dimensional analyses additional equations are required. In scalar form the following six equations apply:

$$\Sigma F_X = 0 \qquad \Sigma M_{XX} = 0$$
$$\Sigma F_Y = 0 \qquad \Sigma M_{YY} = 0$$
$$\Sigma F_Z = 0 \qquad \Sigma M_{ZZ} = 0$$

The vector forms of the equilibrium equations are the same,

$$\mathbf{F}_R = \mathbf{0} \qquad \mathbf{M}_R = \mathbf{0}$$

but the resultant vectors represent three-dimensional force and moment vectors.

2-3 SIGN CONVENTIONS

An essential part of analysis and design is the adoption of an appropriate sign convention for various types of force and deflection. As the various methods of analysis are developed it will become apparent that there are advantages in not using the same sign convention for all methods. The student will find it advantageous in employing a particular method to establish clearly in his mind the sign convention to be used.

The conventions used in this text for the various types of forces and their corresponding deflections are discussed in the following paragraphs.

Bending moment

There are two conventions used for bending moment: the *beam convention,* shown in Fig. 2-9*a*, and what is often referred to as the *joint convention,* shown in Fig. 2-9*b*. The beam convention is probably familiar from a previous course in strength of materials. Positive moment is commonly described by the beam convention as moment on a beam producing a deflected shape whose curvature is concave up, or the moment producing compression in the upper fiber. The joint convention describes moment which tends to rotate a joint counterclockwise, as shown in Fig. 2-9*b*. A clockwise direction is sometimes taken as positive; however, in this text counterclockwise moment on a joint will be considered positive. The merits of the joint sign convention for moment will become evident in later treatments of frames and continuous beams.

Axial force

As shown in Fig. 2-9*c*, an axial force is considered positive when it produces tension in the member and therefore tends to extend the member. A compressive force is therefore negative.

Shear

Shearing forces which tend to shear a member in the manner shown in Fig. 2-9*d* are considered positive. It is helpful to remember that this force is in effect a force acting upward to the left of the section and downward to the right.

Twist

Twisting moment is considered positive when it acts on a member as shown in Fig. 2-9*e*. The convention thus corresponds to the right-hand screw rule; that is, if the thumb of the right hand is pointed in the direction coincident with the twist vector, the arcing motion in closing the other four fingers of the right hand represents the direction of positive twist. The twist vector will be distinguished from other force vectors by double lines for the shaft, as shown.

For some analyses it is convenient to adopt a sign convention for forces in terms of a framework coordinate system. For an XYZ coordinate system, as shown in Fig. 2-9*f*, the forces and moments acting on the

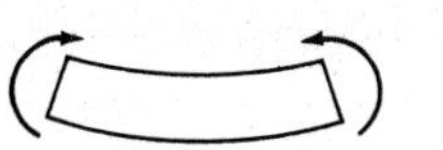

(a) Bending moment (beam convention)

(b) Bending moment (joint convention)

(c) Axial force

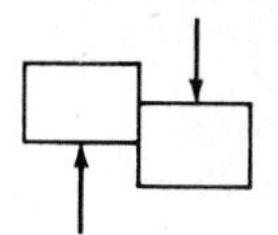

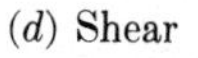

(d) Shear

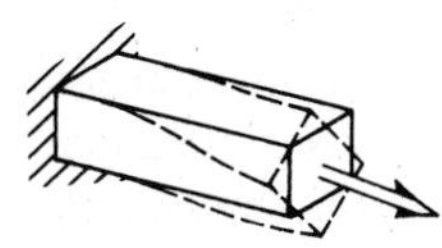

(e) Twisting moment

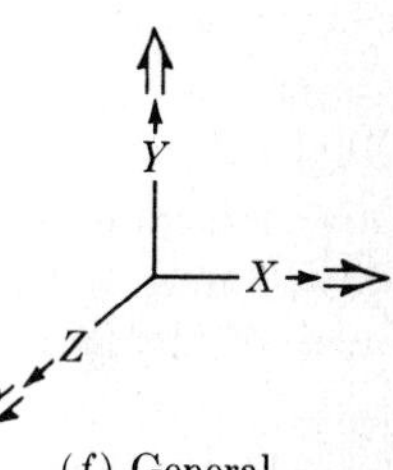

(f) General

Sign conventions (positive)

Fig. 2.9

structure can be considered positive when they act in a direction coincident with the coordinate axes, with the moments positive according to the right-hand screw rule. In this case the moments are represented by vectors with double-lined shafts. This sign convention for force and moment, which can also be used for deflection, is particularly useful in the matrix methods to be discussed later in the text.

EXAMPLE 2-5 An analysis using the beam sign convention has resulted in moments of $M_B = -3.2$ kN-m and $M_A = -3.8$ kN-m for the continuous beam of Fig. 2-10. Free-body diagrams of members AB and BC are to be constructed and the sign of the moments in members AB and BC at joint B are to be indicated according to the joint convention.

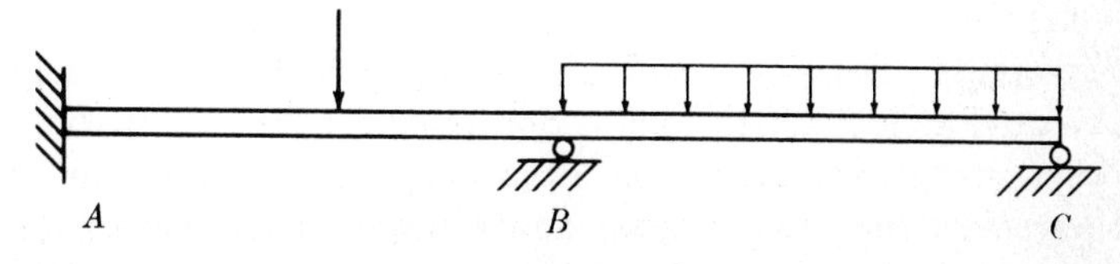

Fig. 2.10

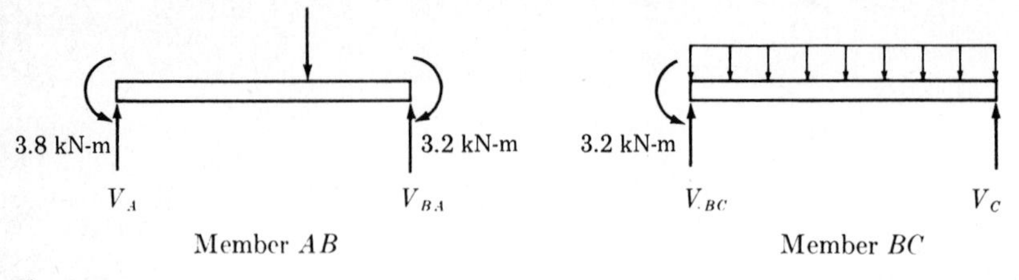

Fig. 2.11

From Fig. 2-9*a*, the negative moments by the beam sign convention result in compression of the lower fibers. Therefore the moments at A and B act on the free bodies as shown in Fig. 2-11. To make the free-body diagrams complete, the reactions on the individual members are shown. The values of the reactions can be obtained from equations of equilibrium for each member. It is apparent that the reactions at A and C are equal to V_A and V_C, respectively. At B, however, the reaction is a combination of the shear values V_{BA} and V_{BC}. Because it is obvious that the axial forces in the members are zero, they have not been shown on the free-body diagrams.

Note in Fig. 2-9*b* that, according to the joint convention, the given moment at B of -3.2 kN-m results in $M_{BA} = 3.2$ kN-m and $M_{BC} = -3.2$ kN-m. It is helpful in interpreting the joint sign convention to visualize the effect of the moment on the direction of concavity, and hence the direction in which the joint tends to be rotated. The resulting signs of moment can be verified by checking the equilibrium of the joint. Taking the summation of moments on joint B,

$$M_{BA} + M_{BC} = 3.2 - 3.2 = 0$$

we see that the joint is in equilibrium.

2-4 SHEAR AND MOMENT DIAGRAMS

The ability to construct shear and moment diagrams is an essential part of structural analysis. It should be apparent from previous courses in mechanics and strength of materials that the values of shear and moment in a structure are functions of the position on the members. The statement that a shear or moment diagram is to be constructed for a structure simply means that the value of shear or moment is to be plotted as a function of position on the members of the structure. The plot of a function $f(x)$ is often encountered in mathematics. For our purposes, the function $f(x)$ now becomes shear $V(x)$ or moment $M(x)$, which are functions of x. It is convenient to construct the plot on the

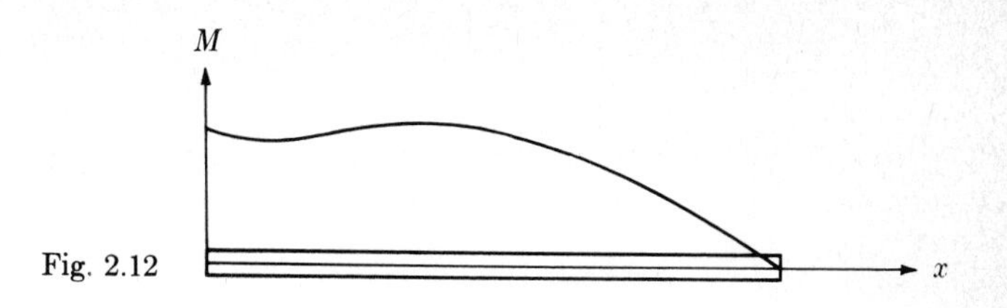

Fig. 2.12

members and to make the x axis coincident with the centroidal axis of the member, as shown in Fig. 2-12. It is not essential to label the centroidal axis as the x axis, but the label is a helpful reminder of its significance.

Having established the significance of shear and moment diagrams, we now need a sign convention. Since the diagrams are considered functions of the position on the member, we must first establish what are considered positive values of shear and moment. One of the most frequently used and convenient sign conventions for shear and moment is given in Figs. 2-9*a* and *d*. For shear this can be stated that shear is positive on a section of a member if the resultant of the external forces to the left of the section acts in an upward direction and negative if the resultant acts in a downward direction. For moment this can be stated that moment is positive if at the section under consideration the resulting moment in the member causes compression in the upper fiber. Because of the "up" and "down" references, this convention is appropriate only for beams. Application of this procedure is illustrated in the following example.

EXAMPLE 2-6 The shear and moment diagrams are to be constructed for the simply supported beam of Fig. 2-13. First the reactions at A and D are determined. Using $\Sigma M_D = 0$ and $\Sigma F_V = 0$, we find these values to be $R_A = 12.5$ kips (upward) and $R_D = 17.5$ kips (upward). Then the shear diagram is constructed, with the values of shear at various points on the beam obtained from free-body diagrams. The unknown shear and the unknown moment should be drawn on the free-body diagrams in positive directions so that resulting values will have the correct sign. Beginning at the left end of the beam and considering first the free-body diagram of a section just to the right of the reaction at A, we see from

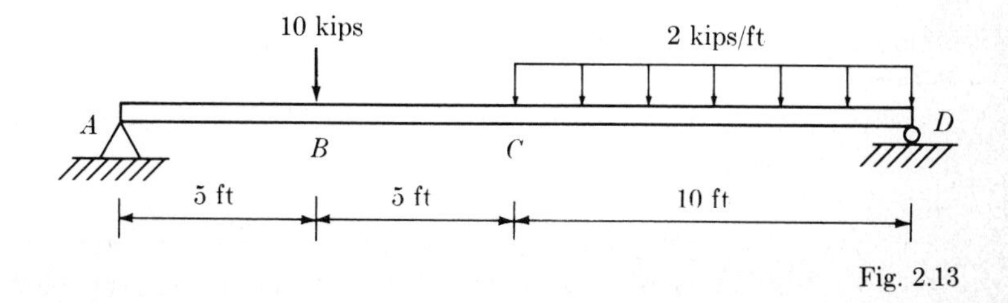

Fig. 2.13

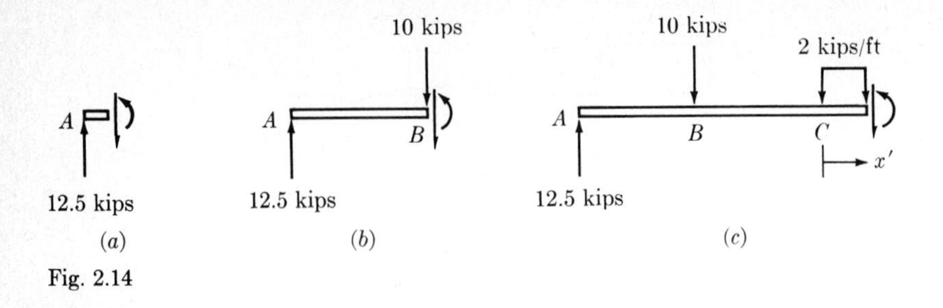

Fig. 2.14

Fig. 2-14*a* that the value of shear is 12.5 kips; that is, the resultant of external forces to the left of the section acts up with a magnitude of 12.5 kips. The shear remains constant out to point B. From the free-body diagram of Fig. 2-14*b* we see that the shear in the beam just to the right of point B is $12.5 - 10.0 = 2.5$ kips. The shear is again constant between points B and C. At point C the shear is 2.5 kips, and to the right of C we see from Fig. 2-14*c* that the shear decreases linearly with the distance x'. The general expression for shear between C and D is

$$V = 2.5 - 2x' \qquad 0 \leq x' \leq 10 \tag{2-1}$$

At the point where $x' = 1.25$ ft the value of shear is zero, and from that point to D it becomes more negative. At point D, where $x' = 10$ ft, the value of shear, from Eq. (2-1), is -17.5, in agreement with the previously determined value of the reaction at D. The resulting shear diagram for the member is shown in Fig. 2-15*a*. As stated earlier, the abscissa of the plot represents the position on the centroidal axis of the member. The student is encouraged to label the shear diagram and to denote the magnitude of shear as shown in Fig. 2-15*a*. The units used on the diagrams can be indicated next to the diagram title, as shown.

The moment diagram can similarly be constructed from free-body diagrams of sections of the member. The moment at point A is zero. From Fig. 2-14*b* we see that the moment increases from A to B by an amount equal to 12.5 multiplied by the distance from A. Thus at point B the moment is $(12.5)(5) = 62.5$ ft-kips. It is positive because the upper fibers of the member at point B are in compression. From point B to C the moment increases at a reduced rate, becoming 75.0 ft-kips at C. As shown in Fig. 2-14*c*, the general expression for moment can be written to the right of point C as

$$\begin{aligned} M &= (12.5)(10 + x') - (10)(5 + x') - 2x' \frac{x'}{2} \\ &= 75 + 2.5x' - x'^2 \qquad 0 \leq x' \leq 10 \end{aligned} \tag{2-2}$$

By differentiating this expression with respect to x' and setting the

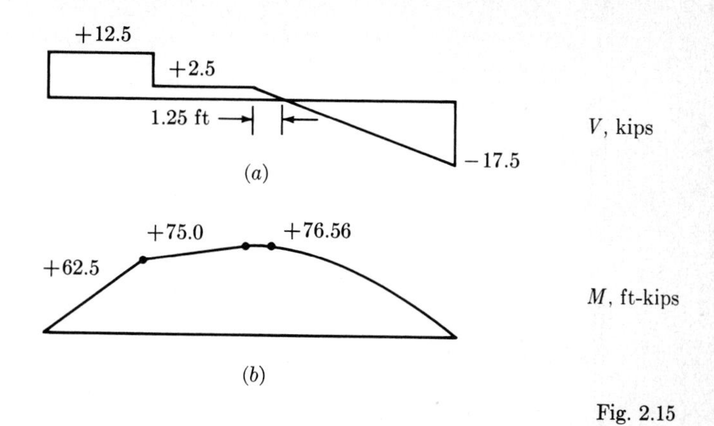

Fig. 2.15

resulting expression equal to zero,

$$\frac{dM}{dx'} = 2.5 - 2x' = 0 \tag{2-3}$$

we find that the maximum moment occurs at the point where $x' = 1.25$ ft. From Eq. (2-2), this maximum value of moment is 76.56 ft-kips, which corresponds to the point of zero shear obtained earlier. Substitution of $x' = 10$ ft into Eq. (2-2) yields $M = 0$ at the right support. The resulting moment diagram for the member is shown in Fig. 2-15*b*. Some interesting and useful relationships between the shear and moment diagrams will be discussed in the following section.

Unless it is properly applied, the above procedure can lead to difficulties in constructing shear and moment diagrams for structures whose members are oriented in a position other than horizontal; the difficulty is in the concept of which direction is "up" for such members. Consider again the (V,x) and (M,x) coordinate axes to be used for each member. If need be, a small sketch of these coordinate axes indicating positive quantities can be made next to the members. The following example illustrates this point.

EXAMPLE 2-7 Shear and moment diagrams are to be constructed for the three members of the frame of Fig. 2-16. From free-body considerations and equations of equilibrium, the external forces and moments on the individual members are found to have the values shown in Fig. 2-17. The student should verify that he understands the free-body diagrams and can obtain the values shown. In particular, he should understand the transformation of force components at C. A free-body diagram of joint C will be helpful.

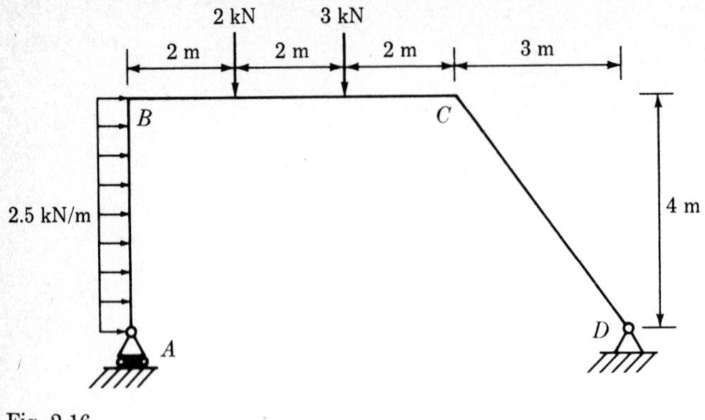

Fig. 2.16

With the free-body diagrams of Fig. 2-17, the shear and moment diagrams for the members are each constructed on a line diagram of the structure. Let us consider the shear diagram first. Positive shear for each member is indicated by the coordinate axes next to each member, as shown in Fig. 2-18*a*. The positive direction of the V axis corresponds to the upward direction of the positive-shear definition in Fig. 2-9*d*. In constructing shear and moment diagrams for total structures, the diagrams for adjacent members will often overlap. Possible confusion can be avoided if, as shown in Fig. 2-18, the diagrams are lightly lined normal to the axis of the member to which the diagram pertains.

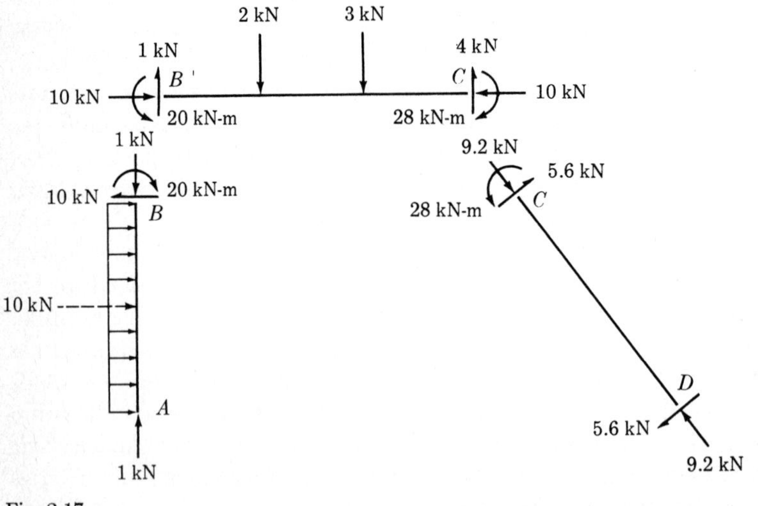

Fig. 2.17

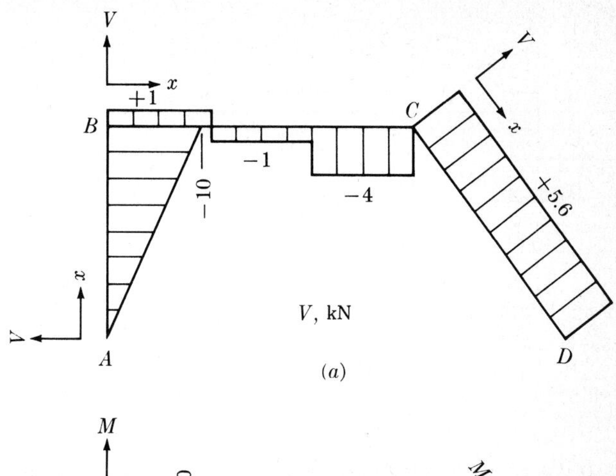

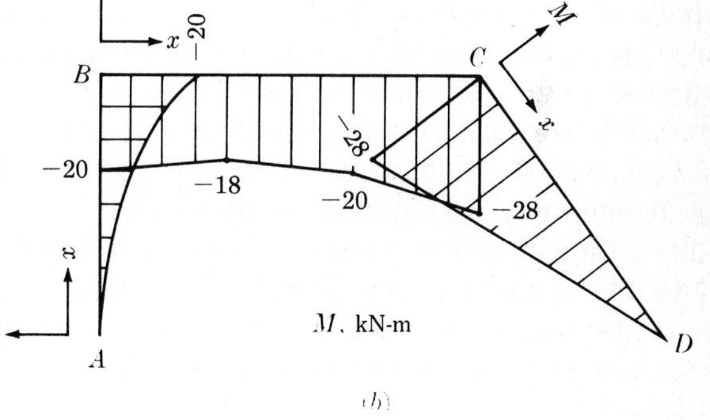

Fig. 2.18

The moment diagrams for the members are as shown in Fig. 2-18*b*. The positive direction of the M axis on each member corresponds to the upward direction of the positive moment definition in Fig. 2-9*a*.

Two other sign conventions that are often used, particularly for frames, are those where the moment diagram is plotted on either the tension side or the compression side of the members. Such conventions do not depend on an "up" direction when the diagram is constructed. If plotted on the tension side, for example, the value of moment at a point on a member is plotted on the side of the member that is in tension. Such a convention does not have negative values; the diagram indicates the value of moment, and the side of the member on which it is plotted establishes the direction of moment. For the structure of Fig. 2-16, it is seen that the moment diagram would be plotted on the outer sides of each member if constructed on the tension side of the members.

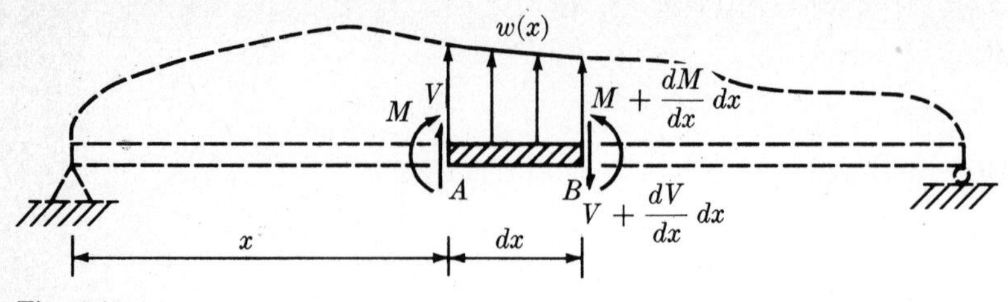

Fig. 2.19

2-5 RELATIONSHIPS OF LOAD, SHEAR, AND BENDING MOMENT

As stated earlier, there are some interesting and useful relationships among load, shear, and bending moment in a member. These relationships are in the form of integral and differential equations.

Consider a differential element taken from a member subjected to a variable load $w(x)$, as shown in Fig. 2-19. The load is positive when it acts in the direction shown. Positive shear and moment are shown acting on the differential element at section A, which is a distance x from the left end of the beam, and at section B, which is a distance dx farther along the beam. The value of shear and moment vary with x. Strictly speaking, they should be written as $V(x)$ and $M(x)$, but for convenience they will be written simply as V and M. The load will also be denoted as w rather than $w(x)$.

Let us first consider the shearing forces acting on the element. There is a shearing force of V at A and a shearing force $V + (dV/dx)\,dx$ at B. Taking the summation of vertical forces on the element, we have

$$V + w\,dx - (V + dV) = 0$$

which reduces to

$$dV = w\,dx \tag{2-4}$$

This can also be written in the form

$$\frac{dV}{dx} = w \tag{2-5}$$

Now consider the bending moment on the element. There is a bending moment of M acting on the section at A and a moment of $M + (dM/dx)\,dx$ acting at B. Taking the summation of moments about B, we see that

$$M + V\,dx + w\,dx\,\frac{dx}{2} - (M + dM) = 0$$

Neglecting the higher-order term including dx^2, we find that this expression reduces to

$$dM = V\,dx \tag{2-6}$$

Equation (2-6) can also be written in the form

$$\frac{dM}{dx} = V \tag{2-7}$$

The significance of these relationships is apparent in the structure of Example 2-7. The loading on the structure was shown in Fig. 2-16, and the shear and moment diagrams were shown in Fig. 2-18. Consider member AB. We see from Fig. 2-16 that the expression for load is

$$w = -2.5 \tag{2-8}$$

In this case it is constant. From Eq. (2-4) we therefore have

$$dV = -2.5\,dx$$

Integrating both sides, we obtain

$$V = -2.5x + C_1$$

where C_1 is a constant of integration to be determined from the boundary conditions. We see from the free-body diagram of member AB in Fig. 2-17 that the shear is zero when x is zero, that is, $V(0) = 0$. It is therefore obvious that $C_1 = 0$. The general expression for shear in member AB becomes

$$V = -2.5x \tag{2-9}$$

When $x = 4$, $V = -10$, which agrees with the value obtained earlier in Fig. 2-18.

With the general expression for shear from Eq. (2-9), Eq. (2-6) becomes

$$dM = -2.5x\,dx$$

Integrating, we obtain

$$M = -1.25x^2 + C_2$$

From the free body of Fig. 2-17 we see that $M(0) = 0$. Therefore $C_2 = 0$. The general expression for moment in member AB becomes

$$M = -1.25x^2 \tag{2-10}$$

When $x = 4$, $M = -20$, as in Fig. 2-18*b*.

The same procedure can be applied to member BC with some slight modifications. Member BC is subjected to concentrated loads, and these must be treated in a manner similar to the treatment of discontinuous functions in calculus; that is, the above relationships among load, shear, and moment remain valid at all points on the member except right at the concentrated loads.

Consider the segment of member BC from point B to the 2-kN load. We see that $w = 0$. Integrating Eq. (2-4) and observing from the free-body diagram of Fig. 2-17 that $V(0) = 1$, we find that the general expression for shear is $V = 1$. Then, integrating Eq. (2-6) and using the boundary condition $M(0) = -20$, we have the general expression for moment,

$$M = x - 20$$

When $x = 2$, $M = -18$.

To continue with these relationships between the 2-kN and the 3-kN loads, the origin of x is shifted just to the right of the 2-kN load. The boundary conditions for this section are $V(0) = -1$ and $M(0) = -18$. The expression for the load remains $w = 0$.

From the above developments and discussions, the following general statements can be made, exclusive of the points where concentrated loads and couples are applied to the beam.

1. The change in the value of shear between two points on a member is equal to the area under the load diagram between the points [Eq. (2-4)]. Also, the slope of the shear diagram at any point is equal to the value of load at that point [Eq. (2-5)].
2. The change in the value of moment between two points on a member is equal to the area under the shear diagram between these points [Eq. (2-6)]. Also, the slope of the moment diagram at any point is equal to the value of shear at that point [Eq. (2-7)].

The student should verify the statements concerning the slope of the shear and moment diagrams for the diagrams in Fig. 2-18. An understanding of these relationships will be helpful in determining the direction of curvature for nonlinear shear and moment diagrams.

2-6 PRINCIPLE OF SUPERPOSITION

The principle of superposition is used continually in mechanics of materials and structural analysis. For example, in strength-of-materials studies the total stress at a point in a material resulting from various types of applied forces can be obtained by summing the stresses due to the forces considered individually. In determining the reactions of a simple beam subjected to a number of loads we can obtain the total reaction by summing the reactions due to the loads considered individually. The principle of superposition can be stated as follows:

PRINCIPLE OF SUPERPOSITION: *The total effect at some point in a structure due to a number of loads applied simultaneously is equal to the sum of the effects for the loads applied individually.*

For the principle of superposition to be valid there must be a linear relationship among forces, stresses, and deflections. There are two conditions for which superposition is *not* valid:

1. When the structural material does not behave according to Hooke's law; that is, when the stress is not proportional to the strain
2. When the deflections of the structure are so large that computations cannot be based on the original geometry of the structure

Unless otherwise stated, the principle of superposition is assumed to be valid in subsequent discussions.

2-7 GRAPHIC STATICS

Graphical methods of analysis are not used so extensively as they have been in the past, primarily because of the potential offered by high-speed computers. However, an understanding of the principles of graphics is a valuable aid in visualizing the force effects on a structure. Even a rough sketch of a graphical solution will provide deeper insight into the behavior of the structure.

The basic concepts of force have been treated in courses in mechanics and will therefore be discussed here only with regard to graphical representation. A force is defined by direction, magnitude, and point of application. Arrows, drawn to an appropriate scale, are used to represent forces in graphical analysis. Thus a force Q of 10 kips acting to the right would be represented as shown in Fig. 2-20*a*, with a scale of 5 kips = 1 inch.

Two planar and intersecting forces Q_1 and Q_2 can be combined into a resultant force R, as shown in Fig. 2-20*b*. These forces are combined in Fig. 2-20*b* by the parallelogram law; that is, the diagonal of the parallelogram whose sides are formed from the two given forces represents the resultant of these forces. The resultant can also be obtained

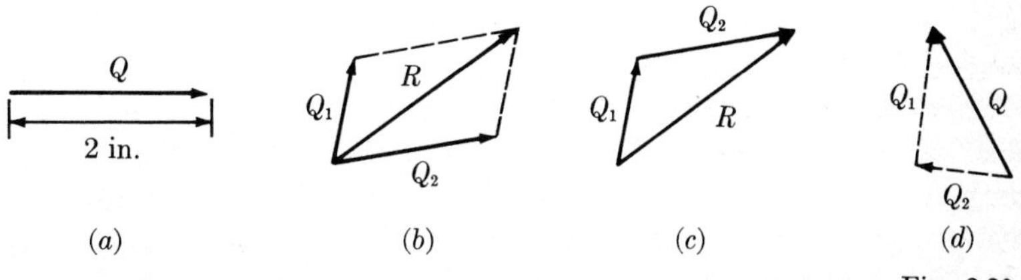

Fig. 2.20

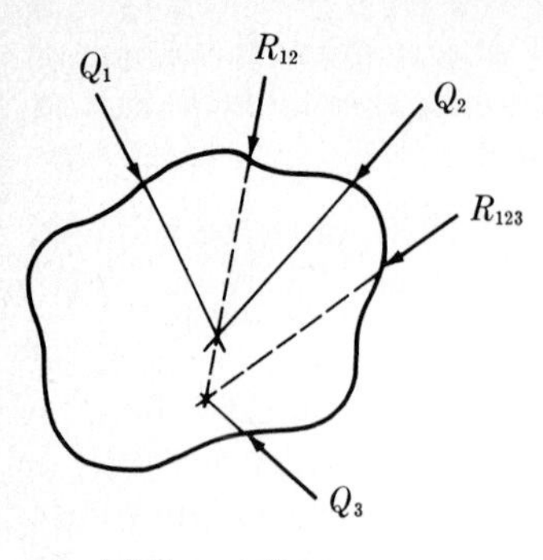

(a) Space diagram

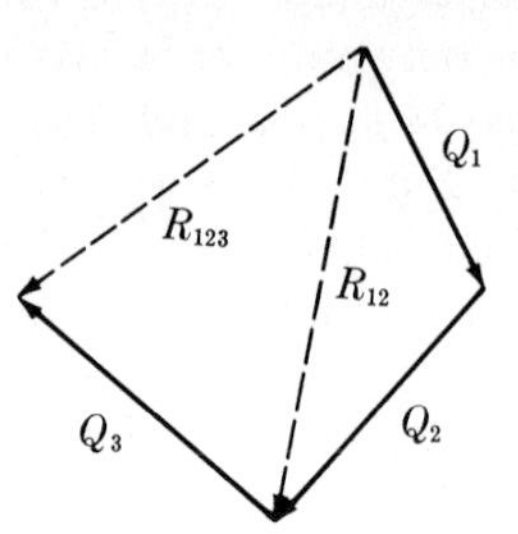

(b) Force polygon

Fig. 2.21

by placing the forces tip to tail, as shown in Fig. 2-20c. The procedure of combining two forces into their resultant force is referred to as the *composition* of forces Q_1 and Q_2. The reverse procedure, replacing a single force by two components, is referred to as the *resolution* of a force Q. Force Q is resolved in Fig. 2-20d into components Q_1 and Q_2.

Let us next consider a rigid body acted upon by three forces, Q_1, Q_2, and Q_3, as shown in Fig. 2-21a. A diagram such as that in Fig. 2-21a, where the points of application of the forces are drawn to scale and the directions of the forces are accurately drawn, is referred to as a *space diagram*. The space diagram can be used with the diagram in Fig. 2-21b to obtain the resultant of the three forces. In Fig. 2-21b the forces on the rigid body are placed tip to tail in the order in which they are encountered in going around the rigid body. We proceeded around the body clockwise in this case, but we could have proceeded counterclockwise. The forces in Fig. 2-21b are drawn to an appropriate scale. The scales for the space diagram and diagram of forces need not be the same. Each scale should be chosen to represent most conveniently the geometric dimensions and the magnitude of the forces.

The magnitude and direction of the resultant of forces Q_1 and Q_2 is denoted as R_{12} in Fig. 2-21b. The point of application on the rigid body of R_{12} is determined on the space diagram by the intersection of the lines of action of Q_1 and Q_2. R_{12} can next be combined with Q_3 in Fig. 2-21b to give the resultant of the three given forces, R_{123}. The point of application of R_{123} on the rigid body is determined on the space diagram by the intersection of the lines of action of R_{12} and Q_3. Thus the resultant force on the rigid body is represented as R_{123} in Fig. 2-21a. For equilibrium of the rigid body a force equal and opposite to R_{123} must be applied along the line of action of R_{123}.

The arrangement of the forces shown in Fig. 2-21b is referred to as a *force polygon*. If a force equal but opposite to R_{123} is applied to the body, the force polygon closes. A closed polygon implies that the equilibrium equations $\Sigma F_X = 0$ and $\Sigma F_Y = 0$ are satisfied. The third equilibrium equation, $\Sigma M = 0$, is satisfied on the space diagram by applying the required equilibrium force along the line of action of R_{123}. This fact can be substantiated by considering the summation of moments about the point of intersection of R_{12} and Q_3, which is also the point through which R_{123} passes.

If the directions of the three forces Q_1, Q_2, and Q_3 are parallel, or nearly parallel, as shown in Fig. 2-22*a*, the intersection points of the forces cannot be obtained. To obtain the location of the resultant for this condition a different approach is used. As an illustration of the technique, let us determine the resultant for forces Q_1, Q_2, and Q_3 in Fig. 2-22*a*. The force polygon for these three forces is as shown in Fig. 2-22*b*. The resultant of the forces is R_{123}. A point P is located in the vicinity of the force polygon, and lines are drawn to the extremities of the given forces. Point P is referred to as a *pole*. The most appropriate location of the pole P will become evident as the method is developed. The lines connecting the pole P to the extremities of the forces are referred to as *rays*. The end points of the rays are numbered 1 to 4, as shown.

Consider force Q_1 in Fig. 2-22*b* to be resolved into components coincident with the directions of lines $1P$ and $P2$. The directions of these components are shown next to the rays. At some arbitrary point A on the line of action of Q_1 in Fig. 2-22*a* the directions of the components $1P$ and $P2$ are constructed. The line representing $P2$ is extended to intersect force Q_2 at B.

Force Q_2 is considered next and resolved into components $2P$ and $P3$,

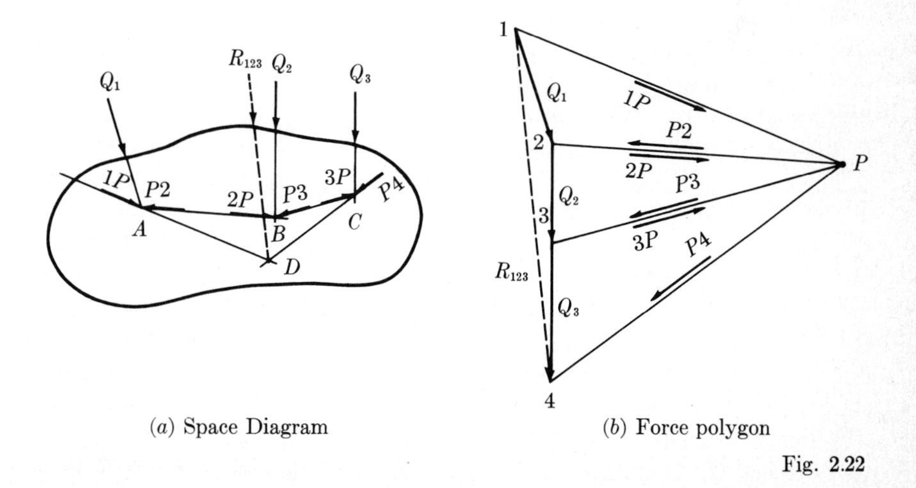

(*a*) Space Diagram

(*b*) Force polygon

Fig. 2.22

as shown in Fig. 2-22*b*. The direction of *P*3 is drawn through point *B* of Fig. 2-22*a*, the location of *B* having been established from *P*2. *P*3 is extended to intersect Q_3 at point *C*. This procedure is repeated for Q_3, with *P*4 drawn through point *C*, as shown.

Considering the results, we see in Fig. 2-22*b* that *P*2 and 2*P* cancel, as do *P*3 and 3*P*. We have therefore essentially replaced the resultant of forces Q_1, Q_2, and Q_3 by the components 1*P* and *P*4. The intersection of the lines of action of 1*P* and *P*4 at point *D* on the space diagram locates the position of the resultant R_{123} on the rigid body.

The polygon *ABCDA* formed on the space diagram is referred to as a *funicular polygon*. The sides of this polygon are often referred to as *strings*. Note that the funicular polygon shown is not unique, as the choice of the location of point *A* along the line of action of Q_1 is arbitrary as is the location of the pole *P*. It should now be apparent that the general rule for selecting the location of point *P* is to select its location such that the strings of the funicular polygon will intersect the lines of action of the given forces at near 90° angles. Thus less space is required for the diagram, and greater accuracy can be attained.

For equilibrium of the body a force equal and opposite to R_{123} must be applied to the body; it must be applied through point *D*. In summary, it can be stated that for equilibrium of the body the force polygon and the funicular polygon must close.

The principles of the funicular and force polygons can be used to determine the reactions of a statically determinate structure. Consider the structure of Fig. 2-23*a*. The reaction at *A* and the reaction of the cable *BC* are to be determined by graphical analysis. It should be noted for later use that the direction of the reaction in cable *BC* is known. Knowing the direction of one reaction is the key to determining the reactions graphically. The force polygon for the given forces Q_1 and Q_2 is constructed in Fig. 2-23*b*. The forces are placed in the order in which they are encountered in going counterclockwise around the structure, beginning with Q_1. The pole *P* is located as shown, and rays are drawn to the extremities of the given forces.

The construction of the funicular polygon begins at a particular point in Fig. 2-23*a*. Although we do not know the magnitude and direction of the reaction at *A*, we do know that it passes through point *A*. We therefore begin the construction of the funicular polygon at this point. A line parallel to ray 1*P* is passed through *A* and extended to intersect the line of action of Q_1. The student should verify with reference to the developments in Fig. 2-22 that this first line through *A* represents the component common to the unknown reaction R_A and the force Q_1.

A line parallel to ray 2*P* is drawn from the intersection point on Q_1 to intersect Q_2. Similarly, a line parallel to 3*P* is drawn through this

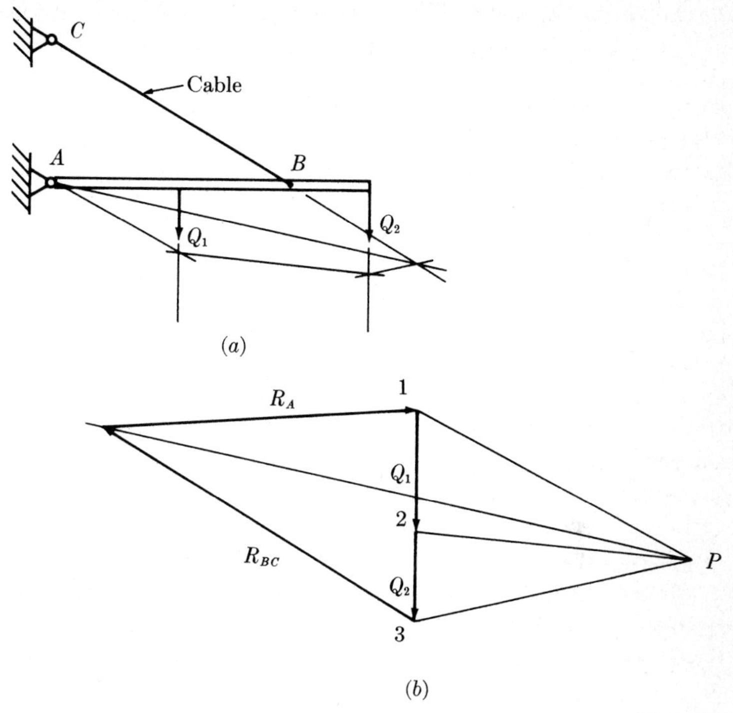

Fig. 2.23

intersection point until it intersects the line of action of the cable BC. It is at this point that we see the need for knowing the direction of one of the reactions. From this point of intersection we can draw the closing line of the funicular polygon through point A.

A line parallel to the closing line of the funicular polygon is drawn through the pole P. From point 3 on the force polygon, the direction of the reaction in the cable R_{BC} is drawn and extended until it intersects the line just drawn through P. The intersection of these two lines determines the magnitude of the reaction R_{BC}. The vector from this intersection to point 1 represents the reaction at A. Thus we see that the force polygon is closed and the structure is in equilibrium.

To successfully determine reactions by the above method we must know the direction of one of the reactions. Commonly encountered forms of such reactions for which the directions are known are roller supports, pinned-pinned members serving as supports, and cable supports. An additional factor to remember is that construction of the funicular polygon is begun at the support other than the one for which the direction is known.

PROBLEMS

2-1. Draw complete free-body diagrams for members AC and BD. Determine the horizontal and vertical components of reaction at C and D and the normal reaction at A.

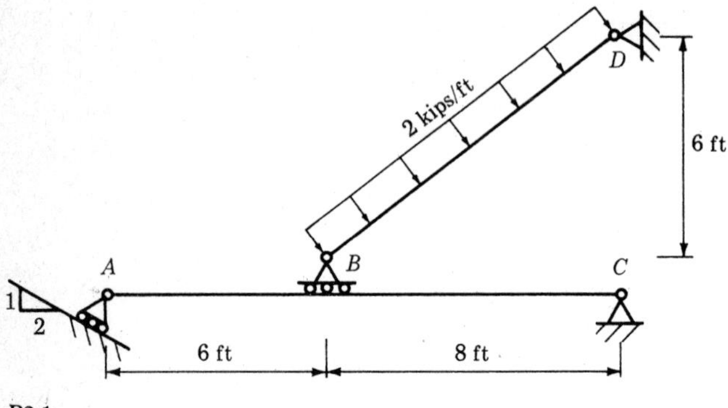

P2.1

2-2. Draw complete free-body diagrams for members AC and CE. Determine the horizontal and vertical components of reaction at B, D, and E.

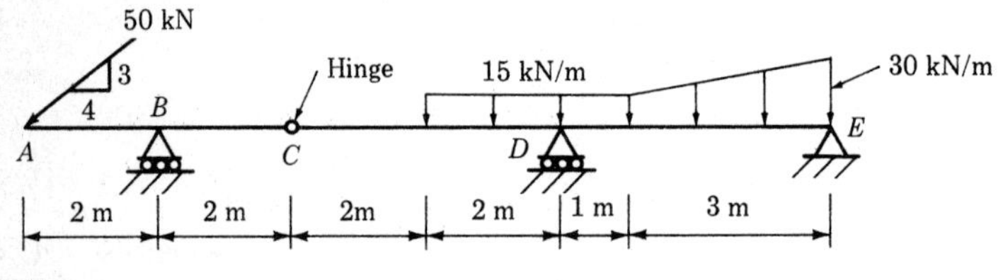

P2.2

2-3. Determine the horizontal and vertical components of reaction at A and the reaction normal to the incline at B.

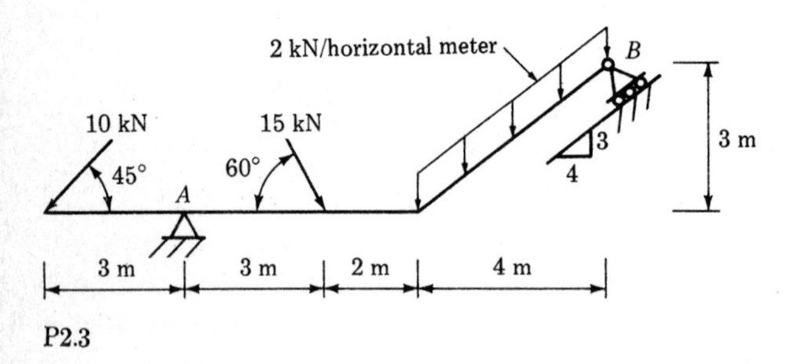

P2.3

2-4. Determine the horizontal and vertical components of reaction at A and D for the frame shown.

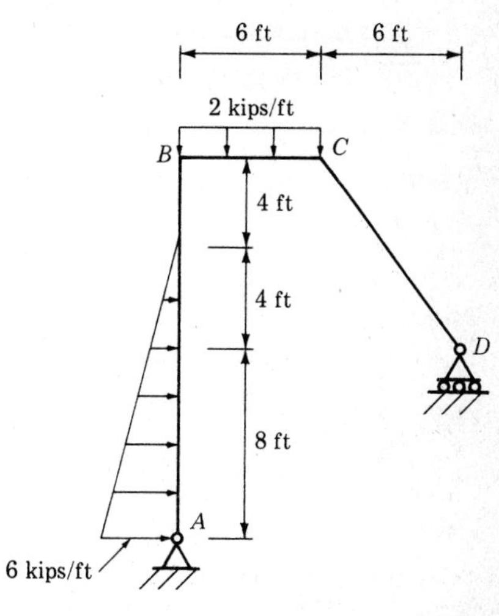

P2.4

2-5 through 2-8. For the beams shown, (*a*) determine the reactions; (*b*) construct complete shear and moment diagrams. Use the beam sign convention.

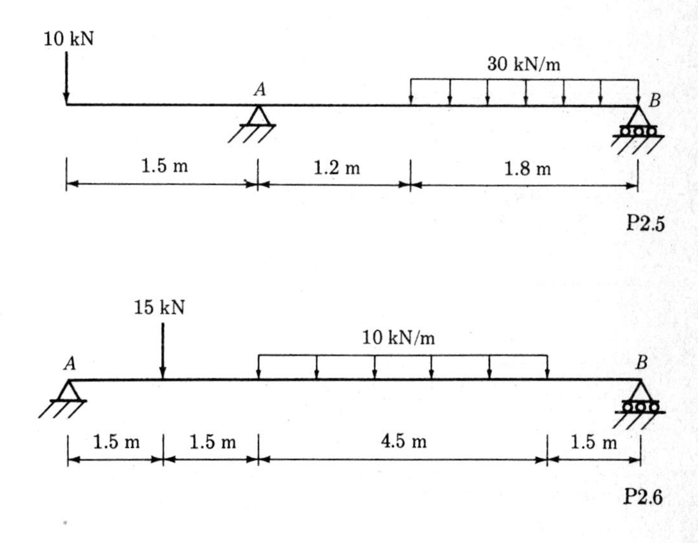

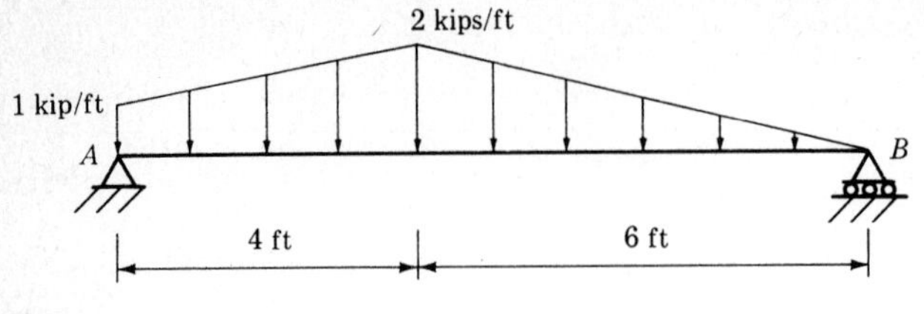

P2.7

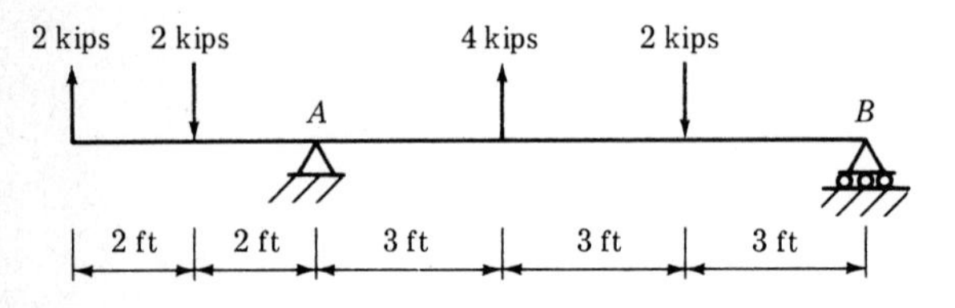

P2.8

2-9 through 2-14. Determine the horizontal and vertical components of reaction for the structures shown, and construct complete shear and moment diagrams for the members indicated. Use the beam sign convention.

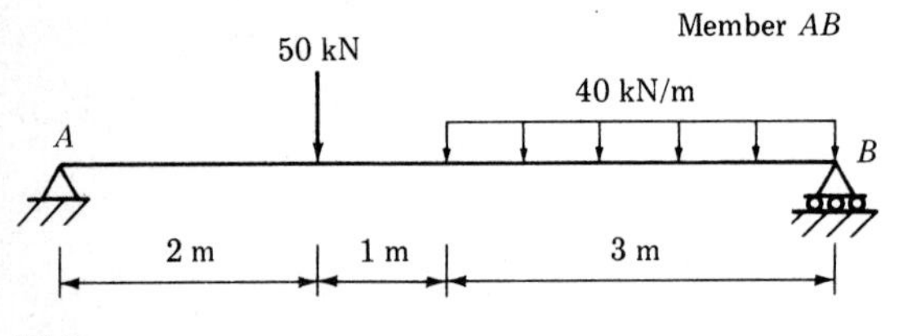

P2.9

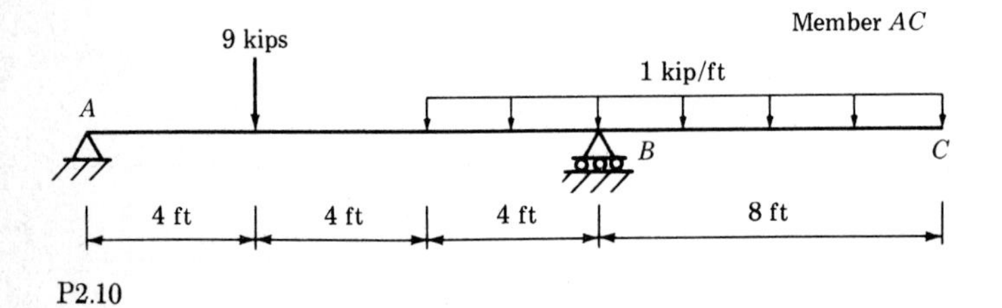

P2.10

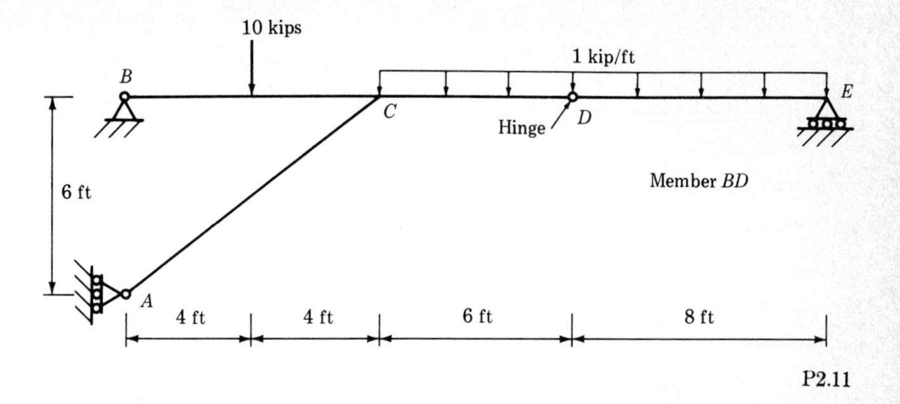

P2.11

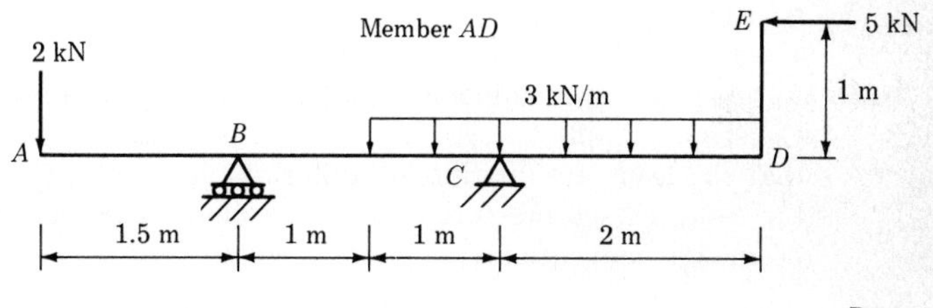

P2.12

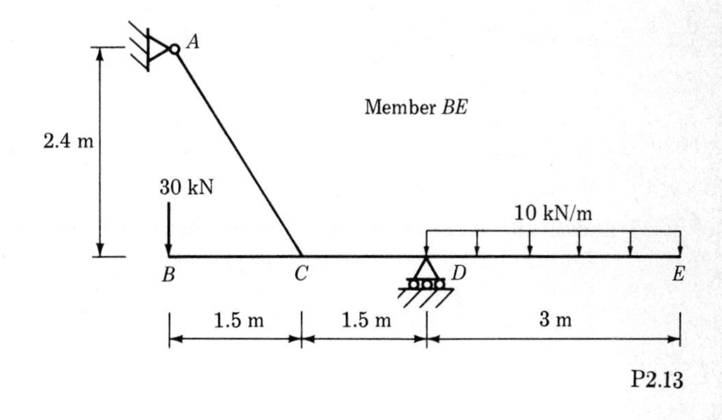

P2.13

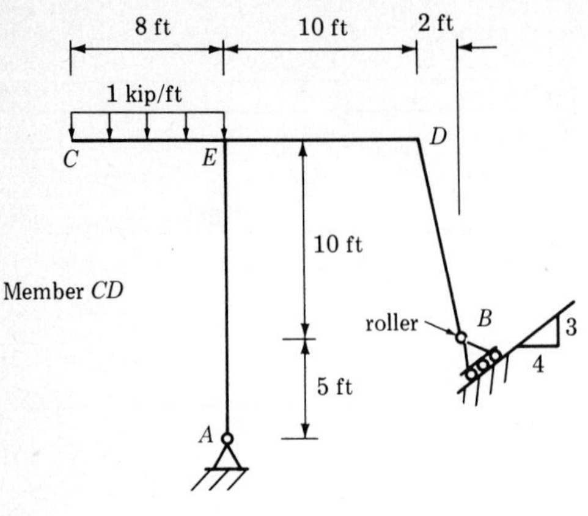

P2.14

2-15. Construct complete shear and moment diagrams for the beam of Prob. 2-2. Use the beam sign convention.

2-16. A spread footing supports two 1.2-MN loads as shown in Fig. P2-16. Construct complete shear and moment diagrams for the footing. Assume that the loads are distributed uniformly across the 1-m widths and that the resulting soil pressure on the base of the footing is uniform. Use the beam sign convention.

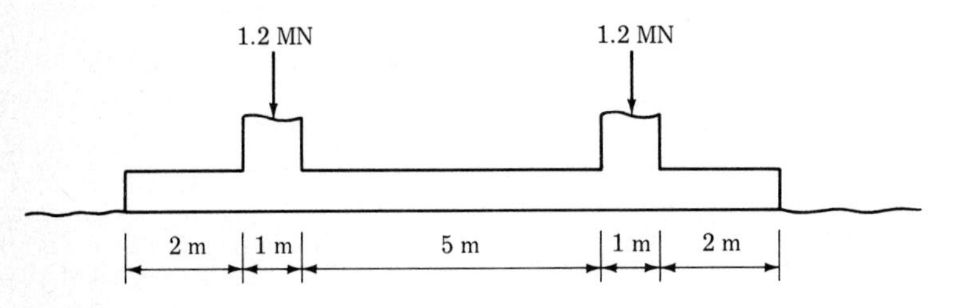

P2.16

2-17 and 2-18. Construct the moment diagrams for the members of Probs. 2-10 and 2-11, respectively, on the tension side of the members.

2-19. Construct the moment diagram for the frame of Example 2-7 on the tension side of the members, and express the values in U.S. Customary units.

2-20 and 2-21. For the frames shown, (*a*) determine the horizontal and vertical components of reaction; (*b*) construct the moment diagram for the frames on the tension side of the members.

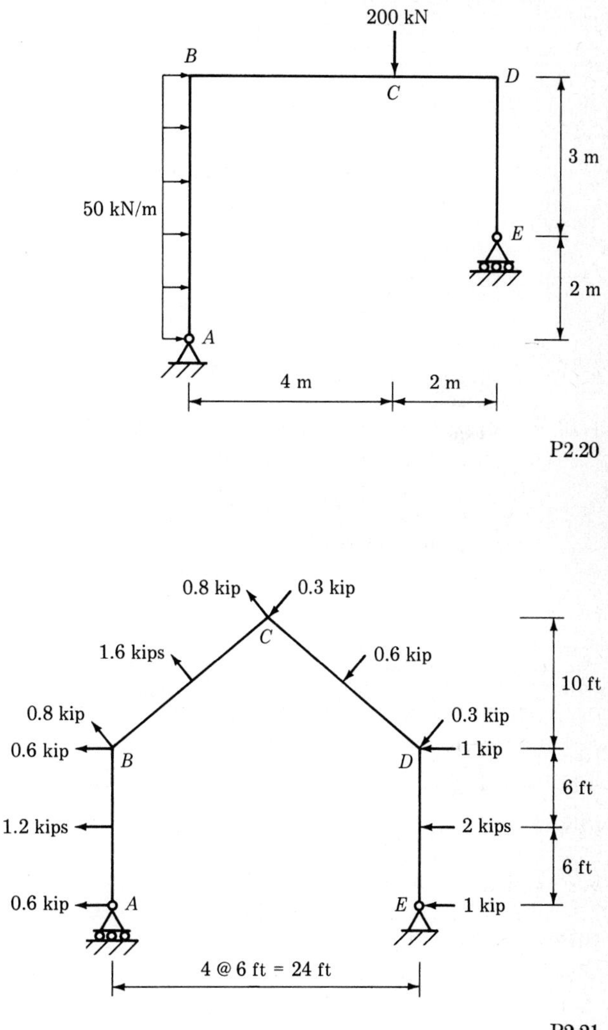

2-22. For the frame of Fig. P2-22, (*a*) determine the horizontal and vertical components of reaction at A and C; (*b*) construct the moment diagram for the frame on the compression side of the members.

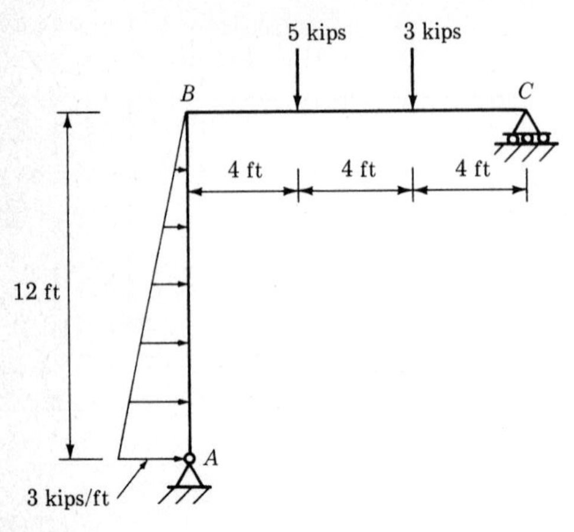

P2.22

2-23. For the frame of Fig. P2-23, (*a*) determine the horizontal and vertical components of reaction at C and D; (*b*) construct the moment diagram for the frame on the tension side of the members.

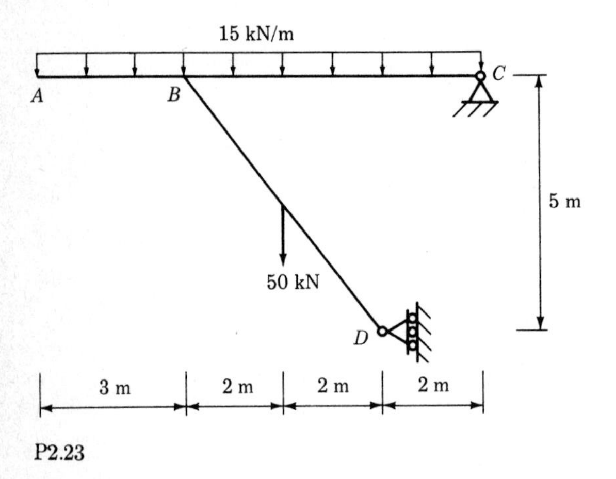

P2.23

2-24. For the frame of Fig. P2-24, (*a*) determine the horizontal and vertical components of reaction at *A* and *D*; (*b*) construct the moment diagram for the frame on the tension side of the members.

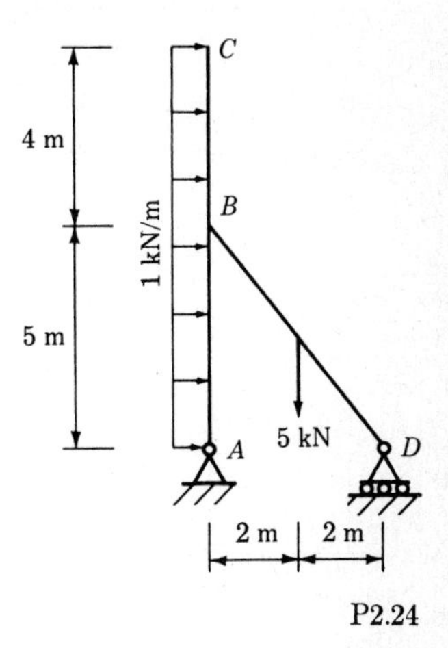

P2.24

2-25. Graphically determine the reactions at *A* and *B* for the frame of Fig. P2-14.

2-26. Graphically determine the reactions at ① and ③ for the truss of Fig. P3-6.

CHAPTER

THREE

PLANE TRUSSES

3-1 INTRODUCTION

A *truss* is a structure whose members are connected and arranged in such a way that they are subjected primarily to axial loading. When all members of a truss lie in one plane, the truss is referred to as a *plane truss*. A three-dimensional truss is referred to as a *space truss*. Space trusses will be discussed in the following chapter.

The basic element of plane truss is a triangle. Three members arranged as shown in Fig. 3-1*a* and pinned together at their ends form a stable structure for supporting the load Q_1. The truss of Fig. 3-1*a* can be enlarged by adding two more members to form a truss composed of two triangular elements, as shown in Fig. 3-1*b*. Various combinations of the basic triangular elements produce general truss structures such as that shown in Fig. 3-1*c*.

In analyzing trusses the following assumptions are made:

1. The members are joined together at their ends by frictionless pins.
2. The loads and reactions are applied to the structure only at the joints.
3. The individual members of the truss are straight.

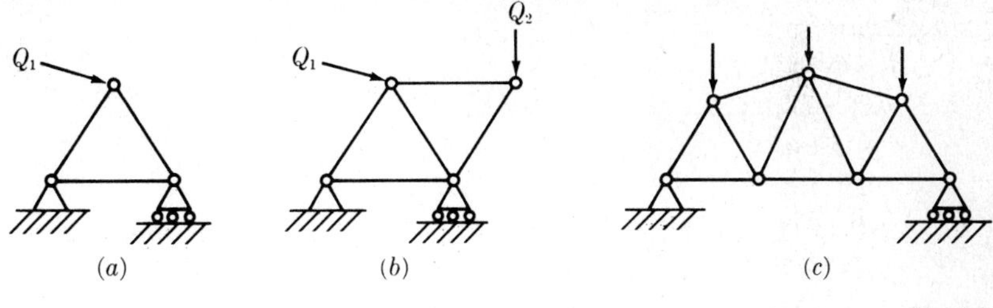

Fig. 3.1

The first of these assumptions is seldom completely satisfied in an actual truss. For example, the welded or bolted gusset plates commonly used to join the member ends do not exemplify a pinned condition. In many cases, however, where the members of the truss are long and slender and eccentricities in the joints are small, little moment is transmitted by the members, and the assumption of a pinned connection produces acceptable results. Later in the text methods will be developed for determining the magnitude of the moments that actually can be present in a truss. Such moments are referred to as *secondary moments*.

A general free-body diagram of a truss member is shown in Fig. 3-2*a*. Because the ends of the member are pinned, there are no moments acting on the ends. The possible forces transmitted by the pins are denoted by horizontal and vertical components at A and B. These pin forces can also be represented by force vectors $\mathbf{F}_A$ and $\mathbf{F}_B$. Applying the vector expression for equilibrium, $\mathbf{F}_R = \mathbf{0}$, we see that $\mathbf{F}_A = -\mathbf{F}_B$; that is, the forces must be equal and oppositely directed. The member is therefore subjected to either axial tension or axial compression, as shown in Fig. 3-2*b*.

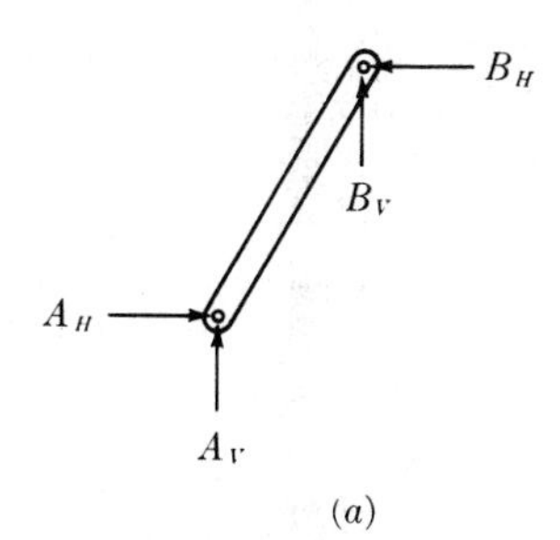

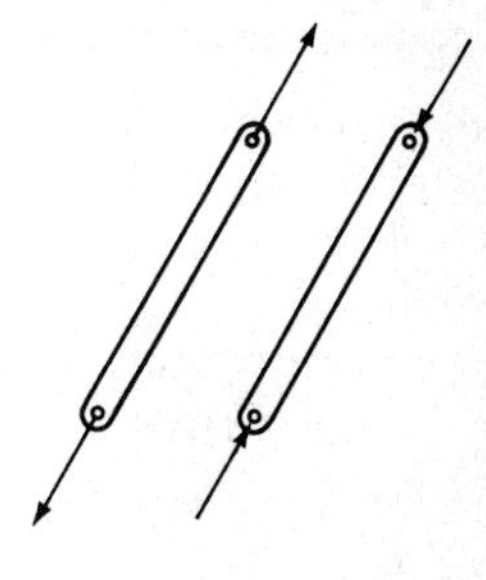

Fig. 3.2

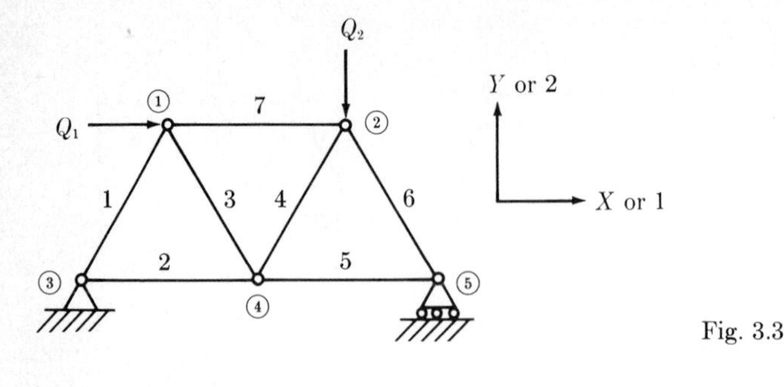

Fig. 3.3

3-2 TRUSS NOTATION

The truss notation used in this text is intended to clearly denote the quantity under consideration and is of a form that lends itself well to possible use in machine computations. The notation for the analysis of a truss is shown in Fig. 3-3. The same general form of notation will be used for other types of structures later in this text. The members of the structure are arbitrarily numbered as shown, with the joints and supports denoted by numbers enclosed in circles. Numbering the members and joints permits the use of subscripts and superscripts which are easily incorporated into computer languages. External loads on the structure are denoted by the letter Q, subscripted with the number of the joint at which the load acts.

The axial forces in the members are denoted by the letter P with the proper subscript; thus P_4 is the axial force in member 4 of Fig. 3-3.

Because it is often necessary to work in terms of components of force, it is convenient to adopt a coordinate system, as shown next to the structure in Fig. 3-3. The X component of reaction at joint ③ can then be written as R_{3X} or, if numbers are to be retained, as R_{31}, where the second subscript refers to the component of force. The number 1 refers to the X component and the number 2 to the Y component. It will probably be less confusing to use X and Y subscripts while learning the theory. The change to numbers can readily be made when desired.

Tension in a member is denoted by a plus sign (+), and compression is denoted by a minus sign (−). Some engineers prefer to use the letter T for tension and C for compression, but for consistency in using notation adaptable for computer languages we shall use plus and minus signs.

3-3 METHOD OF JOINTS

One of the methods used in analyzing a truss for the member forces is the method of joints. This analysis entails the use of free-body diagrams

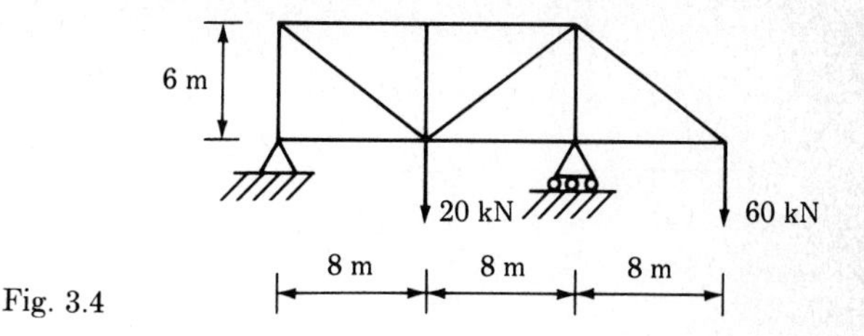

Fig. 3.4

of the joints, with the equilibrium equations $\Sigma F_X = 0$ and $\Sigma F_Y = 0$ written for each one. An aspect of this method that sometimes causes difficulty is determining the joint at which the analysis should begin and the order in which the succeeding joints should be taken. The governing criterion for the choice of a joint is that there must not be more than two unknown member forces. The method is illustrated in the following example.

EXAMPLE 3-1 The member forces are to be determined by the method of joints for the truss shown in Fig. 3-4. The first step in the analysis is to assign appropriate notation to the joints and members. With the notation discussed in the previous section, the joints and members are indicated as shown in Fig. 3-5. The unknown reactions at joints ④ and ⑥ are also shown. For the general solution of this truss by the method of joints we must know the values of the reactions, so we determine these first. From equations of equilibrium for the entire truss, the reactions are found to be

$$R_{4X} = 0 \qquad R_{4Y} = -20 \text{ kN} \qquad R_{6Y} = +100 \text{ kN}$$

The next important step is to study the structure and determine which path can be taken so that the joints will have only two unknowns. One path that will work is the order of joints ④-①-②-⑤-⑥-③- check ⑦. There are other paths, but this will illustrate the method.

A free-body diagram of joint ④ is drawn as shown in Fig. 3-6*a*. In drawing the free-body diagrams of the joints, it is convenient to show the

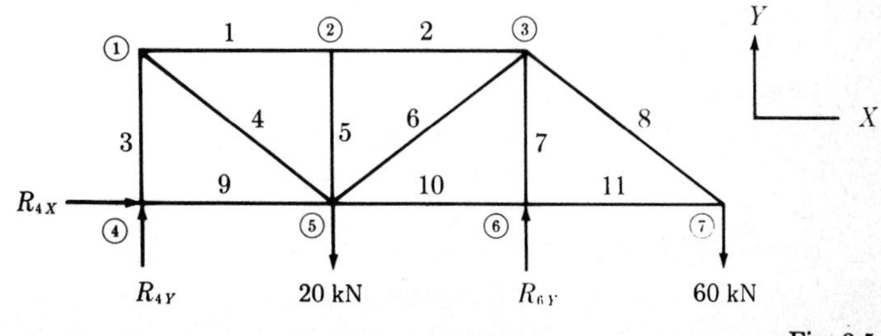

Fig. 3.5

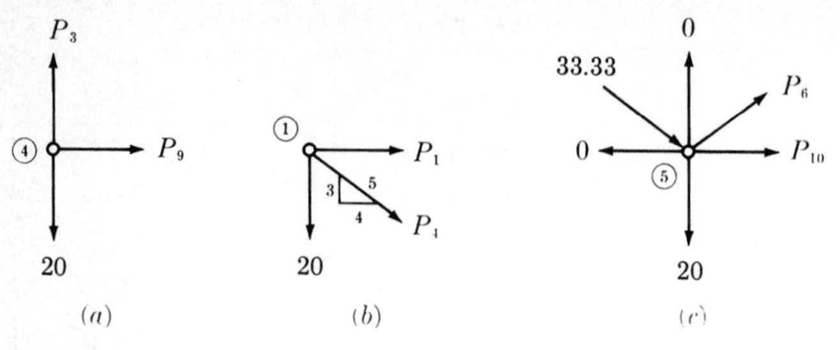

Fig. 3.6

unknown member forces acting in tension. If the answer then comes out negative, the member is in compression. From Fig. 3-6*a*, $\Sigma F_Y = 0$ results in

$$P_3 - 20 = 0 \qquad \text{or} \qquad P_3 = +20 \text{ kN}$$

P_3 is therefore in tension. It is obvious from Fig. 3-6*a* that $P_9 = 0$. The results obtained at each joint can be recorded on a summary diagram of the truss, as shown in Fig. 3-7. Such a diagram of the sign and magnitude of the member forces enables us to proceed to the succeeding joints more expeditiously.

Proceeding to joint ①, we draw the previously determined value of P_3 on the free-body diagram in its determined direction (Fig. 3-6*b*). Taking the summation of forces in the Y direction, we obtain

$$-20 - P_{4Y} = 0 \qquad \text{or} \qquad P_{4Y} = -20 \text{ kN}$$

From the geometric relationships of the components of P_4 we see from Fig. 3-6*b* that

$$P_4 = \tfrac{5}{3}P_{4Y} = (\tfrac{5}{3})(-20) = -33.33 \text{ kN}$$

From the summation of forces in the X direction we obtain

$$P_1 + P_{4X} = 0$$
$$P_1 + \tfrac{4}{3}(-20) = 0$$
$$P_1 = +26.67 \text{ kN}$$

P_4 is therefore in compression, and P_1 is in tension. These values are recorded in Fig. 3-7.

The remaining member forces are determined by the same procedure, with the joints taken in the order suggested above. The resulting member forces are summarized in Fig. 3-7.

Instead of using scalar equilibrium equations at the joints, as was done above, we can use vector expressions. The choice of which to use in two-dimensional analysis is largely a matter of personal preference. There

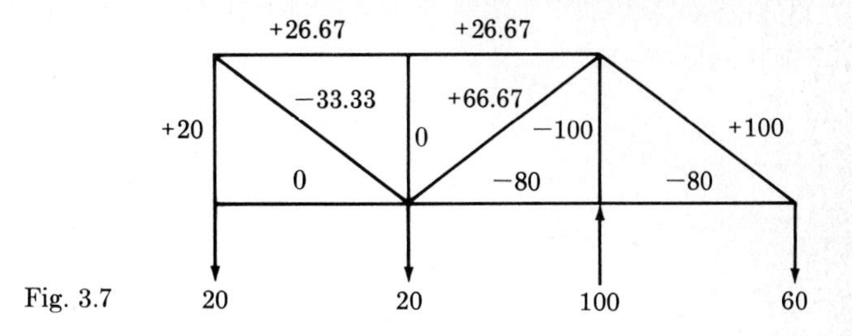

Fig. 3.7

are cases, particularly with space trusses, in which vector expressions can be used to good advantage.

Consider the analysis of joint ⑤ as shown in Fig. 3-6*c*. After the truss joints have been analyzed in the order given, the only unknowns at joint ⑤ are P_6 and P_{10}. To write the vector equilibrium equation at joint ⑤ we must determine the unit vector **u** of the members at the joint. The unit vector of a member is considered positive when directed away from the joint, so that the force results correspond to the sign convention being used. The unit vector can be determined from the given components of the lengths of the members. Thus the unit vector for member 4 at joint ⑤ is

$$\mathbf{u}_4 = \frac{-8\mathbf{i} + 6\mathbf{j}}{(8^2 + 6^2)^{1/2}} = \frac{-4\mathbf{i} + 3\mathbf{j}}{(4^2 + 3^2)^{1/2}} = -0.8\mathbf{i} + 0.6\mathbf{j}$$

Similarly, for the other members

$$\mathbf{u}_6 = 0.8\mathbf{i} + 0.6\mathbf{j}$$
$$\mathbf{u}_{10} = \mathbf{i}$$

The base vectors **i** and **j** correspond, respectively, to the X and Y directions.

For equilibrium of the joint $\mathbf{F}_R = \mathbf{0}$, which, in terms of the force vectors, becomes

$$-33.33(-0.8\mathbf{i} + 0.6\mathbf{j}) - 20\mathbf{j} + P_6(0.8\mathbf{i} + 0.6\mathbf{j}) + P_{10}\mathbf{i} = \mathbf{0}$$

First we equate the coefficients of **j** to obtain

$$-20 - 20 + 0.6P_6 = 0$$
$$P_6 = +\ 66.67 \text{ kN}$$

Equating the coefficients of **i**, we obtain

$$26.67 + (66.67)(0.8) + P_{10} = 0$$
$$P_{10} = -80 \text{ kN}$$

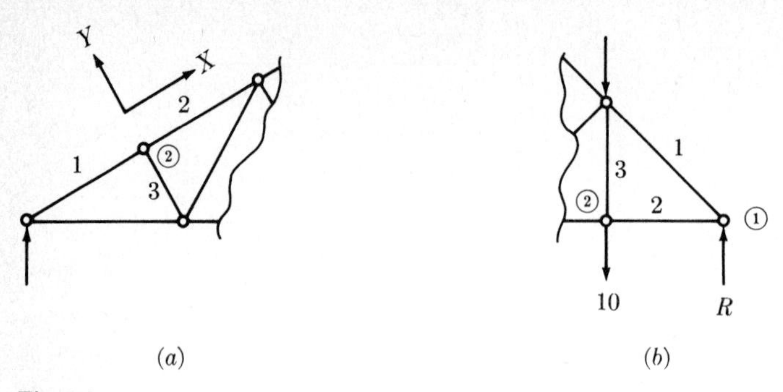

Fig. 3.8

The equilibrium of forces at certain joints is easily handled. A few of these conditions are shown in Fig. 3-8. In analyzing a joint such as ② in Fig. 3-8*a* the computations are simplified if the XY coordinate system is oriented as shown rather than in the horizontal and vertical directions. Furthermore, it should be obvious that the force in member 3 is zero and that $P_1 = P_2$. A support joint such as that in Fig. 3-8*b* is often encountered. It should be apparent that $P_{1Y} = -R$, and that P_2 is then found from $P_2 = -P_{1X}$. It is also obvious that at joint ② $P_3 = +10$.

As the student becomes more adept at analyzing trusses, he may find less need for complete free-body diagrams of the joints, particularly for joints at which the member forces are obvious. However, the value of free-body diagrams during the learning stages, and for the more complicated joints, cannot be overemphasized.

3-4 METHOD OF SECTIONS

The distinguishing feature of the method of sections is that it enables us to determine the force in a particular member of a truss without first having to determine many other member forces, as we might have to do with the method of joints. An appropriate portion of the structure is taken as a free-body diagram, and the unknown is found from an equation of equilibrium for the free body. The critical aspect of this method is the choice of the proper free-body diagram. The student is encouraged when first learning this method to concentrate only on selecting an appropriate free-body diagram, and to forego the computations until this has been mastered. Exercises at the end of the chapter will aid in developing this ability.

The following examples illustrate the method of sections.

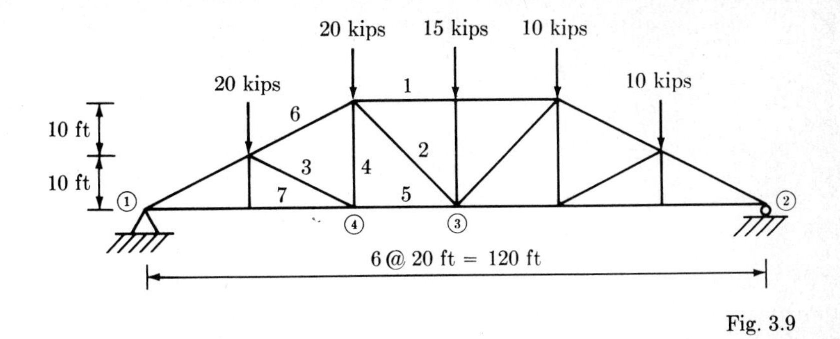

Fig. 3.9

EXAMPLE 3-2 The forces are to be determined in members 1, 2, and 3 of the truss in Fig. 3.9. The reactions are found from a free-body diagram of the entire structure to be $R_{1Y} = 42.5$ kips and $R_{2Y} = 32.5$ kips.

To determine the force in member 1 we cut the structure through members 1, 2, and 5 and draw a free-body diagram of the portion to the left of the cut, as shown in Fig. 3-10*a*. The member forces at the cuts are denoted as shown. For convenience, these unknowns are assumed to be acting in tension. It is evident that if the summation of moments is taken about the point coincident with joint ③ in the given structure, the only unknown in the equilibrium equation is P_1. Thus the value of P_1 is found to be

$$(42.5)(60) - (20)(40) - (20)(20) + 20P_1 = 0$$
$$P_1 = -67.5 \text{ kips}$$

Member 1 is therefore in compression.

The same free-body diagram can be used to determine the force in member 2. From Fig. 3-10*a* we see that the summation of forces in the Y

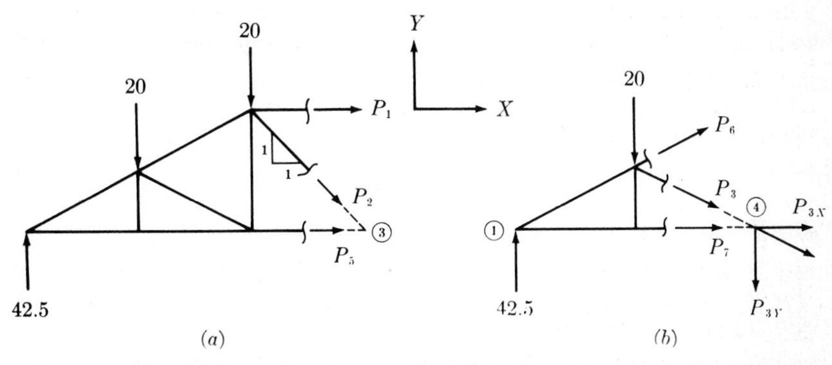

Fig. 3.10

direction results in only one unknown, the Y component of P_2. Thus

$$42.5 - 20 - 20 - P_{2Y} = 0$$
$$P_{2Y} = +2.5 \text{ kips}$$

From the given geometric components of member 2,

$$P_2 = \sqrt{2}\, P_{2Y} = +3.54 \text{ kips}$$

so the member is in tension.

To determine the force in member 3 we cut members 6, 3, and 7 and draw the free-body diagram to the left of this cut, as shown in Fig. 3-10*b*. The equilibrium equation to be used in this case is not quite so obvious. Note that use of the summation of forces in the Y direction would result in two unknowns. If, however, the summation of moments is taken about joint ①, the only unknown is P_3. The lines of action of P_6 and P_7 pass through joint ①. To further facilitate evaluation of P_3 we move P_3 along its line of action to the location of joint ④, where it is resolved into its components P_{3X} and P_{3Y}. Taking moments about joint ① will then involve only P_{3Y}, as the line of action of P_{3X} passes through joint ①. The value of P_{3Y} is found to be

$$(20)(20) + 40P_{3Y} = 0$$
$$P_{3Y} = -10 \text{ kips}$$

The force in member 3 is

$$P_3 = \sqrt{5}\, P_{3Y} = -22.36 \text{ kips}$$

and is therefore compressive.

EXAMPLE 3-3 The force in member 1 of the truss in Fig. 3-11 is to be determined by the method of sections. The first step is to decide where the structure can be cut so that the member force will appear as the unknown in an equilibrium equation for the free-body diagram. We cut members 1, 2, and 3, and the free-body diagram to the right of the cut is as shown in Fig. 3-12. If the lines representing members 2 and 3 are

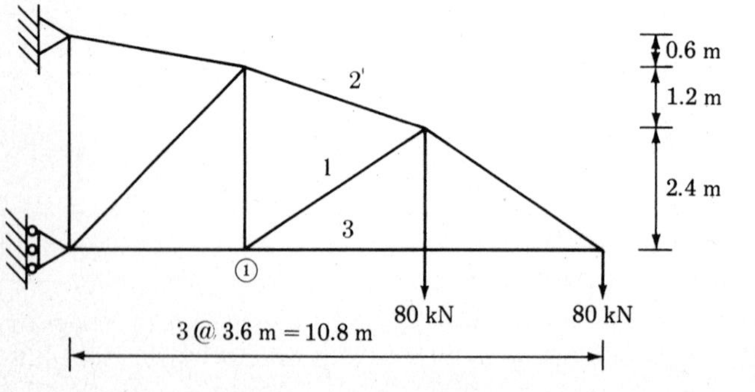

Fig. 3.11

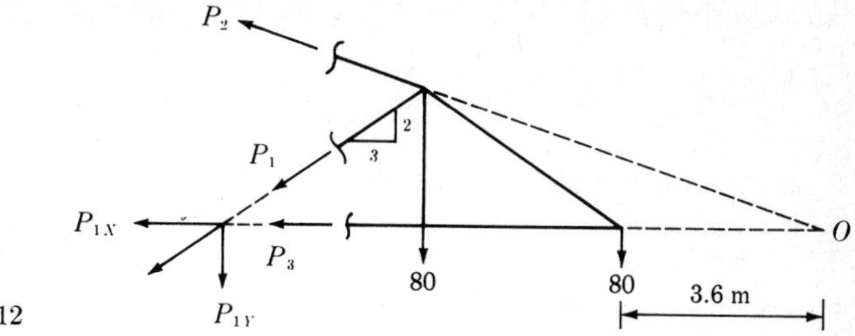

Fig. 3.12

extended until they intersect at O, the equation for the summation of moments about O will contain only the desired unknown P_1. To simplify the operation we move P_1 along its line of action to the location of joint ① and then resolve it into its X and Y components. The line of action of P_{1X} therefore passes through point O. Taking the summation of moments about point O, we obtain

$$-10.8P_{1Y} - (80)(7.2) - (80)(3.6) = 0$$
$$P_{1Y} = -80 \text{ kN}$$

From the given geometric components of P_1 we see that

$$P_1 = \frac{\sqrt{13}}{2} P_{1Y} = -144.2 \text{ kN}$$

Member 1 is therefore subjected to a compressive force.

These examples have illustrated the general procedure for analyzing a truss by the method of sections. Basically, this procedure entails the selection of a free-body diagram of a portion of the structure such that the desired unknown will appear in an equilibrium equation. Occasionally it may be necessary to solve simultaneous equilibrium equations of the free body, and sometimes we must solve first for some other member force before the desired member force can be found. The method of sections can also be used in combination with the method of joints for the most efficient analysis of some trusses.

3-5 GEOMETRIC STABILITY AND STATIC DETERMINACY OF TRUSSES

In previous examples we have assumed that the configurations of the trusses were capable of supporting the loads, and furthermore that the member forces could be determined from equations of statics. This section is devoted to the study of the criteria necessary to establish whether or not a truss is geometrically stable and statically determinate. We shall develop the criteria here for planar trusses and in the following chapter for space trusses.

As stated in Sec. 3-1, the basic element of a truss is a triangle composed of three straight members pinned together at their ends. Such an element is geometrically stable. Geometric stability is maintained if the basic element is expanded, as in Fig. 3-1; that is, if two members are added to the existing system for each new joint established. It is of interest to note that a variation of this procedure can be used to make an unstable configuration stable. For example, four pin-connected members in the form of a rectangle can be made into a stable configuration by adding a member on one of the diagonals.

The stability of a truss also depends on proper support conditions. For example, if the left support of the truss in Fig. 3-1*a* were a roller support rather than a pinned support, obviously the structure would not be stable. In general, we can state that for stability the structure must be supported by at least three reaction forces, all of which are neither parallel nor concurrent.

A planar truss can be thought of as a structural device which constrains j pinned joints in a plane. The forces which act on the joints are the member forces, the external forces, and the reactions. For each joint we can write two equilibrium equations, $\Sigma F_X = 0$ and $\Sigma F_Y = 0$. Thus for the entire truss we can write $2j$ equations. The unknown forces to be determined are the member forces and the reaction components. Therefore, if the structure is statically determinate, we can write the relation

$$2j = m + r \tag{3-1}$$

where m is the number of members and r is the number of reaction components. The following general statements can be made concerning the relations between j, m, and r:

1. $2j < m + r$. There are too many unknowns to be determined from the equations available. The structure is therefore statically indeterminate. To analyze statically indeterminate structures we need additional relationships, such as compatibility in deflections. Statically indeterminate structures will be treated in later chapters.
2. $2j = m + r$. The structure is statically determinate, and the unknowns can be obtained from the $2j$ equations.
3. $2j > m + r$. There are not enough unknowns. The structure is statically unstable.

Equation (3-1) can be examined more closely in terms of Fig. 3-13*a*. This truss is statically determinate according to Eq. (3-1), with $j = 3$, $m = 3$, and $r = 3$. It can be analyzed by first determining its reactions from the equations of equilibrium for the entire structure and then deter-

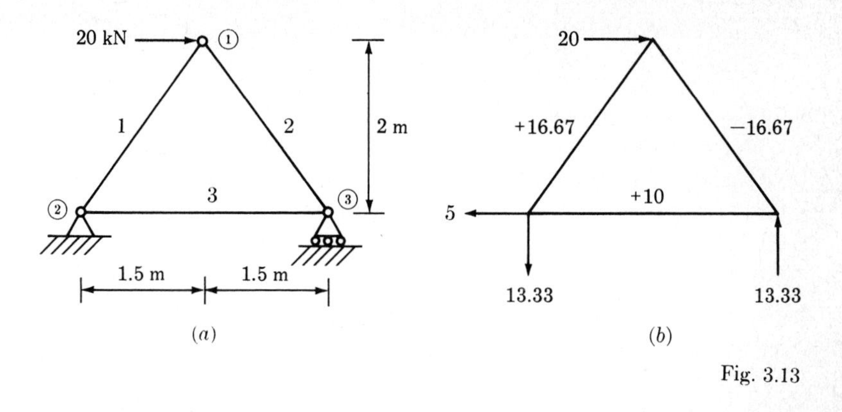

Fig. 3.13

mining the member forces by the method of joints. The results of such an analysis are shown in Fig. 3-13*b*. Now consider the various effects shown in Fig. 3-14. In Fig. 3-14*a* the 20-kN force and the r unknown components of reaction are shown acting on a general rigid body. The configuration of the body has been changed to emphasize the point that the equilibrium equations for the rigid body are not dependent on the internal composition of the body. The external forces act at the same points in space as those in Fig. 3-13*a*. Essentially, we are establishing the fact that we have a system in equilibrium with regard to external forces. We have previously stipulated that trusses be composed of straight members, pinned at their ends, and arranged in triangular configurations. Thus the rigid body of Fig. 3-14*a* is composed of a system of geometrically stable members with m unknown forces as shown in Fig. 3-14*b*. The resulting effect on the joint pins is shown in Fig. 3-14*c*. For these pins we can write $2j$ equilibrium equations which will contain the $m + r$ unknowns.

As an illustration, the vector equations of equilibrium for the three joints are seen from Fig. 3-13*a* to be

Joint ①: $20\mathbf{i} + P_1(-0.6\mathbf{i} - 0.8\mathbf{j}) + P_2(0.6\mathbf{i} - 0.8\mathbf{j}) = \mathbf{0}$

Joint ②: $R_{2X}\mathbf{i} + R_{2Y}\mathbf{j} + P_1(0.6\mathbf{i} + 0.8\mathbf{j}) + P_3\mathbf{i} = \mathbf{0}$

Joint ③: $R_{3Y}\mathbf{j} - P_3\mathbf{i} + P_2(-0.6\mathbf{i} + 0.8\mathbf{j}) = \mathbf{0}$

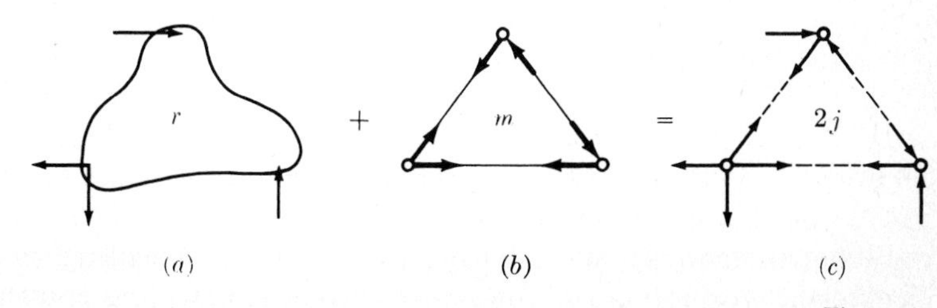

Fig. 3.14

Equating the coefficients of like base vectors in each expression, we obtain the following set of equations

$$
\begin{aligned}
-0.6P_1 + 0.6P_2 &= -20\\
-0.8P_1 - 0.8P_2 &= 0\\
0.6P_1 + P_3 + R_{2X} &= 0\\
0.8P_1 + R_{2Y} &= 0\\
-0.6P_2 - P_3 &= 0\\
0.8P_2 + R_{3Y} &= 0
\end{aligned}
$$

These equations can be written in the general matrix form for simultaneous equations as

$$\mathbf{AX} = \mathbf{B}$$

where **X** is a column matrix of the unknowns, **A** is a square matrix of the coefficients of the unknowns, and **B** is a column matrix of the constants appearing as the right side of the equations. The above equations can thereby be written as

$$
\begin{bmatrix}
-0.6 & 0.6 & 0 & 0 & 0 & 0\\
-0.8 & -0.8 & 0 & 0 & 0 & 0\\
0.6 & 0 & 1 & 1 & 0 & 0\\
0.8 & 0 & 0 & 0 & 1 & 0\\
0 & -0.6 & -1 & 0 & 0 & 0\\
0 & 0.8 & 0 & 0 & 0 & 1
\end{bmatrix}
\begin{Bmatrix} P_1\\ P_2\\ P_3\\ R_{2X}\\ R_{2Y}\\ R_{3Y} \end{Bmatrix}
=
\begin{Bmatrix} -20\\ 0\\ 0\\ 0\\ 0\\ 0 \end{Bmatrix}
$$

This matrix equation can be solved for the unknowns **X**, using a computer program such as that of Appendix B, to obtain

$$
\mathbf{X} = \begin{Bmatrix} P_1\\ P_2\\ P_3\\ R_{2X}\\ R_{2Y}\\ R_{3Y} \end{Bmatrix}
=
\begin{Bmatrix} 16.67\\ -16.67\\ 10.0\\ -20.0\\ -13.33\\ 13.33 \end{Bmatrix}
$$

The results are the same as those obtained earlier by solving for the reactions and then for the member forces by the method of joints.

For the truss under consideration there is only one set of loads which resulted in **B** being a column matrix. It should be noted that more than a single set of loads can be considered, which would result in additional columns in the **B** matrix, one column for each set of loads. The **A** matrix would not change. When solved, each column in **X** would represent the results of the corresponding set of loads. It will be recalled from matrix algebra that a solution for simultaneous equations such as those above is

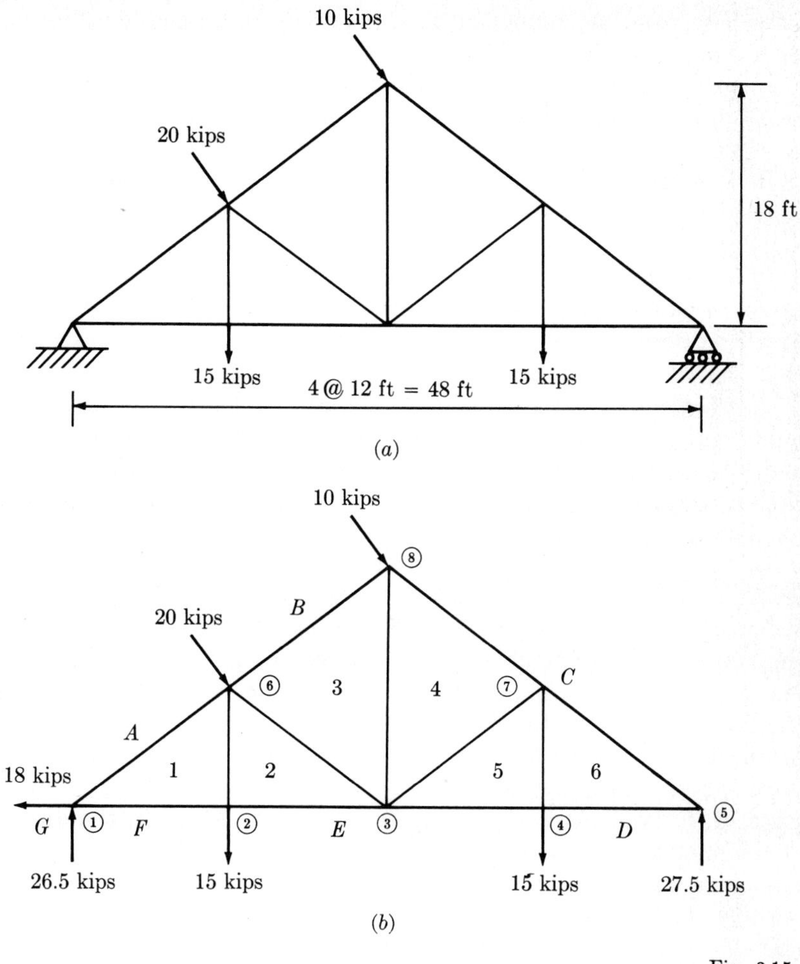

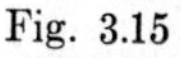

Fig. 3.15

possible only if the **A** matrix is nonsingular, i.e., the value of its determinant is not equal to zero. Thus the evaluation of the determinant of the **A** matrix can be used as a criterion for the stability of a truss.

3-6 GRAPHICAL ANALYSIS OF TRUSSES

The member forces in a statically determinate truss can be obtained by graphical analysis. To see the procedure, let us consider the truss in Fig. 3-15*a*. The analysis begins with the values of the reactions already determined. These values can be obtained either graphically or analytically. The reaction components for the truss under consideration have the values shown in Fig. 3-15*b*.

It is convenient in graphical analyses of trusses to adopt a system of notation such as that shown in Fig. 3-15*b*. The spaces between the external loads and the reaction components are lettered, and the space inside each triangle is numbered. With this notation, known as *Bow's notation*, we can conveniently describe external forces and member forces by the letters or numbers adjacent to them. The notation also provides a convenient designation of forces on the force polygon. A similar designation was used earlier in Chap. 2 in constructing force polygons. The joints are denoted by circled numbers, as before. It is necessary to designate the joints so that they can be referred to during the analysis. Some analysts prefer to denote the joints of a truss by the subscripted letters U_i and L_i, where joints on the lower chord are denoted as L with a numbered subscript and all other joints are denoted as U with a numbered subscript. We shall not use such notation in this text, for consistency with the circled-number convention adopted earlier.

To begin the general procedure for determining the member forces, let us first determine the member forces at an individual joint. As in the analytic method, we must begin at a joint where there are no more than two unknown member forces. A force polygon is then constructed for the joint. For reasons that will soon become apparent, we construct the force polygon in the order in which the forces are encountered in proceeding clockwise around the joint. Beginning with joint ①, we draw force *fg* and then *ga*, as shown in Fig. 3-16*a*. Next we encounter member force *a*1, for which we know only the direction. A line in the direction of this member is drawn through point *a* on the force polygon and extended down and to the left. Looking ahead to member force 1*f*, we see that its direction is horizontal and that the line must pass through point *f* on the force polygon. By drawing this horizontal line through *f* until it intersects the line through *a*, we determine point 1. This gives us the magnitude of the member forces *a*1 and 1*f*. By proceeding around the joint in a prescribed direction, we can determine whether the force in *a*1 and 1*f* is compression or tension. In Fig. 3-16*a* force *a*1 is in a direction

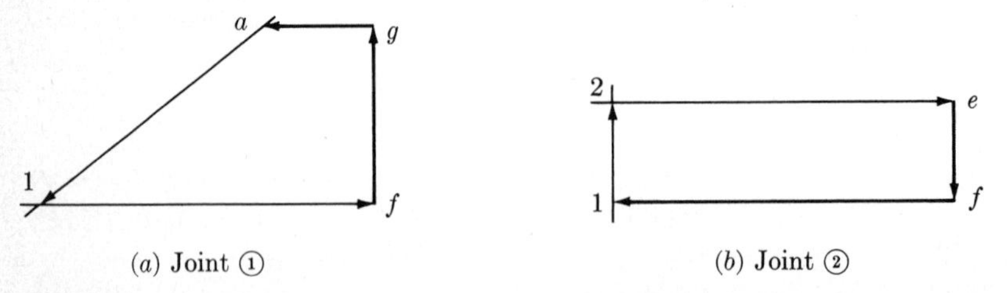

(*a*) Joint ①

(*b*) Joint ②

Fig. 3.16

down and to the left. A member force acting in this direction on joint ① is compressive. This force is referred to as *a*1 rather than 1*a* because this is the order in which the symbols are encountered in a clockwise direction around joint ①. Force 1*f* on the force polygon acts to the right. A member force acting in this direction on joint ① is tensile.

The force polygon thus determines the magnitude of the member forces; the type of force in a member is determined from the Bow's notation by proceeding around the joint in a prescribed direction. We have used a clockwise direction here; the same results could be obtained with a counterclockwise convention, but the force polygon would have a different shape.

Once we have determined the forces in the members at joint ①, we see from Fig. 3-15*b* that joint ② can next be analyzed. Construction of the force polygon is begun with the 15-kip force *ef*. The previously determined tensile force *f*1 is drawn next, as shown in Fig. 3-16*b*. Since we know the directions of the member forces 12 and 2*e*, we can establish point 2 on the force polygon. Thus the magnitudes of the two unknown forces are determined, and we see that member force 12 is tensile, as is member force 2*e*.

The procedure can be continued for the other joints of the structure. We see from Fig. 3-15*b* that joint ⑥ could be analyzed next, and then the remaining joints in the order ⑧-③-④-⑦.

Instead of constructing individual force polygons for the joints, we can perform the entire analysis on a single diagram known as a *Maxwell diagram*. As an illustration, let us construct the Maxwell diagram for the truss in Fig. 3-15. For easy reference, the truss is shown again in Fig. 3-17*a*. The first step in constructing the Maxwell diagram is to construct the force polygon of all external forces and reactions acting on the truss. The forces in the polygon are placed in the order in which they are encountered in going clockwise around the truss. In Fig. 3-17*b* the polygon began with *fg* and then continued in the order *ga-ab-bc-cd-de-ef*. The force polygon must close. Arrows need not be drawn on the forces, as their directions are evident from the notation.

The next step is to begin at a joint where there are not more than two unknowns. Joint ① is selected. Forces *fg* and *ga* have already been constructed, so it is a simple matter to locate point 1 as we did in Fig. 3-16*a*. Having located point 1, and hence the magnitude of member forces *a*1 and 1*f*, we determine whether the forces are compressive or tensile as was done earlier.

Next we consider joint ②. Force *ef* has already been established, and point 1 was established in the previous step. Point 2 can therefore be located from the directions of 12 and 2*e*, as in Fig. 3-16*b*.

The procedure is continued for the other joints of the structure, and

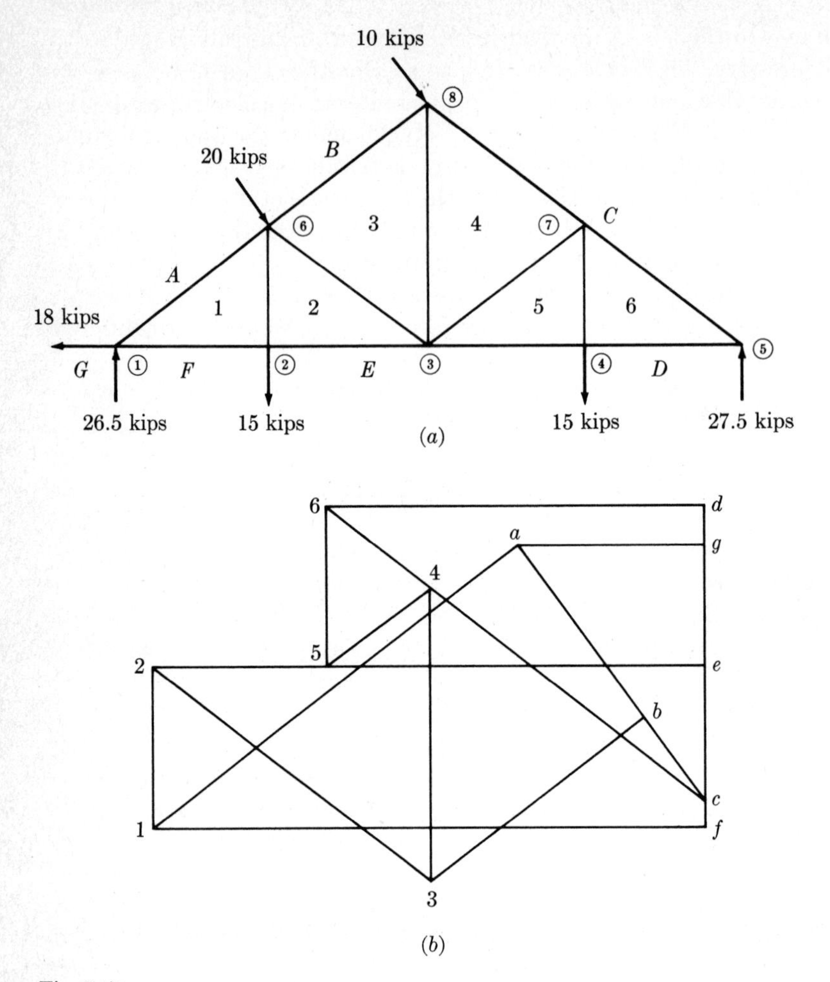

Fig. 3.17

the resulting Maxwell diagram is shown in Fig. 3-17*b*. The Maxwell diagram must close within a reasonable degree of accuracy. The magnitude and type of any member force can be obtained from the completed Maxwell diagram. For example, at joint ⑦ we see by proceeding around the joint clockwise and noting the direction in which the forces act on the Maxwell diagram that 54 acts toward the joint and is therefore compressive, 4*c* is compressive, *c*6 is compressive, and 65 is tensile. The resulting member forces are generally recorded on a sketch of the truss, as in Fig. 3-7.

Occasionally in the analysis of more complex structures a situation is

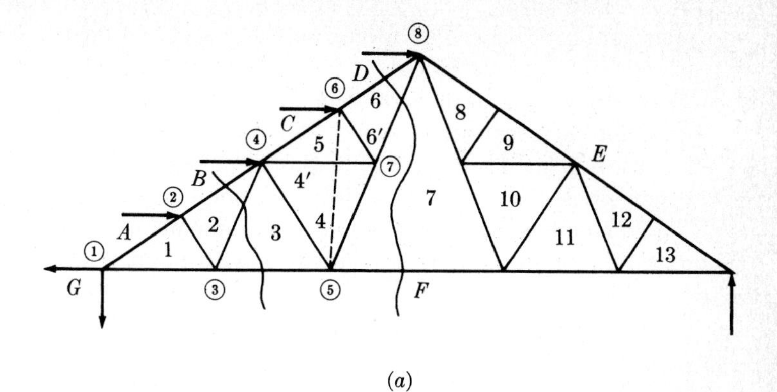

(*a*)

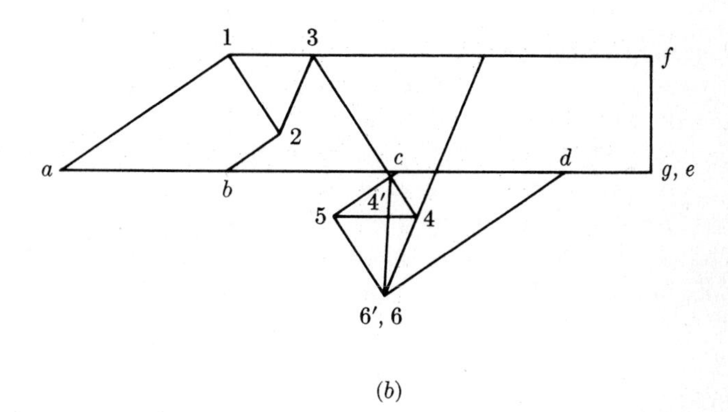

(*b*)

Fig. 3.18

encountered that requires additional considerations. The truss of Fig. 3-18*a*, known as a *Fink truss*, illustrates such a situation. The reactions have been determined for the given loads, and therefore the graphical analysis of member forces can proceed. As shown in Fig. 3-18*b*, the graphical analysis can begin at joint ① and proceed to ②, and then to ③. At this point, however, we see that joints ④ and ⑤ both have more than two unknown member forces. To overcome this difficulty a *substitute member* can be employed. In Fig. 3-18*a* members 45 and 56 can be replaced by a substitute member denoted as 4′6′ and indicated by a dashed line. Note that where there were previously three triangles, there are now two, defined by the joints ⑤-④-⑥ and ⑤-⑥-⑧. The spaces inside these triangles are denoted as 4′ and 6′, respectively. The structure remains stable with the substitute member. In addition, the true forces in the members adjacent to the zone of the substitute member have not been altered. This can be verified by considering the member forces at the sections cut as shown in Fig. 3-18*a*. The forces at these sections do not reflect the change in member arrangement. If there is any doubt, free-body diagrams should be drawn for each arrangement and the equilibrium equations examined for each section.

With the new member arrangement we can now analyze joint ④ and then joint ⑥. In the analysis at joint ⑥ point 6′ is established in Fig. 3-18*b*. The member force $d6'$ is equal to the member force $d6$ in the real structure. Therefore point 6 is coincident with point 6′ on the Maxwell diagram. Having determined member force $d6$, we can determine member forces 65 and $5c$ at joint ⑥ by locating point 5 on the Maxwell diagram.

Joint ④ can now be solved, since there are only two unknown member forces, 54 and 43. The analysis can then be continued to joint ⑤ and then ⑦. The analysis for the other side of the truss is not shown; it would proceed in the same manner.

It should be apparent that the selection of a substitute member in this procedure is not completely arbitrary. It should be kept in mind in selecting the substitute member that its prime function is to provide a means of getting beyond unsolvable joints and at the same time to enable the true forces to be determined in the members beyond these joints. A prerequisite to accomplishing this is to have the members revised in a stable configuration.

PROBLEMS

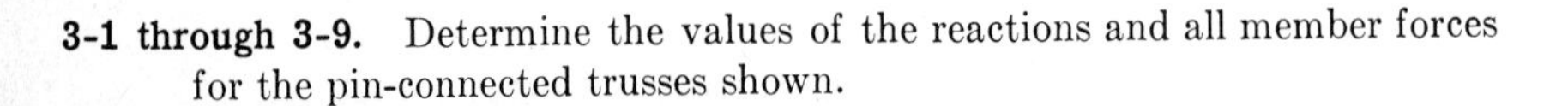

3-1 through 3-9. Determine the values of the reactions and all member forces for the pin-connected trusses shown.

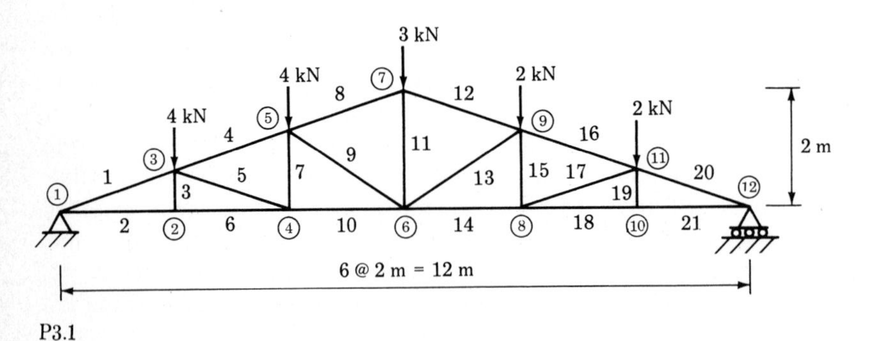

P3.1

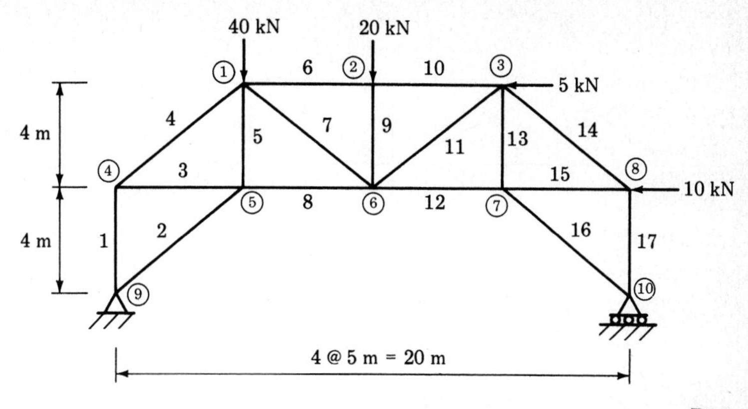
40 kN
20 kN
5 kN
10 kN
4 m
4 m
4 @ 5 m = 20 m

P3.2

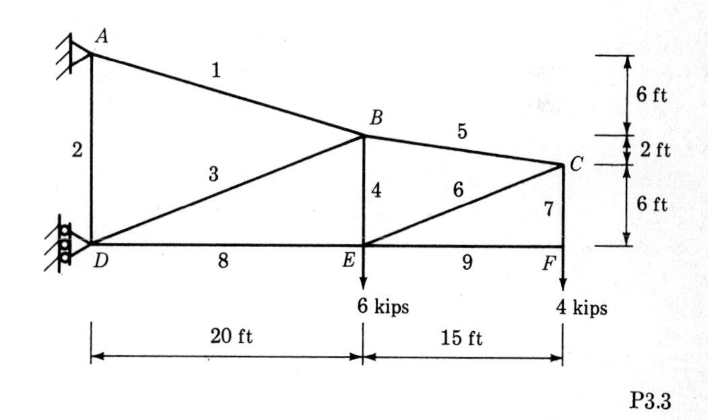
A
B
C
D
E
F
6 ft
2 ft
6 ft
6 kips
4 kips
20 ft
15 ft

P3.3

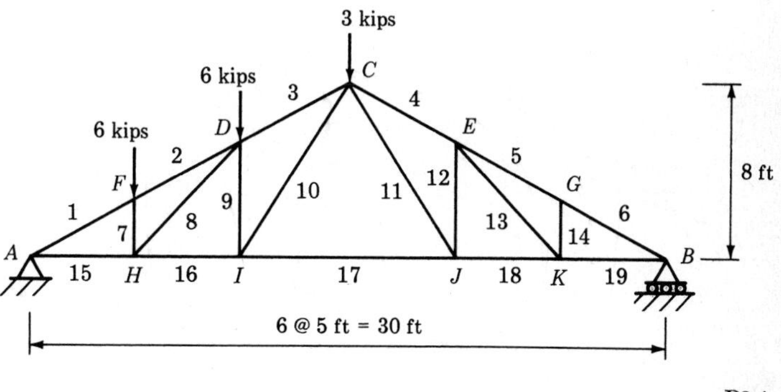
3 kips
6 kips
6 kips
A
B
C
D
E
F
G
H
I
J
K
8 ft
6 @ 5 ft = 30 ft

P3.4

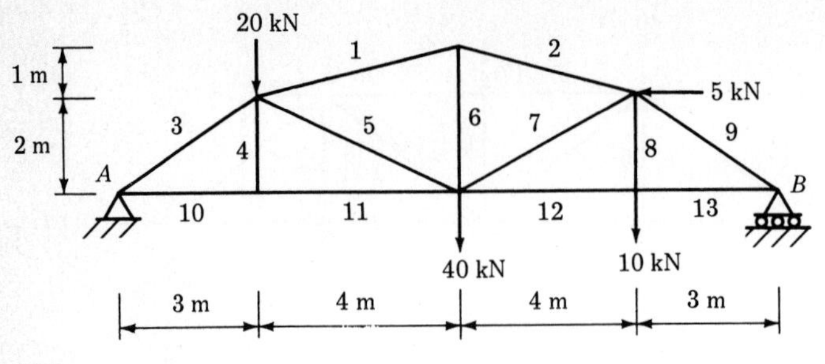
20 kN
1
2
5 kN
1 m
2 m
3
4
5
6
7
8
9
A
B
10
11
12
13
40 kN
10 kN
3 m
4 m
4 m
3 m

P3.5

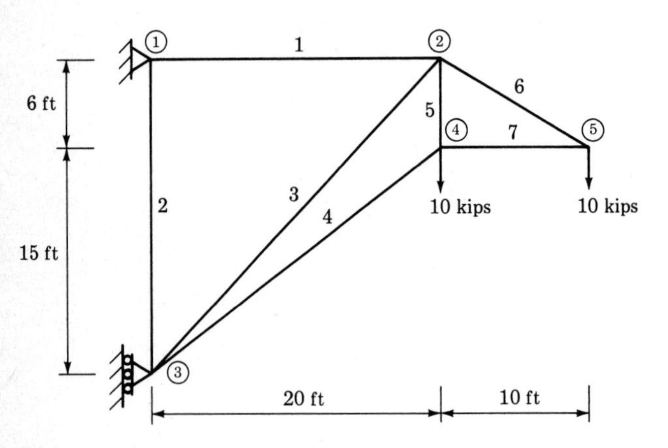
1
2
1
6
6 ft
5
4
7
5
2
3
4
10 kips
10 kips
15 ft
3
20 ft
10 ft

P3.6

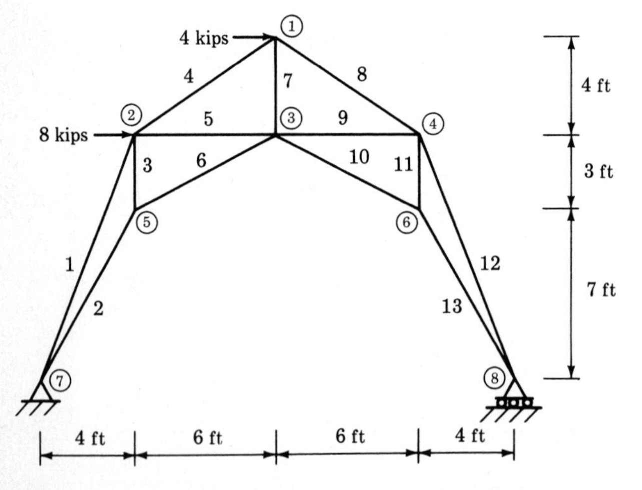
4 kips
1
4
7
8
4 ft
8 kips
2
5
3
9
4
3
6
10
11
3 ft
5
6
1
12
2
13
7 ft
7
8
4 ft
6 ft
6 ft
4 ft

P3.7

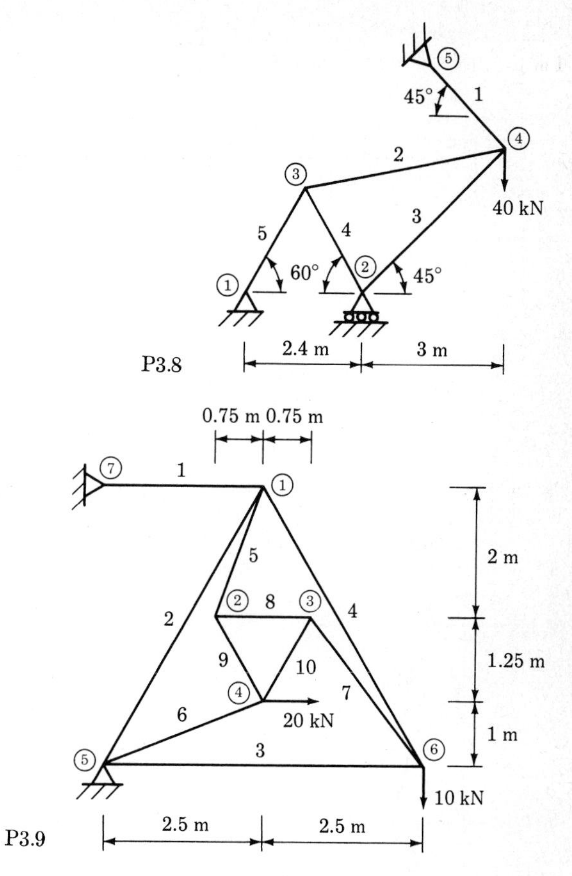

P3.8

P3.9

3-10. The Fink truss shown is to be analyzed for a snow load of 30 psf of horizontal roof projection. The snow load is transferred by the roof system to the truss as concentrated loads at the joints. The spacing between trusses is 5 m. Determine the values of the reactions and all member forces in kilonewtons.

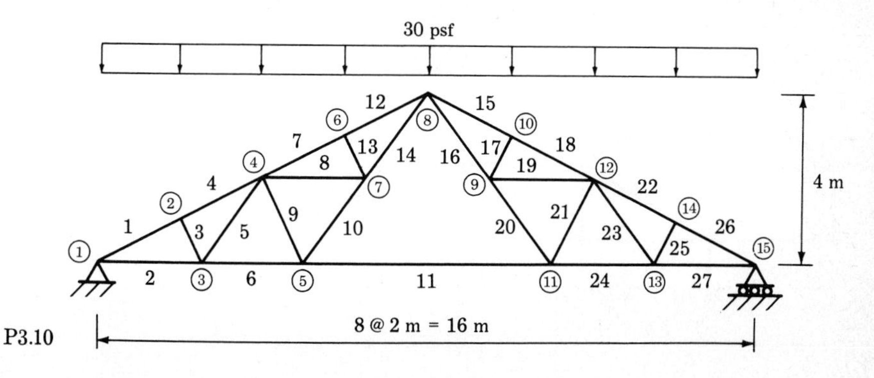

P3.10

3-11 through 3-17. For the pin-connected trusses shown, determine the axial force P in the numbered members.

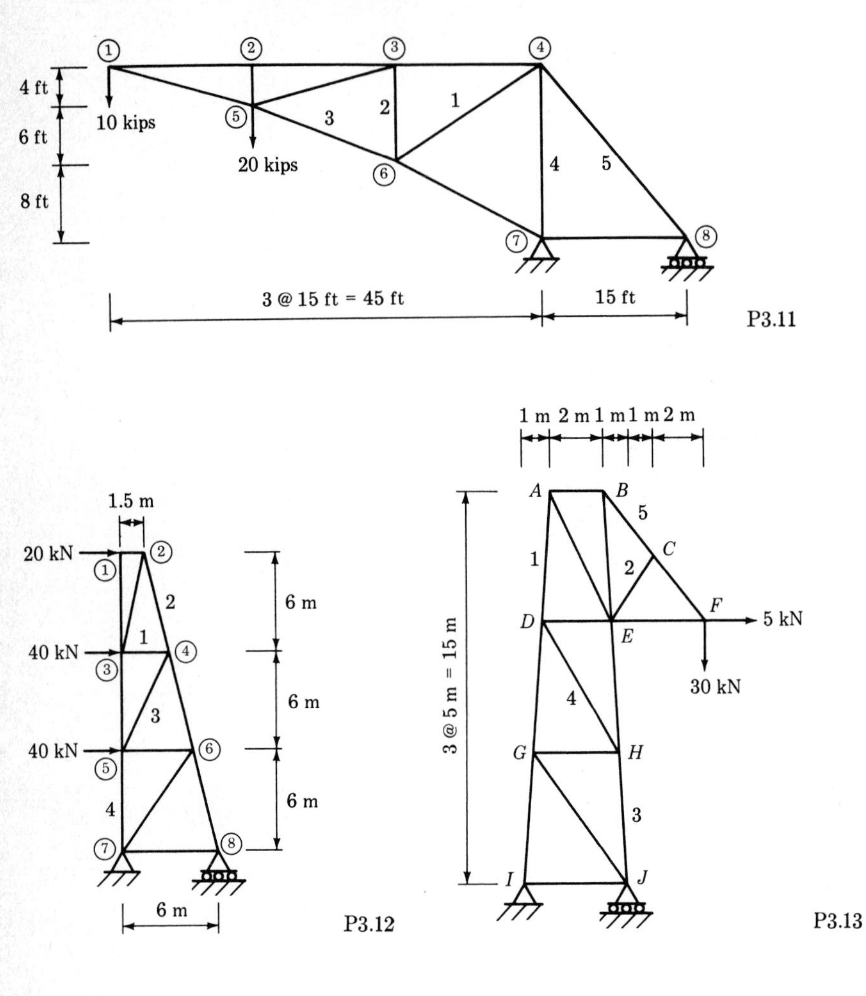

P3.11

P3.12

P3.13

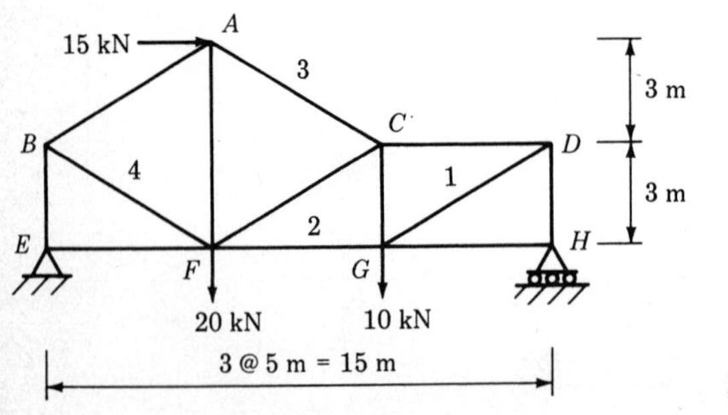

P3.14

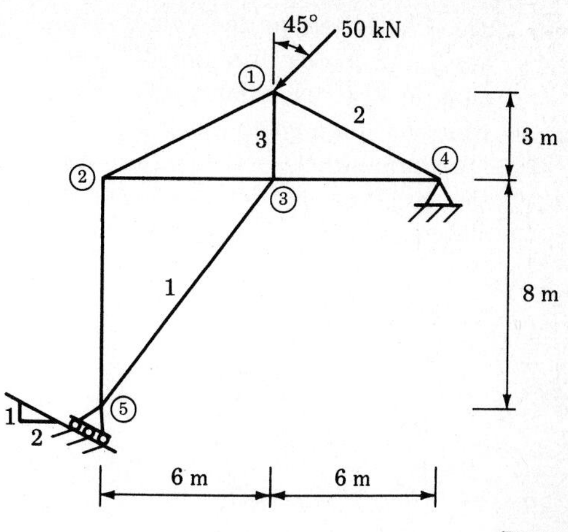

P3.15

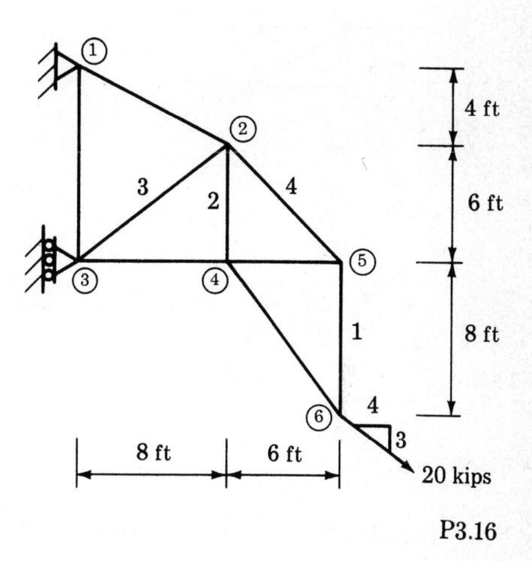

P3.16

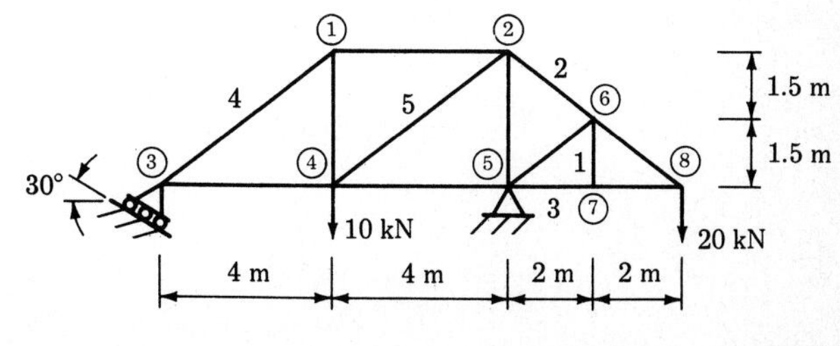

P3.17

3-18 and 3-19. Determine the values of the reactions and member forces in the trusses of Probs. 3-8 and 3-9, respectively, by writing the equilibrium equations at each joint and solving the resulting set of simultaneous equations on a computer.

3-20. Determine whether the trusses shown are stable or unstable and whether they are statically determinate or indeterminate. If they are indeterminate, state the degree of indeterminacy.

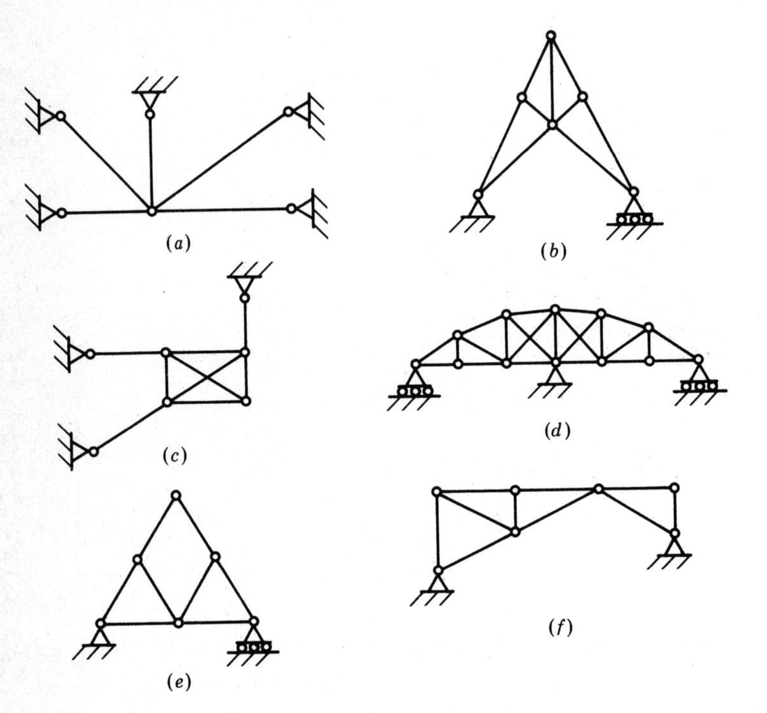

P3.20

3-21 and 3-22. Graphically solve Probs. 3-5 and 3-6, respectively, using the notation described in Sec. 3-6.

CHAPTER

FOUR

SPACE TRUSSES

4-1 EQUILIBRIUM EQUATIONS AND STATIC DETERMINACY

The analysis of space trusses can be regarded as an extension of the principles previously discussed in the analysis of plane trusses. Because of the third dimension, it is necessary to reconsider the notation to be used for the reactions and to reexamine the equilibrium equations available.

The connections of the members in a space truss are considered to be ball-and-socket joints. It is assumed that the joints transmit no moment. The members are therefore subjected only to axial forces.

There are various forms of support for space trusses. It is necessary to understand the support condition in all three dimensions, and it is convenient to describe the components of reaction and the components of member forces in terms of an XYZ coordinate system for the entire truss. Three types of supports and the corresponding components of reaction that are possible are shown in Fig. 4-1. Also shown are the descriptive terms commonly used in referring to the supports. The ball-and-socket support of Fig. 4-1a prevents movement of the support in each of the three directions. The components of reaction at a support denoted as

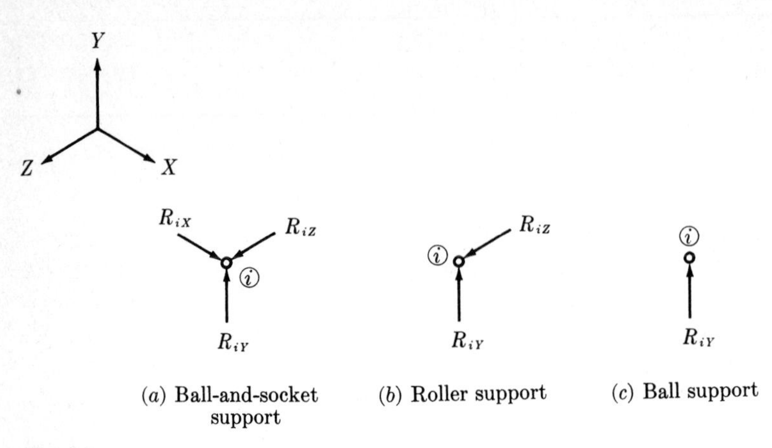

Fig. 4.1

(i) can conveniently be described with subscripts, as shown. The reaction components are assumed positive when they act in the positive direction of the reference coordinate axes. The use of such a sign convention for reaction components simplifies the interpretation of the directions for values found in analysis.

The roller support of Fig. 4-1*b* prevents movement of the support in two directions; movement is permitted in the third direction. The support condition in the *YZ* plane is comparable to a pinned support in planar analysis. The support condition in the *XY* plane is comparable to a roller support in planar analysis. Other combinations of support are also possible for a roller support. For example, R_{iZ} can be zero, and R_{iX} and R_{iY} will exist.

The ball support of Fig. 4-1*c* prevents movement in only one direction, the *Y* direction for the support shown. Movement is permitted in each of the other two directions.

Equilibrium of an entire space truss or equilibrium of sections of a space truss is described by the six scalar equations

$$\Sigma F_X = 0 \qquad \Sigma M_{XX} = 0$$
$$\Sigma F_Y = 0 \qquad \Sigma M_{YY} = 0$$
$$\Sigma F_Z = 0 \qquad \Sigma M_{ZZ} = 0$$

or, in vector form,

$$\mathbf{F}_R = \mathbf{0} \qquad \mathbf{M}_R = \mathbf{0}$$

The resultant vectors $\mathbf{F}_R$ and $\mathbf{M}_R$ represent three-dimensional force and moment vectors.

Equilibrium of a ball-and-socket joint is described by the three scalar equations

$$\Sigma F_X = 0 \qquad \Sigma F_Y = 0 \qquad \Sigma F_Z = 0$$

or by the vector equation

$$\mathbf{F}_R = \mathbf{0}$$

A space truss can be thought of as a structural device which constrains j ball-and-socket-connected joints in space. The forces which act on the joints are the member forces, the external forces, and the reaction components. Thus, from the three equilibrium equations for each joint, we can write $3j$ equilibrium equations for the entire truss. Denoting the number of members by m and the number of reaction components by r, the condition for the space truss to be statically determinate is

$$3j = m + r \tag{4-1}$$

For conditions where $3j$ is not equal to $m + r$ the criteria are the same as for planar trusses. If $3j < m + r$, the truss is statically indeterminate; that is, there are more unknowns than equations of equilibrium. If $3j > m + r$, the structure is statically unstable.

We shall see later in the chapter that the previously discussed method of sections and the method of joints can be applied to the analysis of space trusses. However, the methods must be modified to enable us to consider the forces in the third dimension. Often member forces and reactions for a given space truss can be determined with less computational effort if certain conditions of force at the joint are recognized. These special conditions of force at a joint are presented below in the form of theorems.

For certain conditions of loading at a joint in a space truss and for certain arrangements of the members meeting at a joint, the forces in the bars can be determined by observation. Such conditions are stated in the following theorems.

THEOREM 1: *If all but one of the members meeting at a joint lie in a plane, and if an external load is applied to the joint, the components of the member force and the external load, normal to the plane, are equal. If no external load is applied at the joint, the force in the member is zero.*

THEOREM 2: *If all but two of the members meeting at a joint have zero forces, and the two members are not collinear, and if no external loads are applied to the joint, the force in each of the two members is zero.*

These theorems can be justified by equilibrium considerations of a joint under the conditions stated. Their application will be illustrated in Example 4-2.

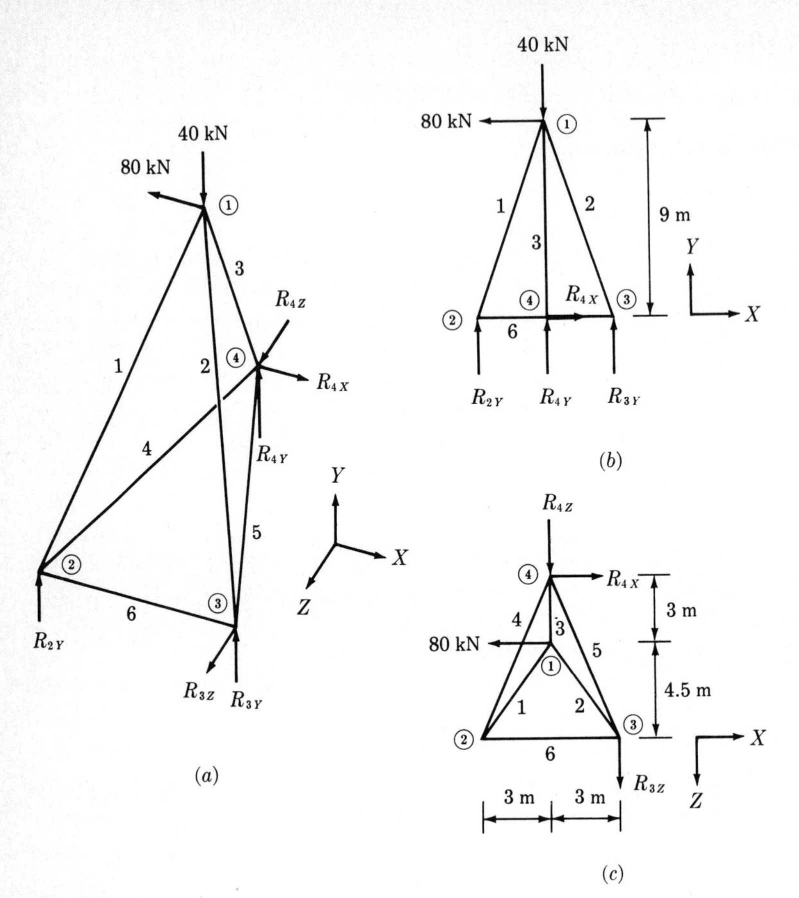

Fig. 4.2

4-2 ANALYSIS OF SPACE TRUSSES

The following examples illustrate the procedures involved in the analysis of space trusses.

EXAMPLE 4-1 The reaction components and the member forces are to be determined for the space truss of Fig. 4-2. Force quantities are to be described in terms of the XYZ coordinates of Fig. 4-2a. For additional clarity, the truss is also shown in two-dimensional views in Figs. 4-2b and c. The unknown components of reaction at support points ②, ③, and ④ are shown acting in the direction of the positive X, Y, and Z axes.

The truss is first checked by Eq. 4-1 for statical determinacy. From

Fig. 4-2 we see that

$$j = 4 \qquad m = 6 \qquad r = 6$$

Therefore in Eq. (4-1)

$$(3)(4) = 6 + 6$$

The truss is statically determinate.

Now let us consider the entire truss. There are six equilibrium equations available. There are six unknown reaction components, so we shall determine the reaction components first. In obtaining the reaction components, we look for an equilibrium equation that involves only one unknown. The reaction component R_{4Y} can be obtained by taking the summation of moments about an X axis through points ② and ③,

$$-(40)(4.5) + 7.5R_{4Y} = 0$$
$$R_{4Y} = 24 \text{ kN}$$

In taking the summation of moments we use the right-hand screw rule, with the thumb in the positive direction of the axis. Taking the summation of moments about a Z axis through point ② and using the value of R_{4Y} determined above, we obtain the value of R_{3Y},

$$(80)(9) - (40)(3) + (24)(3) + 6R_{3Y} = 0$$
$$R_{3Y} = -112 \text{ kN}$$

Taking the summation of forces in the Y direction and using the above determined values of reaction, we obtain the value of R_{2Y},

$$-40 + 24 - 112 + R_{2Y} = 0$$
$$R_{2Y} = 128 \text{ kN}$$

The value of R_{3Z} can be obtained by taking the summation of moments about a Y axis through point ④,

$$-(80)(3) - 3R_{3Z} = 0$$
$$R_{3Z} = -80 \text{ kN}$$

R_{4Z} is then obtained by taking the summation of forces in the Z direction,

$$-80 + R_{4Z} = 0$$
$$R_{4Z} = 80 \text{ kN}$$

The remaining reaction component, R_{4X}, is obtained from the summation of forces in the X direction,

$$-80 + R_{4X} = 0$$
$$R_{4X} = 80 \text{ kN}$$

The member forces are determined from equilibrium considerations of the joints. In general, we have three scalar equilibrium equations for each joint, and we must therefore begin at a point where there are no more than three unknowns. For the given structure we could begin at any point once we have determined the values of the reaction components. Let us analyze joint ② first, using vector expressions in determining the member forces. At joint ②, and at any other joint to be analyzed, the unknown member forces are assumed to be acting in a direction away from the joint. Thus the resulting signs of the member forces will correspond with plus for tension and minus for compression. The unit vectors for the members at joint ②, directed away from the joint, are found to be

$$\mathbf{u}_6 = \mathbf{i}$$
$$\mathbf{u}_4 = 0.371\mathbf{i} - 0.928\mathbf{k}$$
$$\mathbf{u}_1 = 0.286\mathbf{i} + 0.857\mathbf{j} - 0.428\mathbf{k}$$

where **i**, **j**, and **k** represent the unit base vectors in the X, Y, and Z directions, respectively. The vector equation for equilibrium of joint ②, $\mathbf{F}_R = \mathbf{0}$, is

$$128\mathbf{j} + P_6\mathbf{i} + P_4(0.371\mathbf{i} - 0.928\mathbf{k}) + P_1(0.286\mathbf{i} + 0.857\mathbf{j} - 0.428\mathbf{k}) = \mathbf{0}$$

Equating the coefficients of **j**, we obtain

$$128 + 0.857P_1 = 0$$
$$P_1 = -149.4 \text{ kN}$$

The minus sign indicates that member 1 is in compression. The member forces P_4 and P_6 are obtained by equating the coefficients of **k** and **i**, respectively. The resulting values are

$$P_4 = 68.9 \text{ kN} \qquad P_6 = 17.2 \text{ kN}$$

Next we consider joint ③. The unit vectors for the members, directed away from the joint, are

$$\mathbf{u}_6 = -\mathbf{i}$$
$$\mathbf{u}_5 = -0.371\mathbf{i} - 0.928\mathbf{k}$$
$$\mathbf{u}_2 = -0.286\mathbf{i} + 0.857\mathbf{j} - 0.428\mathbf{k}$$

The equilibrium equation at joint ③, with the previously determined value of P_6, is

$$-80\mathbf{k} - 112\mathbf{j} + 17.2(-\mathbf{i}) + P_5(-0.371\mathbf{i} - 0.928\mathbf{k}) + P_2(-0.286\mathbf{i} + 0.857\mathbf{j} - 0.428\mathbf{k}) = \mathbf{0}$$

Equating the coefficients of like base vectors, we obtain

$$P_2 = 130.7 \text{ kN} \qquad P_5 = -147.1 \text{ kN}$$

The remaining member force, P_3, is found from equilibrium considerations at joint ④ to be

$$P_3 = -25.3 \text{ kN}$$

It would have been possible to determine the member forces of this truss without first determining the reactions. If we had begun the analysis at joint ①, there would have been three unknowns. However, the equilibrium expression at ① would have resulted in simultaneous equations. Having determined the member forces at ①, we could have analyzed joint ②, joint ③, and then joint ④. In solving for the reactions first we encountered no simultaneous equations.

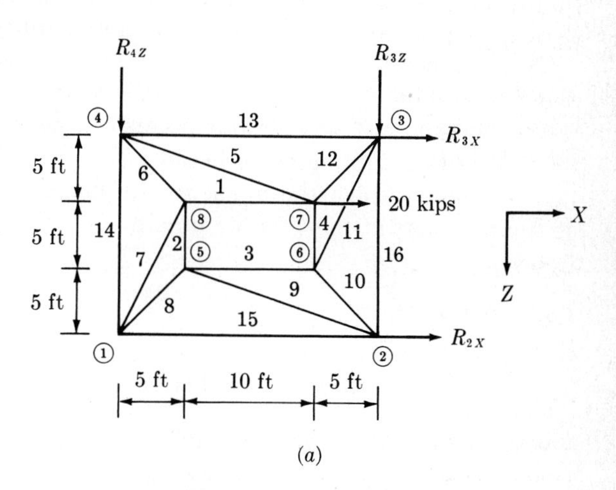

(a)

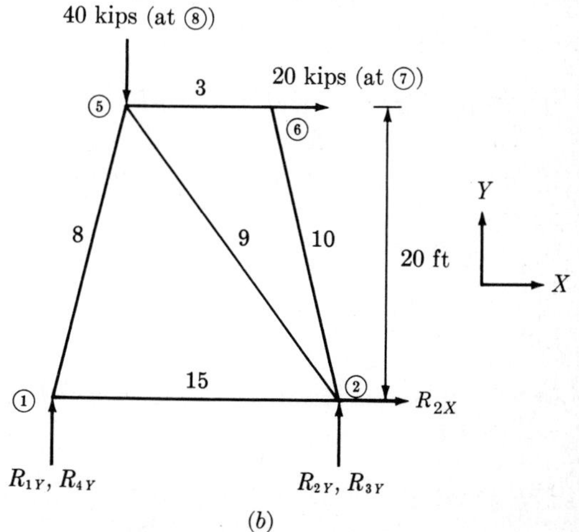

(b)

Fig. 4.3

EXAMPLE 4-2 Theorems 1 and 2 above are to be used in determining the member forces for the space truss of Fig. 4-3. Fig. 4-3*a* represents a plan view of the truss and Fig. 4-3*b* an elevation view. The X, Y, and Z reference axes are oriented as shown.

From Fig. 4-3 we see that $j = 8$, $m = 16$, and $r = 8$. Therefore, from Eq. (4-1), the truss is statically determinate. Note, however, that there are eight unknown reaction components. Thus the values of all of the reactions cannot be determined first.

At joint ⑤ all members except member 2 lie in the same plane, and there are no external loads applied at the joint. Therefore, according to Theorem 1, the member force P_2 is zero. Similarly, at joints ⑥ and ⑦ the member forces P_3 and P_4 are zero. From Theorem 2, we then see that at joint ⑤ P_8 and P_9 are zero and at joint ⑥ P_{10} and P_{11} are zero.

The member force P_1 can be obtained by considering the free-body diagram of joint ⑧, as shown in Fig. 4-4. If we take the summation of moments about a Z axis through points ④ and ①, we obtain an equation with P_1 as the only unknown because the other two unknowns intersect the axis about which the moments are being taken. Recalling that the 40-kip load acts at joint ⑧, we obtain

$$-(5)(40) - 20P_1 = 0$$
$$P_1 = -10 \text{ kips}$$

Bar force P_7 can be obtained by taking the summation of moments about an X axis through points ④ and ③ on the free-body diagram of Fig. 4-4,

$$(40)(5) + 20P_{7Z} + 5P_{7Y} = 0$$

However, from the given dimensions of the truss,

$$P_{7Y} = 2P_{7Z} \qquad P_7 = 2.29P_{7Z}$$

The value of the member force P_7 is thus found to be -15.3 kips.

The member forces P_5 and P_{12} can be obtained from the summation of forces acting on joint ⑦. The remaining member forces and the reaction

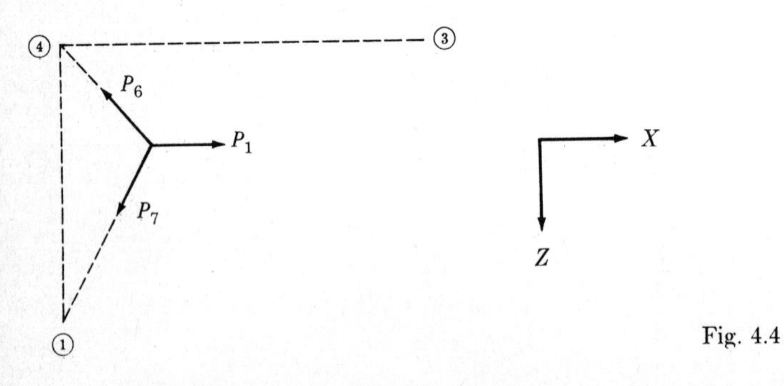

Fig. 4.4

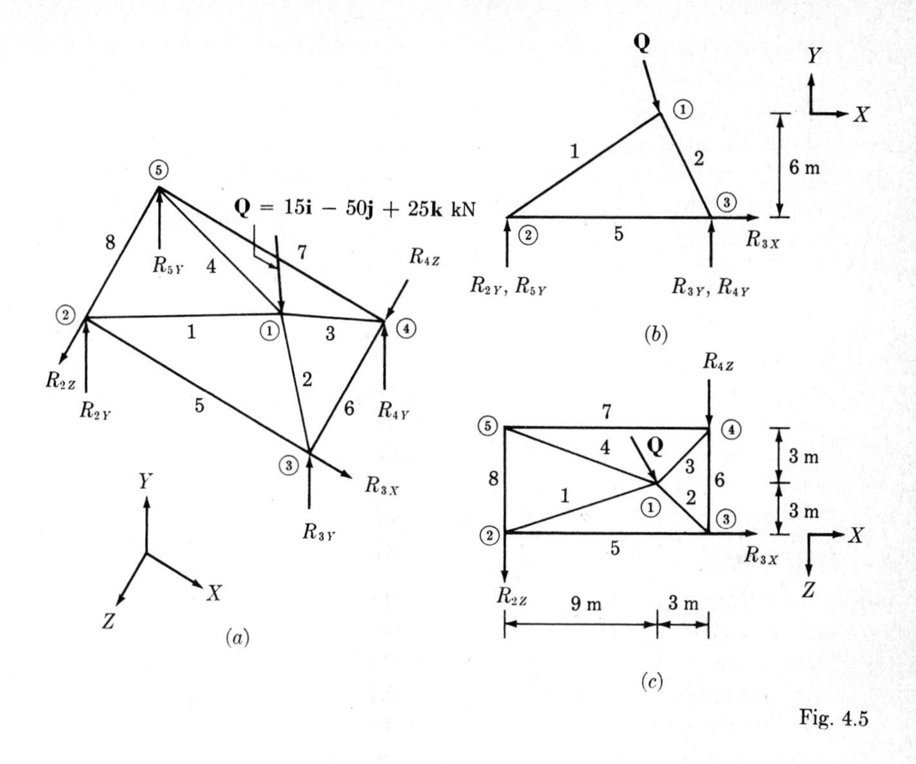

Fig. 4.5

components can then be determined from equilibrium expressions for joints ①, ②, ③, and ④.

EXAMPLE 4-3 The member forces and the reaction components are to be determined for the space truss of Fig. 4-5. The loading on the truss at joint ① is described by the force vector **Q** as shown in Fig. 4-5*a*. We first check to see if the truss is statically determinate. From Fig. 4-5 it is seen that $j = 5$, $m = 8$, and $r = 7$. Therefore, according to Eq. (4-1), the truss is statically determinate.

The number of unknown reaction components exceeds the number of equilibrium equations available. Furthermore, it is not possible to start an analysis at joint ① because there are four unknown member forces and there are only three equations of equilibrium for the joint. We know, however, that the truss is statically determinate from the check performed above.

One approach to such a problem is to assume the value of a reaction component or a member force to have a value denoted as X. The analysis is then performed in terms of X until a point is reached where the value of X is obtained.

Another approach is to write all the equilibrium equations for the joints in terms of the unknowns and the given loads and then solve the resulting simultaneous equations. For the truss of Fig. 4-5 there will be $3j = 15$ equations. In writing equilibrium equations at each joint we must take care to use a consistent sign convention for the unknowns. If, as in previous examples, we assume the unknown reaction components to act in the direction of the positive coordinate axes and if we assume unknown member forces to act away from a joint, then the signs of the results can readily be interpreted.

We shall use vector expressions to develop the equilibrium equations at the joints. Beginning at joint ②, we find the unit vectors for the members to be

$$\mathbf{u}_5 = \mathbf{i}$$
$$\mathbf{u}_8 = -\mathbf{k}$$
$$\mathbf{u}_1 = 0.802\mathbf{i} + 0.535\mathbf{j} - 0.267\mathbf{k}$$

The equilibrium equation $\mathbf{F}_R = \mathbf{0}$ at joint ② is therefore

$$R_{2Y}\mathbf{j} + R_{2Z}\mathbf{k} + P_5\mathbf{i} + P_8(-\mathbf{k}) + P_1(0.802\mathbf{i} + 0.535\mathbf{j} - 0.267\mathbf{k}) = \mathbf{0}$$

Equating the coefficients of like base vectors, we obtain the following three equilibrium equations at joint ②:

$$P_5 + 0.802P_1 = 0$$
$$R_{2Y} + 0.535P_1 = 0$$
$$R_{2Z} - P_8 - 0.267P_1 = 0$$

Proceeding in the same manner at joint ③, we obtain the three equilibrium equations

$$R_{3X} - P_5 - 0.408P_2 = 0$$
$$R_{3Y} + 0.816P_2 = 0$$
$$-P_6 - 0.408P_2 = 0$$

At joint ④ we obtain the equations

$$-P_7 - 0.408P_3 = 0$$
$$R_{4Y} + 0.816P_3 = 0$$
$$R_{4Z} + P_6 + 0.408P_3 = 0$$

At joint ⑤ we obtain the equations

$$P_7 + 0.802P_4 = 0$$
$$R_{5Y} + 0.535P_4 = 0$$
$$P_8 + 0.267P_4 = 0$$

At joint ① we obtain the equations

$$-0.802P_1 + 0.408P_2 + 0.408P_3 - 0.802P_4 = -15$$
$$-0.535P_1 - 0.816P_2 - 0.816P_3 - 0.535P_4 = 50$$
$$0.267P_1 + 0.418P_2 - 0.408P_3 - 0.267P_4 = -25$$

These resulting equilibrium equations can be written in the general form for simultaneous equations as

$$\mathbf{AX} = \mathbf{B}$$

There are a total of 15 unknown member forces and reaction components. Therefore, **A** is a 15 × 15 matrix. **B** is a 15 × 1 matrix of the constants that appear on the right side of the equilibrium equations. The matrix equation for the truss is

$$\begin{bmatrix}
-0.802 & 0.408 & 0.408 & -0.802 & . & . & . & . & . & . & . & . & . & . & . \\
-0.535 & -0.816 & -0.816 & -0.535 & . & . & . & . & . & . & . & . & . & . & . \\
0.267 & 0.408 & -0.408 & -0.267 & . & . & . & . & . & . & . & . & . & . & . \\
. & . & . & 0.802 & . & . & 1.0 & . & . & . & . & . & . & . & . \\
. & . & . & 0.535 & . & . & . & . & . & . & . & . & . & . & 1.0 \\
. & . & . & 0.267 & . & . & . & 1.0 & . & . & . & . & . & . & . \\
. & . & -0.408 & . & . & . & -1.0 & . & . & . & . & . & . & . & . \\
. & . & 0.816 & . & . & . & . & . & . & . & . & . & 1.0 & . & . \\
. & . & 0.408 & . & . & 1.0 & . & . & . & . & . & . & . & 1.0 & . \\
. & -0.408 & . & . & -1.0 & . & . & . & . & . & 1.0 & . & . & . & . \\
. & 0.816 & . & . & . & . & . & . & . & . & . & 1.0 & . & . & . \\
. & -0.408 & . & . & . & -1.0 & . & . & . & . & . & . & . & . & . \\
0.802 & . & . & . & 1.0 & . & . & . & . & . & . & . & . & . & . \\
0.535 & . & . & . & . & . & . & . & 1.0 & . & . & . & . & . & . \\
-0.267 & . & . & . & . & . & . & -1.0 & . & 1.0 & . & . & . & . & .
\end{bmatrix}
\begin{Bmatrix} P_1 \\ P_2 \\ P_3 \\ P_4 \\ P_5 \\ P_6 \\ P_7 \\ P_8 \\ R_{2Y} \\ R_{2Z} \\ R_{3X} \\ R_{3Y} \\ R_{4Y} \\ R_{4Z} \\ R_{5Y} \end{Bmatrix}
=
\begin{Bmatrix} -15 \\ 50 \\ -25 \\ . \\ . \\ . \\ . \\ . \\ . \\ . \\ . \\ . \\ . \\ . \\ . \end{Bmatrix}$$

Solving this matrix equation by a computer program such as that of Appendix B, we obtain

$$\mathbf{X} = \begin{Bmatrix} P_1 \\ P_2 \\ P_3 \\ P_4 \\ P_5 \\ P_6 \\ P_7 \\ P_8 \\ R_{2Y} \\ R_{2Z} \\ R_{3X} \\ R_{3Y} \\ R_{4Y} \\ R_{4Z} \\ R_{5Y} \end{Bmatrix} = \begin{Bmatrix} -9.35 \\ -55.15 \\ 0 \\ 0 \\ 7.50 \\ 22.50 \\ 0 \\ 0 \\ 5.00 \\ -2.50 \\ -15.00 \\ 45.00 \\ 0 \\ -22.50 \\ 0 \end{Bmatrix} \text{ kN}$$

A negative value for member forces P indicates compression, and a positive value indicates tension. A negative value of a reaction component R indicates that the component is acting in the negative direction of the coordinate axis, and a positive value indicates that it is acting in the positive direction of the coordinate axis.

PROBLEMS

4-1 and 4-2. Show that the space trusses are statically determinate. Determine all reaction components and member forces.

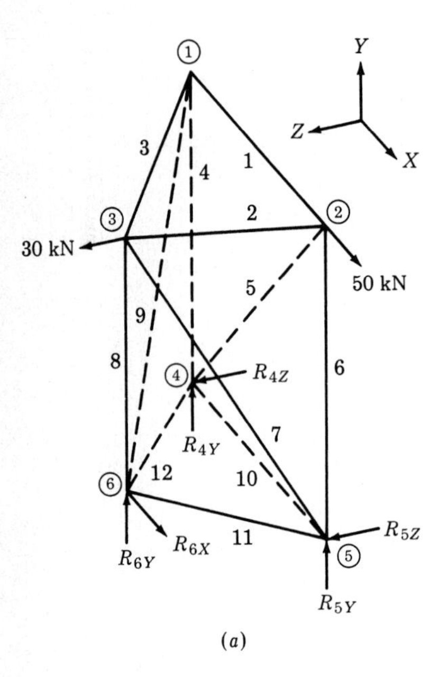

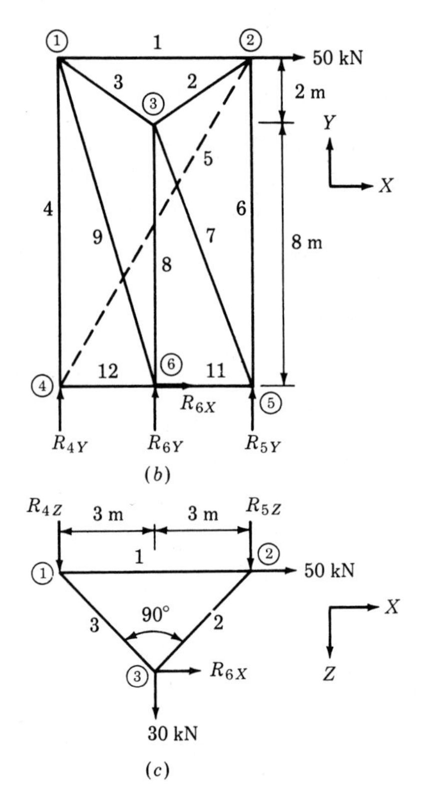

P4.1

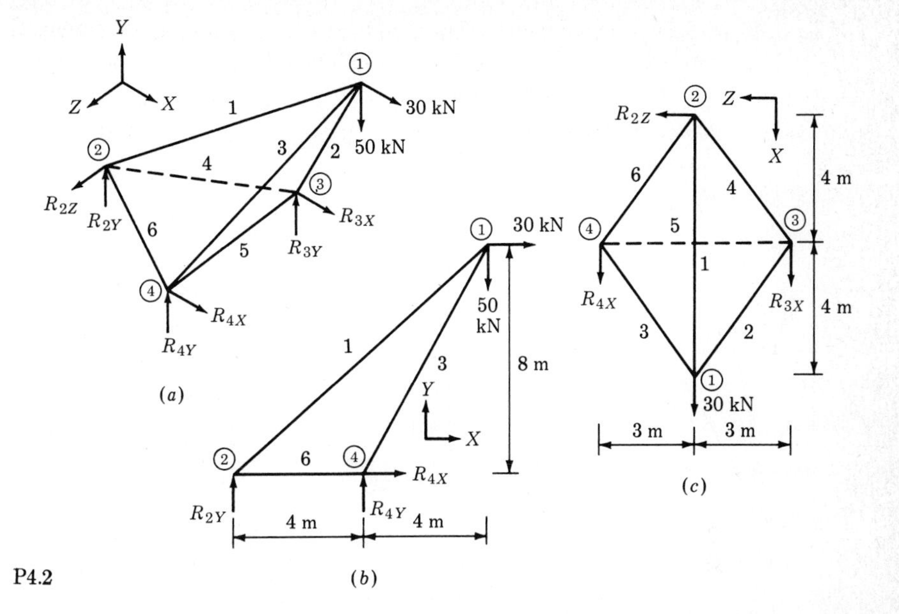

P4.2

4-3. For the space truss of Prob. 4-1, determine all reaction components and member forces for the external load condition of $Q_{1Z} = 40$ kN and $Q_{2Z} = -60$ kN.

4-4. For the space truss of Prob. 4-2, determine all reaction components and member forces for the external load condition of $Q_{1X} = 10$ kN, $Q_{1Y} = -30$ kN, and $Q_{1Z} = 40$ kN.

4-5. For the space truss shown, determine (*a*) the members that obviously have zero force; (*b*) the values of reaction components R_{5Z} and R_{6Y}; (*c*) the member forces P_4 and P_7.

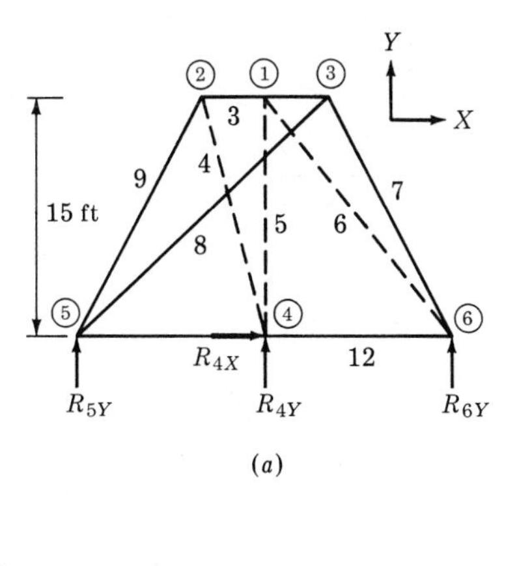

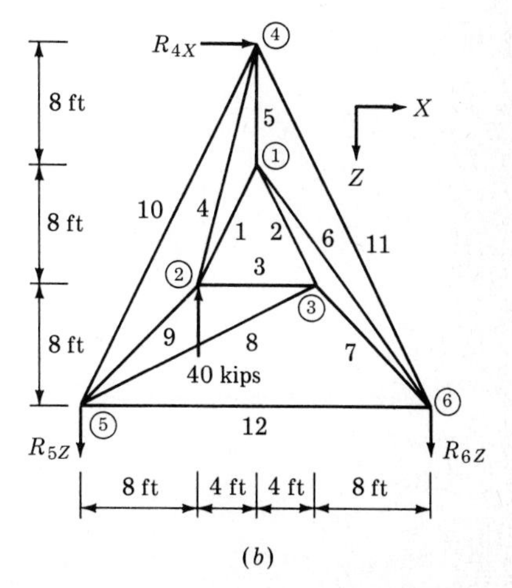

P4.5

4-6. For the space truss shown, determine (*a*) the members that obviously have zero force; (*b*) the member forces P_4 and P_6; (*c*) the reaction components R_{7Y}, R_{6Z}, and R_{8X}.

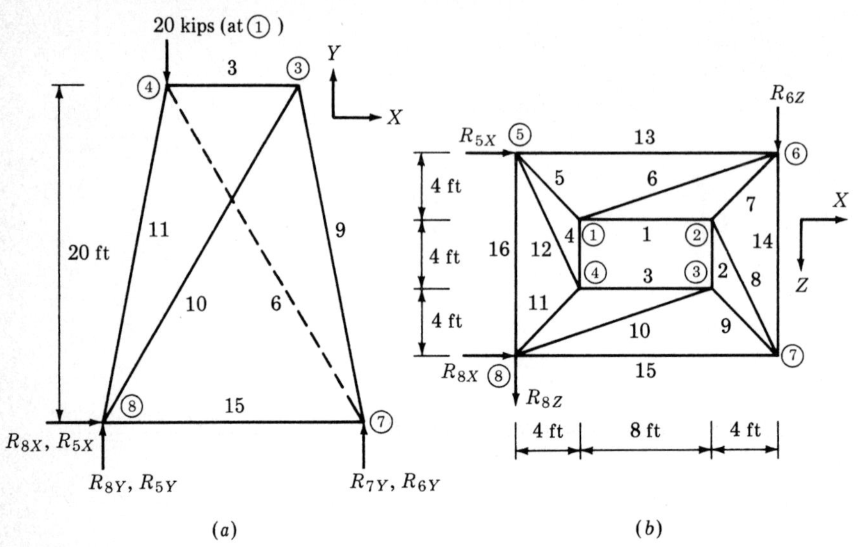

P4.6

4-7. For the space truss shown, determine (*a*) the members that obviously have zero force; (*b*) the reaction components R_{3X}, R_{2Y}, and R_{2Z}; (*c*) the member forces P_8 and P_5.

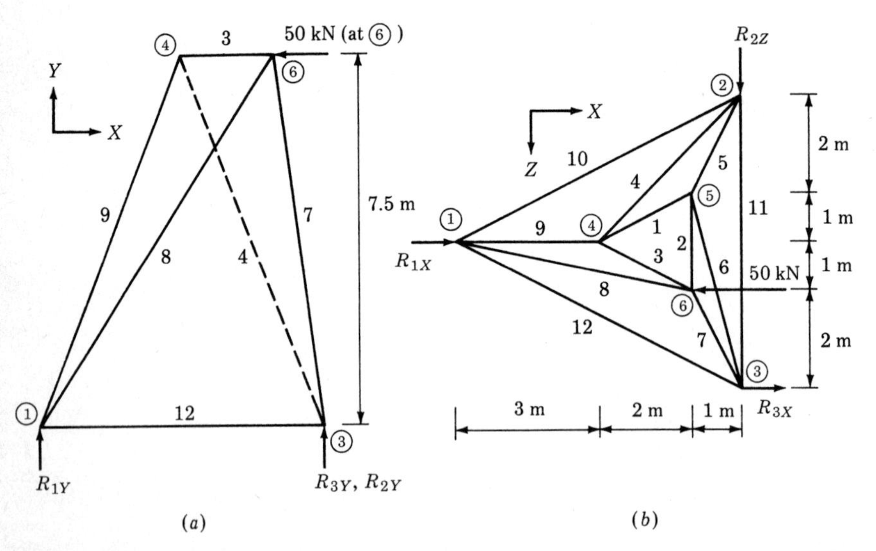

P4.7

4-8. Solve Prob. 4-5 for the external load condition of $Q_{3X} = -50$ kips.

4-9. Solve Prob. 4-6 for the external load condition of $Q_{1X} = -10$ kips and $Q_{1Z} = -5$ kips.

4-10. Determine the values of the reaction components and the member forces for the space truss shown by writing the equilibrium equations at each joint and solving the resulting set of simultaneous equations on a computer. Use an external load at joint ① of $\mathbf{Q} = 2\mathbf{i} - 10\mathbf{j} + 4\mathbf{k}$ kN.

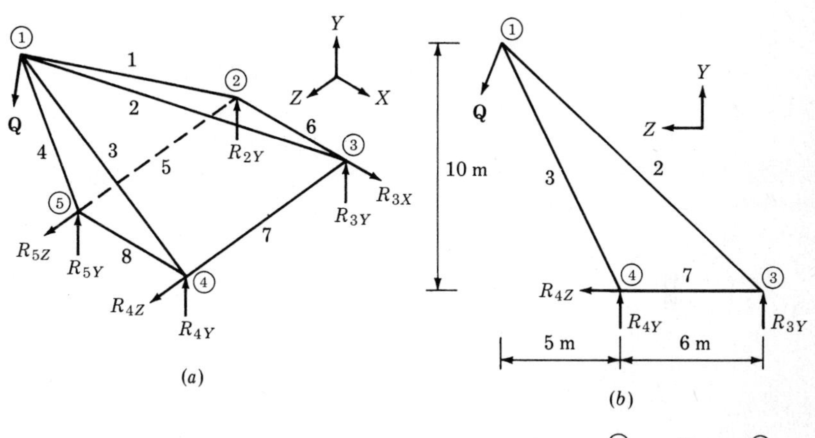

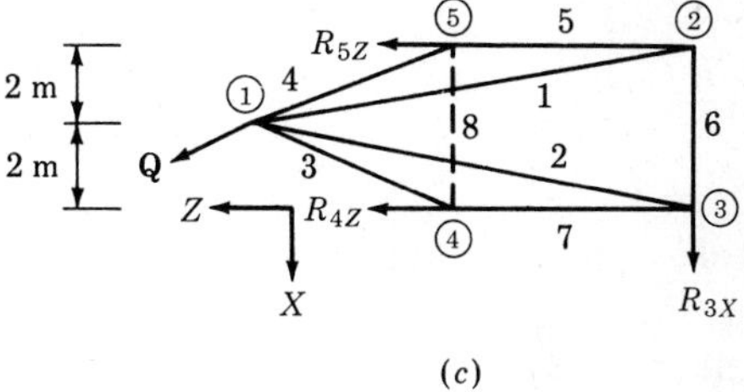

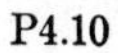

P4.10

CHAPTER

FIVE

ARCHES AND CABLE STRUCTURES

5-1 ARCH CHARACTERISTICS

Arches are distinguished from other types of structures by their shape, which results in a unique manner of supporting the loads. The several types of arches shown in Fig. 5-1 vary in the manner in which they are supported. Figure 5-1*a* shows a *hingeless arch,* one with fixed supports and no hinges along its length. Also shown are the names by which two points on an arch are commonly referred to in analysis. In descriptive terms, the *crown* is the high point on an arch and the *springings* are the support points. The arch of Fig. 5-1*b*, referred to as a *two-hinged arch,* is supported at its springings by hinges. Another form of two-hinged arch is shown in Fig. 5-1*c*. Although the arch of Fig. 5-1*c* resembles a truss, it can be considered an arch because of the manner in which it supports loads. The arch of Fig. 5-1*d* is a *three-hinged arch*. The three-hinged arch is statically determinate, and is thus the only type analyzed in this chapter. The others are statically indeterminate, and their analysis must be deferred until we have developed the principles of statically indeterminate structures.

The distinguishing features of an arch can be explained by comparing

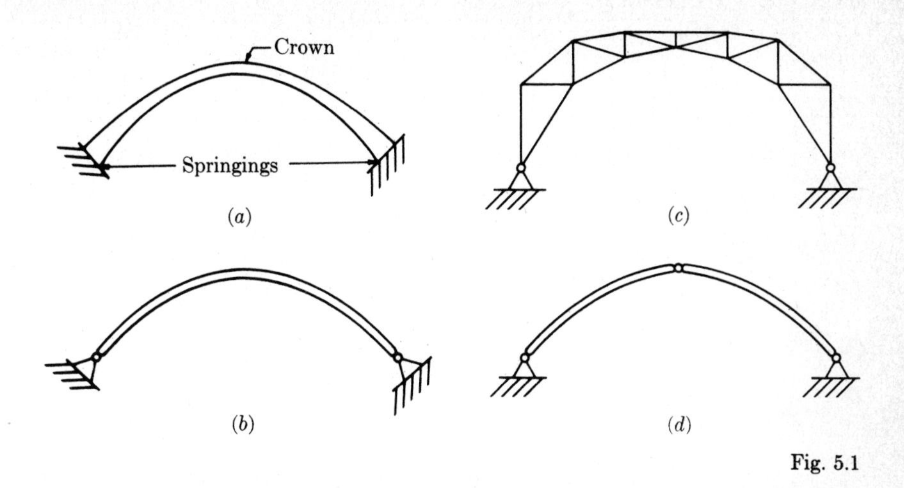

Fig. 5.1

its response to loading to that of a beam, as shown in Fig. 5-2*a*. A concentrated load P applied to the beam results in vertical reactions and moment in the beam, as shown by the moment diagram in Fig. 5-2*b*. Consider the same load supported by a two-hinged arch, as shown in Fig. 5-2*c*. The arch resists the load P by vertical reactions, but also by horizontal components of reaction. The horizontal components of reaction in effect reduce the

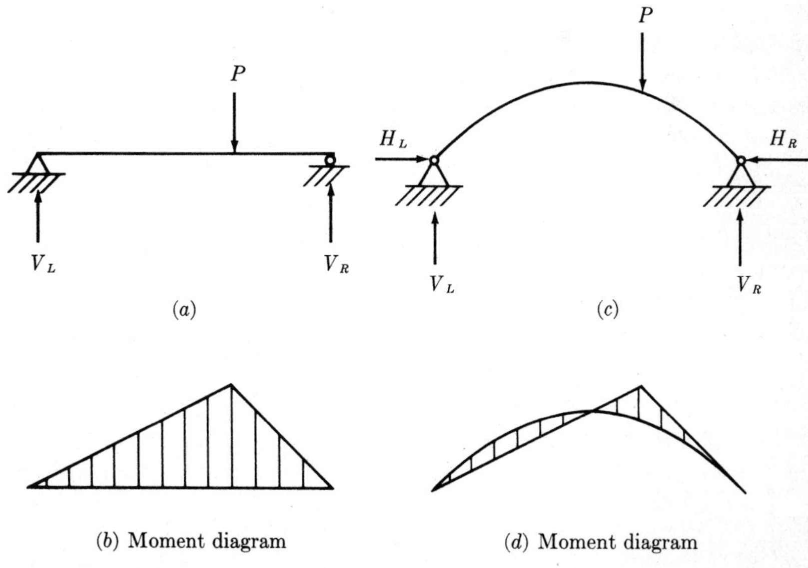

Fig. 5.2

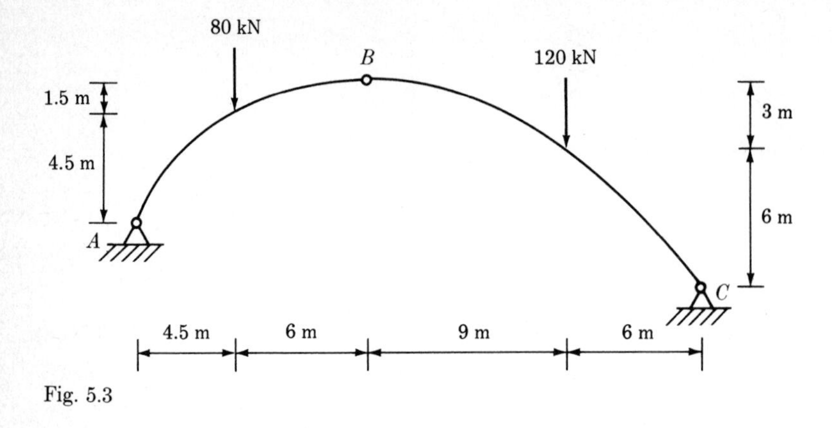

Fig. 5.3

moment from that in the simple beam. The resulting moment in the arch is shown by the moment diagram in Fig. 5-2*d*. Thus we see that owing to its configuration, and with proper support, an arch supports loading with less moment than a simple beam configuration. It must be recognized, however, that in the process considerable compressive stresses are generated in the arch. An arch is intended to function primarily by transmitting forces through compression, not by bending, which is secondary.

5-2 ANALYSIS OF THREE-HINGED ARCHES

A three-hinged arch, which is statically determinate, is analyzed by the basic equations of statics. The key to this analysis is the fact that there is zero moment at the hinge located out in the span of the arch. The procedure is illustrated in the following example.

EXAMPLE 5-1 The reactions are to be determined and the moment diagram constructed for the three-hinged arch in Fig. 5-3. A free-body diagram of the entire structure is shown in Fig. 5-4. The pinned supports at A and C have been replaced by the horizontal and vertical components of reaction. We see that there are four unknowns, considered to be positive in the directions shown. Recall that we have only three equations of equilibrium. An additional equilibrium equation must therefore be found if the problem is to be statically determinate. The additional equation is obtained from the condition of connection at B. Because the connection at B is a hinge, the summation of moments about that point must be zero.

Taking first the summation of moments about point A, we obtain

$$(80)(4.5) + (120)(19.5) + 3C_H - 25.5C_V = 0$$
$$3C_H - 25.5C_V = -2700$$

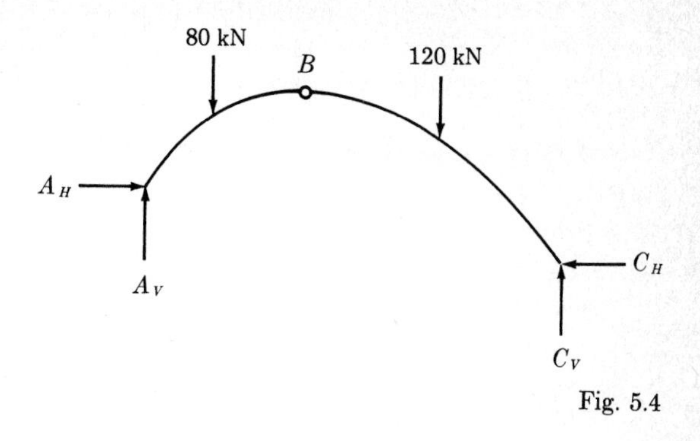

Fig. 5.4

Another equation containing C_H and C_V is obtained by considering the segment of the arch between B and C. Taking the summation of moments about B for this segment, we obtain

$$(120)(9) + 9C_H - 15C_V = 0$$
$$9C_H - 15C_V = -1080$$

Solving these two equations for the unknown values of C_H and C_V results in

$$C_H = 70.24 \text{ kN} \qquad C_V = 114.15 \text{ kN}$$

For equilibrium of forces in the horizontal direction

$$A_H = C_H = 70.24 \text{ kN}$$

The remaining unknown is found by considering the equilibrium of forces in the vertical direction

$$-80 - 120 + 114.15 + A_V = 0$$
$$A_V = 85.85 \text{ kN}$$

Moment diagrams for arches are commonly constructed on the axis of the arch, with the magnitude of moment plotted in a vertical direction. The moment diagram for the arch under consideration is shown in Fig. 5-5. The moment diagram has been plotted on the compression side of the members.

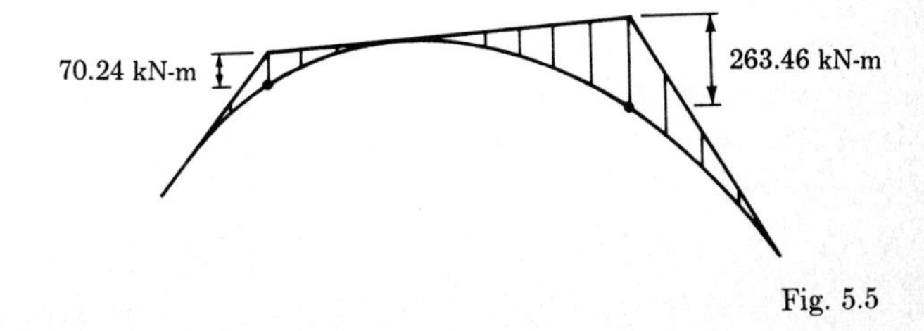

Fig. 5.5

5-3 GRAPHICAL ANALYSIS OF THREE-HINGED ARCHES

The procedure presented in Sec. 2-7 for graphically determining reactions can be applied to the analysis of three-hinged arches. Once we have found the reactions, the graphical procedure can be extended to obtain additional quantities needed for design.

As an illustration of the procedure, let us consider the analysis of the three-hinged arch shown in Fig. 5-6*a*. Such an arch is commonly used in buildings as the primary framing system. The arch is composed of two curved members joined at the top and supported at the base. For our purposes, all connections are considered to be pinned. The external loads are transmitted to the arch by purlins and girts which span between the arches. The loading on the arch is therefore in the form of concentrated loads, as shown.

For convenience in treating the large number of loads on the arch, the spaces between the loads are lettered as shown in Fig. 5-6*b*. It is convenient to use such a line diagram of the outer shape of the arch for the analysis.

The reactions of the arch are determined in two steps. In each step the loads are removed from one segment of the arch. This makes it possible to determine the reactions, because the reaction on the unloaded side of the arch must pass through the two pins, and its direction is therefore known. The results obtained in each step are superimposed to give the actual condition.

The loads are removed from the right segment of the arch in step 1. The space diagram for this condition is shown in Fig. 5-7*a*. The load at the crown has been included with the loads on the left segment. It could just as well have been included with those on the right. The direction of the

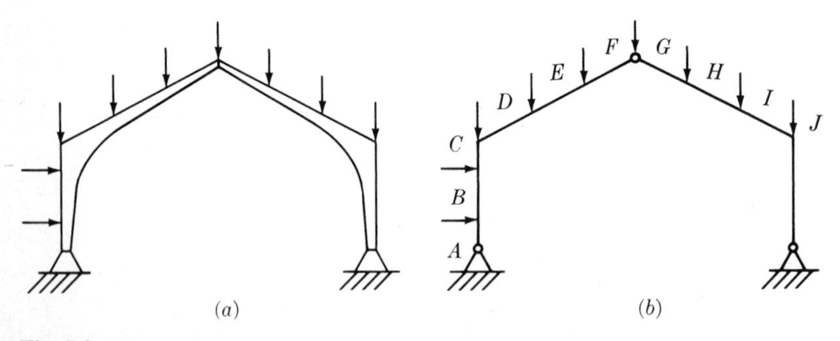

Fig. 5.6

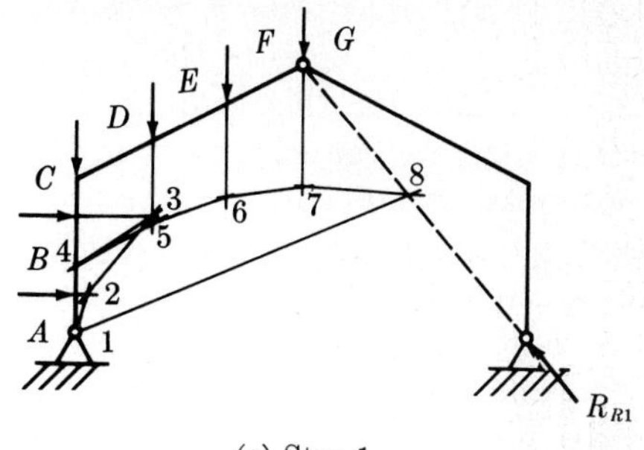

(a) Step 1

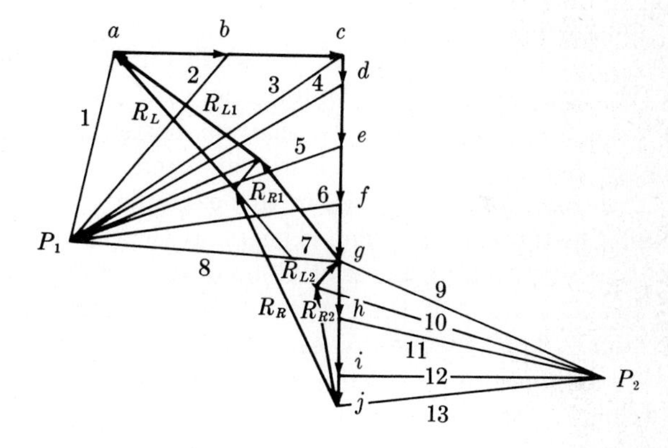

(b) Force polygon

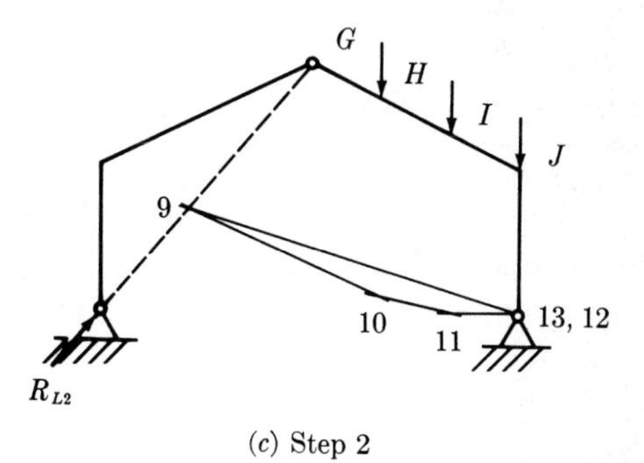

(c) Step 2

Fig. 5.7

reaction at the right support is known and is denoted as R_{R1}, where the second subscript 1 refers to the step 1 loading. To determine the values of the reactions for this loading we use the same procedure as in Sec. 2-7. The force polygon is shown in Fig. 5-7*b*. The forces are denoted by the lowercase letters corresponding to the lettering in the space diagram. Pole P_1 is selected as shown, and the rays are drawn to the extremities of the given forces. To avoid confusion in constructing the funicular polygon we number the spaces between the rays as shown. We then construct the funicular polygon, beginning at the left support. The strings of the polygon are denoted by their corresponding numbers from the force polygon. With the direction of R_{R1} known, the reactions for the loading of step 1 are determined. These are denoted as R_{R1} and R_{L1} on the force polygon.

The loading for step 2 is shown in Fig. 5-7*c*. These loads are added onto the previous loads on the force polygon. Although the same pole could be used for both steps of the analysis, better intersection of the strings with the lines of action of the loads can often be obtained if two poles are used. A second pole P_2 is used in this example.

The funicular polygon is begun at the right support. It should be noted that ray 13-12 is the first to be drawn. However, this ray intersects the load *I-J* right at the support, and therefore point 12 is coincident with point 13. The student is cautioned against overlooking cases such as this, as they can quite often occur. The remainder of the funicular polygon is then constructed, with the direction of the left reaction known. The resulting reactions for the loading of step 2 are denoted on the force polygon as R_{R2} and R_{L2}.

The total reactions are found by superimposing the results of step 1 and step 2. On the force polygon this entails adding R_{R1} to R_{R2} and R_{L1} to R_{L2}. As shown in Fig. 5-7*b*, this is accomplished by constructing a parallelogram with sides R_{R1} and R_{L2}. The resulting reactions of the structure are found to be R_R and R_L.

Before we continue with additional uses of the graphical results, mention should be made of the reactions at the crown hinge. These reactions are available from the force polygon in Fig. 5-7*b*. The reaction of the pin at the crown on the left segment of the arch is represented by a force vector from point *f* to the intersection of R_R and R_L. Similarly, the reaction of the pin on the right segment of the arch is represented by a force vector from the point of intersection of R_R and R_L to point *g*. Thus the force polygon for each arch segment is closed, and there is a closed force polygon for the pin, with one side the vector *fg* and the other two sides vectors equally and oppositely directed to those for the segments.

The graphical analysis can be extended for additional information on moment, shear, and thrust in the arch. Consider the left segment of the arch in Fig. 5-6. Figure 5-8*b* shows the force polygon for the left segment as obtained from Fig. 5-7*b*. The space diagram is shown in Fig. 5-8*a*. Note that for the following analysis the reactions at the left support and at the crown could have been determined analytically rather than graphically. The point is that we begin the analysis with knowledge of the reactions acting on the segment of the arch.

On the force polygon in Fig. 5-8*b* lines are drawn from point P to the extremities of the forces acting on the arch segment. The spaces between these lines are numbered as shown. Point P is considered to be a particular pole, and the lines just drawn are the rays. The string representing ray 1-2 is drawn through the left support and extended to intersect force A-B. The remaining strings of the funicular polygon are drawn on the space diagram as shown. The final string 6-7 passes through the crown and serves as a check on the construction.

The reason for selecting the location of the pole P as we did will now become apparent. We see in Fig. 5-8*b* that the true reaction at the left support of the arch is represented by ray 1-2. The point of application of the reaction is known on the space diagram. We see that the next ray 2-3 represents the reaction on a free-body diagram of the arch segment, with a hypothetical cut made between loads A-B and B-C on the space diagram. The point of application of this resultant is represented by string 23 of the funicular polygon. A similar interpretation can be made for the other rays of the force polygon. We can therefore state that the moment at any point on the arch is equal to the perpendicular distance from the appropriate string of the funicular polygon to that point multiplied by the magnitude of the force represented by that string on the force polygon. For example, the moment at point M on the arch is equal to the distance d_M multiplied by the magnitude of ray 2-3 on the force polygon. The distance d_M is, of course, measured according to the scale of the space diagram and the force in ray 2-3 according to the scale of the force polygon.

The graphical evaluation of moment serves as a means of visualizing the moment pattern in such a structure. The resulting funicular polygon readily indicates the type of moment at a point in the structure, compression in the inner fiber of the member or in the outer fiber. The graphical presentation also adds to an understanding of how the shape of the arch and the moment in the arch are related. The student should recognize what happens to the magnitude of the moment as the funicular polygon becomes coincident with the centroidal axis of the arch.

The arch cross-section can be designed on the space diagram. A trial shape of the inner face of the arch is lightly drawn on the space diagram. The moment at the centroid of any cross-section can then readily be obtained by measuring out to the funicular polygon and multiplying the distance by the appropriate force on the force polygon. The bending

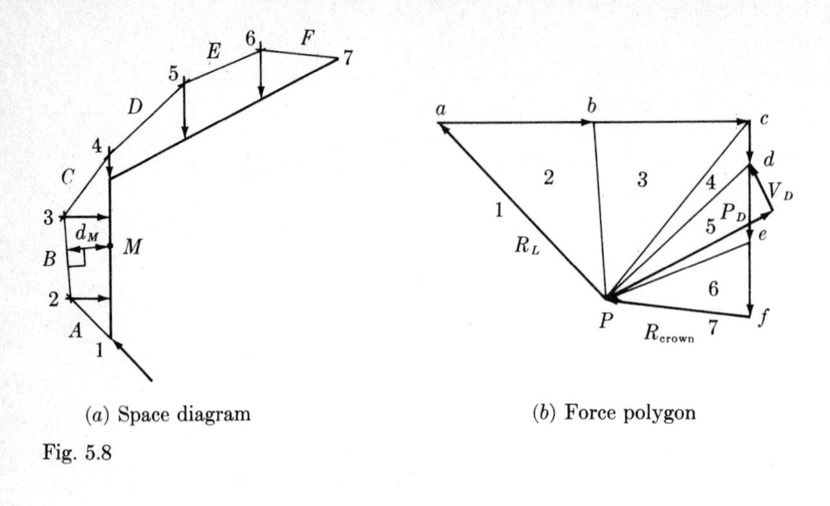

(a) Space diagram

(b) Force polygon

Fig. 5.8

stress can be calculated from simple sliderule calculations and checked against the allowable stress. If necessary, the depth of the arch can easily be adjusted, and the designer can at the same time visualize how the change in depth affects the appearance of the arch.

Shear and thrust can also be obtained from the graphical solution. For example, the shear and thrust in the section of the arch between loads *C-D* and *D-E* of Fig. 5-8 are obtained by resolving ray 4-5 of the force polygon into components parallel and perpendicular to the axis of the arch at the section under consideration. The shear and thrust for this section are indicated as V_D and P_D, respectively, in Fig. 5-8*b*.

5-4 CABLE CHARACTERISTICS

In analyzing cable structures it is generally assumed that the cable is flexible and therefore does not resist bending moment. By assuming that there are zero moments at all points on the cable, it is possible to analyze the cable by basic statics.

Let us first consider a cable that is subjected to a series of concentrated vertical loads, as shown in Fig. 5-9. The weight of the cable is assumed negligible. The horizontal and vertical components of reaction at *A* and *B* are considered to be positive in the directions shown. If we take the summation of horizontal forces on a free-body diagram of a segment of the cable and on a free-body diagram of the entire structure, we find that for the vertical loading

$$A_H = B_H = H$$

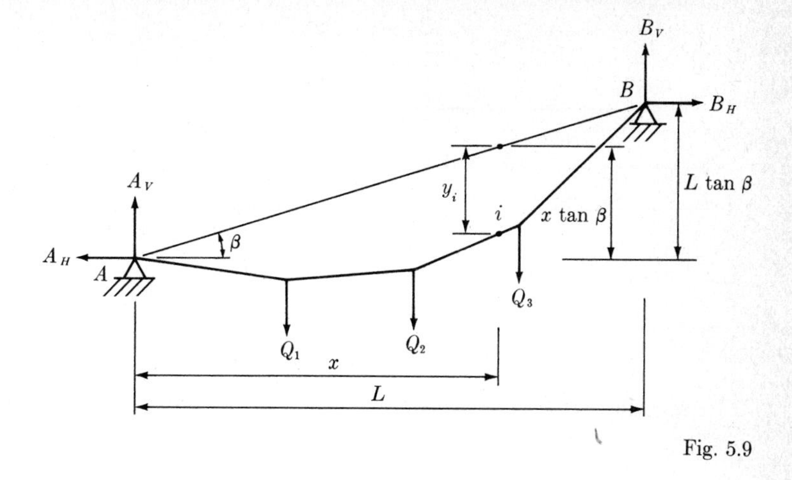

Fig. 5.9

where H denotes the constant value of the horizontal component of force in the cable.

Taking the summation of moments about B in Fig. 5-9, we obtain

$$H(L \tan \beta) + A_V L - \sum_{j=1}^{3} Q_j(L - x_j) = 0$$

or
$$A_V = \frac{\sum_{j=1}^{3} Q_j(L - x_j)}{L} - H \tan \beta \tag{5-1}$$

The term x_j is the distance of the jth load from the left support. If we next take the summation of moments about some point on the cable, such as point i in Fig. 5-9, we obtain for the forces acting to the left of i,

$$H(x \tan \beta - y_i) + A_V x - M_Q = 0 \tag{5-2}$$

M_Q denotes the moment of the loads Q about point i. The term x describes the location of the point i, and y_i is the vertical distance from i to the chord of the cable. With the expression for A_V of Eq. (5-1), Eq. (5-2) becomes

$$H(x \tan \beta - y_i) + \frac{x}{L} \sum_{j=1}^{3} Q_j(L - x_j) - Hx \tan \beta - M_Q = 0$$

or
$$Hy_i = \frac{x}{L} \sum_{j=1}^{3} Q_j(L - x_j) - M_Q \tag{5-3}$$

Let us consider the same set of loads applied to a simple beam having the same span L, as shown in Fig. 5-10. If we take the summation of

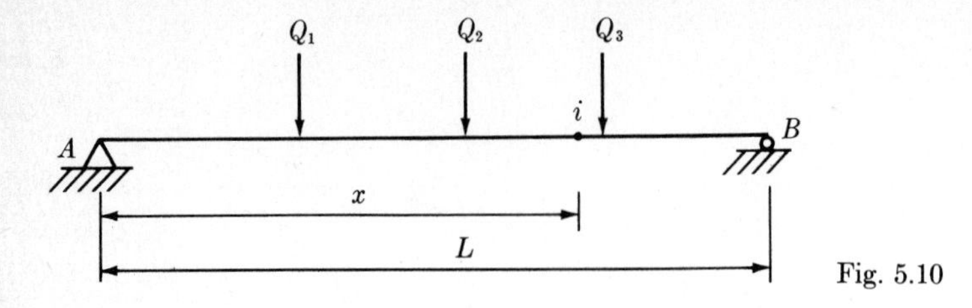

Fig. 5.10

moments about support point B, we obtain

$$R_A L - \sum_{j=1}^{3} Q_j(L - x_j) = 0$$

The reaction at A is therefore

$$R_A = \frac{\sum_{j=1}^{3} Q_j(L - x_j)}{L}$$

Taking the summation of moments to the left of point i, for moment in the beam at i we have

$$M_i = \frac{x}{L} \sum_{j=1}^{3} Q_j(L - x_j) - M_Q \tag{5-4}$$

We see that the right sides of Eqs. (5-3) and (5-4) are identical. From these results we can therefore make the following statement: *For a vertically loaded cable, the product of the horizontal component of force in the cable and the vertical distance from a point on the cable to the chord of the cable is equal to the moment at that point in a simple beam of the same span supporting the same loads.*

Applications of this statement and other aspects of cable analysis are illustrated in the following example.

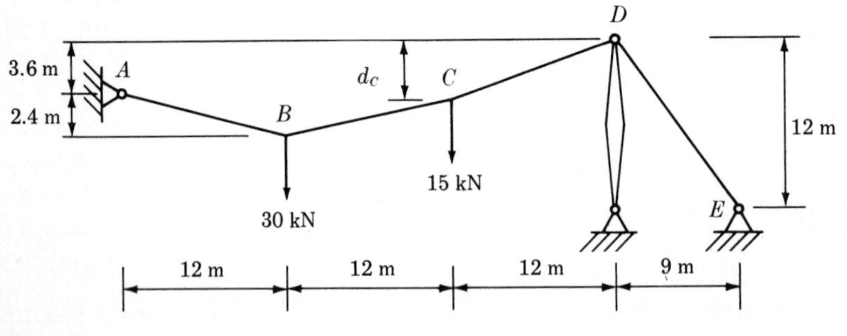

Fig. 5.11

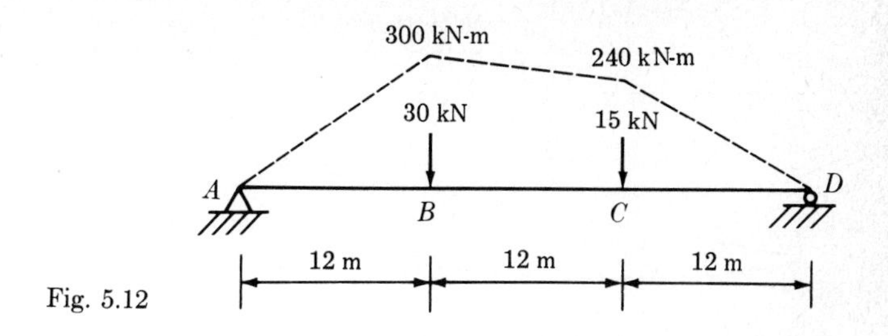

Fig. 5.12

EXAMPLE 5-2 For the vertically loaded cable structure of Fig. 5-11 we are to determine the value of the horizontal and vertical components of reaction at A, the distance d_C, the maximum tension in cable AD, and the tension in cable DE. The simple-beam analogy will be used to determine the value of H, and hence the horizontal component of reaction at A. The simple beam for cable AD and the moment diagram for the given loads are shown in Fig. 5-12. Noting that the vertical distance from point B on the cable to the chord is 3.6 m, we can state that

$$3.6H = 300$$
$$H = 83.33 \text{ kN}$$

The horizontal component of reaction at A is therefore 83.33 kN. The vertical component of reaction at A can be determined by taking the summation of moments about point B on the cable:

$$-(83.33)(2.4) + 12A_V = 0$$
$$A_V = 16.67 \text{ kN}$$

The distance d_C is obtained by first evaluating the distance y_C, which is the vertical distance from point C on the cable to the chord of the cable. With the value of moment in the simple beam at C, y_C is found to be

$$83.33y_C = 240$$
$$y_C = 2.88 \text{ m}$$

From the given dimensions of the cable structure, the desired distance is therefore

$$d_C = 2.88 + 1.20 = 4.08 \text{ m}$$

The maximum tension in a cable such as AD will occur in the section where the vertical component of force is the largest because the horizontal component of force is constant from A to D. The vertical component of force in section CD is equal to $45 - 16.67 = 28.33$ kN. The maximum tension therefore occurs in section CD and is equal to

$$T_{max} = (83.33^2 + 28.33^2)^{1/2} = 88.01 \text{ kN}$$

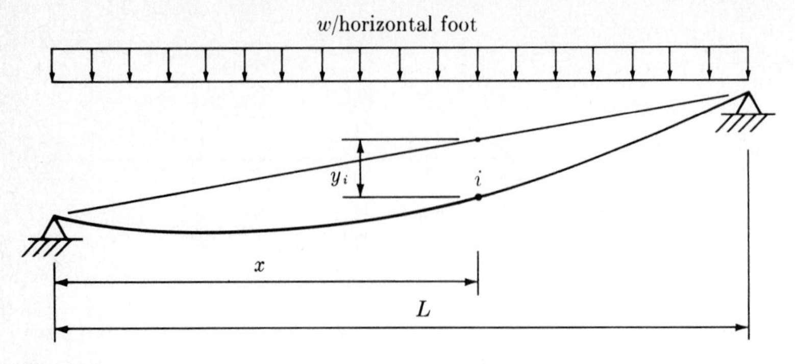

Fig. 5.13

The tension in cable DE is obtained by considering equilibrium conditions at pin D. Because the tower is pinned at each end, we know that the horizontal components of force in CD and DE are equal. Therefore

$$T_{DE} = \frac{(5)(83.33)}{3} = 138.88 \text{ kN}$$

Let us next consider a cable subjected to uniform loading per horizontal foot, as shown in Fig. 5-13. Using the simple-beam analogy, we find that

$$Hy_i = \frac{wLx}{2} - \frac{wx^2}{2}$$

$$= \frac{w}{2}(Lx - x^2)$$

If we let $y_i = h$ at $x = L/2$, we obtain a convenient expression for the horizontal component of force in the cable,

$$H = \frac{wL^2}{8h}$$

The expression for y_i is therefore

$$y_i = \frac{4h}{L^2}(Lx - x^2)$$

which is recognized as an expression for a parabola.

The weight of a cable can generally be approximated by a uniform load per horizontal foot, or even a series of concentrated loads. Analyzing a cable for loading per foot of cable results in a more complicated expression for y_i in the form of hyperbolic functions. The resulting curve is referred to as a *catenary*.

In this chapter only the basic concepts of arch and cable structures have been developed. For a more extensive discussion of arches and cables the student is referred to texts that include the advanced topics.[1]

PROBLEMS

5-1. For the three-hinged arch shown, determine (*a*) the horizontal and vertical components of reaction at A and C; (*b*) the values of moment at points D and E.

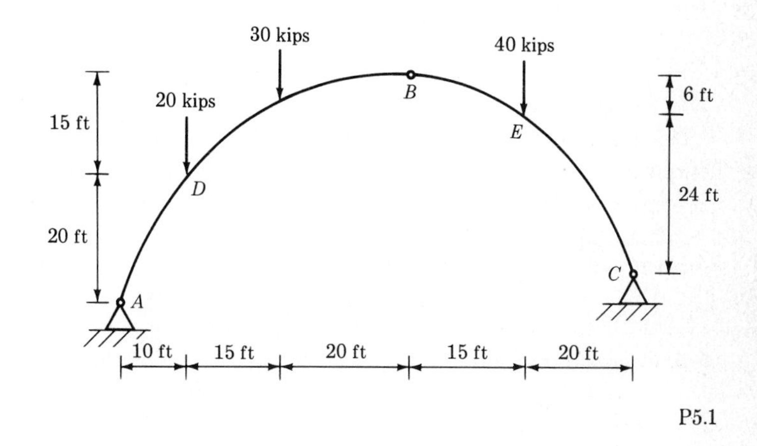

P5.1

5-2. For the symmetric three-hinged arch loaded as shown, determine (*a*) the horizontal and vertical components of reaction at A and C; (*b*) the thrust, shear, and moment at point D.

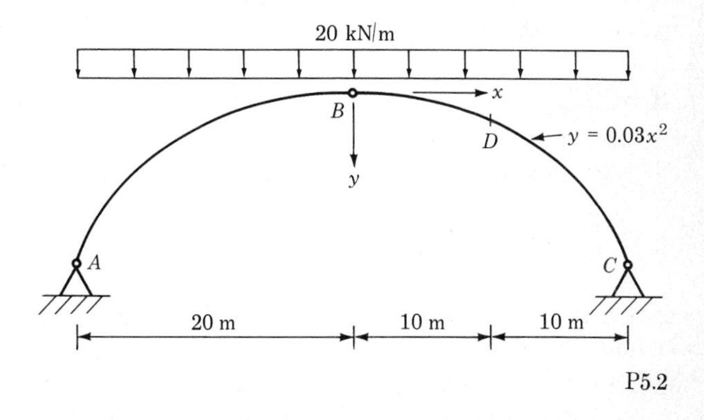

P5.2

[1] J. I. Parcel and R. B. B. Moorman, "Analysis of Statically Indeterminate Structures," John Wiley & Sons, Inc., New York, 1955; S. T. Carpenter, "Structural Mechanics," John Wiley & Sons, Inc., New York, 1960.

5-3. Solve Prob. 5-2 with the uniform load on only one-half of the arch from A to B.

5-4. For the three-hinged frame loaded as shown, (*a*) determine the horizontal and vertical components of reaction at A and E; (*b*) construct the moment diagram for $ABCD$ on the tension side of the members.

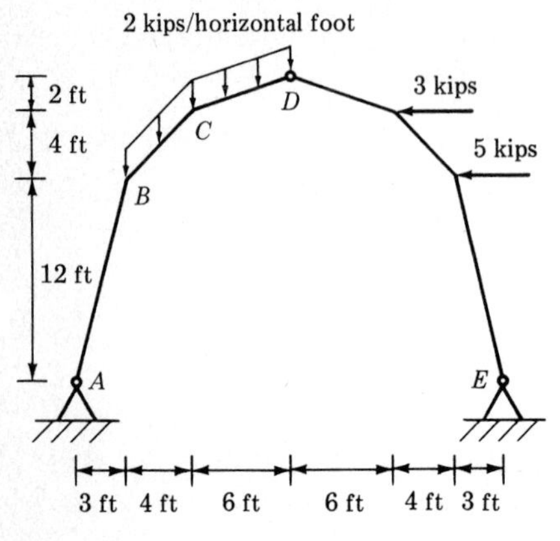

P5.4

5-5. For the three-hinged frame shown, (*a*) determine the horizontal and vertical components of reaction at A and E; (*b*) construct the moment diagram for segment CDE on the tension side of the members.

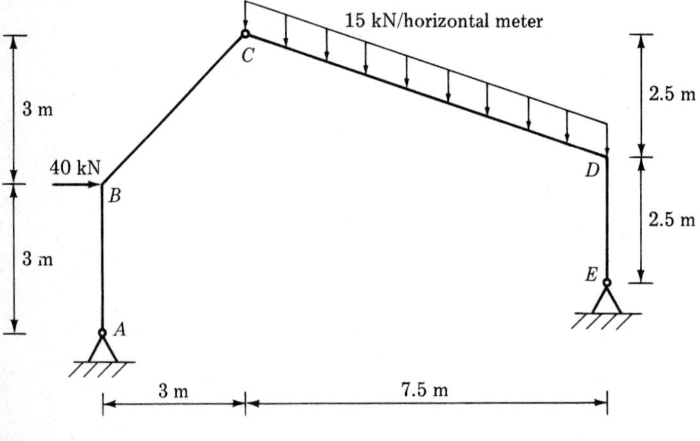

P5.5

5-6. For the three-hinged truss structure shown, determine (*a*) the horizontal and vertical components of reaction at *A* and *C*; (*b*) the forces in members *DE*, *BF*, and *GH*.

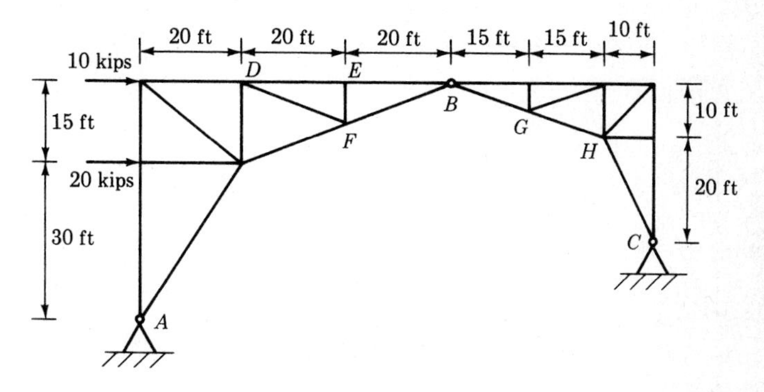

P5.6

5-7. The three-hinged frame shown is subjected to wind pressures as shown. The wind loads are transmitted to the frame as concentrated loads by purlins and girts spanning between frames. The spacing of the frames is 5 m. Convert the given wind pressures to SI units of newtons per square meter (Pascals), and (*a*) determine the resulting horizontal and vertical components of reaction at *A* and *E*; (*b*) construct the moment diagram for the frame on the compression side of the members.

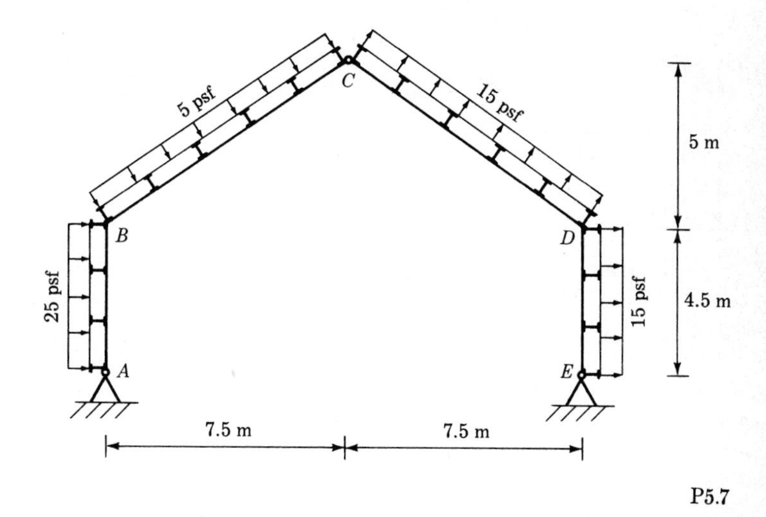

P5.7

5-8. For the cable structure shown, determine (*a*) the vertical distance d_E; (*b*) the maximum tension in cable *CDEF*; (*c*) the horizontal and vertical components of reaction at *B*.

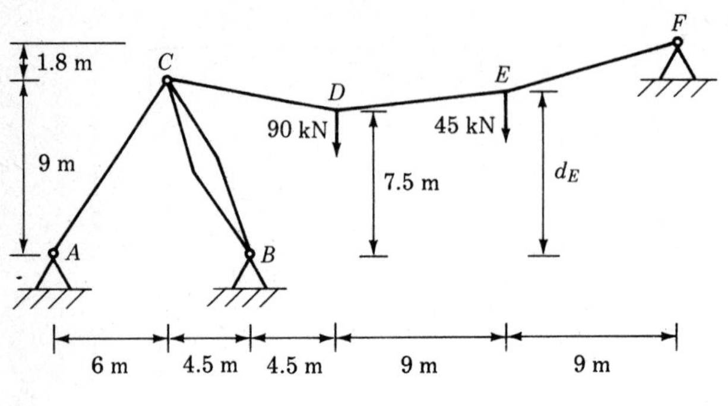

P5.8

5-9. For the cable structure shown, determine (*a*) the maximum tension in cable *AC*; (*b*) the value of d_D; (*c*) the maximum tension in cable *CDE* in kips. *B* is a point on the cable of known elevation.

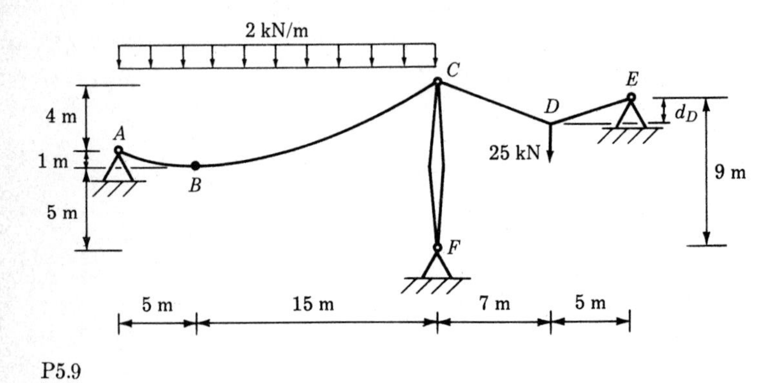

P5.9

5-10. For the cable and truss structure shown, determine (*a*) the horizontal and vertical components of reaction at A; (*b*) the maximum tension in cable AC; (*c*) the axial force in truss member DE. B is a point on the cable of known elevation.

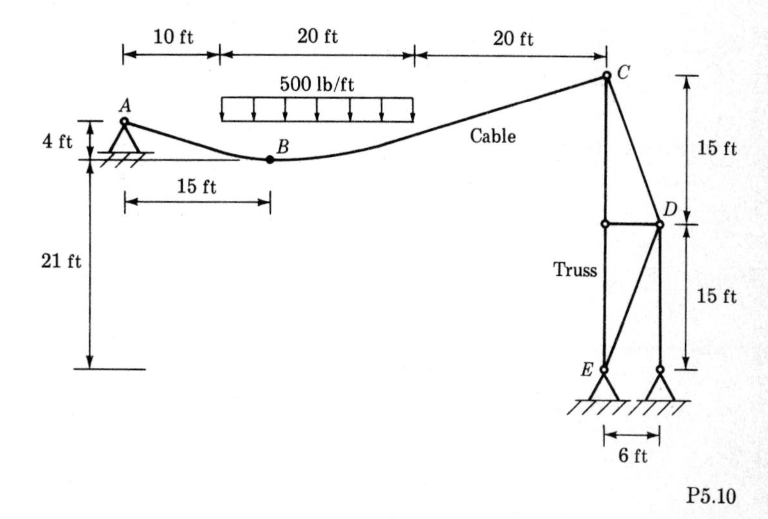

P5.10

5-11. For the frame and cable structure shown, (*a*) determine the horizontal and vertical components of reaction at E; (*b*) construct the moment diagram for ABC on the tension side of the members. D is a point on the cable of known elevation.

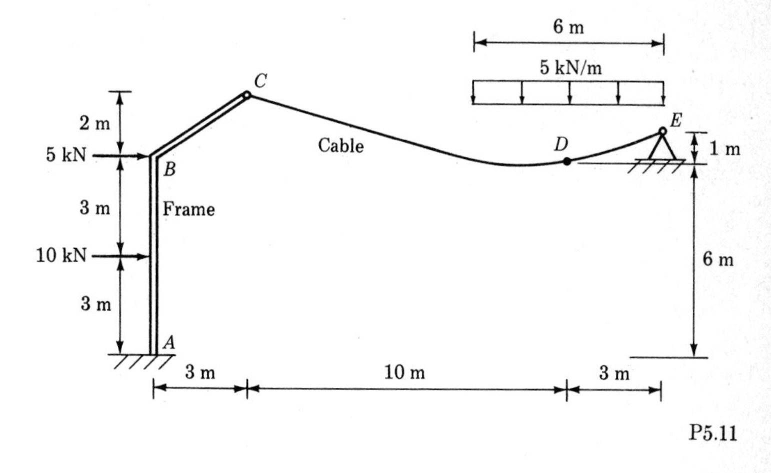

P5.11

5-12. For the cable structure shown, determine (*a*) the vertical distance d_F; (*b*) the maximum tension in cable *BDEFG*; (*c*) the tension in cable *AB*; (*d*) the tension in cable *GI*.

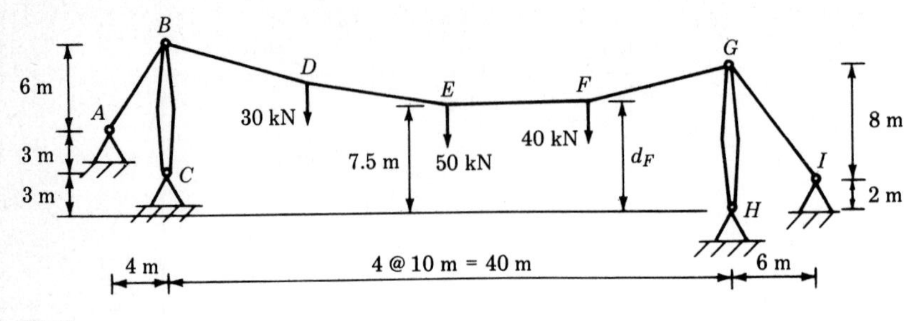

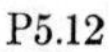
P5.12

5-13. Solve Prob. 5-6 by the graphical method.

CHAPTER

SIX

APPROXIMATE METHODS OF ANALYSIS

6-1 FRAME CHARACTERISTICS

Because frames are often statically indeterminate, we shall not discuss their analysis in detail until later in the text. However, it is well to point out here the basic characteristics of a frame and compare these characteristics to those of other structures. It is also appropriate at this point to discuss some of the approximate methods that can be used for the analysis of statically indeterminate structures.

As we have discussed, a truss is formed of triangular arrangements of pin-connected members. A frame, by comparison, is formed by arranging moment-resisting members in the desired configuration and providing moment-resisting connections between some or all of the members. These moment-resisting connections are often referred to as *rigid connections.* A frame can generally be distinguished from an arch structure by its configuration; the members of frames are usually arranged in nearly horizontal and vertical directions.

The structures of Fig. 6-1 are examples of typical frames. Loads can be applied at any point on the structure. Each member is therefore subjected to possible axial force, bending moment, and shear. The supports

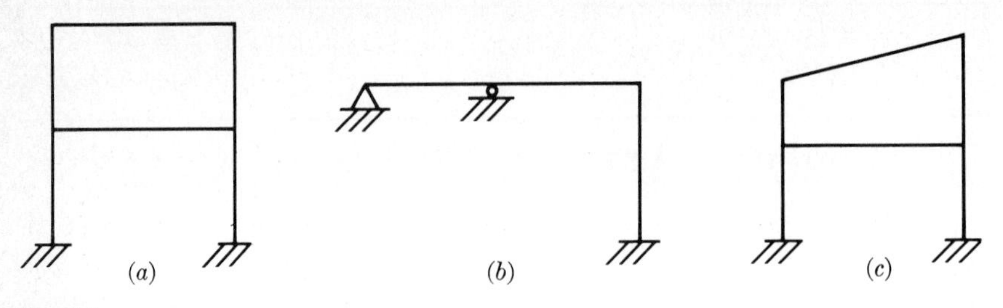

Fig. 6.1

of frames can vary, as long as they provide a stable configuration for the structure. The frames of Fig. 6-1 lie in one plane, but frames can also be three-dimensional structures. To analyze a frame simultaneously in three dimensions complicates the analysis considerably. It is therefore common to consider two dimensions at a time in analyzing three-dimensional frames. The major portion of this text will be devoted to planar frames, with an occasional treatment of three-dimensional frames.

Basically, a frame is statically indeterminate when the number of unknown reactions and member forces is in excess of the number of available equilibrium equations. For some types of frames the degree of indeterminacy is not difficult to ascertain. For example, there are six unknown components of reaction for a frame such as that shown in Fig. 6-2*a*. There are three equations of equilibrium available, so it is apparent that this frame is statically indeterminate to the third degree. By comparison, let us consider the frame of Fig. 6-2*b*. This frame also has six components of reaction. However, even if the reaction components are known, the upper section of the frame remains statically indeterminate because the number of member forces exceeds the number of equations

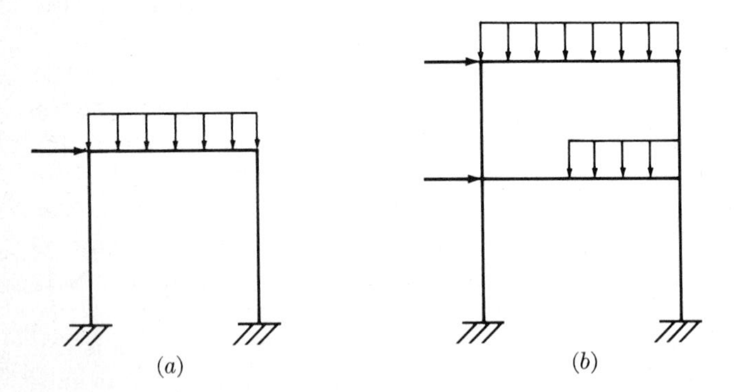

Fig. 6.2

available. The upper section is also statically indeterminate to the third degree. This can be verified by considering a free-body diagram of the upper section. At the base of each upper column there are three unknowns: axial force, moment, and shear. Therefore the entire frame is statically indeterminate to the sixth degree. It is possible to develop criteria for static determinacy of frames, as we did for trusses in Eqs. (3-1) and (4-1), but such criteria for frames are often difficult to apply. It will be found that the degree of static indeterminacy becomes evident as the analysis is attempted, provided the analyst establishes a sound practice of using free-body diagrams and acquiring from these the number of unknown forces and reactions.

For some purposes, such as in preliminary design, it is desirable to obtain a quick approximation of the member forces in statically indeterminate structures. With appropriate assumptions, statically indeterminate structures can be reduced to statically determinate systems, which can then be analyzed for approximate member forces and reactions using statics. Some of the commonly used approximate methods of analysis are discussed in this chapter. It must be remembered that the results obtained are approximate, and their nearness to the true values is dependent upon how good the assumptions are.

6-2 APPROXIMATE ANALYSES

In analyzing a statically indeterminate structure by approximate procedures one assumption is made for each degree of indeterminacy. These assumptions are based on logical interpretations of how the structure will react to the given loading. Let us consider an approximate analysis of the statically indeterminate truss of Fig. 6-3. The truss is statically indeterminate because of the additional diagonal in each panel. If the diagonals of the truss in Fig. 6-3 are considered to be long and slender, we can assume that their capacity to resist compressive force is negligible. Therefore the

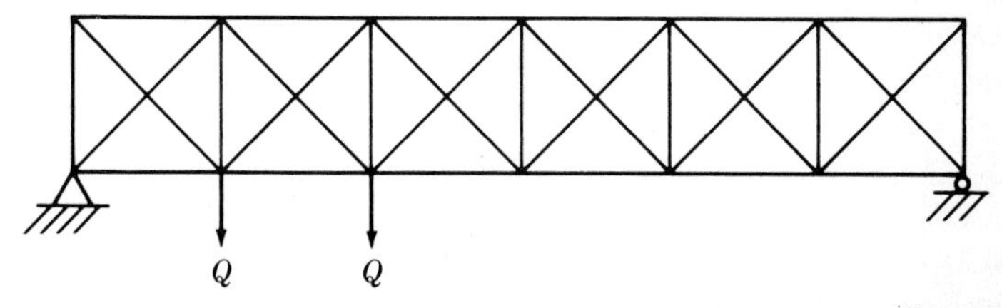

Fig. 6.3

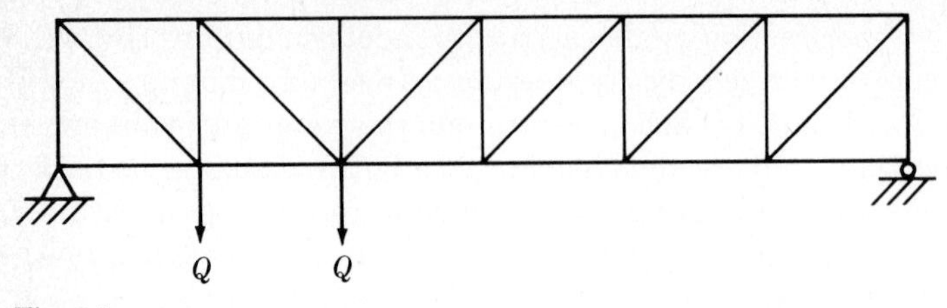

Fig. 6.4

diagonals acting in compression are removed, resulting in a statically determinate truss. For loading such as that in Fig. 6-3 the approximated statically determinate truss is as shown in Fig. 6-4. The question of which diagonal in a panel is in tension is answered by noting that the shear in each panel is carried by the diagonal in tension.

If the diagonals in the truss of Fig. 6-3 are assumed to have considerable stiffness, we can perform an approximate analysis by assuming a certain distribution of the shear in each panel between the two diagonals. One diagonal will be in tension and the other in compression. For convenience, the shear is often assumed to be distributed equally between the diagonals. In either of these approximate analyses of the truss the number of assumptions is equal to the degree of indeterminacy.

Another problem that is commonly encountered is lateral loading on a *mill bent*. Two configurations of mill bents with lateral loading, such as that due to wind, are shown in Fig. 6-5. The mill bents are composed of trusses for the roof section, supported by vertical columns. The trusses are considered to be pin connected, and the columns are continuous from the supports to the top chord.

To make rational assumptions for approximate analyses of such structures we must consider the manner in which the bents deflect when they are subjected to lateral loading. For the mill bent of Fig. 6-5*a* the left column will deflect in the manner shown in Fig. 6-6*a*. At some point on

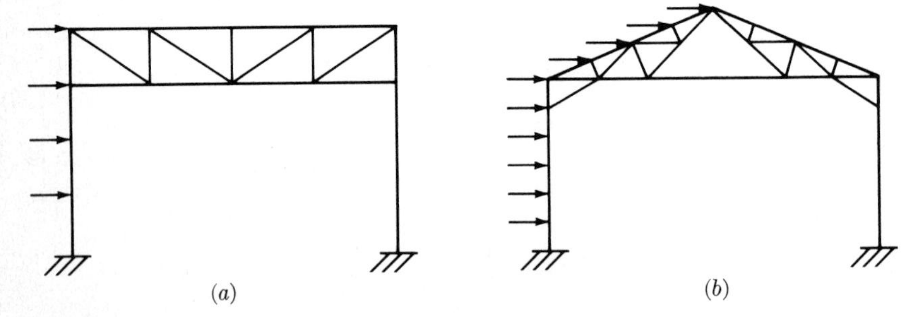

Fig. 6.5

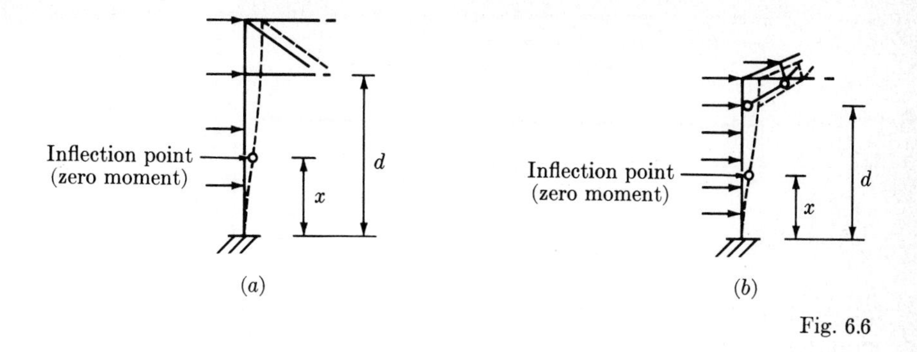

Fig. 6.6

the column there will be an *inflection point*, a point of zero moment. The distance of the inflection point from the base is denoted as x in Fig. 6-6*a*. For a rigid support at the base of the column it is common to assume that the inflection point is located midway up to the bottom of the truss, and therefore $x = d/2$. For a less rigid support the inflection point is lower, being at the base for a pinned support. For pinned supports the bent is less indeterminate by two degrees. The same reasoning is applied to the location of the inflection point on the right column. Locating the inflection point on each column, and hence the point of zero moment, entails two assumptions for the bent. Because the bent is statically indeterminate to the third degree, a third assumption must be made. A common third assumption is that the shear in the columns is distributed equally at the inflection points. The shear in the columns is equal to the total horizontal force on the structure above the level under consideration.

For the mill bent of Fig. 6-5*b* the left column will deflect as shown in Fig. 6-6*b*. The diagonal member between the truss and the column, commonly referred to as a *knee brace*, is considered to be pin connected at its

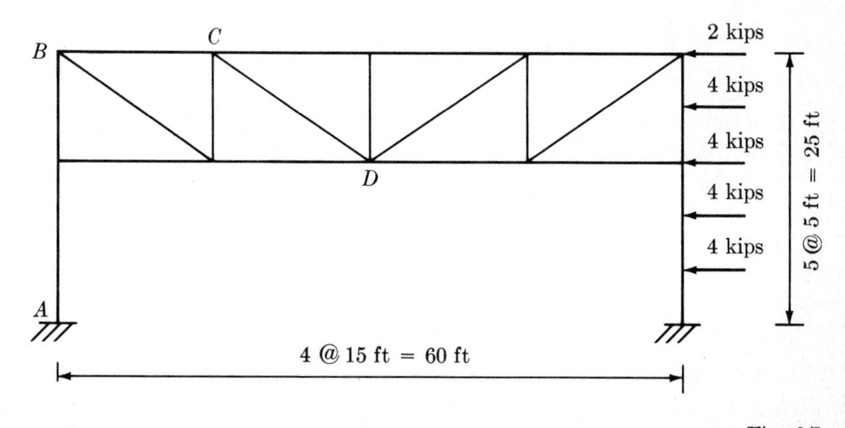

Fig. 6.7

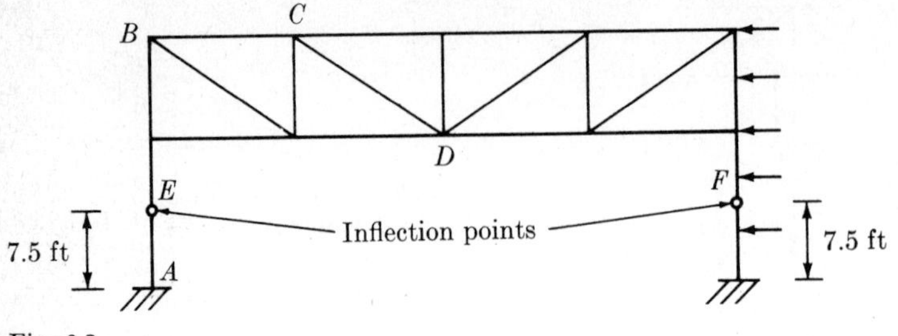

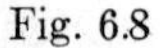

Fig. 6.8

ends to the column and the truss. The column is a continuous member from the support to the bottom of the truss. For a fixed support it is common to assume that the inflection point is located midway up to the point of bracing, that is, $x = d/2$. The assumption of inflection points in each column and an equal distribution of shear in the columns at the inflection points thus permits an analysis of such a structure.

EXAMPLE 6-1 With appropriate assumptions, we are to determine for the mill bent of Fig. 6-7 the components of reaction at support point A; the moment diagram for member AB, constructed on the tension side of the member; the force in member CD.

Because the supports are fixed, the inflection points in each column, denoted as E and F in Fig. 6-8, are assumed to be 7.5 ft up from the supports. With the shear in the columns assumed to be divided equally at the inflection points, the values of shear are

$$V_E = V_F = 14/2 = 7 \text{ kips}$$

The axial force in column AB at point E is obtained by taking the summation of moments about inflection point F for the forces acting on the free body of the structure above the inflection points,

$$60E_V - (2)(17.5) - (4)(12.5) - (4)(7.5) - (4)(2.5) = 0$$
$$E_V = 2.08 \text{ kips}$$

The desired components of reaction at A are obtained from equilibrium considerations of column segment AE. The results are shown in Fig. 6-9a.

The remaining forces on member AB are obtained from the free-body diagram of Fig. 6-9b. Taking the summation of moments about point B, we obtain

$$52.5 + 10X - (25)(7) = 0$$
$$X = 12.25 \text{ kips}$$

The net horizontal force on AB at point B is therefore

$$12.25 - 7 = 5.25 \text{ kips}$$

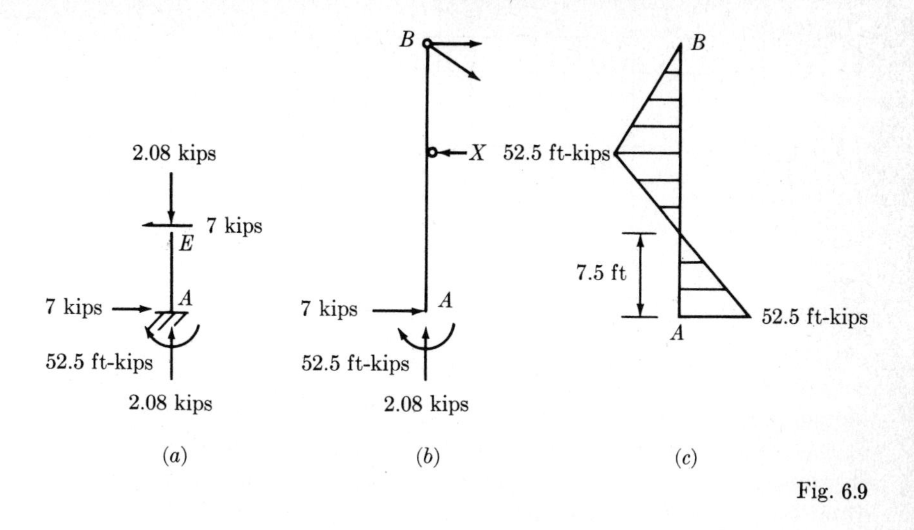

Fig. 6.9

acting to the right. The resulting moment diagram for AB, drawn on the tension side of the member, is shown in Fig. 6-9*c*.

The force in member CD is obtained by considering a free-body diagram of the truss to the left of a vertical cut made in the truss between C and D. The vertical component of force in CD is found from the summation of vertical forces on the free body to be 2.08 kips. The desired force in member CD is therefore

$$P_{CD} = \left(\frac{3.61}{2}\right)(2.08) = 3.75 \text{ kips}$$

The force in CD is a tensile force.

Laterally loaded portals, such as that of Fig. 6-10*a*, can also be analyzed by the approximations used above for mill bents. With fixed supports, the portal is statically indeterminate to the third degree. The deflected shape of the portal and an assumed location of the inflection points on the

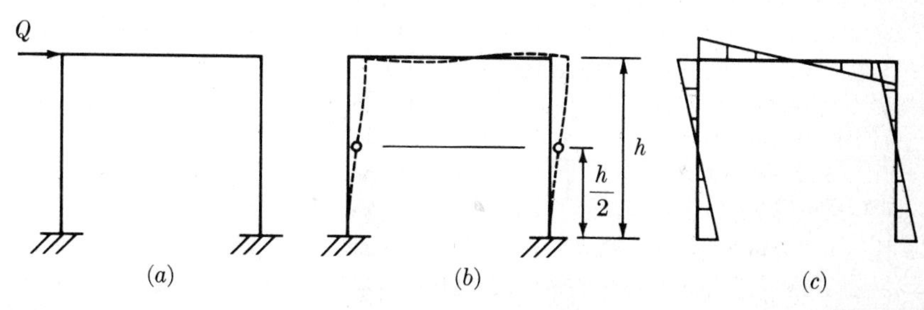

Fig. 6.10

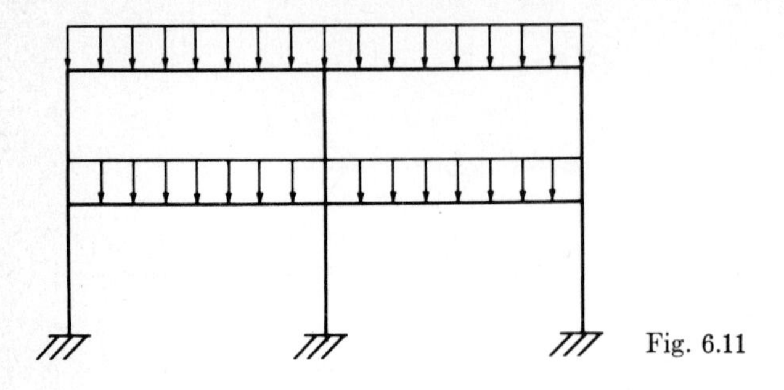

Fig. 6.11

columns are shown in Fig. 6-10*b*. As in the analysis of mill bents, zero moments at the two inflection points and an equal distribution of shear in the columns at the inflection points constitute the three assumptions required to perform an approximate analysis. The resulting moment diagram for the portal, drawn on the compression side of the members, is shown in Fig. 6-10*c*.

It is often useful to be able to determine the approximate forces in a statically indeterminate frame subjected to vertical loading. Such a condition of vertical loading is shown in Fig. 6-11, where the horizontal members of the frame are subjected to uniform vertical loading. To obtain an approximate solution for the forces in such a frame let us first examine the conditions of rotational restraint at the ends of a typical uniformly loaded beam. If the ends of the beam are rigidly supported, it can be

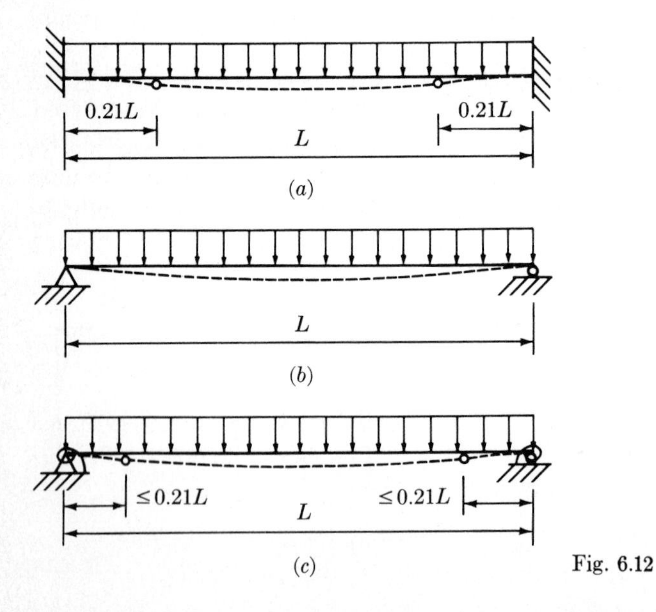

Fig. 6.12

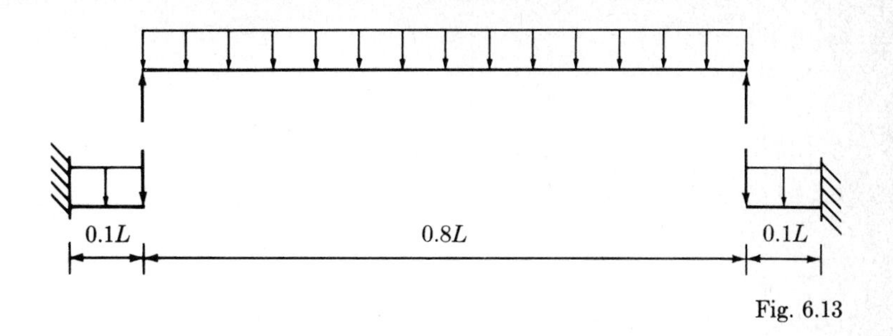

Fig. 6.13

shown by methods of analysis to be discussed in later chapters that the inflection points (zero moment) of the beam are located a distance $0.21L$ from each end of the beam, as shown in Fig. 6-12*a*. If no resistance to rotation exists at the ends of the beam, we have the simple-beam condition of Fig. 6-12*b*, where the points of zero moment are at the supports. In a frame such as that of Fig. 6-11 the ends of the beam are restrained against rotation by the columns and adjacent beams, which in effect serve as elastic restraints against rotation. The resistance to rotation can be represented by springs, as shown in Fig. 6-12*c*. The location of the inflection points for such an elastically restrained beam depends on the stiffness of the springs, but it must be somewhere between the end of the beam and the location for the fixed condition in Fig. 6-12*a*. A generally assumed location for the inflection points for beams of a frame is $0.1L$ from the ends of the beam. Thus for such an assumed location of the inflection points the beam can be analyzed by considering the free-body diagram of the beam segments shown in Fig. 6-13. Although the loading in our discussion has been considered to be uniform, satisfactory results can also be obtained for some forms of concentrated loading with the same location for the inflection points.

In addition to the approximate procedures for analysis discussed above, there are general methods for the approximate analysis of building frames subjected to lateral loading. Two commonly used approximate methods, the *portal method* and the *cantilever method*, are presented in the following sections.

6-3 PORTAL METHOD

The portal method is an approximate method for determining the forces in laterally loaded frames such as that of Fig. 6-14. In such an analysis the following assumptions are made:

1. An inflection point is located at midheight on each column.
2. An inflection point is located at midspan on each beam.

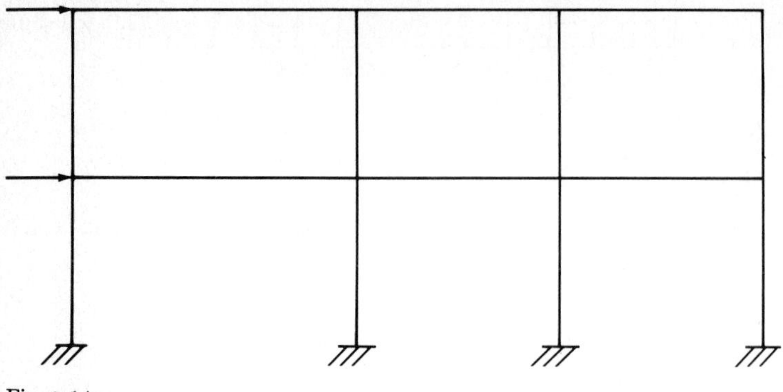

Fig. 6.14

3. The horizontal shear is divided among the columns in the ratio of one part to exterior columns and two parts to interior columns.

The third assumption is a result of considering each level of the frame to be composed of individual portals, as shown in Fig. 6-15. Thus, by this assumption, an interior column is in effect resisting the shear of two columns of the individual portals.

Each level of the frame in Fig. 6-14 is statically indeterminate to the ninth degree. Thus the entire frame is statically indeterminate to the eighteenth degree. With the assumptions of the portal method, we assume eight inflection points on the columns, six inflection points on the beams, and in effect three assumptions on each level for shear; that is, the shear in three columns is related to that in the fourth. The total number of assumptions is therefore 20, which is more than enough for an approximate analysis.

The following example illustrates the procedure of analysis by the portal method.

EXAMPLE 6-2 The approximate bending moment, shear, and axial force in each member of the frame of Fig. 6-16 are to be determined by the portal method. Inflection points are assumed at midheight on each upper column. We obtain the value of shear in each upper column from the

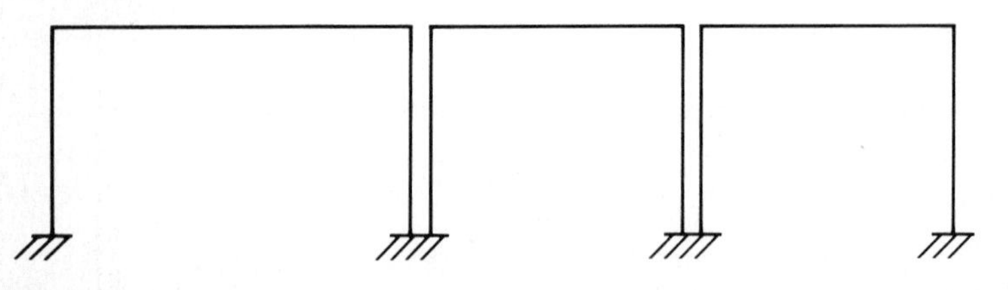

Fig. 6.15

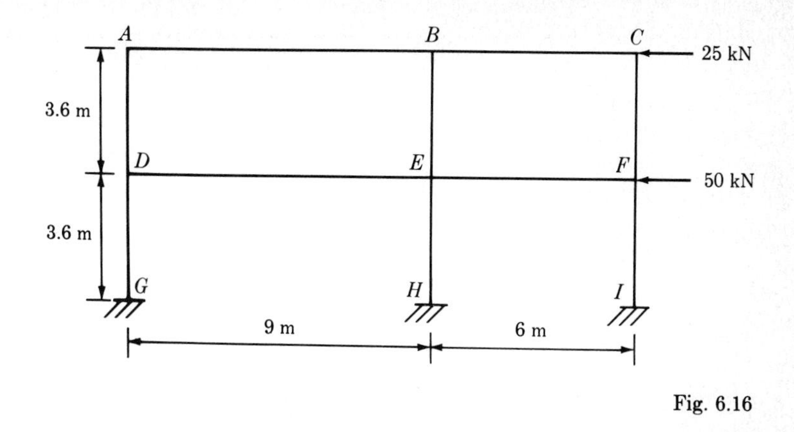

Fig. 6.16

free-body diagram of Fig. 6-17*a* and the assumption of twice as much shear in the interior column as the exterior columns. Therefore

$$H + 2H + H = 25$$
$$H = 6.25 \text{ kN}$$

Inflection points are also assumed at midspan on beams AB and BC. The member forces in the upper level of the frame are thus obtained from the free-body diagrams of Figs. 6-17*b*, *c*, and *d*, beginning either with the free body for A or for C and working across the structure. The resulting forces must check on the free-body diagram at the opposite end.

Given inflection points at midheight of the lower columns, the shear in each lower column is obtained from the free-body diagram of Fig. 6-18*a*.

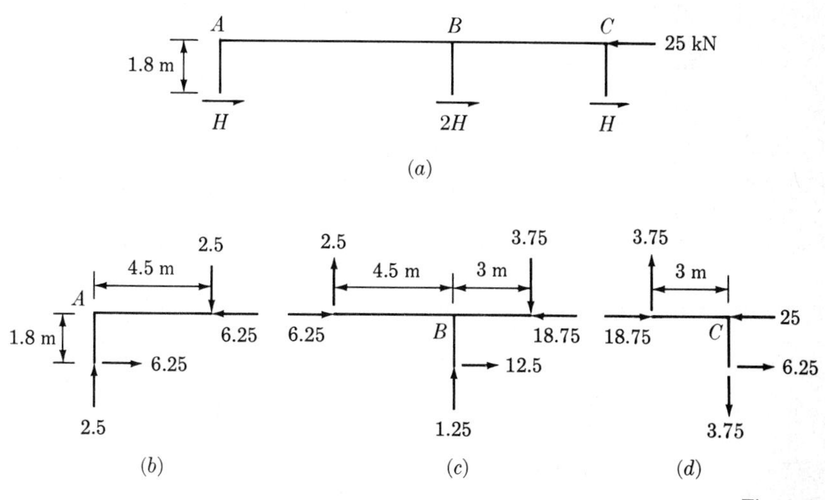

Fig. 6.17

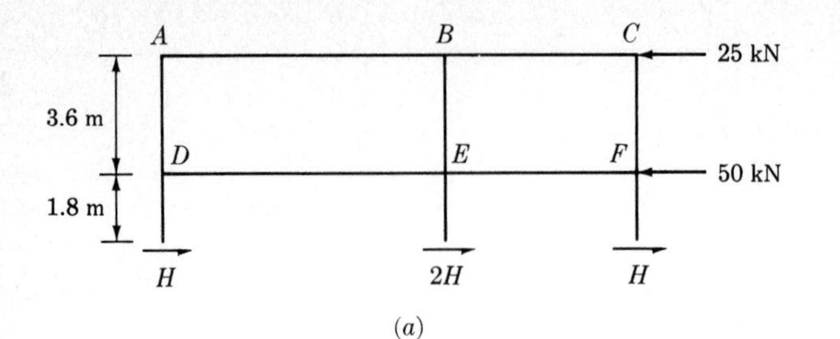

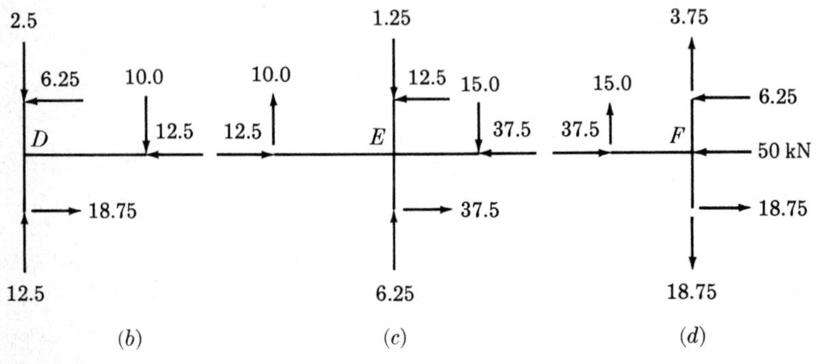

Fig. 6.18

Thus

$$H + 2H + H = 25 + 50$$
$$H = 18.75 \text{ kN}$$

The forces in the members of the lower level of the structure are obtained from the free-body diagrams of Figs. 6-18*b*, *c*, and *d*.

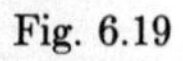

Fig. 6.19

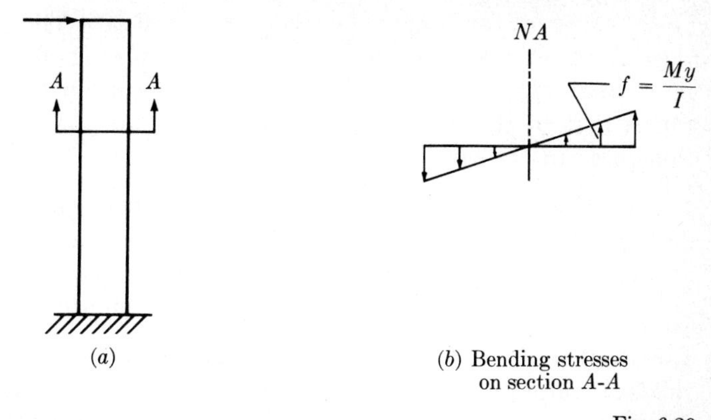

(a)

(b) Bending stresses on section A-A

Fig. 6.20

The maximum moment in each member of the structure is readily obtained once the values of shear at the inflection points have been determined. The maximum moments, shears, and axial forces in the members can conveniently be summarized on a diagram of the structure as shown in Fig. 6-19. The bending moments and shears are expressed without regard to sign. Negative values of axial force indicate compression and positive values indicate tension.

6-4 CANTILEVER METHOD

The basis for the assumptions used in the cantilever method can be explained by comparing a laterally loaded frame to a laterally loaded cantilever beam, as shown in Fig. 6-20*a*. From strength-of-materials studies we know that the resulting bending-stress distribution on a typical

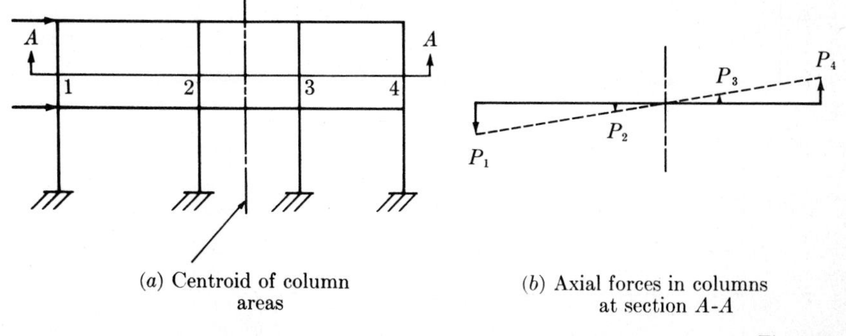

(a) Centroid of column areas

(b) Axial forces in columns at section A-A

Fig. 6.21

cross-section of the beam is as shown in Fig. 6-20*b*. The analogous situation for a laterally loaded frame is shown in Fig. 6-21. The bending moment at a typical section *A*-*A* is resisted by concentrated column forces rather than by distributed stresses in the beam of Fig. 6-20. The assumptions used in the cantilever method are:

1. An inflection point is located at midheight on each column.
2. An inflection point is located at midspan on each beam.
3. The axial force in each column is proportional to its distance from the centroid of all the column cross-sectional areas at that level.

Note that the cantilever method differs from the portal method in the third assumption. Instead of assuming that shear is distributed in the columns, as in the portal method, we assume a distribution of axial forces in the columns. The effects of columns having unequal cross-sectional areas can be included in the third assumption of the cantilever method. The areas of the columns are involved in the location of the centroid.

EXAMPLE 6-3 The approximate bending moment, shear, and axial force in each member of the frame analyzed in Example 6-2 are to be determined by the cantilever method (Fig. 6-22). The cross-sectional areas of the columns are assumed to be equal. The location of the centroid of the column areas is determined first. Taking the summation of moments of the column areas about the left column, we obtain

$$d = \frac{9 + 15}{3} = 8.0 \text{ m}$$

Considering a free-body diagram of the frame above the inflection

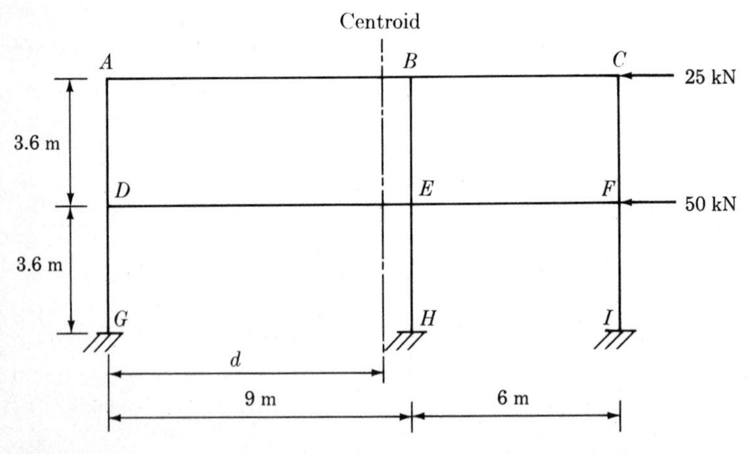

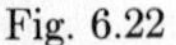
Fig. 6.22

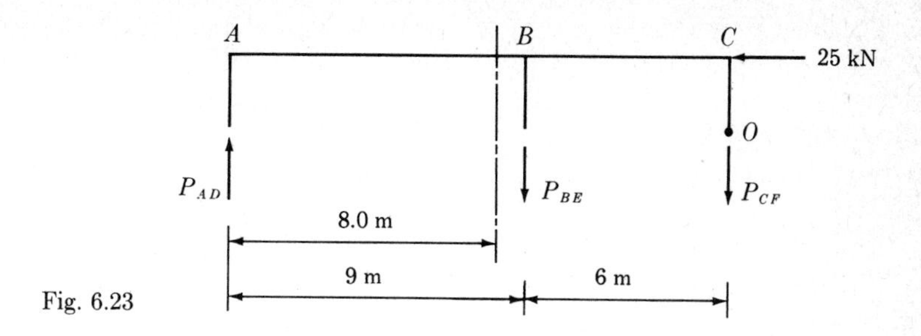

Fig. 6.23

points of the upper columns, as shown in Fig. 6-23, we can write the following relationships for axial forces in the columns according to the third assumption of the cantilever method as

$$P_{BE} = \frac{1.0}{8.0} P_{AD} = 0.125 P_{AD}$$

$$P_{CF} = \frac{7.0}{8.0} P_{AD} = 0.875 P_{AD}$$

Taking the summation of moments about point O in Fig. 6-23, we obtain

$$-(25)(1.8) + 15P_{AD} - (6)(0.125P_{AD}) = 0$$
$$P_{AD} = 3.158 \text{ kN}$$

Therefore

$$P_{BE} = (0.125)(3.158) = 0.395 \text{ kN} \qquad P_{CF} = (0.875)(3.158) = 2.763 \text{ kN}$$

Positive axial forces are considered to be acting in the directions shown in Fig. 6-23. Once we have determined the values of the axial forces in the columns, the remaining member forces in the upper section of the frame are determined from free-body considerations as in the portal method.

The axial forces in the columns of the lower level are obtained by the same procedure used for the upper level. The 50-kN load must be included with the 25-kN load in taking moments about the lower column inflection point. The results of the analysis by the cantilever method are shown in Fig. 6-24.

The results obtained by the cantilever method and those obtained by the portal method are quite similar for some values and differ by an appreciable amount for other values. Remember that these are both approximate methods of analysis and that the results obtained are only as good as the assumptions used. For example, it may not be very accurate for some frames to assume that the inflection points are located at mid-height on the columns and at midspan on the beams. It is also evident that a change in the area of the columns can influence the location of the centroid of the column areas and thus appreciably alter the results.

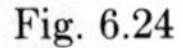

Fig. 6.24

Discussions of the accuracy of portal and cantilever methods, as well as the presentation of other approximate methods, can be found in a text devoted more to this subject.[1] The statically indeterminate methods of analysis developed in later chapters can be used to obtain exact results for lateral loading of frames.

PROBLEMS

6-1 and 6-2. Determine the axial forces in the members of the trusses shown, assuming that (*a*) the diagonals are very slender and therefore carry no compressive forces; (*b*) the shear in each panel of the truss is divided equally between the diagonals.

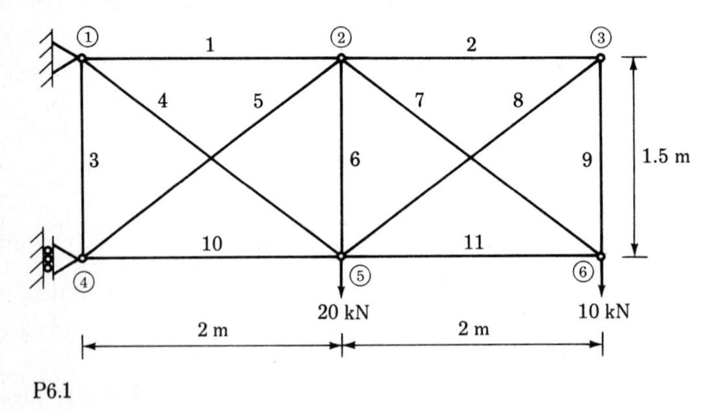

P6.1

[1] J. R. Benjamin, "Statically Indeterminate Structures," McGraw-Hill Book Company, New York, 1959.

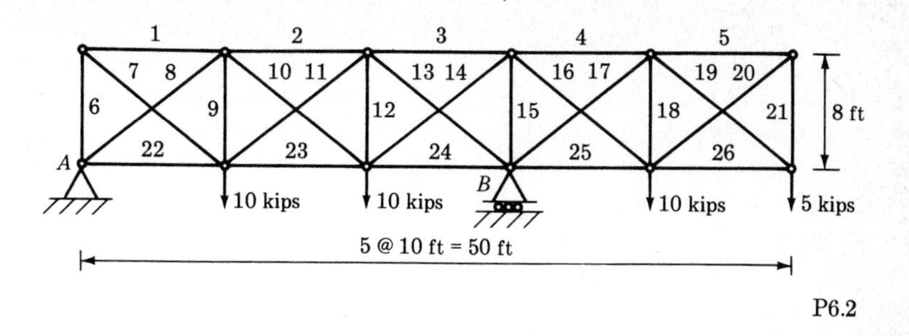

P6.2

6-3. Assuming inflection points to be formed midway between the base of the columns and the lower level of the roof truss, for the mill bent shown determine (*a*) the components of reaction at A and F; (*b*) the axial forces in the members of the roof truss, assuming that the diagonals are long and slender and therefore carry no compressive forces; (*c*) the shear and moment diagrams for member AC with the moment diagram constructed on the compression side of the member.

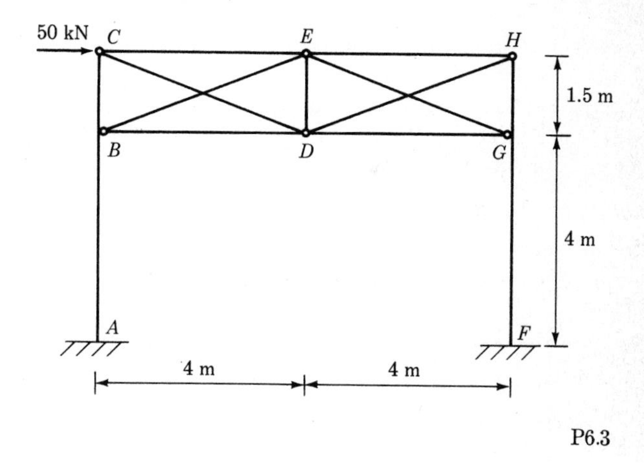

P6.3

6-4. For the mill bent shown, use appropriate assumptions and (*a*) determine the values and directions of the reaction components at A and G; (*b*) construct the moment diagram for member AC on the tension side of the member; (*c*) determine the axial force in members BD, CE, DE, and EF.

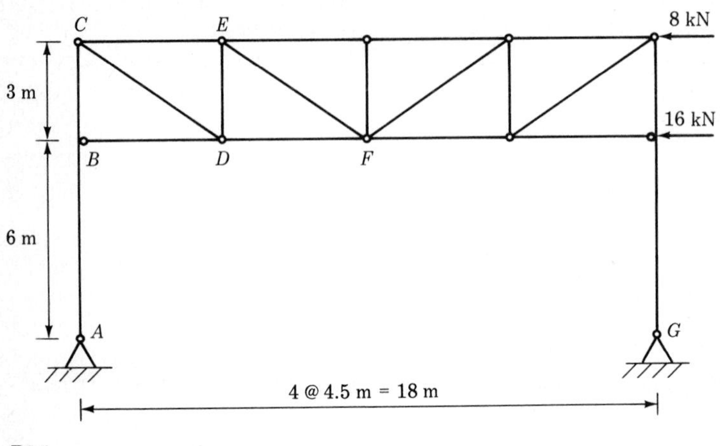

P6.4

6-5. Assuming inflection points to be formed midway between the base of the columns and the lower end of the knee braces, for the mill bent shown determine (*a*) the values and directions of the reaction components at A and B; (*b*) the axial force in members CE, DE, and EF; (*c*) the moment diagram for member AD drawn on the compression side of the member.

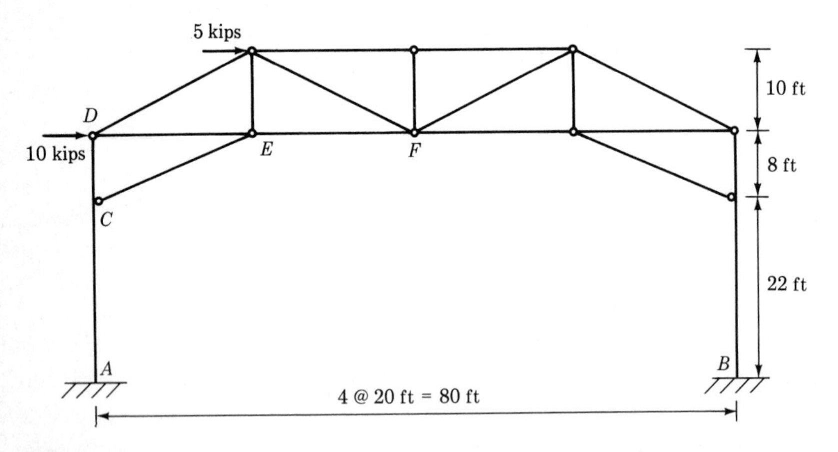

P6.5

6-6. An approximate analysis is to be made for the mill bent subjected to wind pressures as shown. The wind loads are transferred to the bent by purlins and girts, spaced on the bent as shown. The spacing between bents is 5 m. Convert the given wind pressures to SI units of newtons per square meter (Pascals) and, assuming inflection points on the columns midway up from the base to the point of bracing, (*a*) determine the values and directions of the reaction components at A and H; (*b*) construct the shear and moment diagrams for column AC with the moment diagram drawn on the compression side of the member; (*c*) determine the axial forces in members BD, DE, and FG.

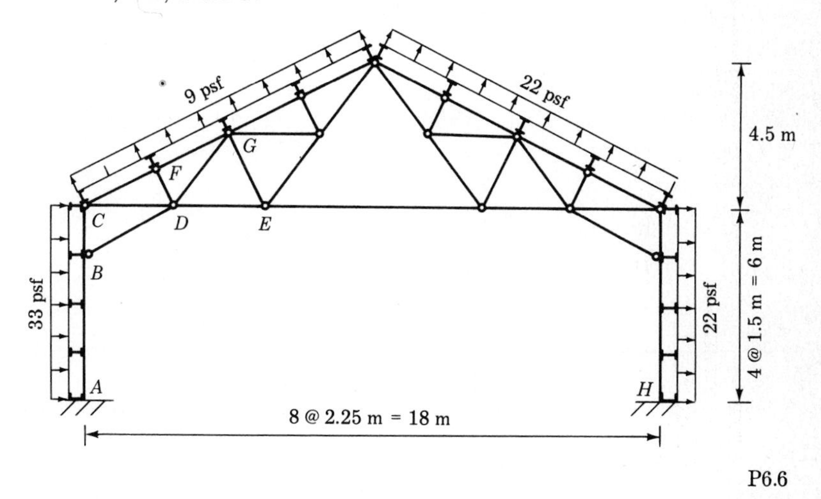

P6.6

6-7. For the portal shown, use appropriate assumptions and (*a*) determine the values and directions of the reaction components at A and D; (*b*) construct the moment diagram for the portal on the tension side of the members.

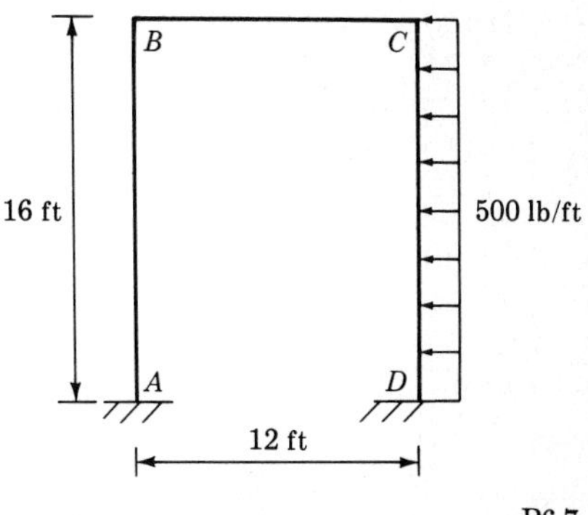

P6.7

6-8 through 6-10. Determine the approximate moment, shear, and axial force in each member of the frames shown, by the portal method of analysis. Put the results on a summary diagram like that of Fig. 6-19.

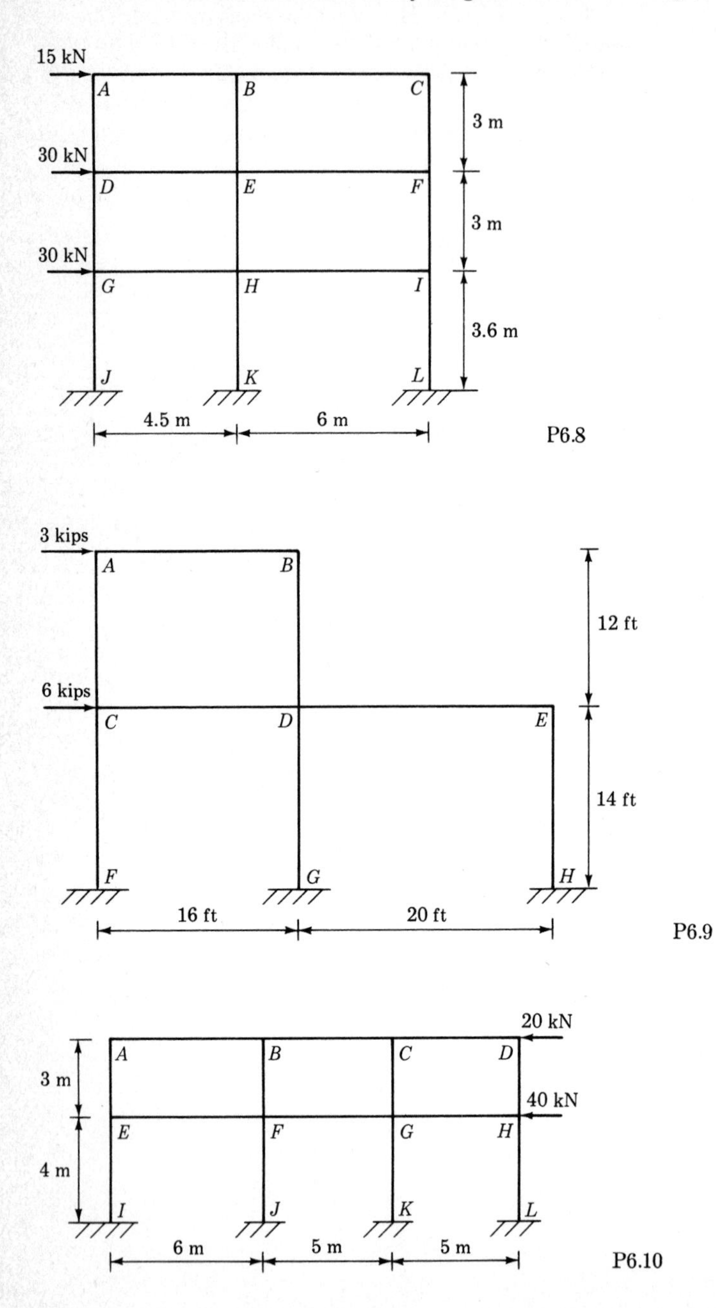

6-11. Solve Prob. 6-8 by the cantilever method. Assume the column areas are equal.

6-12. Solve Prob. 6-9 by the cantilever method. Assume the column areas are equal.

6-13. Solve Prob. 6-10 by the cantilever method. Assume that the cross-sectional areas of the interior columns are twice those of the exterior columns.

CHAPTER
SEVEN

DEFLECTIONS: GEOMETRIC METHODS

7-1 DEFLECTED SHAPES

A necessary part of analyzing structures is the evaluation of deflections. Deflections are evaluated not only in order to keep their magnitudes within design limitations, but also for the essential purpose of using them in the analysis of statically indeterminate structures. It has previously been stated that a structure is statically indeterminate when it contains unknowns in excess of the number of equilibrium equations available. This necessitates acquiring additional equations. Equations representing consistent deflections can be used for this purpose.

In this chapter we shall discuss the geometric methods of computing deflections in a structure. The energy methods of determining deflections will be treated in the following chapter. The intent in these two chapters is to develop the concepts and methods of evaluating deflections. Their use in analyzing statically indeterminate structures will be demonstrated in later chapters.

As a first step in discussing deflections, it is helpful to develop an understanding of the manner in which a structure deflects when loaded. The drawing of deflected shapes of structures serves as a means of understanding the deflections. Deflections can be the result of various kinds of forces in the members, such as bending moment, axial force, and shear. Deflections due to axial forces and shear are often neglected in beams and frames because they are so much smaller than those due to bending moments. In trusses, however, deflections are primarily due to axial distortions of the members. Because the forces in the members are not quantitatively known beforehand, the deflected shape must be drawn from a feeling of how the structure will behave under loading. Often the feeling can be substantiated by visualizing the general nature of the resistive moments or axial forces in the members.

Let us first consider the elementary case of a cantilever beam with a concentrated load, as shown in Fig. 7-1*a*. The deflected shape of the beam is shown by the dashed line. In drawing deflected shapes the student should concentrate on satisfying certain requirements of the deflected structure. First, the deflected shape representing a member is a smooth, continuous curve. It must also be such that it satisfies the connections or supports at the ends of the member. Great exaggeration of the deflections of the structure will often lead to better understanding of the deflected shape. With regard to the cantilever beam of Fig. 7-1*a*, the deflected shape is continuously curved downward. The fixed support at the left end is represented by zero deflection and zero slope of the deflected shape.

Now consider the beam of Fig. 7-1*b*. At points A and C the deflected shape has zero slope and zero deflection. At point B the requirement for

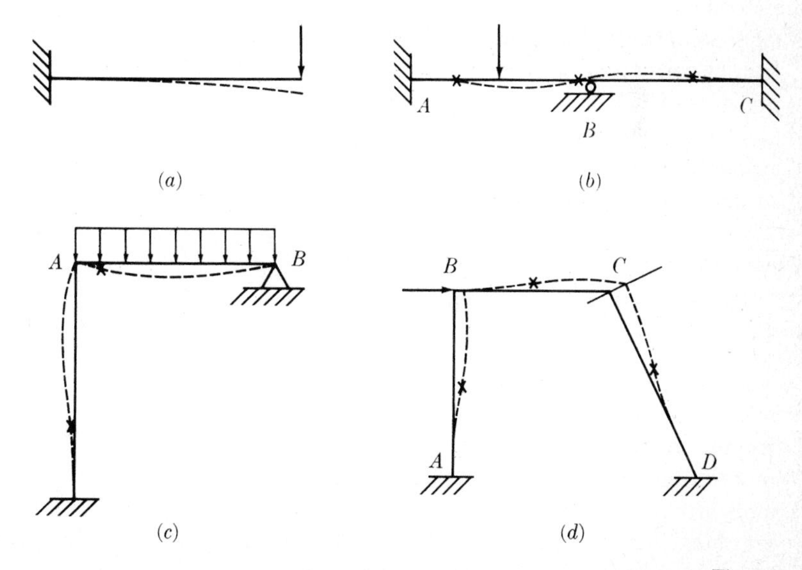

Fig. 7.1

zero deflection is satisfied. Since it is continuous, the beam is represented by a smooth continuous curve from A to C. The curve deflects downward in the left span and upward in the right span for the loading shown. At certain points on the deflected shape the curvature changes sign; these points are referred to as *inflection points*. Inflection points also represent points of zero moment in the beam. The assumed locations of the inflection points on the deflected shapes of Fig. 7-1 are denoted by an x.

For the frames of Figs. 7-1*c* and *d* the axial deflections are considered negligible in comparison with those due to bending. Thus point A does not displace in Fig. 7-1*c*. However, the connection at A does rotate. In drawing the deflected shape of a joint such as A it should be remembered that at the immediate point of connection the members maintain the same angular orientation to each other in the deflected position.

Neglecting axial deformations, we see in Fig. 7-1*d* that joint C must displace along the line drawn perpendicular to member CD at joint C. This displacement is, in fact, along the arc of a circle with its center at D, but because we are concerned only with small deflections, the arc can properly be represented by a straight line. Point B moves to the right the same distance as C moves horizontally because axial deformations in BC are neglected. Furthermore, point B displaces only horizontally. There will be rotation of joints B and C. The members meeting at B and at C will retain the same angular orientation *to each other* at the immediate point of the connection.

Learning to draw the deflected shapes of structures qualitatively will provide a more thorough understanding of the methods for analyzing such structures. For example, when it becomes necessary to use a condition of consistent displacements to acquire an additional equation for the analysis of statically indeterminate structures, equating the slope at the end of one member to the slope at the end of an adjoining member or equating the slope or deflection at a point to zero will be much more meaningful if the deflected shape of the structure is visualized first.

7-2 MOMENT-AREA METHOD

The moment-area method of determining deflections provides a means of determining the slopes and deflections due to bending in beams and frames. As will be seen in subsequent developments and examples, the moment-area method can readily be used in the often troublesome problems of varying moment of inertia and discontinuous loadings. The semigraphical form of the moment-area method makes the solution of these more complicated problems possible.

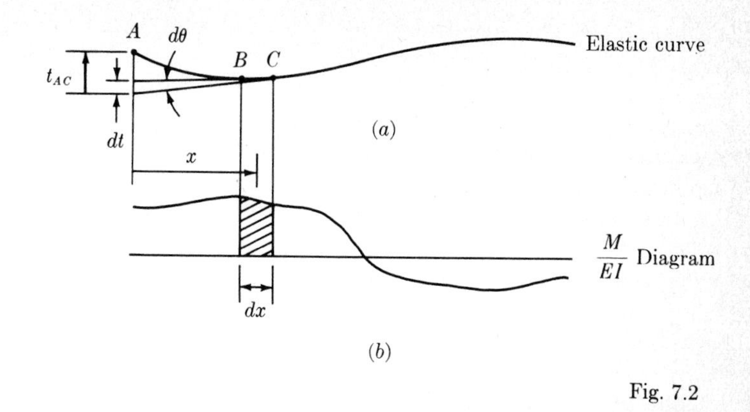

Fig. 7.2

This method entails two basic theorems, one having to do with the variation in slope and the other with the variation in deflection. To develop these theorems let us consider a segment of an elastic curve of a member deflected by bending moment, as shown in Fig. 7-2*a*. The moment producing the deflected shape is shown in Fig. 7-2*b* in the form of the moment diagram divided by the product of the modulus of elasticity E and the moment of inertia I of the beam; it is thus referred to as the *M/EI diagram*. Such a diagram represents a plot of the moment in the beam divided by the corresponding value of EI.

From previous studies in strength of materials it will be recalled that the differential equation of such an elastic curve is

$$\frac{d^2y}{dx^2} = \frac{M}{EI} \tag{7-1}$$

The sign of the bending moment corresponds to the beam convention of Fig. 2-9*a*, and the positive y direction is upward. With the slope of the elastic curve denoted by the angle θ, it is given by the expression

$$\theta = \frac{dy}{dx}$$

Equation (7-1) can therefore be written as

$$\frac{d\theta}{dx} = \frac{M}{EI}$$

or $$d\theta = \frac{M}{EI}\,dx \tag{7-2}$$

Considering a differential length of the beam between points B and C, we see that the change in slope is represented by the angle $d\theta$ of Fig. 7-2*a*, defined by the tangents to the elastic curve at B and C. From Eq. (7-2) and Fig. 7-2*b* we see that the value of the change in slope between B and

C is equal to the area under the M/EI diagram. To determine the change in slope between two points such as A and C, we integrate Eq. (7-2) to give

$$\int_A^C d\theta = \theta_C - \theta_A = \int_A^C \frac{M}{EI}\,dx \tag{7-3}$$

From the results of Eq. (7-3), the first of the moment-area theorems can be stated as follows:

THEOREM 1: *The change in slope between two points on the elastic curve is equal to the area of the M/EI diagram between these points.*

The sign of the change in slope results directly from the sign obtained from the area of the M/EI diagram evaluated by Eq. (7-3). At point A in Fig. 7-2*a* the slope of the tangent to the elastic curve is negative. From Eq. (7-3) and the M/EI diagram, the slope increases from A to C. It is evident from Fig. 7-2*a* that the slope is a more positive quantity at C than at A; in fact, it is a positive slope, as shown. Proceeding to the right of C, we see that the slope continues to increase for some distance and then begins to decrease. With consistent dimensions for M, E, I, and the length of the member under consideration, the resulting change in slope obtained by Theorem 1 will be in radians.

Next consider the use of the moment-area concept to evaluate deflections. In Fig. 7-2*a* the vertical deviation between the two tangents to the curve at B and C can be obtained from the product of the angle between the two tangents and the distance out to the point under consideration. Thus at point A, which is located a distance x from the differential element, the deviation dt between the tangents at B and C is

$$dt = x\,d\theta \tag{7-4}$$

Remember that we are concerned with small deflections and slopes, and hence the tangents are in fact nearly parallel with the axis of the member. The deflections have been greatly exaggerated in Fig. 7-2 for illustrative purposes.

The deviation of point A from the tangent to the elastic curve at C, denoted as t_{AC} in Fig. 7-2*a*, can be obtained by repeating the procedure for each differential slice between A and C and summing the resulting values of dt. This procedure can be accomplished by integrating Eq. (7-4), which leads to the expression

$$t_{AC} = \int_A^C x\,d\theta \tag{7-5}$$

With the expression for $d\theta$ of Eq. (7-2), t_{AC} becomes

$$t_{AC} = \int_A^C \frac{M}{EI}\,x\,dx \tag{7-6}$$

From this expression the second of the moment-area theorems can be stated as follows:

THEOREM 2: *The tangential deviation of a point A from a tangent to the elastic curve at point C is equal to the moment of the area of the M/EI diagram between A and C, taken about point A.*

If the area of the M/EI diagram between points A and C is denoted as A, Eq. (7-6) can be written as

$$t_{AC} = A\bar{x} \tag{7-7}$$

where $\bar{x}$ is the distance from point A to the centroid of the area. Similar expressions were used in determining the centroids of cross-sectional areas, and the various techniques used for that purpose can be applied to the evaluation of t_{AC}. The order of the subscripts used with t is very important. The first subscript denotes the point about which the moments are being taken and the point *to which* the deviation from the tangent to the elastic curve is being determined. The second subscript denotes the point at which the tangent is drawn to the elastic curve. As shown in Fig. 7-2 and as obtained from Eq. (7-6), t_{AC} is a positive quantity; that is, point A is above, or in the positive y direction from, the tangent to the elastic curve.

EXAMPLE 7-1 A cantilever beam of varying moment of inertia is subjected to a 5-kN load, as shown in Fig. 7-3. The slope and deflection due to bending in the beam are to be determined at point B by the moment-area method. The modulus of elasticity E for the material is 207 GPa. The elastic curve for the deflected shape of the beam is shown in Fig. 7-4*a*. Because the slope of the elastic curve at A is zero, the change in slope between A and B represents the slope at point B. The slope at B is denoted by the angle θ_B in Fig. 7-4*a*. Similarly, because the slope of the elastic curve at A is zero, the deviation of the elastic curve from the tangent drawn at A represents the deflection of the beam. The deflection at point B is denoted by t_{BA}.

To evaluate θ_B and t_{BA} by the moment-area concepts we first draw the moment diagram for the beam as shown in Fig. 7-4*b*. The M/EI diagram

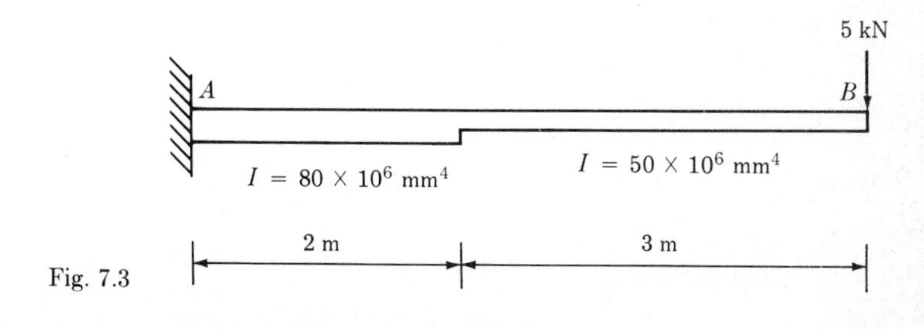

Fig. 7.3

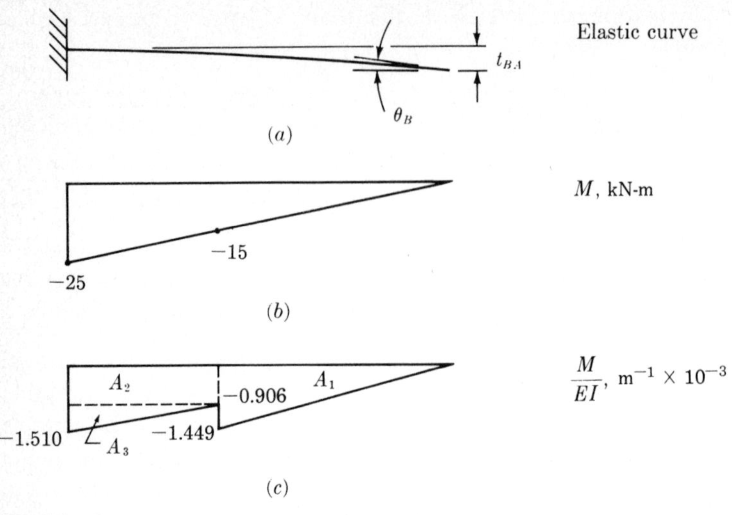

Fig. 7.4

is developed from this moment diagram, with the values of the moment diagram divided by the corresponding values of EI. Using units of meters^4 for moment of inertia, that is, 1 mm^4 = 10^{-12} m^4, the value at A, for example, is

$$\frac{-25 \text{ kN-m}}{(207 \times 10^6 \text{ kN/m}^2)(80 \times 10^{-6} \text{ m}^4)} = -1.510 \times 10^{-3} \text{ m}^{-1}$$

The resulting M/EI diagram is shown in Fig. 7-4c.

The slope at B is determined according to Theorem 1. From Eq. (7-3),

$$\theta_B - \theta_A = \int_A^B \frac{M}{EI}\, dx$$

As stated above, $\theta_A = 0$. The area under the M/EI diagram can conveniently be evaluated by dividing the diagram into the triangular and rectangular configurations shown in Fig. 7-4c. The technique of subdividing the M/EI diagram into figures for which the areas and the locations of centroids are commonly known is particularly useful in the later step of evaluating the deflection. The value of θ_B is therefore given by

$$\theta_B = A_1 + A_2 + A_3$$

which, from Fig. 7-4c, leads to

$$\begin{aligned}\theta_B &= \left[\frac{(3)(-1.449)}{2} + (2)(-0.906) + \frac{(2)(-0.604)}{2}\right]10^{-3} \\ &= (-2.174 - 1.812 - 0.604)10^{-3} \\ &= -0.00459 \text{ rad}\end{aligned}$$

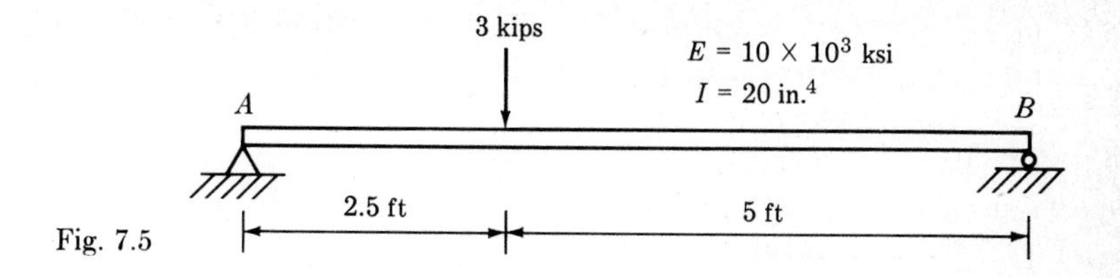

Fig. 7.5

The deflection at B is determined according to Theorem 2. Using the triangular and rectangular components of the M/EI diagram and taking the moments of these areas about point B, we obtain for the deflection at B,

$$t_{BA} = A_1\bar{x}_1 + A_2\bar{x}_2 + A_3\bar{x}_3$$

where $\bar{x}_1$, $\bar{x}_2$, and $\bar{x}_3$ are the distances from B to the centroids of the respective areas. From Fig. 7-4c, the value of deflection is found to be

$$t_{BA} = [(-2.174)(2) + (-1.812)(4) + (-0.604)(4.333)]10^{-3}$$
$$= -0.0142 \text{ m} = -14.2 \text{ mm}$$

EXAMPLE 7-2 The location and magnitude of the maximum deflection due to bending are to be determined for the beam of Fig. 7-5. EI for the beam is constant. For convenience the calculations can be performed in terms of EI. The numerical values of E and I are substituted in the last step, along with a constant, if necessary, to make the results dimensionally correct. The elastic curve for the deflected beam is shown in Fig. 7-6a.

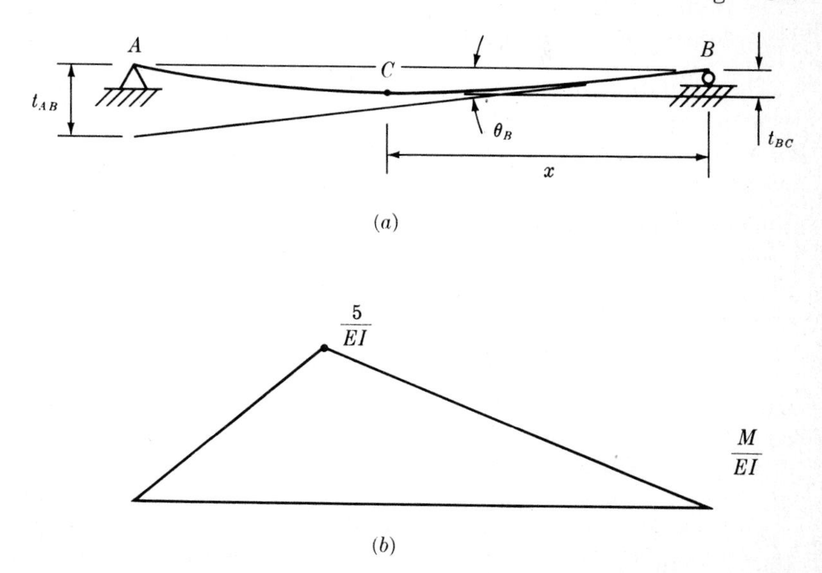

Fig. 7.6

The location of the unknown maximum deflection is denoted as point C, which is a distance x from the right support B. The fact that the slope of the elastic curve at this point is zero is used to determine the value of x. By first determining the value of θ_B we can then determine the value of x for which the value of slope changes by the value θ_B. From Fig. 7-6a we see that θ_B can be found by dividing the deviation t_{AB} by the distance between points A and B.

The M/EI diagram for the beam is shown in Fig. 7-6b with moment expressed in foot-kips. To evaluate t_{AB} we take the moment of the area about A. Thus,

$$t_{AB} = (\tfrac{1}{2})(2.5)\left(\frac{5}{EI}\right)(\tfrac{2}{3})(2.5) + (\tfrac{1}{2})(5)\left(\frac{5}{EI}\right)(2.5 + \tfrac{5}{3})$$
$$= \frac{62.5}{EI}$$

from which θ_B is found to be

$$\theta_B = \frac{t_{AB}}{7.5} = \frac{8.33}{EI}$$

The desired value of x is therefore the length in which the area of the M/EI diagram changes by this value of θ_B,

$$(\tfrac{1}{2})(x)\left(\frac{1x}{EI}\right) = \frac{8.33}{EI}$$

$$x^2 = 16.67$$
$$x = 4.08 \text{ ft}$$

The magnitude of the maximum deflection is found by evaluating t_{BC}, which is equal to the moment about B of the portion of the M/EI diagram between B and C. The numerical values of E and I from Fig. 7-5 are substituted, and a factor of 1728 is used to obtain the deflection in inches.

$$t_{BC} = (\tfrac{1}{2})(4.08 \text{ ft})\left[\frac{4.08 \text{ ft}}{(10 \times 10^3 \text{ k/in.}^2)(20 \text{ in.}^4)}\right](\tfrac{2}{3})(4.08 \text{ ft})(1728 \text{ in.}^3/\text{ft}^3)$$
$$= 0.196 \text{ in.}$$

For some types of loading, such as combinations of concentrated loads and distributed loads, or for varying distributed loads, the moment-area method can become considerably complicated if used directly. The complications arise from the fact that the areas and centroids of the resulting M/EI diagrams are difficult to evaluate. In many cases this difficulty can be overcome by considering the loads separately and superimposing the individual results. For example, the effects of the uniform load w and the concentrated load P on the beam of Fig. 7-7a can be considered to act

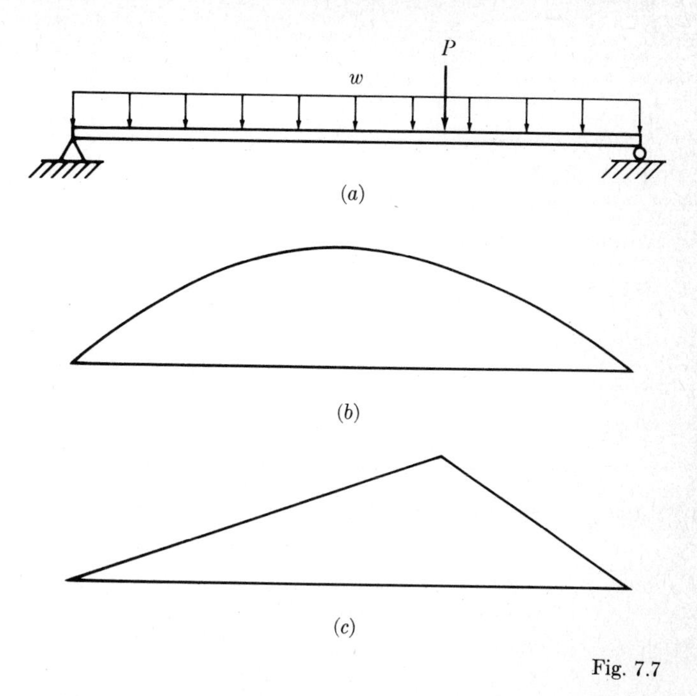

Fig. 7.7

separately. The result is the M/EI diagram of Fig. 7-7*b* for the uniform load and the M/EI diagram of Fig. 7-7*c* for the concentrated load. The areas and centroids can conveniently be determined for each of these diagrams. The results from each diagram are then combined for the total effect of the given loading.

Another manner of superimposing loading conditions can be seen by considering the linearly varying load on the beam of Fig. 7-8*a*. From equations of equilibrium, the reactions are found to be $R_A = 1200$ lb and $R_B = 1800$ lb. The moment in the beam at a distance x from the left support is therefore

$$M = 1200x - 50x^2 - 4.167x^3 \tag{7-8}$$

The M/EI diagram resulting from Eq. (7-8) would be quite difficult to work with. However, if the terms of Eq. (7-8) are considered individually, it is evident that the total moment is represented by the combination of the diagrams in Fig. 7-8*b*. The areas and centroids of each of these diagrams can readily be evaluated from the general properties of such shapes as those given in the general form of Fig. 7-9.

The representation of the moment diagram for the beam of Fig. 7-8*a* by the parts shown in Fig. 7-8*b* is referred to as constructing the moment diagram by *cantilever parts*. From Fig. 7-8*b* we see that each moment diagram is equivalent to that for a cantilever beam loaded with the individual

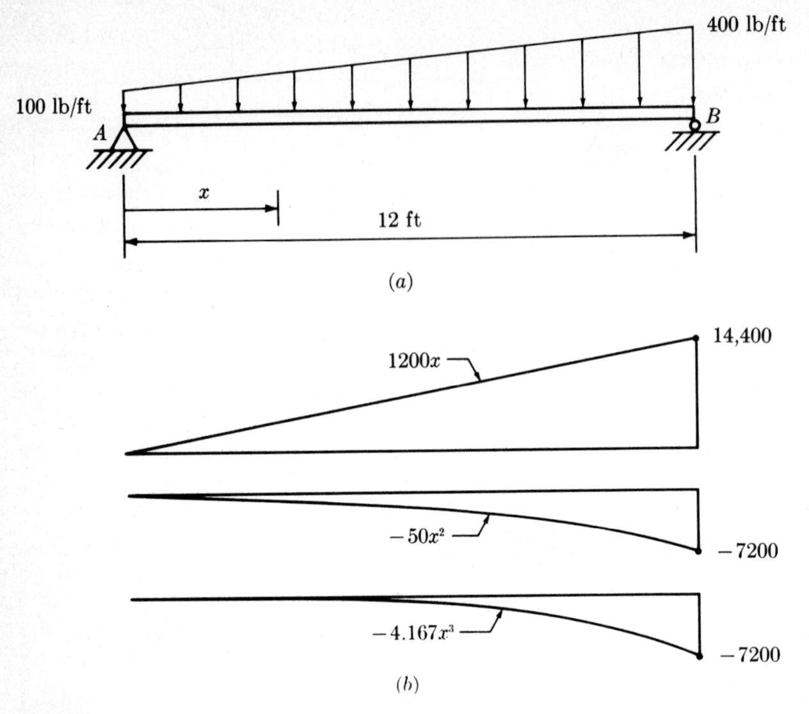
400 lb/ft
100 lb/ft
A
B
x
12 ft
(a)
14,400
1200x
−50x²
−7200
−4.167x³
−7200
(b)

Fig. 7.8

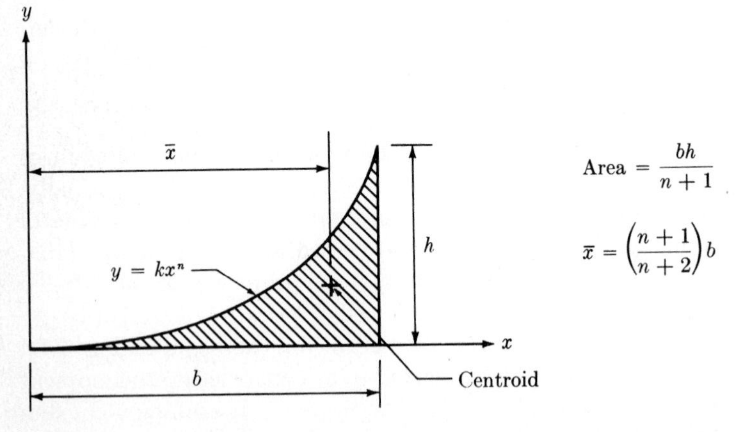
y
x̄
h
y = kxⁿ
x
b
Centroid
Area = bh/(n + 1)
x̄ = ((n + 1)/(n + 2))b

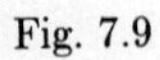
Fig. 7.9

components of force acting on the beam, with the cantilever beam fixed at the right end. The upper moment diagram of Fig. 7-8*b*, for example, represents the moment due to the action of the reaction force at A on the cantilever. The other two diagrams represent the given loading divided into a uniform component and a triangular component.

The general approach to constructing a moment diagram by cantilever parts is to consider a hypothetical cantilever support at any point on the beam. Often the end points serve as the most convenient points. Each load and reaction on the given beam is considered to act individually on the cantilever beam, and the corresponding moment diagrams are constructed. The areas and the centroids of the resulting individual diagrams are generally easy to evaluate. The combination of the individual diagrams represents the total moment diagram for the given beam. Convenient checks on the values of the individual moment diagrams can be made by summing their values at points on the beam where the value of moment in the beam is zero. It is not necessary in drawing the moment diagram by parts to write a general equation such as Eq. (7-8). The application of this method is demonstrated in the following example.

EXAMPLE 7-3 The deflection at point C on the beam of Fig. 7-10*a* is to be determined by the moment-area method. The elastic curve is first drawn as shown in Fig. 7-10*b*, with the desired deflection at point C denoted as Δ_C. A tangent drawn to the elastic curve at B shows that θ_B can be evaluated by first determining t_{AB} in a manner similar to that used in Example 7-2. We then observe that Δ_c' is equal to the product of θ_B and the distance between B and C. We determine the value of t_{CB} by Theorem 2, using the M/EI area between B and C and taking the moment about C. The desired value of deflection at C is therefore

$$\Delta_C = \Delta_C' - t_{CB}$$

The moment diagrams are constructed by cantilever parts, with a cantilever assumed at the right end of the beam. The resulting M/EI diagrams are shown in Fig. 7-10*c*. There are four individual diagrams, two representing the reactions at A and B and one each for the 15-kN load and the uniform load.

The value of t_{AB} is determined first by taking the moments about A of those parts of the M/EI diagrams to the left of B. Thus

$$\begin{aligned} t_{AB} &= \frac{1}{EI}[(\tfrac{1}{2})(3.0)(21.38)(2.0) + (\tfrac{1}{2})(1.5)(-22.5)(2.5) \\ &\qquad + (\tfrac{1}{3})(0.5)(-0.38)(2.88)] \\ &= \frac{21.77}{EI} \end{aligned}$$

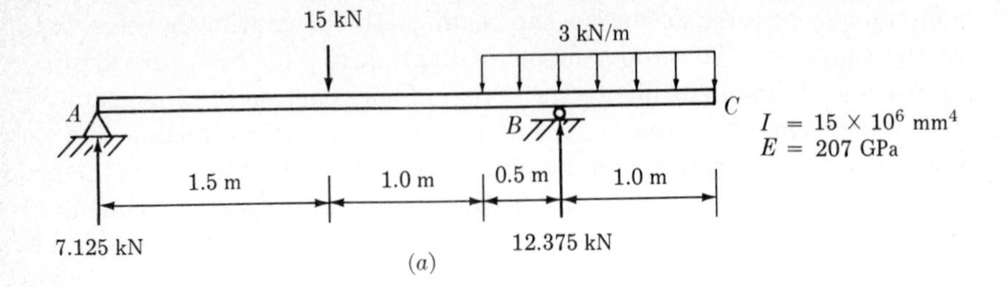
15 kN
3 kN/m
A
B
C
1.5 m
1.0 m
0.5 m
1.0 m
7.125 kN
12.375 kN
$I = 15 \times 10^6$ mm^4
$E = 207$ GPa

(a)

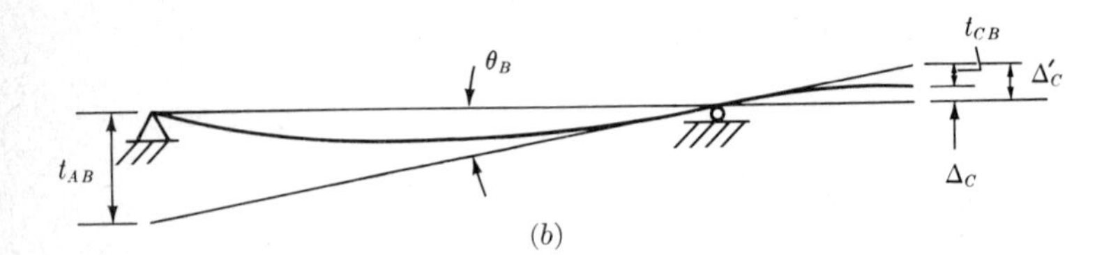
θ_B
t_{AB}
t_{CB}
Δ'_C
Δ_C

(b)

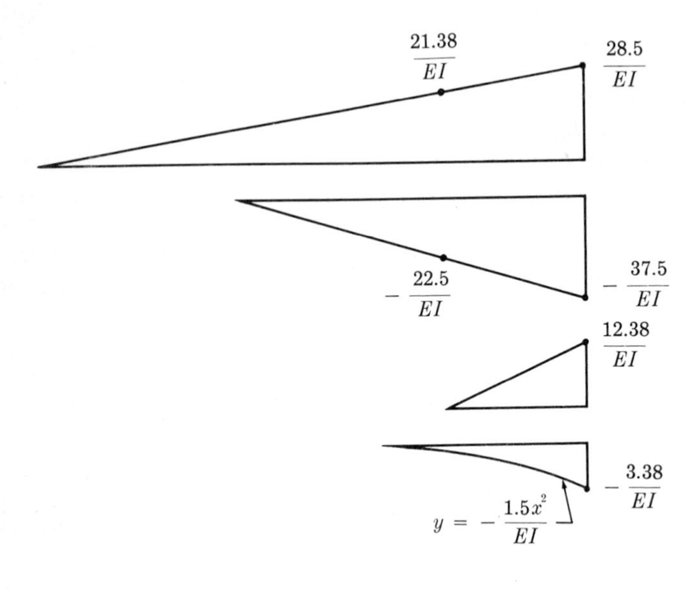
$\frac{21.38}{EI}$
$\frac{28.5}{EI}$
$-\frac{22.5}{EI}$
$-\frac{37.5}{EI}$
$\frac{12.38}{EI}$
$-\frac{3.38}{EI}$

$y = -\frac{1.5x^2}{EI}$

(c)

Fig. 7.10

By proportions, the value of Δ_C' is found to be

$$\Delta_C' = \left(\frac{1.0}{3.0}\right)\left(\frac{21.77}{EI}\right) = \frac{7.26}{EI}$$

The deviation between B and C, t_{CB}, is found by taking the moments about C of those parts of the M/EI diagrams to the right of B:

$$\begin{aligned} t_{CB} &= \frac{1}{EI}\{(1/2)(1.0)(7.12)(1/3)(1.0) + (1.0)(21.38)(1/2)(1.0) \\ &\quad + (1/2)(1.0)(-15.0)(1/3)(1.0) + (1.0)(-22.5)(1/2)(1.0) \\ &\quad + (1/2)(1.0)(12.38)(1/3)(1.0) + [(1/3)(1.5)(-3.38)(1/4)(1.5) \\ &\quad - (1/3)(0.5)(-0.38)(1.0 + 0.125)]\} \\ &= -\frac{0.37}{EI} \end{aligned}$$

The desired value of deflection at C is therefore

$$\begin{aligned} \Delta_C &= \frac{7.26}{EI} - \frac{0.37}{EI} \\ &= \frac{6.89}{EI} \end{aligned}$$

For the given values of $E = 207$ GPa and $I = 15 \times 10^{-6}$ m^4, Δ_C becomes

$$\Delta_C = \frac{6.89\text{kN-m}^3}{(207 \times 10^6\ \text{kN/m}^2)(15 \times 10^{-6}\ \text{m}^4)} = 0.00222\ \text{m} = 2.22\ \text{mm}$$

The moment-area method can also be used to determine the slopes and deflections in a frame. Its use on frames, however, does require more detailed considerations of the deflected shape. The following example will illustrate this point.

EXAMPLE 7-4 The slope and the horizontal and vertical deflections are to be determined at point C for the frame shown in Fig. 7-11. The deflec-

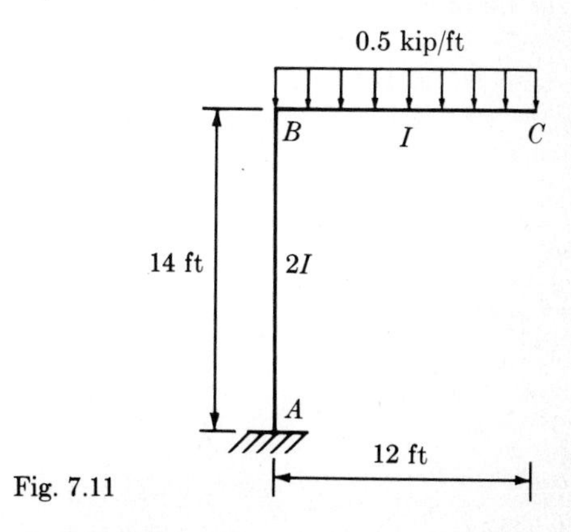

Fig. 7.11

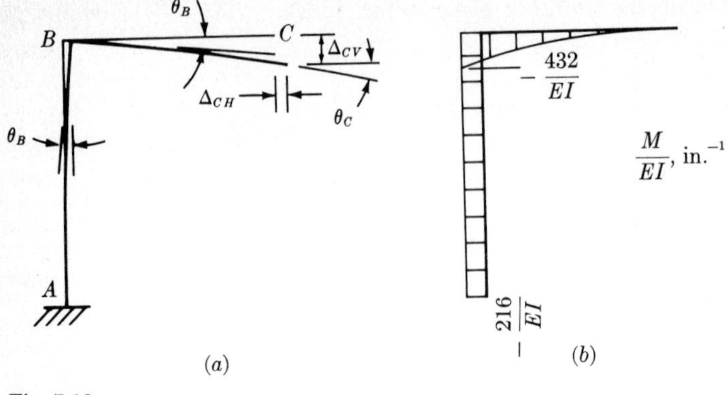

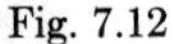

Fig. 7.12

tions to be considered are those due only to bending. The moment of inertia of the vertical member is twice that of the horizontal member. E for each member is the same. The units of E and I are considered to be kips per square inch and inches to the fourth power, respectively.

The deflected shape of the frame is shown in Fig. 7-12*a*. It is seen that the desired slope at C is equal to the slope at B plus the change in slope between B and C. Because axial deformations of the members are being neglected, the horizontal deflection at C, Δ_{CH}, is equal to the horizontal deflection at B. The horizontal deflection at B, t_{BA}, can be determined directly by the second moment-area theorem. The vertical deflection at C is equal to the product of θ_B and the length BC plus the deviation of the elastic curve between B and C.

The M/EI diagrams for the frame are shown in Fig. 7-12*b*. The moment diagrams for each member are drawn according to the beam sign convention. "Up" for beam AB is considered to be to the left. From the first moment-area theorem, θ_C is found to be

$$\begin{aligned}\theta_C &= \theta_B + \theta_{BC} \\ &= \frac{12}{EI}\left[(14)(-216) + \frac{(12)(-432)}{3}\right] \\ &= -\frac{57{,}000}{EI}\end{aligned}$$

The horizontal deflection at C is

$$\Delta_{CH} = t_{BA} = \frac{(144)(14)(216)(7)}{EI} = \frac{3.05 \times 10^6}{EI} \text{ in. (to the right)}$$

In this case the sign of the M/EI diagram has not been used directly to determine the direction in which C deflects horizontally. Its direction has been obtained by observation. This is often the case with frames, because the orientation of the member coordinate axes are generally different from each other and different from the coordinate axes of the entire frame.

The vertical deflection at C is found from the expression

$$\begin{aligned}\Delta_{CV} &= 144\theta_B + t_{CB} \\ &= \frac{(144)(12)(14)(216)}{EI} + \frac{144}{EI}\frac{(12)(432)(9)}{3} \\ &= \frac{7.46 \times 10^6}{EI} \text{ in. (down)}\end{aligned}$$

7-3 CONJUGATE-BEAM METHOD

As a preliminary to developing the conjugate-beam method of determining slopes and deflections in a beam let us consider the general relationships between the deflected shape of a beam, the slope of the deflection curve, the moment in the beam, the shear in the beam, and the load on the beam. If the deflected shape of the beam is described by the function $y(x)$, the following general relationships exist:

$$\begin{aligned}y &= \text{deflection of the elastic curve} \\ \frac{dy}{dx} &= \theta = \text{slope of the elastic curve} \\ \frac{d\theta}{dx} &= \frac{d^2y}{dx^2} = \frac{M_x}{EI} \\ \frac{d^3y}{dx^3} &= \frac{dM_x}{EI\,dx} = \frac{V_x}{EI} \\ \frac{d^4y}{dx^4} &= \frac{dV_x}{EI\,dx} = \frac{w_x}{EI}\end{aligned}$$

The validity of these relationships is dependent on the sign conventions used for the various quantities. The coordinate system used is y directed upward and x to the right, the moment M_x is positive according to the beam convention, the shear V_x is positive as given in Fig. 2-9*d*, and the load w_x is positive when it acts in the direction of positive y (Fig. 2-19). The qualitative nature of these functions for a uniformly loaded beam are demonstrated in Fig. 7-13. The uniform load in Fig. 7-13*a* is positive. Figures 7-13*b*, *c*, *d*, and *e* show the corresponding shear, moment, slope,

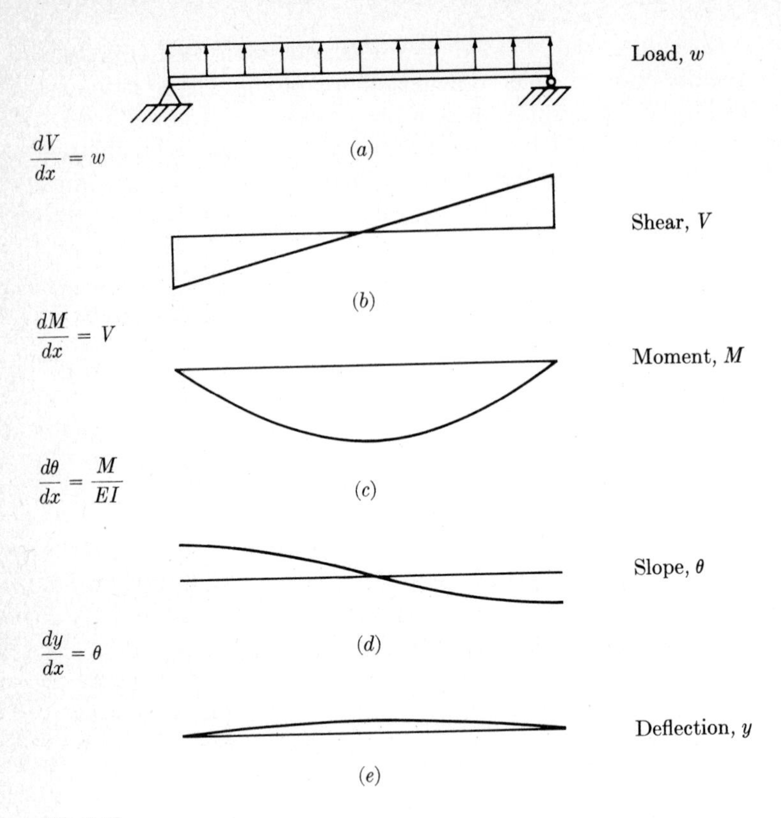

Fig. 7.13

and deflection quantities. In the space between and to the left of these curves are shown the differential relationships between the curves.

From Fig. 7-13 we see that there is a correspondence between the moment-load functions and the deflection-moment/EI functions. For example, the M/EI function can be obtained by successive integration of the load function w. Correspondingly, the deflection function y can be obtained by successive integration of the M/EI function. The intermediate steps of each integration procedure yield the shear function and the slope function. These relationships lead to the conjugate-beam concept of evaluating beam deflections and slopes. If the M/EI diagram for a given beam is considered to be the loading on an imaginary (conjugate) beam, the principles of the conjugate-beam method can be stated as follows:

THEOREM 1: *The shear at any point on the conjugate beam is equal to the slope at the corresponding point on the real beam.*

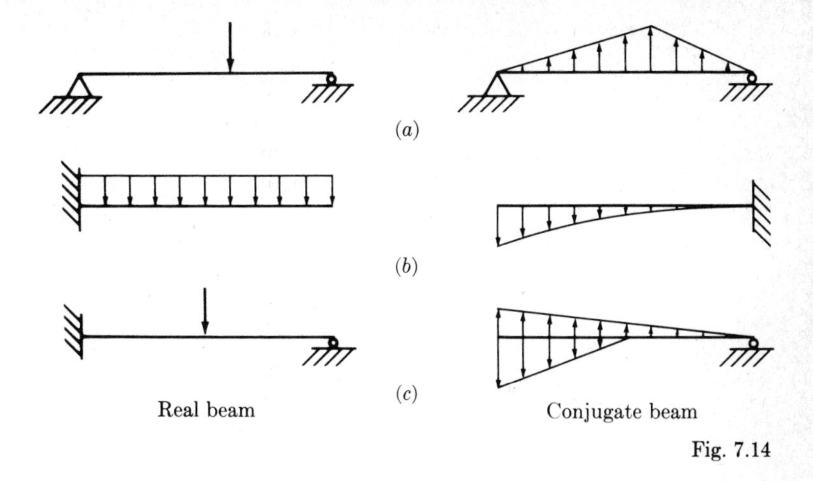

Fig. 7.14

THEOREM 2: *The moment at any point on the conjugate beam is equal to the deflection at the corresponding point on the real beam.*

The supports of the conjugate beam must be such that the shear and moment obtainable in the conjugate beam are consistent with the slope and deflection in the real beam. The conjugate beams and their M/EI loadings for various beams and loadings are shown in Fig. 7-14. With regard to the real beam of Fig. 7-14*a*, we see that slopes are possible at the end points of the beam but deflections are not. Therefore the conjugate beam is supported in such a way that shear is generated at its end points, but no moment is generated. Hence simple supports are used. The moment in the real beam is positive, so the M/EI load on the conjugate beam is positive; positive load is directed upward, as with loading on a real beam. The same reasoning is applied in obtaining the conjugate beams of Figs. 7-14*b* and *c*.

EXAMPLE 7-5 The slope and deflection due to bending are to be determined at point B for the beam of Fig. 7-15 by the conjugate-beam method. This beam was previously analyzed by the moment-area method in Example 7-1. The conjugate beam, loaded with the M/EI diagram of the

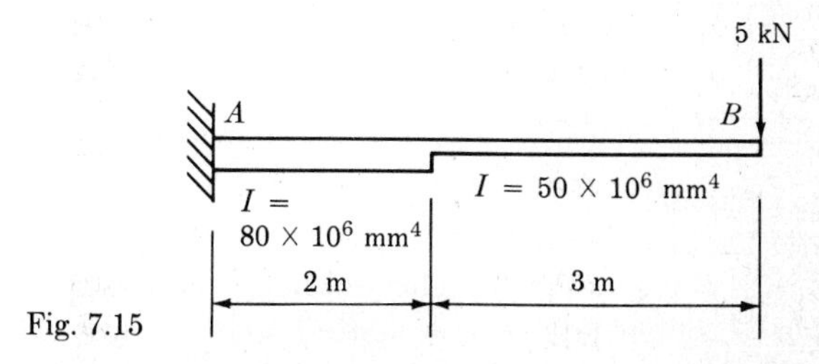

Fig. 7.15

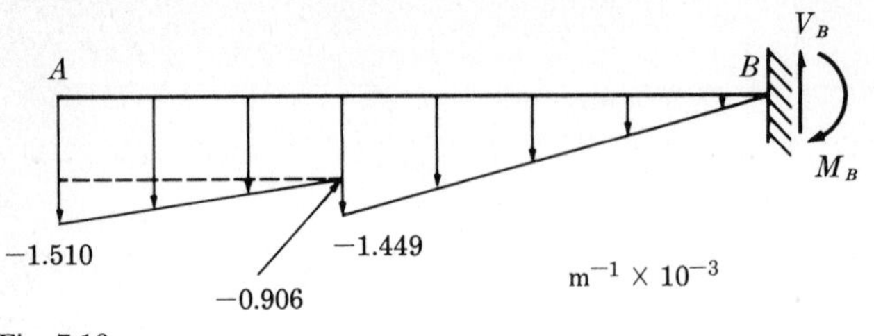

Fig. 7.16

given beam, is shown in Fig. 7-16. Because the moment due to the given loading is negative, the M/EI load on the conjugate beam is negative and hence directed downward. Consider the supports for the conjugate beam. There is no slope or deflection at A in the real beam, so point A on the conjugate beam is free. At point B on the real beam both slope and deflection are possible, which means that at point B on the conjugate beam both shear and moment must be generated. The conjugate beam is therefore supported at this point by a fixed support.

The slope at B in the real beam is determined by evaluating the shear at B in the conjugate beam. The shear V_B as shown in Fig. 7-16 is a negative quantity. Using the rectangular and triangular subdivisions of the M/EI loading in taking the summation of forces in the vertical direction, we find V_B to be

$$V_B - \left[(2)(0.906) + \frac{(2)(0.604)}{2} + \frac{(3)(1.449)}{2}\right]10^{-3} = 0$$

$$V_B = (1.812 + 0.604 + 2.174) \times 10^{-3}$$
$$= 0.00459$$

Because this is negative shear, the slope at B in the real beam is

$$\theta_B = V_B = -0.00459 \text{ rad}$$

The deflection at B in the real beam is determined by evaluating M_B in the conjugate beam. M_B in Fig. 7-16 is a negative quantity. The value of M_B is found by taking the summation of moments about B on the conjugate beam,

$$M_B - [(1.812)(4) + (0.604)(4.333) + (2.174)(2.0)]10^{-3} = 0$$
$$= 14.2 \times 10^{-3}$$

The deflection at B in the real beam is therefore

$$\Delta_B = M_B = -14.2 \text{ mm}$$

EXAMPLE 7-6 The deflection at point C on the beam of Fig. 7-17 is to be determined by the conjugate-beam method. EI for the beam is constant.

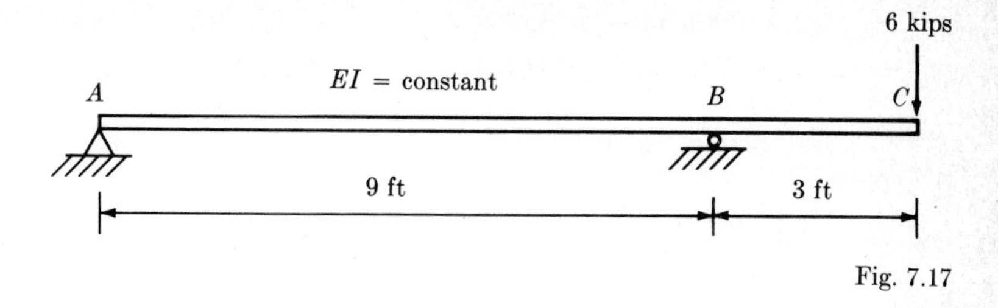

Fig. 7.17

The units of E and I are kips per square inch and inches to the fourth power, respectively. The M/EI loading for the conjugate beam is shown in Fig. 7-18a. The moment is negative over the length of the given beam, so the conjugate beam loading is directed downward. Corresponding to the possible slopes and deflections in the real beam, the conjugate beam is supported on a roller at A, is pinned at B, and is fixed at C. The pin at B results in zero moment, which is consistent with the zero deflection at B in the real beam.

The deflection at C is determined by evaluating the moment at C in the conjugate beam. The value of M_C is found by considering the free-body diagrams of the conjugate beam shown in Fig. 7-18b. The value of force transmitted by the pin at B, R_B, is found from equilibrium considerations of the conjugate-beam segment AB. Taking the summation of moments about A for the segment, we find R_B to be

$$\frac{144}{EI}\frac{(9)(216)}{2}\frac{(2)(9)}{3} - (9)(12)R_B = 0$$

$$R_B = \frac{7780}{EI}$$

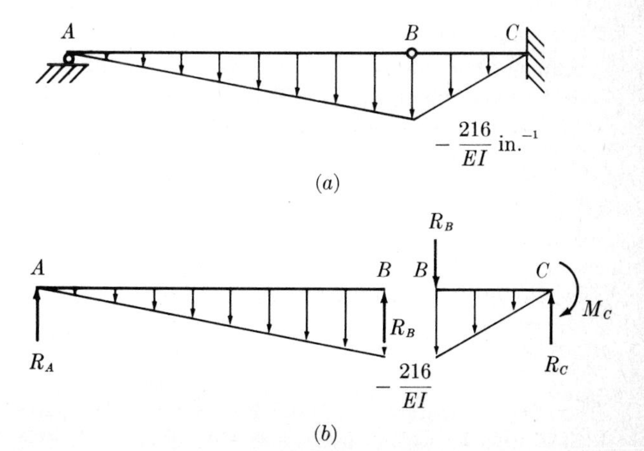

Fig. 7.18

M_C is found by considering segment BC of the conjugate beam. Thus

$$M_C - \frac{(3)(12)(7780)}{EI} - \frac{(3)(216)}{2EI}\frac{(2)(3)(144)}{3} = 0$$

$$M_C = \frac{(2.80 + 0.93) \times 10^5}{EI}$$

$$= \frac{3.73 \times 10^5}{EI}$$

Because this moment is negative, the deflection at C in the real beam is

$$\Delta_C = M_C = -\frac{3.73 \times 10^5}{EI}$$

Note that the slope and deflection at any point on the real beam is readily obtained from the free-body diagrams of the conjugate beam in Fig. 7-18*b*. It is possible, in fact, to write general equations for slope and deflection simply by writing the equations for shear and moment in the conjugate beam.

7-4 NEWMARK'S METHOD

The slopes and deflections in a beam subjected to bending can also be determined by a numerical procedure often referred to as *Newmark's method*.[1] This numerical method is particularly useful for beams with a variable moment of inertia. It is also very useful in determining the critical buckling loads for members with variable moment of inertia. One of the characteristics of Newmark's method is that the distributed M/EI load on the conjugate beam, as discussed in the previous section, is replaced by a series of equivalent concentrated loads. Thus in using this method we have a conjugate beam loaded as shown in Fig. 7-19.

The magnitudes of the equivalent concentrated loads can be expressed in terms of the values of the M/EI diagram at the points of concentrated loading and the distance between the concentrated loads. For convenience, the length of the beam is divided into an equal number of segments.

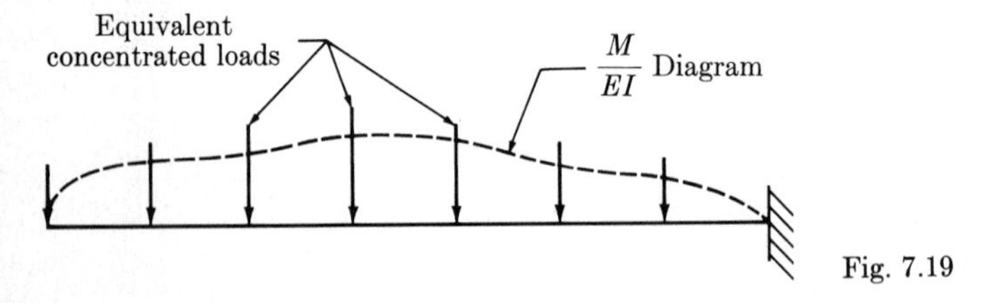

Fig. 7.19

[1] N. M. Newmark, "Numerical Procedure for Computing Deflections, Moments, and Buckling Loads," *Trans. ASCE*, vol. 108, p. 1161, 1943.

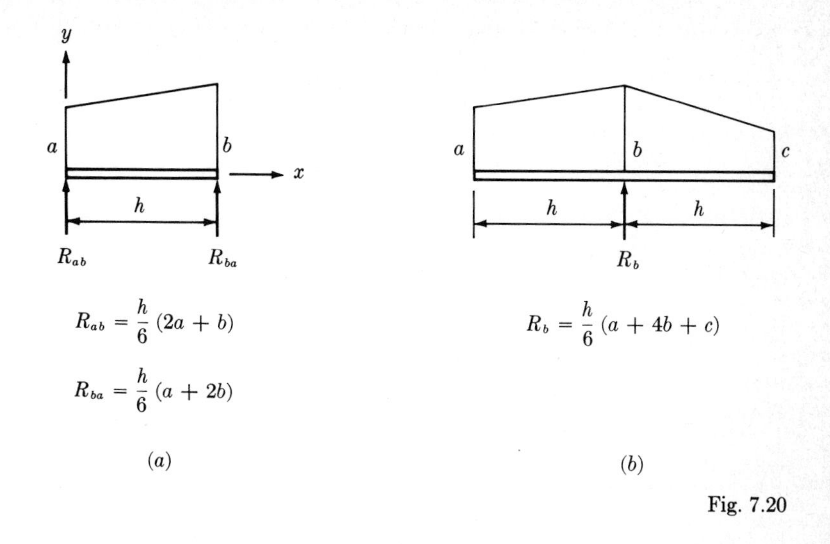

Fig. 7.20

The length of each segment, and hence the distance between concentrated loads, is denoted as h. The general expressions for the magnitude of the equivalent concentrated loads are obtained by representing the shape of the M/EI diagram between the points by segments of polynomials.

The representation of the M/EI diagram between two points by a straight-line segment is shown in Fig. 7-20a. The values of the M/EI diagram at the two points are denoted as a and b. The equivalent concentrated loads are denoted as R_{ab} and R_{ba}. The values of R_{ab} and R_{ba} are obtained from equilibrium considerations. For the xy coordinate system shown the general expression for the straight-line representation of the M/EI diagram between the two points is

$$y_x = Ax + B$$

However,

$$y(0) = a = B \qquad y(h) = b = Ah + B$$

from which we obtain

$$y_x = \frac{b - a}{h} x + a$$

For equilibrium, the following expressions must be satisfied:

$$R_{ab} + R_{ba} = \int_0^h y_x \, dx \qquad hR_{ba} = \int_0^h xy_x \, dx$$

Substituting the above value of y_x into these expressions and integrating, we obtain for the equivalent concentrated loads

$$R_{ab} = \frac{h}{6}(2a + b) \tag{7-9}$$

$$R_{ba} = \frac{h}{6}(a + 2b) \tag{7-10}$$

The value of the concentrated load for points over which the M/EI diagram is continuous, as shown in Fig. 7-20*b*, is obtained from Eqs. (7-9) and (7-10) as

$$R_b = R_{ba} + R_{bc} = \frac{h}{6}(a + 2b) + \frac{h}{6}(2b + c)$$

Thus

$$R_b = \frac{h}{6}(a + 4b + c) \tag{7-11}$$

The representation of the M/EI diagram by a segment of a second-order polynomial is shown in Fig. 7-21*a*. The general expression for the second-order polynomial is

$$y_x = Ax^2 + Bx + C$$

The coefficients A, B, and C are evaluated from the values of y_x at three points. The third value of the M/EI diagram is denoted as c. The value of c need not be an actual value of the M/EI diagram, but can be an extrapolated value.

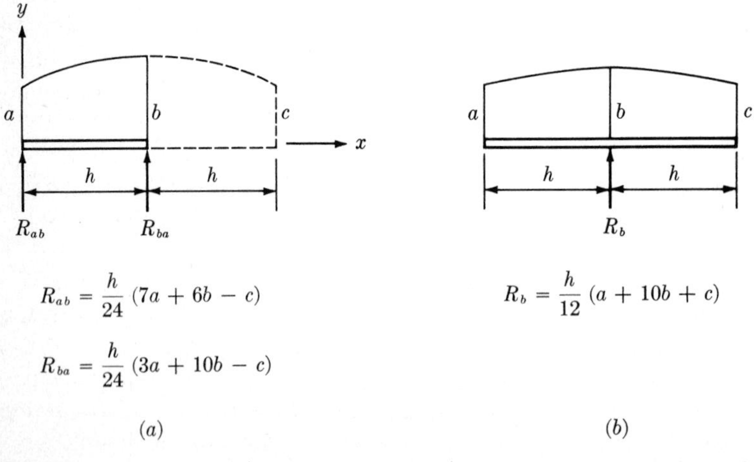

$$R_{ab} = \frac{h}{24}(7a + 6b - c)$$

$$R_{ba} = \frac{h}{24}(3a + 10b - c)$$

(*a*)

$$R_b = \frac{h}{12}(a + 10b + c)$$

(*b*)

Fig. 7.21

The coefficients A, B, and C are found from the conditions

$$\begin{aligned} y(0) &= a = C \\ y(h) &= b = Ah^2 + Bh + C \\ y(2h) &= c = 4Ah^2 + 2Bh + C \end{aligned}$$

which lead to the expression

$$y_x = \frac{a - 2b + c}{2h^2} x^2 + \frac{-1.5a + 2b - 0.5c}{h} x + a$$

For equilibrium in Fig. 7-21a the following expressions must be satisfied:

$$R_{ab} + R_{ba} = \int_0^h y_x \, dx$$
$$hR_{ba} = \int_0^h xy_x \, dx$$

Using the above expression for y_x and integrating results in the following expressions for the equivalent concentrated loads:

$$R_{ab} = \frac{h}{24} (7a + 6b - c) \tag{7-12}$$

$$R_{ba} = \frac{h}{24} (3a + 10b - c) \tag{7-13}$$

The values of Eqs. (7-12) and (7-13) are used to determine the value of the concentrated load for points over which the M/EI diagram is continuous, as shown in Fig. 7-21b. Thus

$$R_b = R_{ba} + R_{bc} = \frac{h}{12} (a + 10b + c) \tag{7-14}$$

Expressions for equivalent concentrated loads can be developed similarly with segments of higher-order polynomials. However, by increasing the number of concentrated loads, and thus decreasing the value of h, these expressions result in sufficient accuracy for most problems. It should be noted that when the M/EI diagram is a straight line, the use of equivalent concentrated loads as given in Eqs. (7-9), (7-10), and (7-11) results in exact values. Similarly, when the M/EI diagram is a second-order curve, the use of equivalent concentrated loads as given in Eqs. (7-12) to (7-14) also yields exact results.

The use of Newmark's method can be illustrated by considering the slope and deflection at point B on the beam of Fig. 7-22a. This is the same beam that was analyzed in Example 7-1 by the moment-area method and in Example 7-5 by the conjugate-beam method. In terms of the work involved, the moment-area or conjugate-beam method would be better in this case. However, this beam will serve to illustrate the general procedure of Newmark's method.

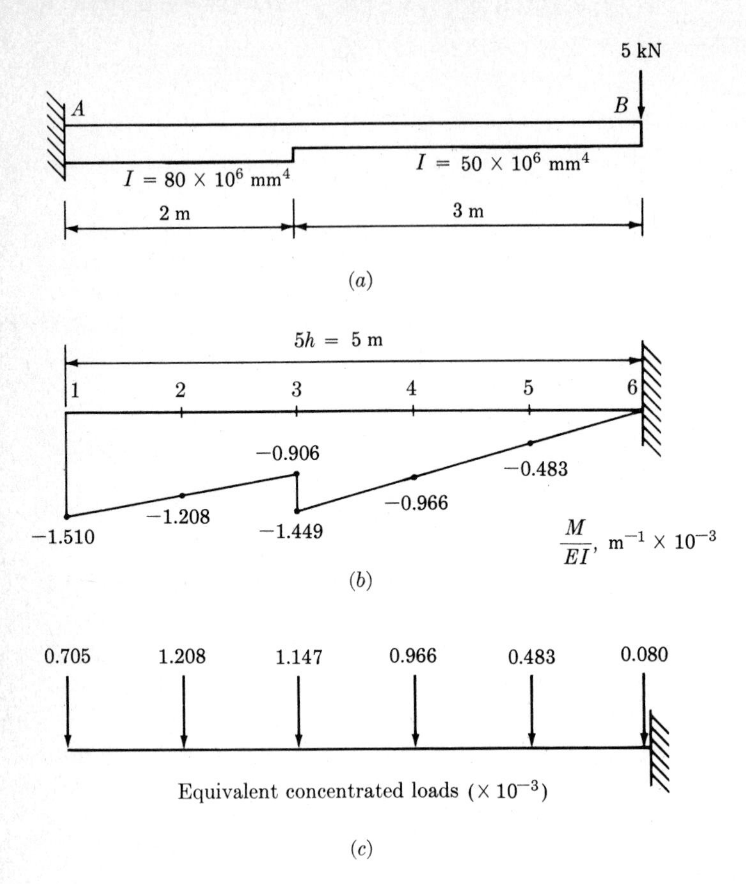

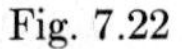

Fig. 7.22

The conjugate beam loaded with the M/EI diagram, as previously developed in Fig. 7-16, is shown in Fig. 7-22*b*. The length of the beam is divided into five equal segments, so that $h = 1$ m. The division points along the beam are numbered 1 to 6, as shown in Fig. 7-22*b*. These points are often referred to as *stations*.

The equivalent concentrated loads acting at each station are obtained by first determining the M/EI values at the stations. These are indicated in Fig. 7-22*b*. Because the M/EI diagram is composed of straight lines, Eqs. (7-9) to (7-11) are used to evaluate the equivalent concentrated loads. From Eq. (7-9), the equivalent load at station 1 is

$$R_{12} = \tfrac{1}{6}[(2)(1.510) + 1.208]10^{-3} = 0.705 \times 10^{-3}$$

The load at station 2 is found from Eq. (7-11) to be

$$R_2 = \tfrac{1}{6}[1.510 + (4)(1.208) + 0.906]10^{-3} = 1.208 \times 10^{-3}$$

Because the M/EI diagram is discontinuous over station 3, the value of load at station 3 is obtained by combining the effects to the left of the point and to the right. Thus, from Eq. (7-10),

$$R_{32} = \tfrac{1}{6}[1.208 + (2)(0.906)]10^{-3} = 0.503 \times 10^{-3}$$

and, from Eq. (7-9),

$$R_{34} = \tfrac{1}{6}[(2)(1.449) + 0.966]10^{-3} = 0.644 \times 10^{-3}$$

The value of load a station 3 is therefore

$$R_3 = R_{32} + R_{34} = 1.147 \times 10^{-3}$$

The equivalent concentrated loads at the remaining stations are similarly found to be

$$R_4 = 0.966 \times 10^{-3} \qquad R_5 = 0.483 \times 10^{-3} \qquad R_6 = 0.080 \times 10^{-3}$$

The conjugate beam loaded with the equivalent concentrated loads is shown in Fig. 7-22*c*. For such loading the shear in the beam, and hence the slope in the real beam, is constant between the loads. The constant values of shear represent average values of slope. The abrupt change in shear at the loads corresponds to an abrupt change in slope in the real beam. The continuous elastic curve is therefore represented by straight-line segments with abrupt angle changes at the points of concentrated loading.

The general procedure of analysis can conveniently be tabulated in the form of Fig. 7-23. The values of the M/EI diagram at the various stations are entered in the first row; from these values the equivalent concentrated loads can be computed. The values of the loads are entered in the second row; because the moment diagram for the given beam is negative, the equivalent loads are negative and directed downward. The column to the right of the table is used for common factors by which the tabular values are to be multiplied. A factor has been included such that the resulting slopes and deflections will have units of radians and millimeters. The shear in the conjugate beam is entered in the third row. For the

	1		2		3		4		5		6	Common factor
$\frac{M}{EI}$	−1.510		−1.208		−0.906 −1.449		−0.966		−0.483		0	$\times 10^{-3}$
R	−0.705		−1.208		−1.147		−0.966		−0.483		−0.080	$\times 10^{-3}$
Average slope		−0.705		−1.913		−3.060		−4.026		−4.509	(−4.589)	$\times 10^{-3}$
Deflection	0		−0.705		−2.618		−5.678		−9.704		−14.213	1.0

Fig. 7.23

conjugate beam under consideration the shear is computed from left to right, with the shear equal to zero at the left end and becoming more negative at each point of loading. The values of shear (average slope in the real beam) are recorded in the spaces between stations.

The moments at the stations on the conjugate beam result from the fact that the change in moment is equal to the change in area under the shear diagram. The zero moment at the left end of the beam is the basis for these computations. The resulting values of moment (deflections in the real beam) are recorded in the fourth row.

The desired value of slope at point B in the real beam is obtained from the value of slope just to the right of station 6 in Fig. 7-23. The desired value corresponds to the shear that exists just to the right of the 0.080 concentrated load, because there is a concentrated angle change right at the load. The slope at B is thus found to be

$$\theta_B = -4.589 \times 10^{-3} = -0.004589 \text{ rad}$$

The deflection at B is found to be

$$\Delta_B = (-14.213)(1.0) = -14.213 \text{ mm}$$

These values correspond to those obtained earlier by the moment-area and conjugate-beam methods.

In the analysis above the values of shear in the conjugate beam were obtained directly because the value of shear was known to be zero at the left end. This would not be the case if the conjugate beam were simply supported. The shear, and hence the moment, can be evaluated in such cases by a procedure referred to as "estimating the shear." As an illustration, consider the simply supported beam of Fig. 7-24 loaded with a series

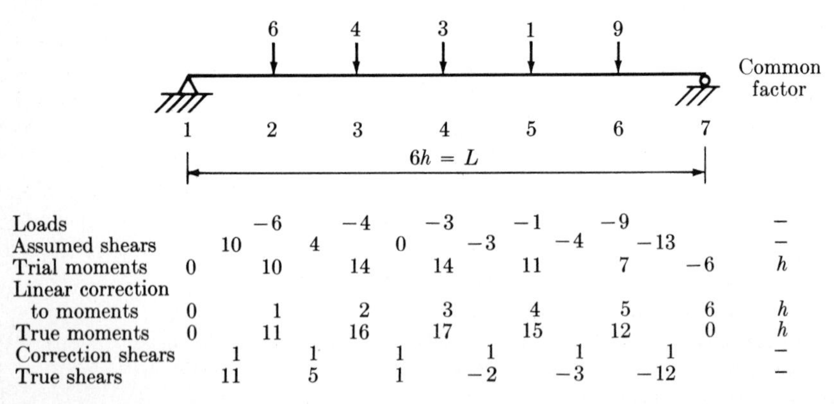

	1		2		3		4		5		6		7	Common factor
Loads			−6		−4		−3		−1		−9			—
Assumed shears		10		4		0		−3		−4		−13		—
Trial moments	0		10		14		14		11		7		−6	h
Linear correction to moments	0		1		2		3		4		5		6	h
True moments	0		11		16		17		15		12		0	h
Correction shears		1		1		1		1		1		1		—
True shears		11		5		1		−2		−3		−12		—

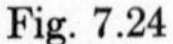
Fig. 7.24

of concentrated loads. The shear and moment in the beam are to be determined. A tabular form is used below the beam to evaluate these quantities. After the loads are recorded in the first row of the table, a value of shear is assumed for the section of the beam between stations 1 and 2. The shears in the remaining sections of the beam are then evaluated on this basis with the load values from the first row. Since zero moment exists at the left end of the beam, trial moments are obtained from the assumed shear values.

We find in the process that the resulting trial moment at the right end has a value of -6 instead of the required value of zero. To correct this we apply a *correction moment* to the beam. This is a moment applied at the right end of the beam with a sign opposite to and of the same magnitude as the resulting trial moment at this point. The application of such a moment results in values of moment over the length of the beam which vary linearly to zero at the left end. These values of correction moments are shown in row 4 of Fig. 7-24. The true moments in the beam, row 5, are obtained by adding together the values of trial moments and correction moments. A correction must also be applied to the assumed values of shear. The corrections, which are shears corresponding to the correction moment, are shown in row 6 of Fig. 7-24. The true values of shear as shown in row 7 are obtained by adding together the values in rows 2 and 6.

An application of the method of estimating shears is illustrated in the following example.

EXAMPLE 7-7 The slopes and deflections are to be evaluated for the beam of Fig. 7-25 by Newmark's method. The length of the beam is divided into six equal segments for analysis. The moments of inertia at the division points are shown below the beam in terms of the moment of inertia at the left end. The units of E and the moment of inertia are kips per square inch and inches to the fourth power, respectively. The moment diagram for the uniform load is as shown in Fig. 7-25, with the values of moment given at the stations along the beam. From these values of moment and the corresponding values of the moment of inertia the M/EI values are obtained and recorded in the first row of the table.

The conjugate beam for the given beam is simply supported. The loads to be placed on the conjugate beam are positive, acting upward, because the moment in the real beam is positive. Because the M/EI diagram is some higher-order curve, Eq. (7-14) is used to approximate the equivalent concentrated loads. Thus

$$R_b = \frac{h}{12}(a + 10b + c)$$

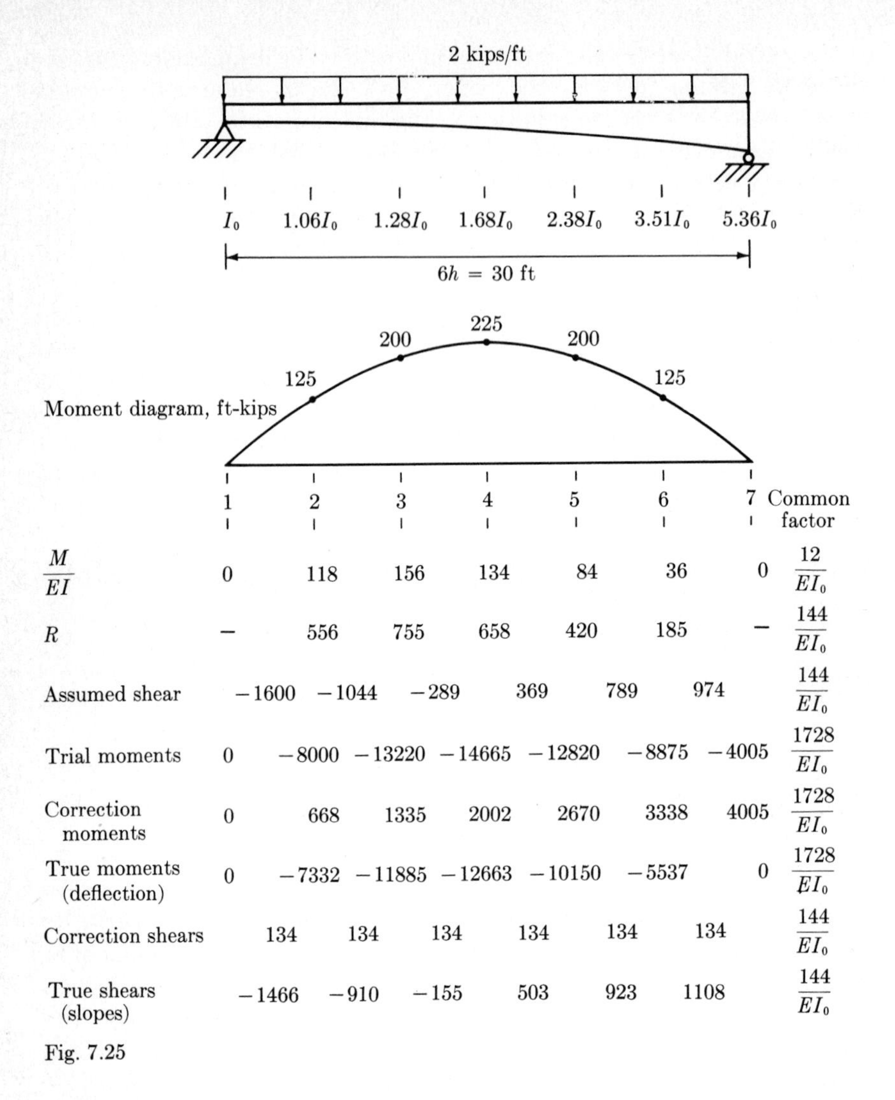

	1		2		3		4		5		6		7	Common factor
$\frac{M}{EI}$	0		118		156		134		84		36		0	$\frac{12}{EI_0}$
R	–		556		755		658		420		185		–	$\frac{144}{EI_0}$
Assumed shear		−1600		−1044		−289		369		789		974		$\frac{144}{EI_0}$
Trial moments	0		−8000		−13220		−14665		−12820		−8875		−4005	$\frac{1728}{EI_0}$
Correction moments	0		668		1335		2002		2670		3338		4005	$\frac{1728}{EI_0}$
True moments (deflection)	0		−7332		−11885		−12663		−10150		−5537		0	$\frac{1728}{EI_0}$
Correction shears		134		134		134		134		134		134		$\frac{144}{EI_0}$
True shears (slopes)		−1466		−910		−155		503		923		1108		$\frac{144}{EI_0}$

Fig. 7.25

The values of the loads are found to be

$$R_2 = \tfrac{5}{12}[0 + (10)(118) + 156] = 556$$
$$R_3 = \tfrac{5}{12}[118 + (10)(156) + 134] = 755$$
$$R_4 = \tfrac{5}{12}[156 + (10)(134) + 84] = 658$$
$$R_5 = \tfrac{5}{12}[134 + (10)(84) + 36] = 420$$
$$R_6 = \tfrac{5}{12}[84 + (10)(36) + 0] = 185$$

The loads at the ends are not needed. These values of positive load are recorded in the second row of the table.

A value of -1600 is assumed for shear in the conjugate beam between stations 1 and 2. This assumed value is obtained from a general inspection of the loads on the beam. The shears in the remaining sections of the beam are then computed from this value and the loads at the stations.

The corresponding values of trial moments are shown in row 4. Note that a correction moment of 4005 must be applied to the right end of the beam. Application of this moment and correction of the trial moments yields the true moments, and hence the deflection at the various stations on the beam. The shears are similarly corrected and yield the average values of slope shown in the last row of the table.

From the values of moments obtained we see that the maximum deflection occurs at station 4 and is approximately equal to

$$\Delta_4 = \frac{(-12{,}663)(1728)}{EI_0}$$

$$= -\frac{21.9 \times 10^6}{EI_0}$$

With E in kips per square inch and I_0 in inches to the fourth power, the resulting value will be in inches because of the common factor used.

It is seen that the method of analysis presented in this section can be particularly useful for determining slopes and deflections when the M/EI diagram is a complicated function.

7-5 WILLIOT-MOHR DIAGRAMS

The graphical method presented in the paragraphs to follow provides a means of obtaining the deflections of a statically determinate truss structure. The deflections of a truss arise from axial shortening or elongation of the members. The two common sources of member deformation are member forces and temperature changes. It will be recalled from strength-of-materials studies that the expression for the change in length e of a member due to a change in temperature ΔT is

$$e = \alpha L\,\Delta T$$

where α is the coefficient of thermal expansion for the member material and L is the length of the member. The expression for the change in length of a member due to axial loading P is given by

$$e = \frac{PL}{EA}$$

where A is the cross-sectional area of the member and E is the modulus of elasticity of the material. Corresponding to the sign convention used for

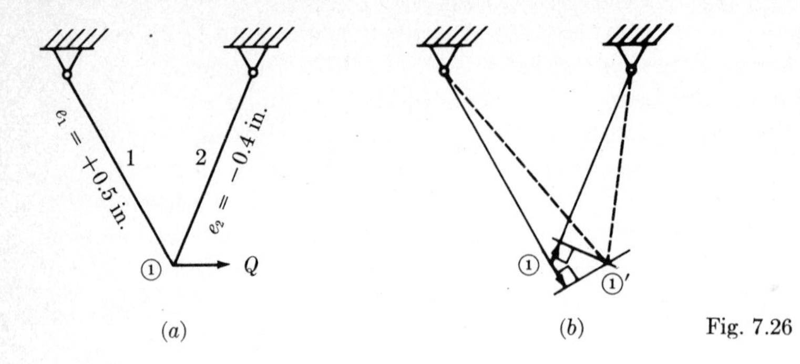

Fig. 7.26

axial forces, extension is considered a positive quantity and shortening is considered a negative quantity.

As an introduction to the deflections of a truss resulting from member deformations, let us consider the simple truss of Fig. 7-26*a*. The member deformations e as shown for members 1 and 2 are values that might result from axial forces generated by the load Q. Member 1 is in tension and undergoes extension, so e_1 is positive. Similarly, member 2 is in compression and shortens, so the value of e_2 is a negative quantity.

The resulting displacement of joint ① can be determined as shown in Fig. 7-26*b*. The amount of extension in member 1 is drawn on member 1 at joint ①. The scale for deformations is greatly exaggerated to show the resulting displacements clearly. The shortening of member 2 is also drawn at joint ①. These two deformations are shown by the vectors in Fig. 7-26*b*. So that these deformations will be consistent with the configuration of the truss, the extended length of member 1 is swung through an arc until it intersects a similar arc for the shortened length of member 2. The centers of these arcs are the opposite ends of the respective members. Because the member deformations are small in comparison with the dimensions of the structure, it is sufficient to represent the arcs by straight lines perpendicular to the original directions of the members. The resulting location of joint ① is found to be at ①′, as shown in Fig. 7-26*b*.

This procedure can be extended to larger trusses. As an example, let us consider the truss of Fig. 7-27*a*. The member deformations to be considered are shown next to each member. To analyze such a structure we must first recognize that two joints are fixed in space; these are joints ① and ②. We can therefore determine the displacement of joint ④ with respect to ① and ②. Proceeding as we did for the truss of Fig. 7-26, we find the displaced location of ④ at ④′, as shown in Fig. 7-27*b*.

The displaced location of joint ③ can be found with points ① and ④′ serving as fixed points. Member 4 is moved in Fig. 7-27*b* so that its right end is coincident with ④′. The amount of deformation in member 4 is then

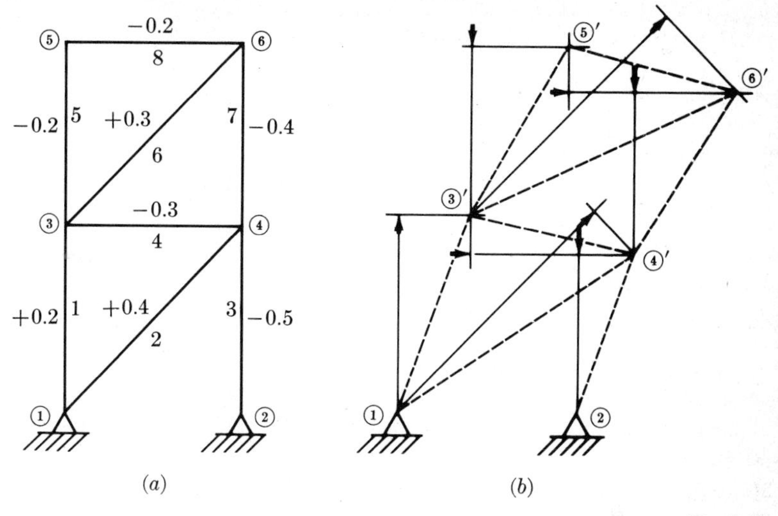

Fig. 7.27

constructed at its left end. The amount of deformation in member 1 is constructed at its upper end. Then the location of ③′ is found by arcing the deformed lengths of members 1 and 4 about points ① and ④′, respectively, with the arcs represented by straight lines perpendicular to the original direction of the members.

The location of ⑥′ is found by considering ③′ and ④′ as the fixed points. As shown in Fig. 7-27*b*, members 7 and 6 are moved to points ③′ and ④′, respectively, and the member deformations are constructed. The intersection of the lines perpendicular to the deformed lengths thus locates ⑥′, which is then used with ③′ to locate the final point ⑤′.

For correct use of this procedure, the member lengths and the member deformations must be drawn to the same scale. It is apparent from a comparison of magnitudes of the member deformations and the member lengths that an enormous drawing would be required for good accuracy. To overcome this difficulty we can use a diagram of only the member deformations. Such a diagram, known as a *Williot diagram*, after the engineer who originated it, affords an accurate graphical means of determining the joint displacements with no need for a large-scale drawing of the truss.

As an illustration of the use of the Williot diagram, let us consider again the truss of Fig. 7-27, shown again in Fig. 7-28*a* for convenient reference. Basically, in the construction of the Williot diagram, the original lengths of the members are assumed to be zero, and only the member deformations are drawn. For the truss under consideration points ① and ② are

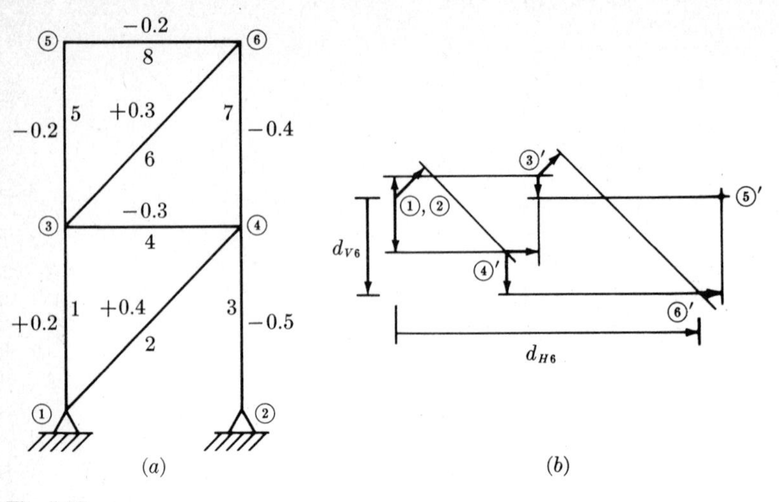

Fig. 7.28

fixed in space and are considered to be coincident on the Williot diagram of Fig. 7-28*b*. As before, we find the displacement of ④ with respect to ① and ②. For the given member deformations ④ moves down with respect to ②, and ④ moves up and to the right with respect to ①. These movements and their magnitudes are drawn from the point (①,②) on the Williot diagram of Fig. 7-28*b*. The intersection of straight lines drawn perpendicular to the end points of these deformation vectors locates ④′.

We now use point ① in Fig. 7-28*b* and point ④′ just established to determine the location of ③′. With respect to joint ①, joint ③ moves up because of the extension in member 1. Point ③ moves to the right with respect to point ④′. These deformation vectors are drawn from points ① and ④′, respectively, and the lines perpendicular to their end points establish ③′. The procedure is repeated to locate ⑥′ and then ⑤′.

The resulting Williot diagram for the truss is shown in Fig. 7-28*b*. Such a diagram gives the actual deflections and the relative deflections of joints. Because points ① and ② are fixed in space, deflections of joints measured from point (①,②) on the Williot diagram represent the actual deflections. For example, in Fig. 7-28*b* the actual deflection of joint ⑥ is downward by the amount d_{V6} and to the right by the amount d_{H6}. The relative deflections between joints on the truss can similarly be obtained by measuring between the primed points on the Williot diagram.

In constructing the Williot diagram for the truss of Fig. 7-28 a straightforward analysis was possible because two joints of the truss remained fixed in space and therefore served as a starting point for the analysis. Note that in constructing a Williot diagram we must know the location

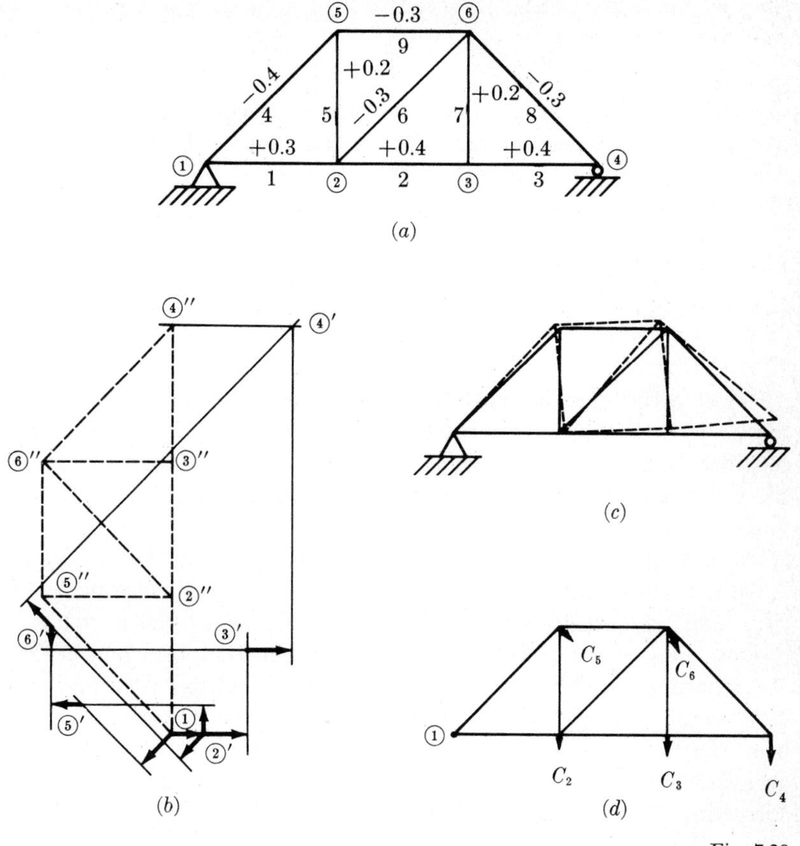

Fig. 7.29

of two adjacent joints in order to establish the displacement of any point. No such condition exists for starting the analysis of a truss such as that shown in Fig. 7-29*a*. For example, the location of joint ① remains fixed, but joint ② will displace. Therefore we would have only one known point from which to start the Williot diagram. As we have seen, this is not sufficient. However, we can temporarily assume the direction of a member and, if necessary, the location of a joint, to construct a Williot diagram which can then be corrected to yield the correct deflections.

As an illustration, consider the truss of Fig. 7-29*a*, with the direction of member 1 assumed fixed. On the Williot diagram of Fig. 7-29*b* we locate point ① first. Since the direction of member 1 is assumed fixed, point ②′ is located to the right of point ① by the amount of extension in member 1. Point ③′, and hence the remaining points of the Williot diagram, can then be established. The points on the Williot diagram are denoted, as before, by single-primed numbers. The resulting deflected shape of the truss is shown in Fig. 7-29*c*. The indicated vertical displacement at the right support of the truss, joint ④, obviously violates the support condition for which only horizontal displacement is permitted. This discrepancy arises from the assumed direction of member 1. To correct for it we rotate the truss clockwise about the left support, joint ①, until there is zero vertical displacement at joint ④.

Such a rotation is equivalent to applying corrections C_i at each joint, as shown in Fig. 7-29*d*. In general, the magnitude of the correction at any joint i can be expressed as

$$C_i = r_i \, d\theta$$

where $d\theta$ is the small angle through which the truss is rotated and r_i is the radius from the center of rotation, joint ① in this case, to the joint under consideration. Since the correction at any point i, C_i, is proportional to the radius r_i, we can use a convenient correction diagram on the Williot diagram.

To understand the basis for such a correction diagram observe in Fig. 7-29*d* that the directions of the corrections are perpendicular to the respective radii, and as stated above, the magnitudes of the corrections are proportional to the radii. The corrections can therefore be obtained by constructing a scaled diagram of the truss through point ① and rotating it 90°, as shown in Fig. 7-29*b*. The scale of the truss diagram is established from the support condition that is to be satisfied. In this case joint ④ is permitted to displace only in the horizontal direction, so point ④″ on the correction diagram is located on a line drawn horizontally through point ④′ on the Williot diagram. Having established the scale, we can locate the other points on the correction diagram. These points are denoted by double-primed numbers in Fig. 7-29*b*.

The resulting diagram, Fig. 7-29*b*, is referred to as a *Williot-Mohr diagram* (Mohr originated the idea of the correction diagram). *The true deflection of a truss joint is the distance measured from the double-primed number to the corresponding single-primed number on the Williot-Mohr diagram.* To verify this, consider the resulting displacements of joint ④ as shown in Fig. 7-30*a*. The displacement $d'_{④}$ is that obtained from the Williot diagram. A correction $C_{④}$ obtained from the correction diagram is added to this. The true deflection $d_{④}$ can be written in vector form as

$$\mathbf{d}_{④} = \mathbf{d}'_{④} + \mathbf{C}_{④} \qquad \text{or} \qquad \mathbf{d}_{④} = \mathbf{C}_{④} + \mathbf{d}'_{④}$$

This last expression is represented graphically in Fig. 7-30*a* and is also the form in which the displacements appear on the Williot-Mohr diagram

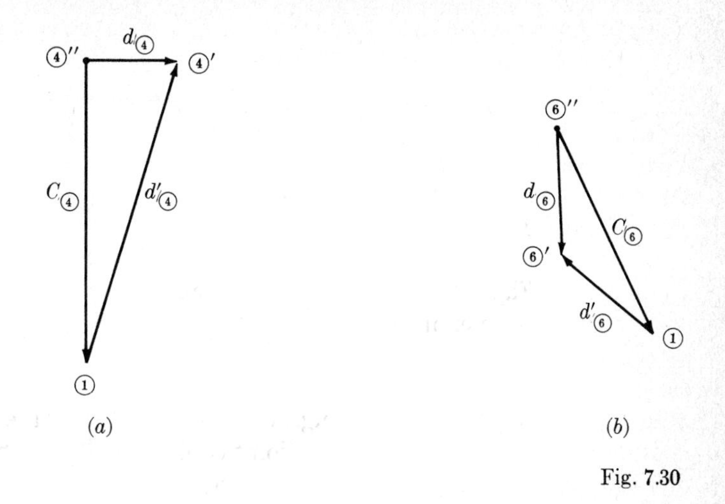

Fig. 7.30

of Fig. 7-29*b*. The true displacement of joint ⑥ is also illustrated by the vector diagram of Fig. 7-30*b*.

The construction of Williot-Mohr diagrams for other types of trusses and other support conditions follows basically the same procedure. Once the basic principles of the method are mastered, it should be a simple matter to apply the method to various conditions. For example, if the roller support surface at joint ④ in Fig. 7-29*a* were inclined at some angle instead of being horizontal, the line drawn through ④′ to locate ④″ in Fig. 7-29*b* would be at the same angle of inclination as the support surface, reflecting the type of displacement possible. Cases where the supports of a truss are located other than at the same height should also cause little difficulty once it is understood that the directions of the corrections are accounted for by constructing the correction diagram 90° to the given orientation of the truss and obtaining its scale from support considerations.

PROBLEMS

Use the moment-area method to solve Probs. 7-2 through 7-15.

7-1. Sketch the deflected shapes for the structures shown. Neglect axial deformations of the members. Indicate the approximate locations of inflection points.

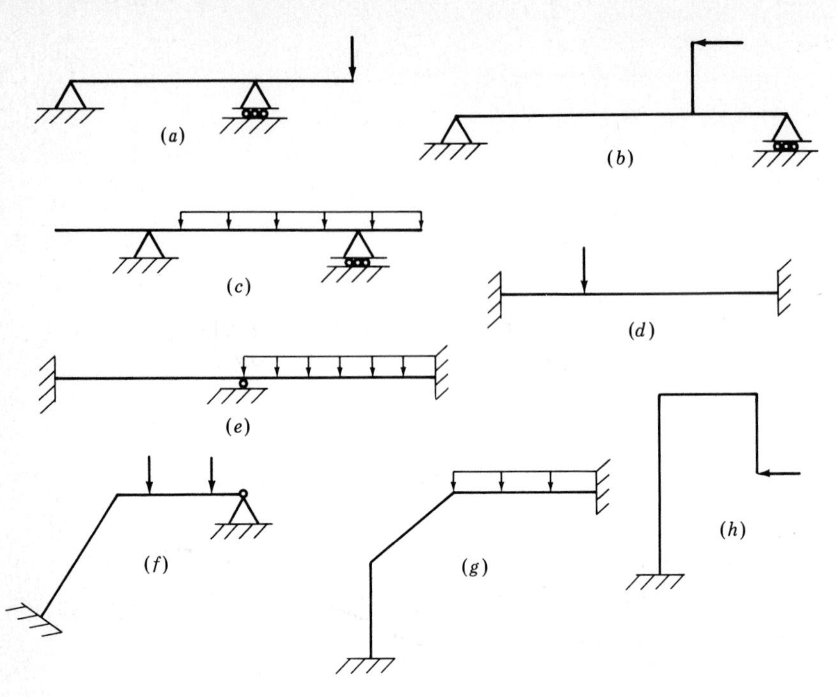

P7.1

7-2. Determine the slope at D in radians and the vertical deflection at D in millimeters. For the beam, $E = 70 \times 10^9\ \text{N/m}^2$ and $I = 40 \times 10^6\ \text{mm}^4$.

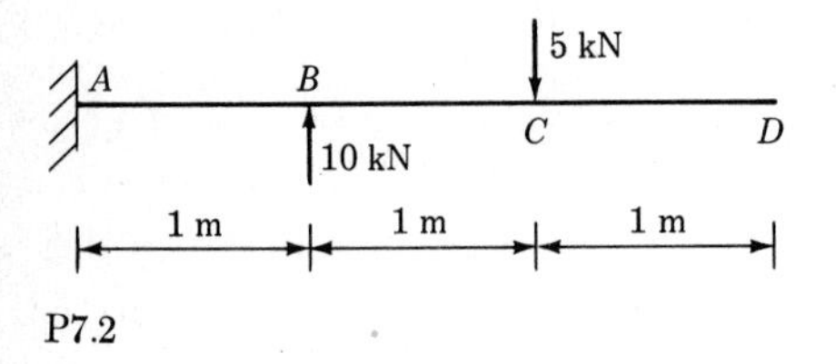

P7.2

7-3. For the beam shown, determine (*a*) the slope at A in radians; (*b*) the vertical deflection at C in inches. For the beam, $E = 30 \times 10^3$ ksi and $I = 100\ \text{in.}^4$.

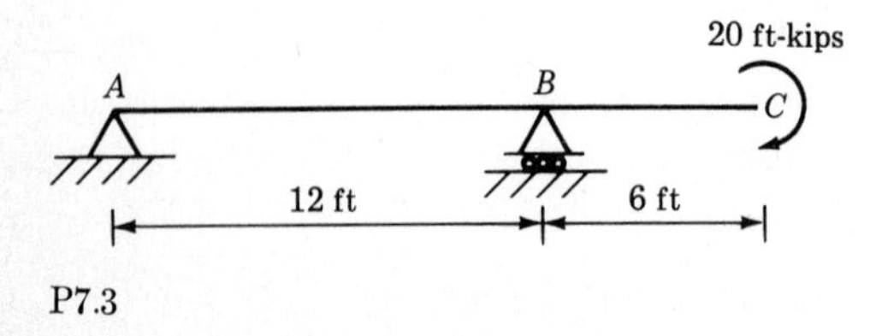

P7.3

7-4. For the cantilever beam shown, determine (*a*) the slope at B in radians; (*b*) the vertical deflection at B in millimeters. For the beam, $E = 200 \times 10^9$ N/m² (Pa) and $I = 30 \times 10^6$ mm⁴.

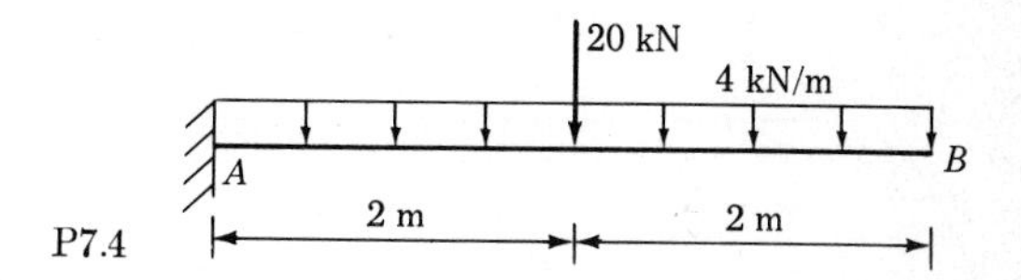

P7.4

7-5. Determine the distance from A in meters and the value of the maximum vertical deflection in millimeters for the beam shown. $I = 40 \times 10^6$ mm⁴ and $E = 200$ GPa.

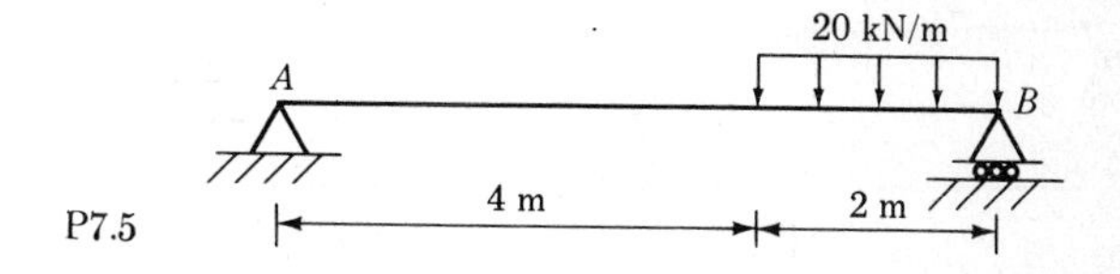

P7.5

7-6. Determine the value of the vertical deflection at point D in inches. Does point D displace up or down? $I = 200$ in.⁴ and $E = 10 \times 10^3$ ksi.

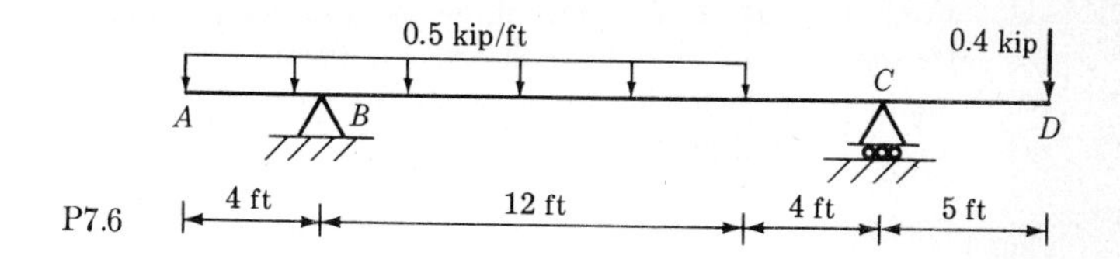

P7.6

7-7. Determine the vertical deflection of point B in inches. $E = 200$ GPa and $I = 30 \times 10^{-6}$ m⁴.

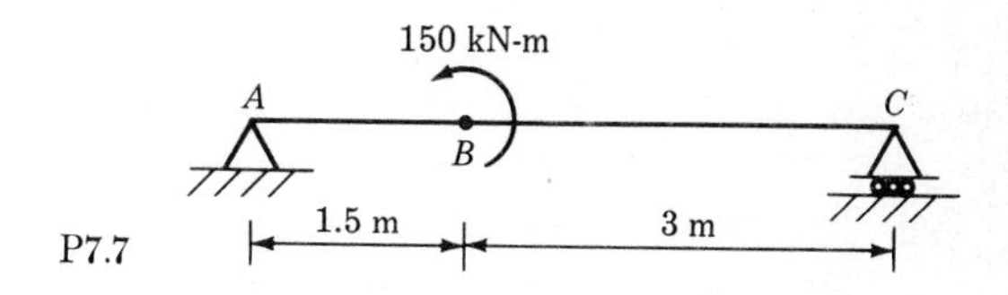

P7.7

7-8. For the beam shown, determine (*a*) the rotation at A in radians; (*b*) the vertical deflection at A in millimeters. $E = 200 \times 10^9$ N/m² (Pa).

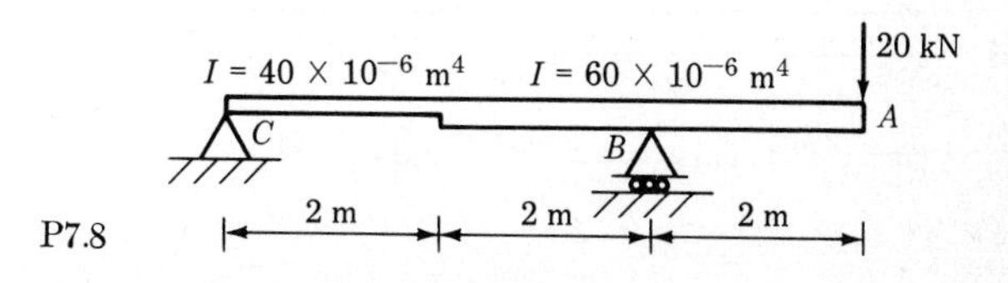

P7.8

7-9. Determine the distance from A in inches and the value of the maximum vertical deflection in inches for the beam shown. I = 100 in.^{4} and E = 30×10^3 ksi.

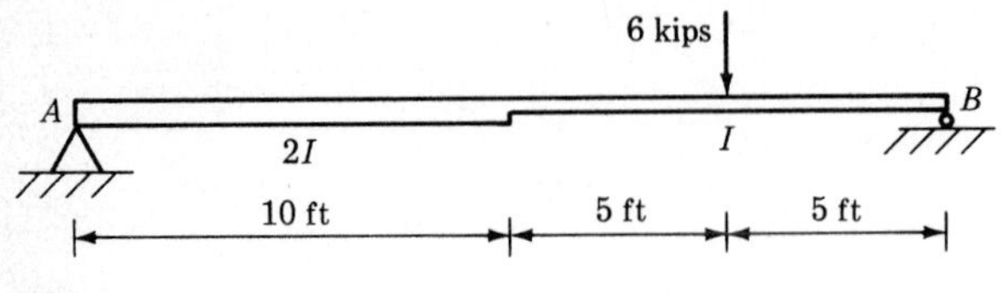

P7.9

7-10. For the beam shown, determine (a) the vertical deflection at A in millimeters; (b) the slope at B in radians. E = 200 GPa and I = 40×10^6 mm^4.

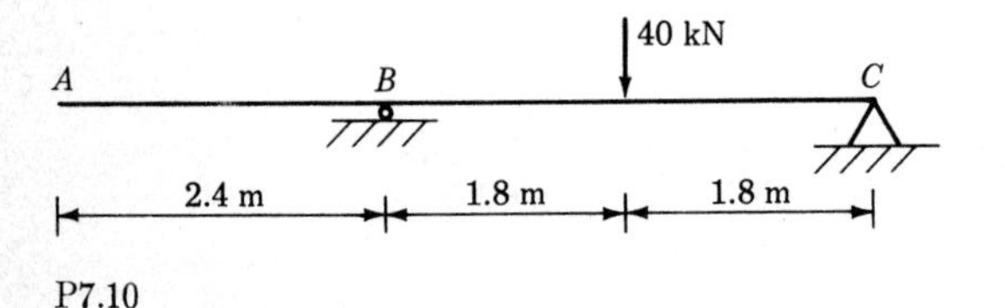

P7.10

7-11. For the beam shown, determine (a) the vertical deflection in inches and the rotation in radians at D; (b) the location and the value of the maximum vertical deflection in inches between A and C. E = 30×10^3 ksi and I = 100 in.^{4}

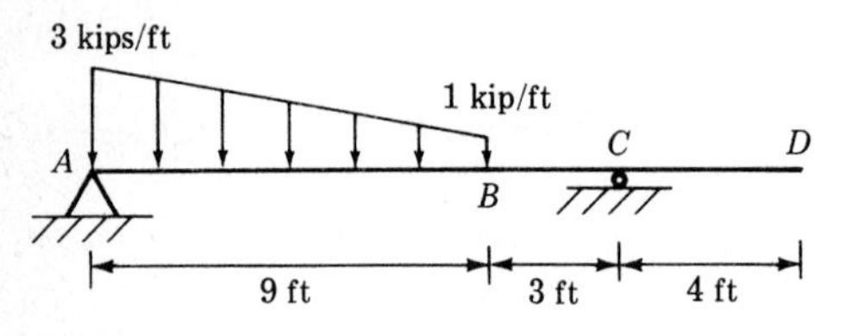

P7.11

7-12. For the beam shown, determine (a) the vertical deflection at C in millimeters; (b) the location and the value of the maximum vertical deflection between A and B in millimeters. E = 25 GPa and I = 200×10^{-6} m^4.

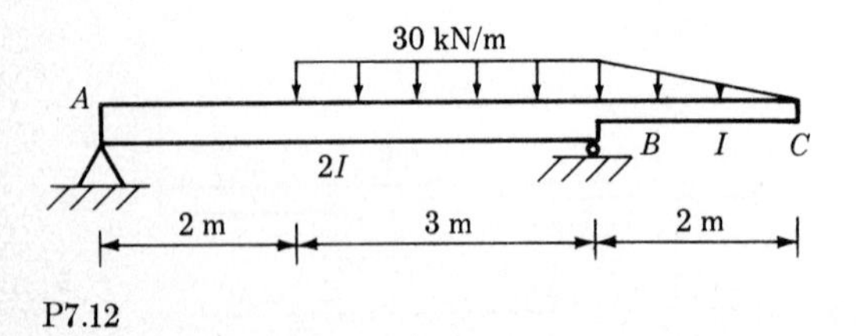

P7.12

7-13. For the frame shown, determine (*a*) the horizontal deflection in inches associated with the 10-kip load; (*b*) the vertical deflection at point *A*. $E = 30 \times 10^3$ ksi.

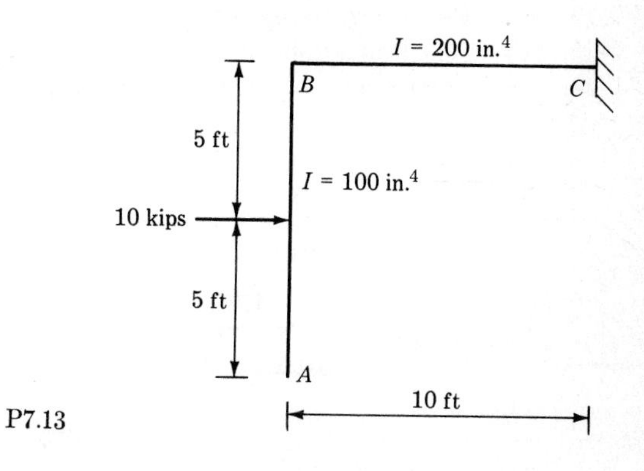

P7.13

7-14. Determine the horizontal deflection in millimeters of point *D*. $E =$ 200 GPa and $I = 300 \times 10^6$ mm⁴ for each member.

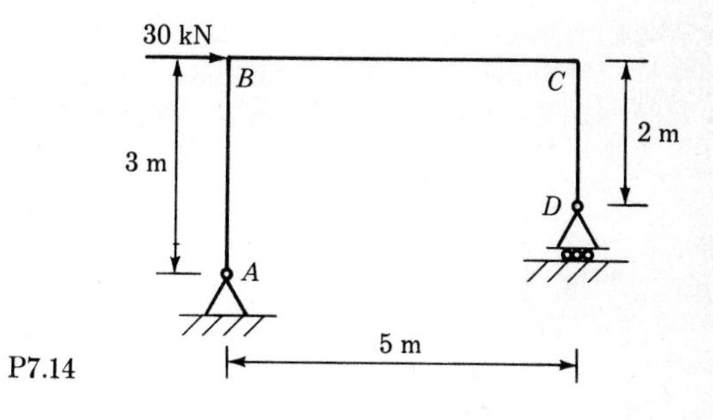

P7.14

7-15. Determine the horizontal deflection in inches of point *D*. $E = 30 \times 10^3$ ksi and $I = 200$ in.⁴ for each member.

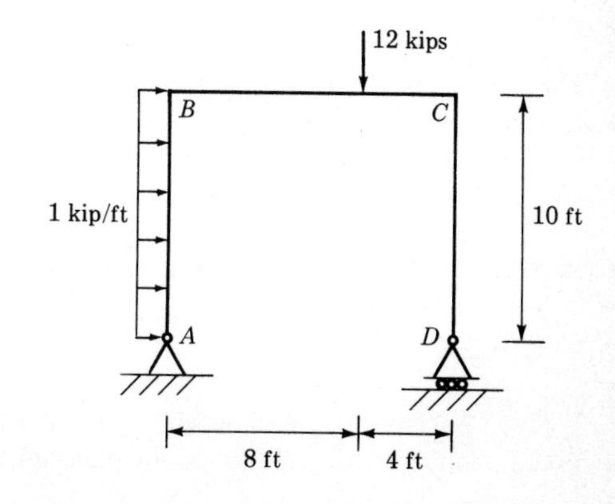

P7.15

7-16. Use the conjugate-beam method to solve Prob. 7-2.
7-17. Use the conjugate-beam method to solve Prob. 7-3.
7-18. Use the conjugate-beam method to solve Prob. 7-5.
7-19. Use the conjugate-beam method to solve Prob. 7-6.
7-20. Use the conjugate-beam method to solve Prob. 7-8.
7-21. Determine the rotation and deflection in inches at point B on the beam shown, using Newmark's method. Divide the length of the beam into six equal segments for the analysis. The beam has a constant width of 1 ft, and $E = 3 \times 10^3$ ksi.

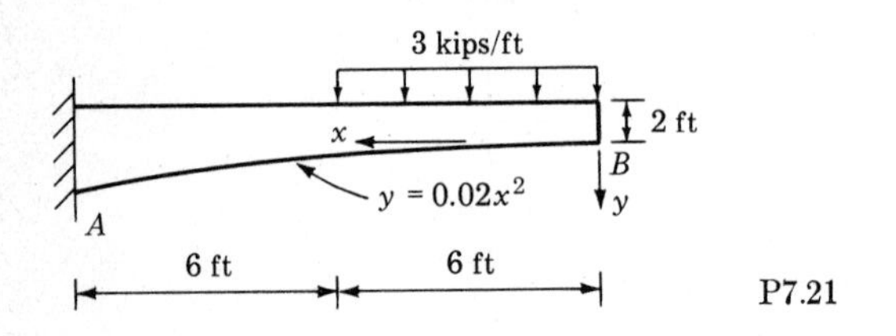

P7.21

7-22. Determine the deflections of the beam of Prob. 7-9 using Newmark's method. Divide the length of the beam into eight equal segments for the analysis.
7-23. Determine the deflections of the beam shown using Newmark's method. Divide the length of the beam into eight equal segments for the analysis. The width of the beam is 400 mm, and $E = 25$ GPa.

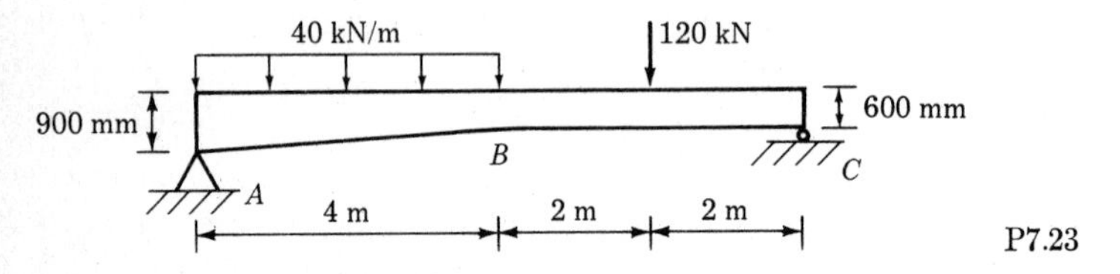

P7.23

7-24. Construct the Williot-Mohr diagram for the truss shown and for the member axial deformations as given. Assume joint ① and the direction of member ①-② to be fixed. On the diagram indicate by a vector the deflection of joint ③.

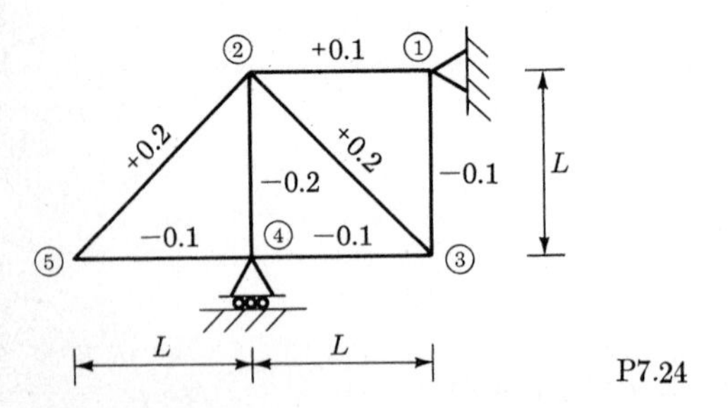

P7.24

7-25. Construct the Williot-Mohr diagram for the truss of Prob. 8-7.
7-26. Construct the Williot-Mohr diagram for the truss of Prob. 8-8.

CHAPTER

EIGHT

DEFLECTIONS: ENERGY METHODS

8-1 INTRODUCTION

In this chapter the energy methods of computing deflections are developed. We shall see that the energy methods provide a highly versatile means for evaluating deflections. Although the geometric methods of the preceding chapter can be used for analyzing many of the more complicated structures, the energy methods are often preferable because of the manner in which the solution can be formulated.

It is important in developing methods for analyzing the more complicated structures, particularly when the energy methods are being used, to adopt notation for deflections that can be used for various types of structures. In the preceding chapter, where we were primarily concerned with rather simple structures, it was sufficient to denote deflections by symbols such as Δ and θ, and if we were concerned with horizontal and vertical components of deflection, the subscripts H and V could be used with Δ.

For the more involved analyses it is often convenient to describe deflections at points on a structure in terms of a framework coordinate system. For example, the deflections at points ① and ② on the frame of Fig. 8-1*a* can conveniently be described in terms of the components shown

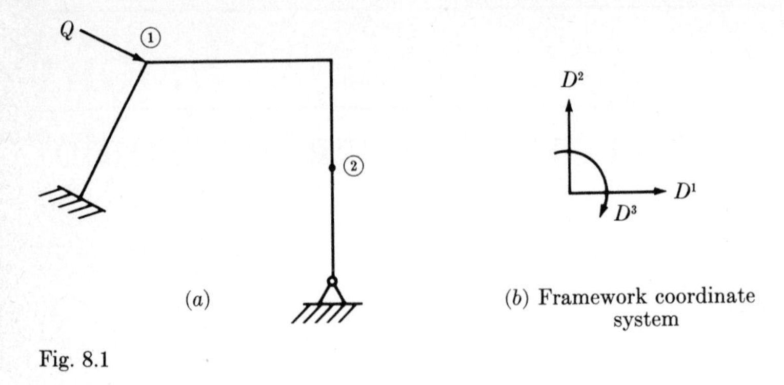

Fig. 8.1

in Fig. 8-1*b*. The horizontal displacement of point ② on the structure can therefore be denoted as $D_2{}^1$, where the subscript indicates the point on the structure under consideration. Similarly, the rotation of point ② can be denoted as $D_2{}^3$. Although the direction of D^3 as shown is not in accordance with the right-hand convention for the Z axis, it is often convenient in planar analysis to consider clockwise joint rotation as positive because that direction corresponds to the often used clockwise convention for member-end rotations. Superscripts such as H, V, and R or X, Y, and Θ could also be used to denote the components of displacement.

Concentrated values of loading can also be represented in terms of superscripted components. Loads in the form of moments can be denoted by a superscript 3 or R. The horizontal and vertical components of a load such as Q in Fig. 8-1*a* can be written as Q^1 and $-Q^2$ for the coordinate system of Fig. 8-1*b*. The negative sign of Q^2 indicates that the component of load acts in the negative direction of the framework coordinate axis. The use of signs on load and deflection quantities can be particularly useful in more complicated problems, where it is not possible to tell by observation the direction in which a point moves. The dependence on the resulting signs, however, requires that careful consideration be given to formulating the problem correctly.

The use of components of force and displacement and the adoption of a positive direction for each is particularly useful in the energy methods of deflection analysis. Therefore, where applicable, such notation will be used in the discussions that follow.

8-2 REAL WORK

The first energy method to be considered is the method of *real work*. The term "real work" implies that the forces and displacements considered are real, as opposed to the method of *virtual work* discussed in a

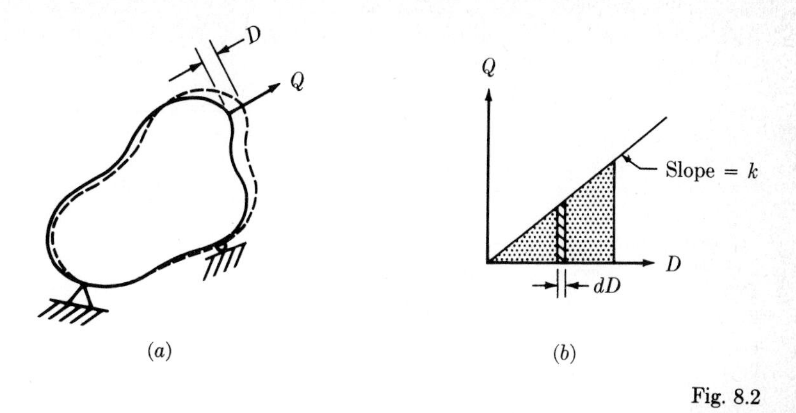

Fig. 8.2

later section, in which the forces and displacements can be virtual or imaginary.

If a nonrigid body is subjected to an external load Q, as shown in Fig. 8-2a, the body will undergo a small change in shape. The displacement associated with the load Q is denoted as D in Fig. 8-2a. D is measured in the direction of the applied load Q. If the material of the body is linearly elastic, the relation between Q and D is represented by a straight line, as shown in Fig. 8-2b.

From the basic definition of work, which is the product of a force acting through a distance, we can obtain an expression for the external work W_E done on the body. Considering a differential quantity of external displacement dD, as shown in Fig. 8-2b, we find the corresponding differential quantity of external work,

$$dW_E = Q\,dD \tag{8-1}$$

However, for the linearly elastic material under consideration

$$Q = kD \tag{8-2}$$

where k is the slope of the load-deflection relationship and therefore represents the stiffness of the body. The expression for the differential quantity of external work can thus be written as

$$dW_E = kD\,dD \tag{8-3}$$

The total amount of external work associated with an external displacement D is obtained by integrating Eq. (8-3),

$$W_E = \frac{kD^2}{2}$$

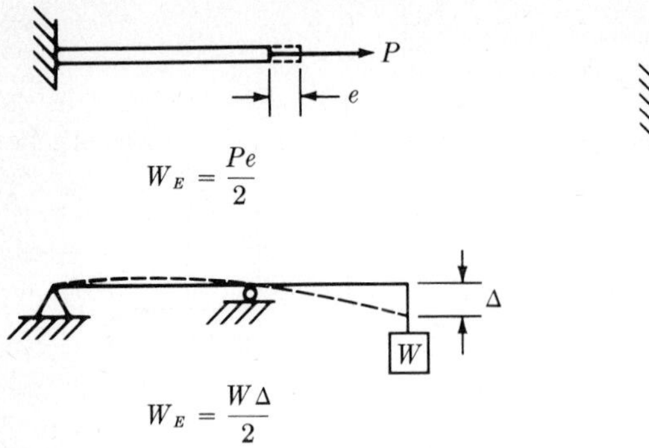

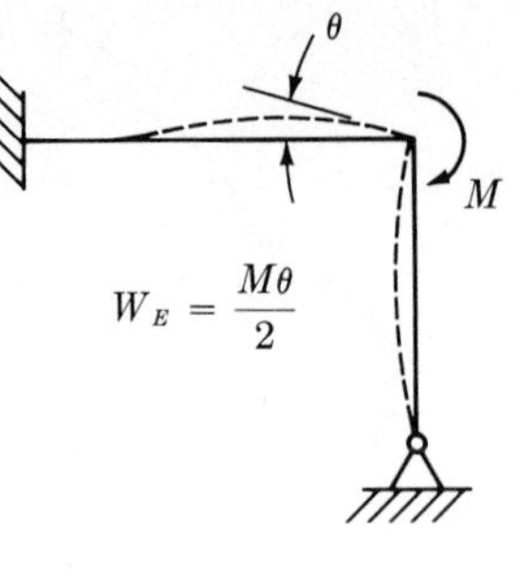

Fig. 8.3

which from Eq. (8-2) becomes

$$W_E = \frac{QD}{2} \tag{8-4}$$

Thus for a linearly elastic body the external work can be stated as the product of the average value of the external force Q and the resulting external displacement D.

Although the load Q of Fig. 8-2a is shown as a force quantity, the above expression for external work applies equally to external moments. In the case of an applied moment the moment Q has units of moment, and the associated displacement D is nondimensional (radians). The resulting units for external work are therefore the same for either type of load. Figure 8-3 shows some forms of external work encountered in analysis of structures.

For a structure subjected simultaneously to n external loads the expression for external work is

$$W_E = \tfrac{1}{2} \sum_{i=1}^{n} Q_i D_i$$

where D_i is the deflection measured in the direction of Q_i. If the external loads **Q** and the external displacements **D** are represented by the column matrices (vectors),

$$\mathbf{Q} = \begin{Bmatrix} Q_1 \\ Q_2 \\ \cdot \\ \cdot \\ \cdot \\ Q_n \end{Bmatrix} \qquad \mathbf{D} = \begin{Bmatrix} D_1 \\ D_2 \\ \cdot \\ \cdot \\ \cdot \\ D_n \end{Bmatrix}$$

the expression for external work can be written in matrix form as

$$W_E = \tfrac{1}{2}\mathbf{Q}'\mathbf{D} = \tfrac{1}{2}\mathbf{D}'\mathbf{Q} \tag{8-5}$$

The result of the matrix multiplication is a 1×1 matrix, which can be considered a scalar quantity.

8-3 FORMS OF INTERNAL WORK

When a structure is subjected to external loading, and hence external work is done, internal work or strain energy is generated and stored in the individual members of the structure. The internal work can be expressed in terms of the stresses and strains developed in the individual members. The forms of stresses and strains commonly considered are those due to axial forces, bending moment, twist, and sometimes shear.

During the loading process the external work will be equal to the internal work (strain energy), by the law of conservation of energy. If the reactions do not displace, then only the loads do external work, and we can state that

$$W_E = W_I \tag{8-6}$$

where W_I denotes the internal work. If the reactions do displace, then their work is part of the external work, and we can state that

$$W_E + W_R = W_I \tag{8-7}$$

where W_R represents the work done by the reactions. Unless otherwise stated, it will be assumed in the following discussions that the reactions do not displace and that the expression for equality of external and internal work is given by Eq. (8-6).

When the loads are removed, the structure will return to its original position, provided the structural material has not been stressed beyond its elastic limit. We shall consider only those loads which produce stresses within the elastic limit of the material.

Equation (8-6) is the basis for evaluating deflections by real work—or Eq. (8-7) in cases where support movements are to be considered. To use Eq. (8-6) we must first develop expressions for internal work. Internal work is a function of the types of stresses and strains generated in the members of the structure during loading. Four types of forces commonly considered in analyzing member deflections are shown acting on a differential slice of a member in Fig. 8-4. The forces are axial force P, moment about the z axis M, shear in the negative y direction V, and twist about the x axis T. The force quantities may vary with the position along the length of the member and should be thought of as functions $P(x)$, $M(x)$,

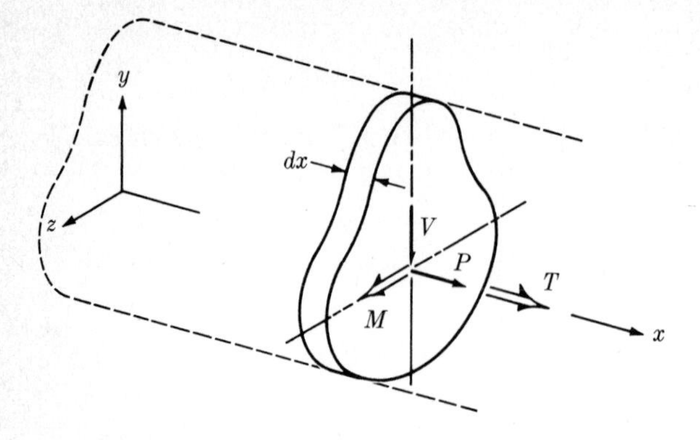

Fig. 8.4

$V(x)$, and $T(x)$. For convenience, however, they will be denoted by the single letters as shown in Fig. 8-4. The moment M and twist T are shown in accordance with the right-hand screw rule. Quantities as shown are considered positive. The forces are considered to act at the centroid of the cross-sectional area A of the member. For the developments that follow we shall assume that the cross-sectional area A and the moment of inertia I are constant. We shall also assume that the member cross-section is symmetric about the y axis.

To develop expressions for internal work, let us consider the effect of each force separately and consider views of each force acting on the differential slice as shown in Fig. 8-5. For axial force effects as shown in the two-dimensional view of Fig. 8-5a, the axial force P acting on the cross-sectional area A causes a change in length de of the length dx. For a linearly elastic material the internal work done on the differential slice due to the axial force P is

$$dW_I = \frac{(\text{force})(\text{displacement})}{2}$$

$$= \frac{P\,de}{2} \tag{8-8}$$

From previous work in strength of materials we recall that change in length divided by original length is defined as strain ϵ, and that strain can be expressed in terms of stress σ and the modulus of elasticity of the material E. The stress σ can be expressed in terms of the axial force P and the member cross-sectional area A. We can thereby write

$$\frac{de}{dx} = \epsilon = \frac{\sigma}{E} = \frac{P}{EA}$$

The expression for internal work done on the differential slice as given

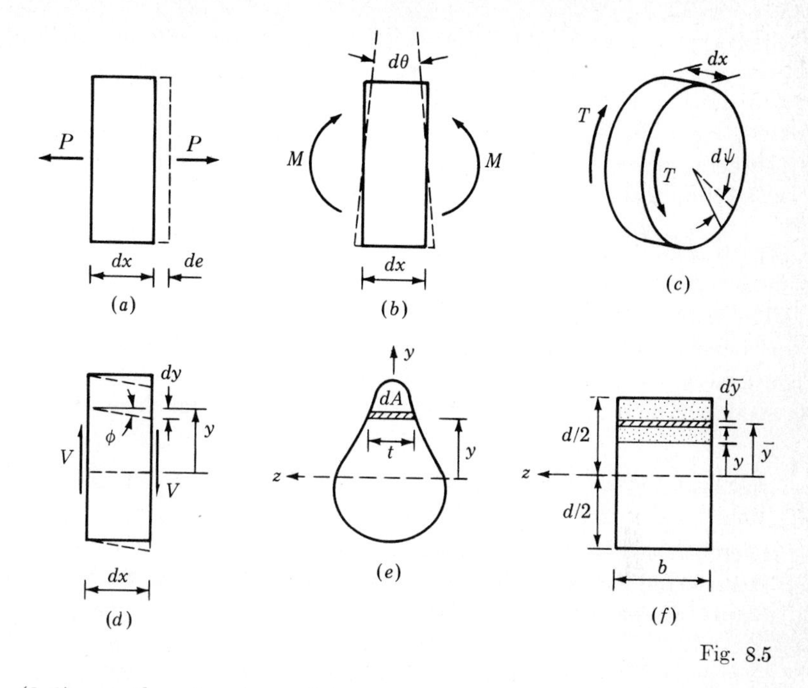

Fig. 8.5

in Eq. (8-8) can therefore be written as

$$dW_I = \left(\frac{P}{2}\right)\left(\frac{P\,dx}{EA}\right) = \frac{P^2}{2EA}\,dx \tag{8-9}$$

The internal work done on the entire member is found by integrating Eq. (8-9) over the length of the member. Thus, for a member of length L, the internal work (strain energy) due to axial force effects is:

$$W_I = \int_0^L \frac{P^2}{2EA}\,dx \tag{8-10}$$

The expression for internal work due to bending can be obtained by considering bending moment M acting on the differential slice as shown in Fig. 8-5*b*. The moment M causes the two vertical sides of the slice to rotate through a relative angle of $d\theta$. The expression for internal work can therefore be written as

$$dW_I = \frac{M\,d\theta}{2} \tag{8-11}$$

From strength of materials we can use the expression for the elastic curve of the beam as used in Eqs. (7-1) and (7-2),

$$\frac{d^2y}{dx^2} = \frac{d\theta}{dx} = \frac{M}{EI}$$

The expression for internal work [Eq. (8-11)] can therefore be written as

$$dW_I = \left(\frac{M}{2}\right)\left(\frac{M\,dx}{EI}\right) = \frac{M^2}{2EI}\,dx$$

Integrating over the length of the member, the expression for internal work due to bending in a member becomes

$$W_I = \int_0^L \frac{M^2}{2EI}\,dx \tag{8-12}$$

The expression for internal work due to twist will be developed for a circular cross-section as shown in Fig. 8-5*c*. It is not possible to develop a general expression for internal work due to twist acting on any cross-section. The internal work on the circular differential slice can be written as

$$dW_I = \frac{T\,d\psi}{2} \tag{8-13}$$

where $d\psi$ is the relative rotation between the cross-sections in length dx. Again using an expression from strength of materials, we can write the expression for rotation in terms of twist T as

$$\frac{d\psi}{dx} = \frac{T}{GJ}$$

where G is the modulus of shear of the material, and J is the polar moment of inertia of the circular cross-section. The expression for internal work in Eq. (8-13) can therefore be written as

$$dW_I = \left(\frac{T}{2}\right)\left(\frac{T\,dx}{GJ}\right) = \frac{T^2}{2GJ}\,dx$$

The expression for internal work due to twist in a member of length L and circular cross-section is therefore

$$W_I = \int_0^L \frac{T^2}{2GJ}\,dx \tag{8-14}$$

To obtain the expression for internal work due to transverse shear, consider the deformation of a differential slice as shown in Fig. 8-5*d*. The shear V causes a vertical deformation dy at position y on the cross-section. The deformation at position y results from shearing stress τ_{xy} acting at that position. It will be recalled that the shearing stress τ_{xy} can be expressed as

$$\tau_{xy} = \frac{VQ}{It} \tag{8-15}$$

where t is the width of the section at position y as shown in Fig. 8-5*e*, and Q is the moment of the area above y about the centroidal axis.

Considering the internal work on a differential strip as shown in Fig. 8-5*e* and integrating over the cross-sectional area A, the internal work on the differential slice can be written as

$$dW_I = \int^A \tfrac{1}{2}(\tau_{xy}\, dA)\, dy \tag{8-16}$$

The shear deformation at y in Fig. 8-5*d* can be written as

$$\frac{dy}{dx} = \phi = \frac{\tau_{xy}}{G}$$

$$\text{or} \qquad dy = \frac{\tau_{xy}}{G}\, dx$$

where G, as before, is the modulus of shear of the material. Using the expression for shearing stress of Eq. (8-15), Eq. (8-16) can be written as

$$dW_I = \int^A \tfrac{1}{2}\left(\frac{VQ}{It}\right) dA \left(\frac{VQ}{ItG}\right) dx$$

$$\text{or} \qquad dW_I = \left[\frac{V^2}{2GI^2}\int^A \left(\frac{Q}{t}\right)^2 dA\right] dx \tag{8-17}$$

A more usable form of Eq. (8-17) can be obtained by considering a rectangular cross-section as shown in Fig. 8-5*f*. For any value of y, the value of Q is seen to be

$$Q = \int_y^{d/2} \bar{y}\, dA = \int_y^{d/2} \bar{y} b\, d\bar{y}$$
$$= \frac{b}{2}\left(\frac{d^2}{4} - y^2\right)$$

Therefore

$$\int^A \left(\frac{Q}{t}\right)^2 dA = \int_{-d/2}^{d/2} \frac{b}{4}\left(\frac{d^2}{4} - y^2\right)^2 dy = \frac{bd^5}{120}$$

Since $I = bd^3/12$ and $A = bd$, the expression for internal work for the differential slice becomes

$$dW_I = \frac{1.2V^2}{2GA}\, dx$$

In general, we can write the internal work due to shear as

$$dW_I = \frac{KV^2}{2GA}\, dA \tag{8-18}$$

where K is a constant whose value is dependent on the shape of the cross-section.[1] As noted above, the value of K for a rectangle is 1.2. The expres-

[1] R. J. Roark and W. C. Young, "Formulas for Stress and Strain," 5th ed., McGraw-Hill Book Company, New York, 1975.

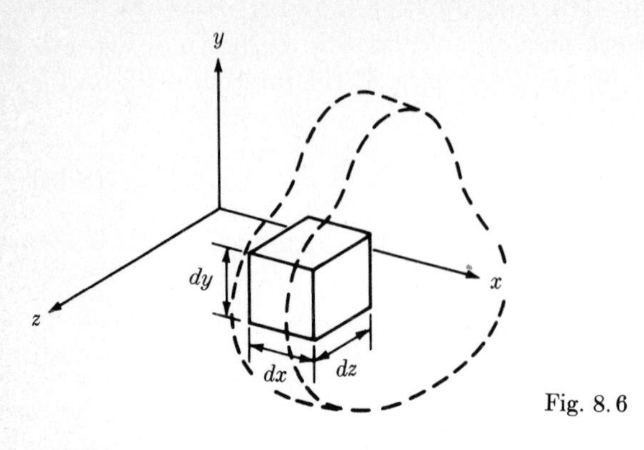

Fig. 8.6

sion for internal work in a member due to shear can be written

$$W_I = K \int_0^L \frac{V^2}{2GA}\,dx \tag{8-19}$$

Expressions for internal work in a member can also be developed by using a generalized matrix form. Consider a differential element of the member as shown in Fig. 8-6. The element is in a differential slice of the member as considered earlier. The element has sides of length dx, dy, and dz. When the member is loaded, the possible stresses on the differential element are as shown in Fig. 8-7. The directions of the stresses shown are considered positive.

Associated with a typical stress such as σ_x there will be strain ϵ_x. For a linearly elastic material the internal work done on the differential element due only to the stress σ_x is

$$\begin{aligned} dW_I &= \frac{(force)(displacement)}{2} \\ &= \frac{(\sigma_x\,dy\,dz)(\epsilon_x\,dx)}{2} \end{aligned} \tag{8-20}$$

Noting that the product of dx, dy, and dz represents the differential volume of the element dV, we can write Eq. (8-20) as

$$dW_I = \frac{\sigma_x \epsilon_x\,dV}{2}$$

Considering all the possible stresses shown in Fig. 8-7, we have the expression for the internal work done on the differential element,

$$dW_I = \tfrac{1}{2}(\sigma_x\epsilon_x + \sigma_y\epsilon_y + \sigma_z\epsilon_z + \tau_{xy}\gamma_{xy} + \tau_{yz}\gamma_{yz} + \tau_{zx}\gamma_{zx})\,dV \tag{8-21}$$

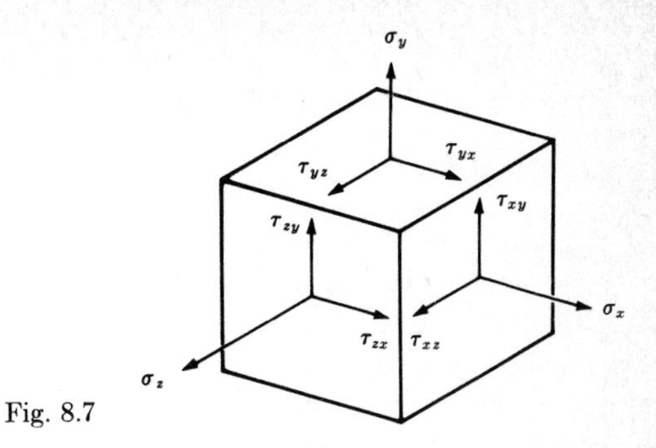

Fig. 8.7

where γ represents shearing strain. The general expressions for strain in terms of stresses for a three-dimensional elastic material are

$$\epsilon_x = \frac{1}{E}[\sigma_x - \nu(\sigma_y + \sigma_z)] \qquad \gamma_{xy} = \frac{\tau_{xy}}{G}$$

$$\epsilon_y = \frac{1}{E}[\sigma_y - \nu(\sigma_z + \sigma_x)] \qquad \gamma_{yz} = \frac{\tau_{yz}}{G}$$

$$\epsilon_z = \frac{1}{E}[\sigma_z - \nu(\sigma_x + \sigma_y)] \qquad \gamma_{zx} = \frac{\tau_{zx}}{G}$$

where ν is Poisson's ratio. By noting that $1/G = 2(1 + \nu)/E$, the expressions for strains in terms of stresses can be written in matrix form as

$$\begin{Bmatrix} \epsilon_x \\ \epsilon_y \\ \epsilon_z \\ \gamma_{xy} \\ \gamma_{yz} \\ \gamma_{zx} \end{Bmatrix} = \frac{1}{E}\begin{bmatrix} 1 & -\nu & -\nu & 0 & 0 & 0 \\ -\nu & 1 & -\nu & 0 & 0 & 0 \\ -\nu & -\nu & 1 & 0 & 0 & 0 \\ 0 & 0 & 0 & 2(1+\nu) & 0 & 0 \\ 0 & 0 & 0 & 0 & 2(1+\nu) & 0 \\ 0 & 0 & 0 & 0 & 0 & 2(1+\nu) \end{bmatrix}\begin{Bmatrix} \sigma_x \\ \sigma_y \\ \sigma_z \\ \tau_{xy} \\ \tau_{yz} \\ \tau_{zx} \end{Bmatrix} \tag{8-22}$$

or $\quad \boldsymbol{\epsilon} = \mathbf{f}\boldsymbol{\sigma}$ (8-23)

where $\boldsymbol{\epsilon}$ = a column matrix of strains
$\boldsymbol{\sigma}$ = a column matrix of stresses
$\mathbf{f}$ = a square symmetric matrix representing the material properties

The expression for the internal work per differential volume of the element as given in Eq. (8-21) can be written in matrix form as

$$\frac{dW_I}{dV} = \tfrac{1}{2}\boldsymbol{\sigma}'\boldsymbol{\epsilon} = \tfrac{1}{2}\boldsymbol{\epsilon}'\boldsymbol{\sigma} \tag{8-24}$$

From Eq. (8-23) and the fact that $\boldsymbol{\epsilon}' = \boldsymbol{\sigma}'\mathbf{f}'$ and $\mathbf{f}' = \mathbf{f}$ owing to symmetry, Eq. (8-24) becomes

$$\frac{dW_I}{dV} = \tfrac{1}{2}\boldsymbol{\sigma}'\mathbf{f}\boldsymbol{\sigma} \tag{8-25}$$

Equation (8-25) expresses the internal work per unit volume in terms of the stresses acting on the element and in terms of the material properties as reflected in the **f** matrix. It is desirable to have the internal work in terms of the forces and moments acting on a section of a member. To develop an expression for internal work in terms of these forces and moments, let us consider again the forces acting on a differential slice of a member, as previously shown in Fig. 8-4. The forces on the slice are related to the stresses on the differential element of the slice by the following expressions:

$$\sigma_x = \frac{P}{A} \qquad \sigma_x = -\frac{My}{I} \qquad \tau_{xy} = -\frac{VQ}{It}$$

As indicated previously, it is not possible to develop a general expression for shearing stress due to twist. For purposes of illustration, however, let us consider the twisting effects for a member with circular cross-section. For a circular cross-section

$$\tau_{xy} = -\frac{Tz}{J} \qquad \tau_{zx} = \frac{Ty}{J}$$

The minus signs on some terms are required for stresses consistent with the positive directions in Fig. 8-7. These expressions for stresses in terms of forces can be written in matrix form as

$$\begin{Bmatrix} \sigma_x \\ \sigma_y \\ \sigma_z \\ \tau_{xy} \\ \tau_{yz} \\ \tau_{zx} \end{Bmatrix} = \begin{bmatrix} 1/A & -y/I & 0 & 0 \\ 0 & 0 & 0 & 0 \\ 0 & 0 & 0 & 0 \\ 0 & 0 & -Q/It & -z/J \\ 0 & 0 & 0 & 0 \\ 0 & 0 & 0 & y/J \end{bmatrix} \begin{Bmatrix} P \\ M \\ V \\ T \end{Bmatrix} \tag{8-26}$$

or $$\boldsymbol{\sigma} = \mathbf{Bq} \tag{8-27}$$

where **q** is a column matrix of the forces acting on the differential slice, and **B** is a rectangular matrix that transforms these forces into stresses on a differential element of the slice. As stated earlier, the terms $-z/J$ and y/J in Eq. (8-26) are valid only for circular cross-sections.

After substitution of Eq. (8-27) into Eq. (8-25), the expression for internal work becomes

$$\frac{dW_I}{dV} = \tfrac{1}{2}\mathbf{q}'\mathbf{B}'\mathbf{f}\mathbf{B}\mathbf{q} \tag{8-28}$$

The above values of **f**, **B**, and **q** and the matrix operations of Eq. (8-28) yield the following expression for internal work:

$$\frac{dW_I}{dV} = \frac{1}{2E}\left[\frac{P^2}{A^2} - \frac{2PMy}{AI} + \frac{M^2y^2}{I^2} + \lambda\left(\frac{V^2Q^2}{I^2t^2} + \frac{2VTQz}{IJt} + \frac{T^2z^2}{J^2} + \frac{T^2y^2}{J^2}\right)\right] \tag{8-29}$$

where $\lambda = E/G$. The expression for internal work per differential length dx of the member can be obtained by integrating the right-hand side of Eq. (8-29) over the cross-sectional area A, resulting in the following expression for internal work per differential length dx:

$$\frac{dW_I}{dx} = \frac{P^2}{2EA} + \frac{M^2}{2EI} + \frac{V^2}{2GI^2}\int^A \left(\frac{Q}{t}\right)^2 dA + \frac{T^2}{2GJ} \tag{8-30}$$

In obtaining Eq. (8-30), the following properties were used:

$$\int^A \frac{P^2}{A^2}\,dA = \frac{P^2}{A} \qquad \int^A y\,dA = 0 \qquad \int^A z\,dA = 0$$

$$\int^A \frac{M^2y^2}{I^2}\,dA = \frac{M^2}{I^2}\int^A y^2\,dA = \frac{M^2}{I}$$

$$\int^A (z^2 + y^2)\,dA = \int^A \rho^2\,dA = J$$

These expressions should be familiar from strength-of-materials courses. In particular, the expression in which the integral of $y\,dA$ is zero requires that the neutral axis be coincident with the centroidal axis for pure bending. The expression in which the integral of $z\,dA$ is zero requires that the cross-section have an axis of symmetry about the y axis. These two conditions have been assumed in these developments.

The internal work done on a member is found by integrating the individual terms of Eq. (8-30) over the length of the member. We obtain the same expressions for internal work as obtained earlier in this section. The internal work (strain energy) due to the various force effects can be summarized as follows:

Axial force: $$W_I = \int_0^L \frac{P^2}{2EA}\,dx \tag{8-31}$$

Bending: $$W_I = \int_0^L \frac{M^2}{2EI}\,dx \tag{8-32}$$

Shear: $$W_I = K\int_0^L \frac{V^2}{2GA}\,dx \tag{8-33}$$

Twist: $$W_I = \int_0^L \frac{T^2}{2GJ}\,dx \qquad \text{(circular cross-section)} \tag{8-34}$$

Again, it should be emphasized that the expression for twist pertains only to circular members.

8-4 DEFLECTIONS BY REAL WORK

To evaluate deflections by real work let us return to the expression for equality of external and internal work given in Eq. (8-6). For a single load Q acting on a structure Eq. (8-4) shows that the external work done is equal to one-half the product of Q and the load displacement D. Forms of internal work were developed and summarized in Eqs. (8-31) to (8-34), where the internal work was expressed as a function of the member forces. The types of force effects to be considered in evaluating the internal work of a particular structure depend on the type of structure under consideration. For example, in evaluating the deflections of a two-dimensional framed structure it is usually necessary to consider only the bending moment effects, since the effects of axial force and shear are so small in comparison with the bending moment effects that they can be neglected. To evaluate any of the forms of internal work we must first determine the forces in the members due to the given loading. For statically determinate structures, to which we are limiting our discussions in this chapter, the expressions for member forces are obtained from statics. The result of equating the internal work to the external work for a given value of Q is an equation containing the unknown value of deflection D. The procedure of evaluating deflections by real work is illustrated in the following examples.

EXAMPLE 8-1 The rotation at B due to an applied moment M_B is to be determined for the beam in Fig. 8-8. EI for the beam is constant. The predominant source of rotation at B is the strain due to bending moment in the beam. The expression for internal work, as given in Eq. (8-32), is therefore

$$W_I = \int_0^L \frac{M^2}{2EI}\,dx$$

Equating external work to internal work, we obtain

$$\frac{M_B\theta_B}{2} = \int_0^L \frac{M^2}{2EI}\,dx$$

where θ_B is the rotation at B. Positive rotation at B is in the same direction as the applied moment.

From statics, the expression for moment in the beam is

$$M = \frac{x}{L}M_B$$

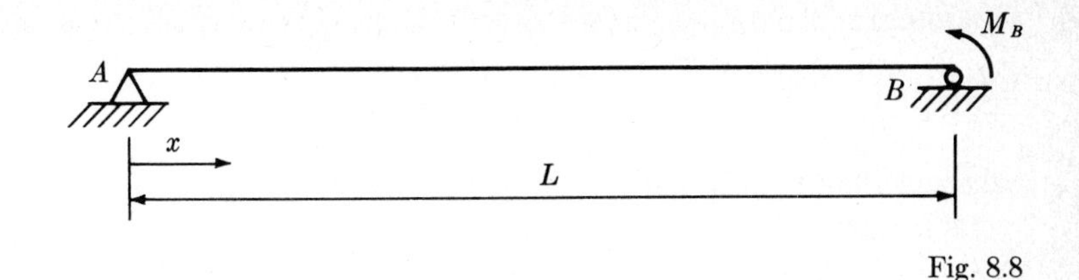

Fig. 8.8

Therefore

$$\frac{M_B\theta_B}{2} = \frac{M_B{}^2}{2EIL^2}\int_0^L x^2\,dx$$
$$= \frac{M_B{}^2}{2EIL^2}\frac{L^3}{3}$$

The value of rotation is thus found to be

$$\theta_B = \frac{M_B L}{3EI}$$

The student should verify that the units of the deflection are in radians.

EXAMPLE 8-2 The deflection at load Q is to be determined for the beam shown in Fig. 8-9. The beam has a rectangular cross-section. EI is constant. As an illustration of the relative importance of shearing strains, in evaluating the deflection we shall consider shearing effects as well as the predominant strains due to bending. From statics, we have

$$M = \frac{Qx}{2} \qquad 0 \leq x \leq \frac{L}{2}$$
$$V = \frac{Q}{2} \qquad 0 \leq x \leq \frac{L}{2}$$

Because of symmetry, the internal work can be expressed as twice that of the left half of the beam. The forms of internal work to be considered are given in Eqs. (8-32) and (8-33). Equating external and internal work,

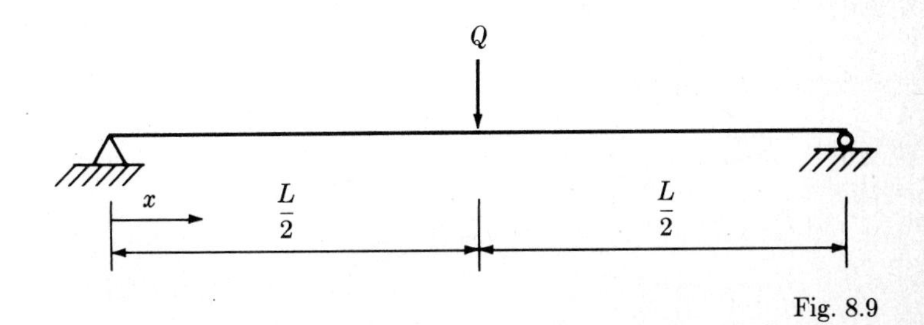

Fig. 8.9

we obtain

$$\frac{Q\Delta}{2} = 2\int_0^{L/2} \frac{M^2}{2EI}\,dx + 2\int_0^{L/2} \frac{1.2V^2}{2GA}\,dx$$

where Δ denotes the deflection of the load Q. Substituting these expressions for M and V and integrating, we find that

$$\Delta = \frac{1}{QEI}\frac{Q^2L^3}{48} + \frac{1.2}{QGA}\frac{Q^2L}{4}$$
$$= \frac{QL^3}{48EI} + \frac{1.2QL}{4GA}$$

which can be written as

$$\Delta = \frac{QL^3}{48EI}\left(1 + \frac{1.2Ed^2}{GL^2}\right)$$

We see from this result that for relatively long beams the second term is quite small. Because of this, the effects of shearing strains are generally neglected in evaluating beam deflections, except for short, deep beams.

The method of real work can be applied to more involved structures, such as that shown in Fig. 8-10. For such structures with more than one member the internal work for each member is evaluated, and the results are summed to give the internal work for the entire structure. For each member we must determine which types of strains contribute appreciably to the total deflection at the point under consideration. In the structure of Fig. 8-10, for example, the deflection at the load Q would be primarily a function of bending and twisting strains in the members.

It should be apparent by now that evaluation of deflections by real work is quite limited. The general procedure shows that only the deflection associated with an applied load can be determined. If more than one load is applied to the structure, more than one unknown value of deflec-

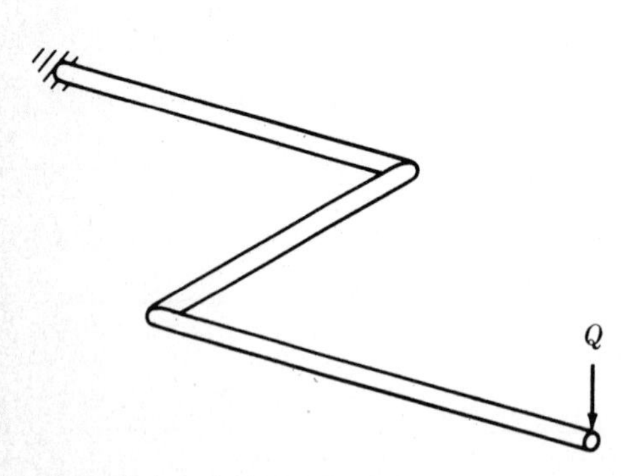

Fig. 8.10

tion will appear in the expression for external work, and the resulting equation cannot be solved. Owing to these limitations, the method of real work is not widely used for deflection analyses. However, it can be used for limited cases of loading and also serves as a means of developing additional principles of deflection.

8-5 VIRTUAL WORK

The concept of virtual work is generally applied in statics courses to the equilibrium of particles and rigid bodies. Detailed discussions of the virtual-work principles can be found in texts devoted primarily to energy concepts.[1] *Virtual work* is the work done by a real force acting through a *virtual displacement* or a *virtual force* acting through a real displacement. The term "virtual" indicates that the quantity is not real, but imaginary. Thus the resulting virtual work is also an imaginary quantity. Virtual quantities are indicated by the symbol δ followed by the usual notation. Thus a virtual displacement is denoted as δD, a virtual force as δQ, and virtual work as δW.

From statics courses we know that *if a rigid body is in equilibrium under a set of real forces, then the virtual work done by the set of forces during a virtual displacement is equal to zero.* The forces can be linear forces or moments, and the virtual displacements can be either linear or rotational. The following example illustrates the use of this principle of virtual work.

EXAMPLE 8-3 The value of the reaction at B is to be determined for the beam of Fig. 8-11 by the principle of virtual work. To determine the value of the reaction R_B we subject the beam to a virtual rigid-body displacement δD at B, as shown in Fig. 8-12. The imaginary displaced position of the beam is shown by the dashed line. Applying the principle of virtual work and noting that the reaction does positive work and the load Q does

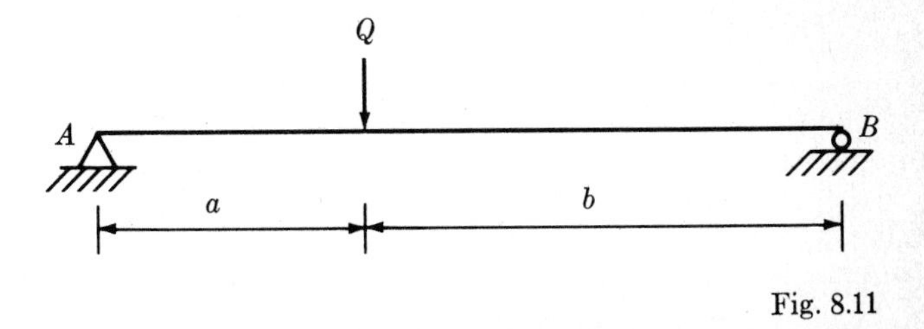

Fig. 8.11

[1] T. M. Charlton, "Energy Principles in Applied Statics," Blackie & Son, Ltd., Glasgow, 1959, and H. L. Langhaar, "Energy Methods in Applied Mechanics," John Wiley & Sons, Inc., New York, 1962.

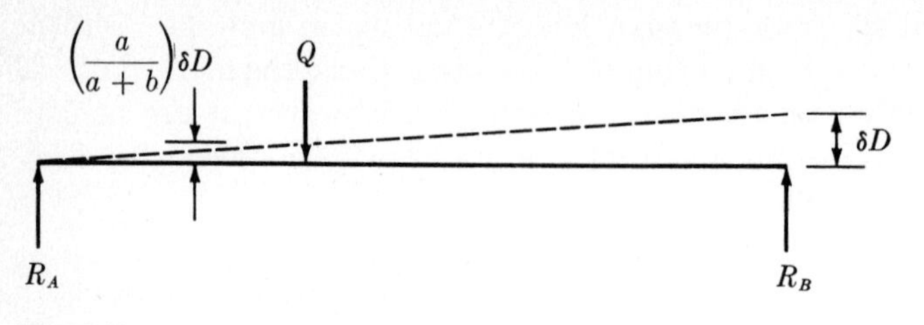

Fig. 8.12

negative work, we obtain

$$-Q\frac{a}{a+b}\,\delta D + R_B\delta D = 0$$

from which we have

$$R_B = \frac{a}{a+b}\,Q$$

The value of the reaction at A could similarly be determined by giving the beam a virtual displacement at A.

The principle of virtual work can be extended to study elastic systems:

PRINCIPLE OF VIRTUAL WORK: *If a structure is in equilibrium under a set of forces and if the structure is given a virtual displacement consistent with the constraints of the structure, then the external virtual work done is equal to the internal virtual work done.*

In equation form this principle states that

$$\delta W_E = \delta W_I \tag{8-35}$$

It is important to realize that the value of the virtual displacement in this statement of virtual work is arbitrary. It is also important to note that the magnitudes and directions of the forces must remain constant during the virtual displacement.

To clarify the concept of virtual work with regard to elastic systems and the distinction between virtual work and real work, let us consider the three-dimensional diagram of Fig. 8-13. Real forces and real displacements are shown on two of the axes. The third axis represents virtual displacement and is thus an imaginary dimension. The real-force–real-displacement relation for a linearly elastic material is represented as the straight line. For a given value of force Q there corresponds a certain value of displacement D, denoted as point (D,Q) on the line. Such a point represents a position of equilibrium for the elastic system. If the force Q is subjected to a virtual displacement δD, the resulting virtual work can be visualized in the plane defined by the real force and the virtual

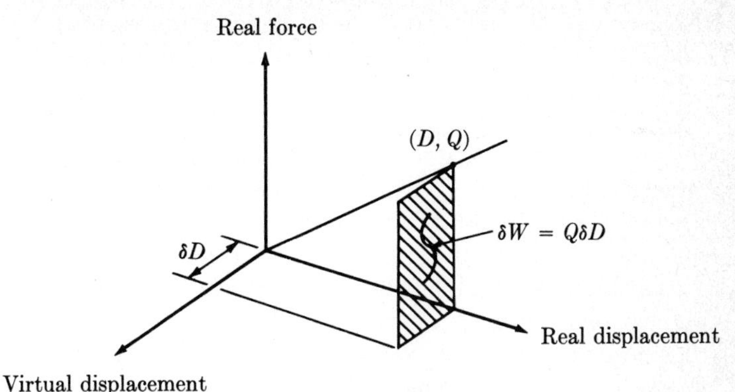

Fig. 8.13

displacement. The amount of virtual work δW is represented by the hatched area in Fig. 8-13. The virtual work done is given by the expression

$$\delta W = Q\delta D \tag{8-36}$$

Remember that the value of δD is arbitrary. As we shall see later, a value of 1 is often used for convenience. It should again be emphasized that the magnitude and direction of the real force are held constant during the virtual displacement δD. Note also that the expression for virtual work in Eq. (8-36) does not include a factor of $\frac{1}{2}$, as does the expression for real work.

Another concept of virtual work is illustrated by the diagram of Fig. 8-14. Figure 8-14 differs from Fig. 8-13 in that the third dimension is a virtual force rather than a virtual displacement. If a structure is first subjected to a virtual force δQ of some arbitrary value, which is imaginary, as was the virtual displacement δD, and is then subjected to a real

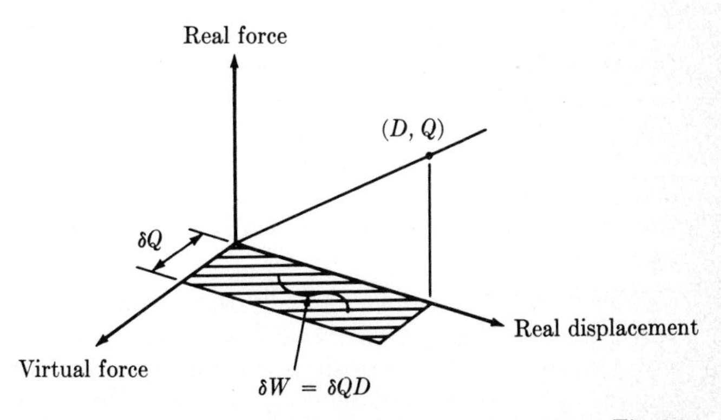

Fig. 8.14

displacement D, the virtual work done is represented by the hatched area of Fig. 8-14. The expression for the virtual work done is

$$\delta W = \delta Q D \tag{8-37}$$

The value of the virtual force is held constant during the displacement D. As we shall see later, a value of 1 is also often used for δQ.

Although the real-force–real-displacement relation of Figs. 8-13 and 8-14 was considered to be linear, these concepts also apply to nonlinear relationships. We shall, however, restrict our consideration of virtual work to linear relationships.

Equations (8-36) and (8-37) represent two forms of virtual work that can be used in developing methods for obtaining deflections by virtual work. In a structure subjected to n simultaneous loads the external virtual work δW_E done by these loads during a set of n virtual displacements is

$$\delta W_E = \sum_{i=1}^{n} Q_i \delta D_i \tag{8-38}$$

If the loads are represented by the column matrix $\mathbf{Q}$ and the virtual displacements by the column matrix $\delta\mathbf{D}$, then the expression for external virtual work can be written as

$$\delta W_E = \mathbf{Q}'\delta\mathbf{D} = \delta\mathbf{D}'\mathbf{Q} \tag{8-39}$$

Similarly, if a set of n virtual forces is considered to act on a structure, and the structure is subjected to a set of real displacements, the expression for external virtual work is

$$\delta W_E = \sum_{i=1}^{n} \delta Q_i D_i \tag{8-40}$$

which can be written in matrix form as

$$\delta W_E = \delta\mathbf{Q}'\mathbf{D} = \mathbf{D}'\delta\mathbf{Q} \tag{8-41}$$

$\delta\mathbf{Q}$ and $\mathbf{D}$ represent column matrices of the virtual forces and real displacements, respectively.

8-6 FORMS OF INTERNAL VIRTUAL WORK

In order to determine deflections by the method of virtual work we need expressions for the internal virtual work done on a structure. Once we have developed these expressions, we can use them with Eq. (8-35),

$$\delta W_E = \delta W_I$$

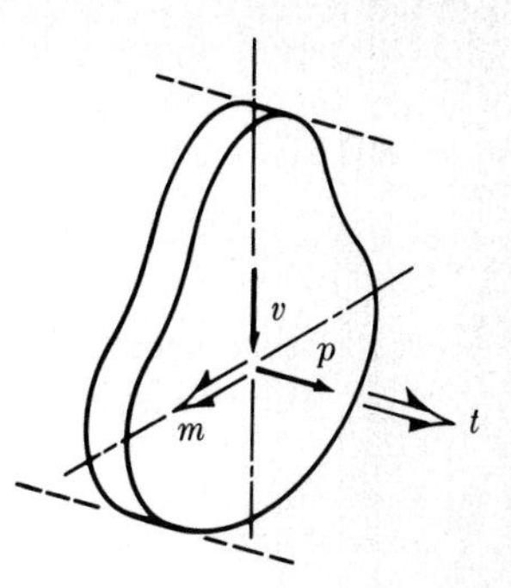

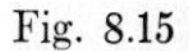
Fig. 8.15

to obtain expressions for deflections. Let us consider the four types of *virtual forces* acting on a differential slice of a member as shown in Fig. 8-15. These are the same types of forces considered earlier in Fig. 8-4 for real-work developments. The virtual forces are denoted by lowercase letters corresponding to the capital letters used for the real forces.

To develop expressions for internal virtual work, let us consider as we did for real work the effect of each force separately, and consider views of each force acting on the differential slice as shown in Fig. 8-16. If a

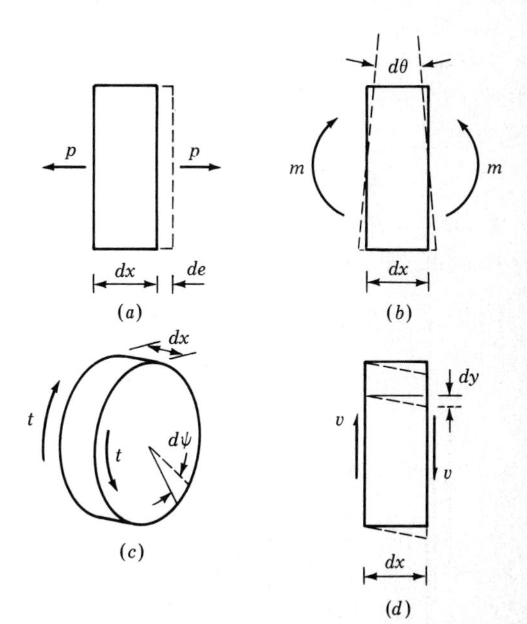

Fig. 8.16

virtual axial force p is applied to the cross-section of the member as shown in Fig. 8-16a, and the slice is subjected to a real displacement de, the virtual work done on the differential slice is

$$\delta W_I = p\, de \tag{8-42}$$

The real displacement de can be expressed, as before, in terms of a real force P, the cross-sectional area A, and the modulus of elasticity E:

$$de = \frac{P\, dx}{EA}$$

The expression for internal virtual work on the differential slice of Eq. (8-42) can therefore be written as

$$\delta W_I = \frac{pP}{EA}\, dx \tag{8-43}$$

The internal virtual work for the entire member is obtained by integrating Eq. (8-43) over the length of the member:

$$\delta W_I = \int_0^L \frac{pP}{EA}\, dx \tag{8-44}$$

The expressions for internal virtual work due to bending, twist, and shear can be developed in a similar manner. A virtual moment m and a real displacement $d\theta$ are shown acting on the differential slice in Fig. 8-16b. The real displacement $d\theta$ can be written in terms of a real moment M as we did in the real-work developments:

$$d\theta = \frac{M\, dx}{EI}$$

The expression for internal virtual work on the differential slice due to bending thereby becomes

$$\delta W_I = m\, d\theta = \frac{mM}{EI}\, dx$$

For the entire member we obtain

$$\delta W = \int_0^L \frac{mM}{EI}\, dx \tag{8-45}$$

Using the virtual force quantities and the real displacements shown in Figs. 8-16c and d for shear and twist, the expression for internal virtual work due to twist is found to be

$$\delta W = \int_0^L \frac{tT}{GJ}\, dx \tag{8-46}$$

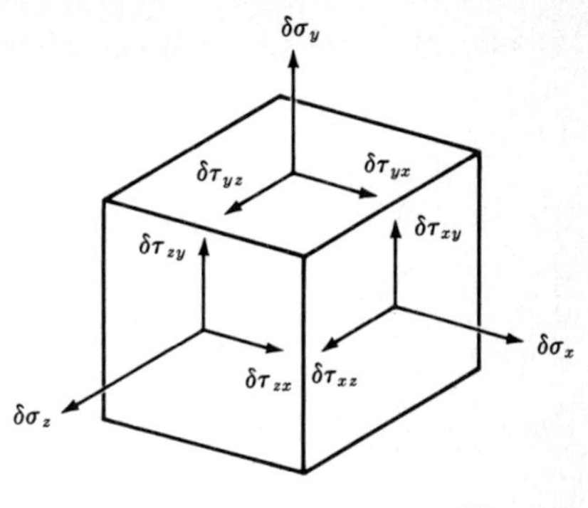

Fig. 8.17

and for shear

$$\delta W = K \int_0^L \frac{vV}{GA}\, dx \tag{8-47}$$

As in the real-work developments, the expression for twist of Eq. (8-46) is applicable only to circular members. The constant K in Eq. (8-47) is the same as that introduced in Eq. (8-18). It should be noted that there is no factor of ½ in the resulting expressions for internal virtual work, as there is in the corresponding real-work expressions.

The expressions for internal virtual work can also be obtained by using a general matrix approach. If the member is subjected to virtual forces, the possible *virtual stresses* acting on a differential element of a slice are shown in Fig. 8-17. If the element is subjected to real displacements in the form of strains, the virtual work done by a virtual stress such as $\delta\sigma_x$ due to strain ϵ_x is

$$\begin{aligned}\delta W_I &= (\delta\sigma_x\, dy\, dz)(\epsilon_x\, dx)\\ &= \delta\sigma_x\epsilon_x\, dV\end{aligned}$$

For all possible types of strains on the element the expression for virtual work is

$$\delta W_I = (\delta\sigma_x\epsilon_x + \delta\sigma_y\epsilon_y + \delta\sigma_z\epsilon_z + \delta\tau_{xy}\gamma_{xy} + \delta\tau_{yz}\gamma_{yz} + \delta\tau_{zx}\gamma_{zx})\, dV$$

This can be written in matrix form as

$$\delta W_I = \boldsymbol{\delta\sigma}'\boldsymbol{\epsilon}\, dV \tag{8-48}$$

where $\boldsymbol{\delta\sigma}$ and $\boldsymbol{\epsilon}$ are column matrices of virtual stresses and real strains, respectively. Recalling that $\boldsymbol{\epsilon}$ can be expressed in terms of $\boldsymbol{\sigma}$ by the $\mathbf{f}$ matrix of Eq. (8-22),

$$\boldsymbol{\epsilon} = \mathbf{f}\boldsymbol{\sigma}$$

we can write the expression for virtual work of Eq. (8-48) as

$$\delta W_I = \delta\boldsymbol{\sigma}'\mathbf{f}\boldsymbol{\sigma}\, dV \tag{8-49}$$

It is desirable to express the internal virtual work in terms of virtual forces acting on a section of the member. The virtual forces considered to be acting on a slice of the member are those shown in Fig. 8-15. The virtual stresses on an element of the slice can be expressed in terms of the virtual forces by the same transformation matrix used in Eq. (8-26),

$$\begin{bmatrix} \delta\sigma_x \\ \delta\sigma_y \\ \delta\sigma_z \\ \delta\tau_{xy} \\ \delta\tau_{yz} \\ \delta\tau_{zx} \end{bmatrix} = \begin{bmatrix} 1/A & -y/I & 0 & 0 \\ 0 & 0 & 0 & 0 \\ 0 & 0 & 0 & 0 \\ 0 & 0 & -Q/It & -z/J \\ 0 & 0 & 0 & 0 \\ 0 & 0 & 0 & y/J \end{bmatrix} \begin{Bmatrix} p \\ m \\ v \\ t \end{Bmatrix} \tag{8-50}$$

or $$\delta\boldsymbol{\sigma} = \mathbf{B}\delta\mathbf{q} \tag{8-51}$$

where $\delta\mathbf{q}$ is a column matrix of the virtual forces acting on the differential slice. The expression for internal virtual work of Eq. (8-49) thus becomes

$$\frac{\delta W_I}{dV} = \delta\mathbf{q}'\mathbf{B}'\mathbf{f}\mathbf{B}\mathbf{q} \tag{8-52}$$

Performing the matrix operations of Eq. (8-52) and integrating the resulting terms over the area of the cross-section yields the expression for virtual work per differential length,

$$\frac{\delta W_I}{dx} = \frac{pP}{EA} + \frac{mM}{EI} + \frac{vV}{GI^2}\int^A \left(\frac{Q}{t}\right)^2 dA + \frac{tT}{GJ} \tag{8-53}$$

Properties similar to those used in obtaining the real-work expressions of Eq. (8-30) are used to obtain Eq. (8-53).

By integrating the individual terms of Eq. (8-53) over the length of the member, it is seen that we obtain the same expressions for internal virtual work as obtained earlier in this section. The forms of internal virtual work can be summarized as

Axial force: $$\delta W_I = \int_0^L \frac{pP}{EA}\, dx \tag{8-54}$$

Bending: $$\delta W_I = \int_0^L \frac{mM}{EI}\, dx \tag{8-55}$$

Shear: $$\delta W_I = K\int_0^L \frac{vV}{GA}\, dx \tag{8-56}$$

Twist: $$\delta W_I = \int_0^L \frac{tT}{GJ}\, dx \qquad \text{(circular cross-section)} \tag{8-57}$$

We see from the above expressions that the internal virtual work is a function of both the real force and the virtual force in a member. Terms such as P/EA represent real displacements through which the virtual forces act. As stated earlier, there is no factor of $\frac{1}{2}$, as there is in the corresponding real-work expressions.

8-7 DEFLECTIONS BY VIRTUAL WORK

In evaluating deflections by means of virtual work we need the two general expressions for external virtual work given in Eqs. (8-36) and (8-37),

$$\delta W_E = Q\delta D \qquad \delta W_E = \delta Q D$$

Equation (8-36) requires that a virtual displacement δD be given to a real external load Q; Eq. (8-37) requires that a real displacement D be given to a virtual external load δQ. Equation (8-37) is the expression appropriate for our purposes in this section.

From the external virtual work of Eq. (8-37), the expression for equality of external and internal virtual work is

$$\delta Q D = \delta W_I \tag{8-58}$$

The internal virtual work δW_I is represented in the forms developed in the previous section. For example, the internal virtual work due to bending, given in Eq. (8-55), is

$$\delta W_I = \int_0^L \frac{mM}{EI}\,dx$$

The expression for equality of external and internal virtual work, considering bending effects, is therefore

$$\delta Q D = \int_0^L \frac{mM}{EI}\,dx \tag{8-59}$$

Similar expressions can be written for other forms of internal virtual work developed in the previous section. In cases where more than one internal-virtual-work effect is being considered, additional terms are included on the right-hand side of the expression. As stated earlier, the value of the external virtual force δQ is arbitrary. The expression for internal virtual force, such as m in Eq. (8-59), is dependent on the value of δQ used. For convenience, the value of δQ can be taken as 1. A unit value of δQ is sometimes referred to as a "dummy unit force." With $\delta Q = 1$, Eq. (8-59) becomes

$$1D = \int_0^L \frac{mM}{EI}\,dx \tag{8-60}$$

By using a unit value of external virtual force we obtain the value of external displacement D directly. In this form the internal virtual moment m is the moment in the structure due to a unit external virtual force. The unit external virtual force can be in the form of either a force or a moment, depending on the form of the external displacement D to be determined. For example, if the external displacement to be determined is a rotational quantity, a unit virtual moment is applied to the structure at the point under consideration.

The following examples illustrate the application of the virtual-work method to various deflection problems. As we shall see, this method of determining deflections is a very versatile one.

Let us consider first the use of the virtual-work method for determining deflections of pin-connected trusses. The deflections to be considered are those due to axial strains in the members. Because the force in each member is constant over its length, p and P of Eq. (8-54) are constant. The internal virtual work of a member is therefore

$$\delta W_I = \frac{pPL}{EA} \tag{8-61}$$

For a truss containing n members the expression for equality of external virtual work can be written as

$$\delta QD = \sum_{i=1}^{n} \frac{pPL}{EA} \tag{8-62}$$

The internal virtual work of the truss is thus found by summing the virtual work of all the members.

EXAMPLE 8-4 The vertical deflection of joint ③ due to the given loads is to be determined for the truss of Fig. 8-18. The cross-sectional area of each member is 1200 mm² and $E = 200$ GPa. To determine the vertical deflection at joint ③, denoted as $D_3{}^V$, we apply a unit virtual load $\delta Q_3{}^V = 1$ kN to the truss in a downward direction at ③. The virtual member forces p resulting from the load on this truss are determined from statics. The member forces P due to the given loads are also evaluated from statics. The results of these analyses are tabulated in Table 8-1, along with the lengths of the members and the product of p, P, and L for each member to be used in determining the total internal virtual work.

The vertical displacement of joint ③ is found from Eq. (8-62), with

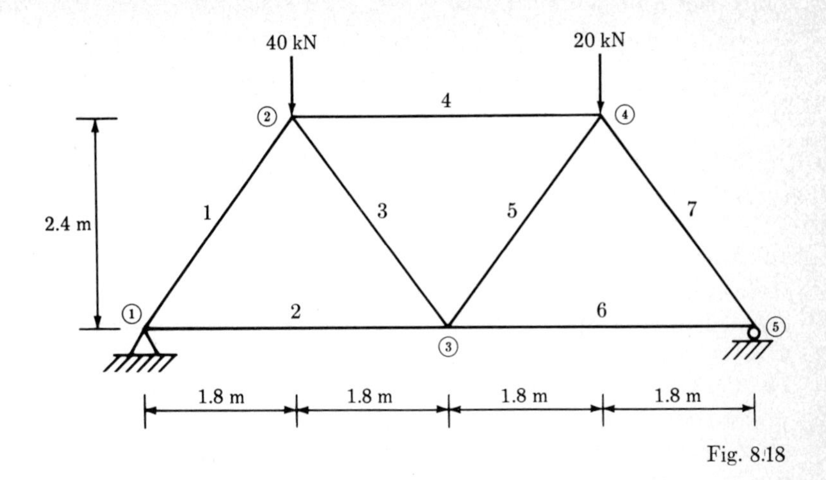

Fig. 8.18

$\delta Q = \delta Q_3{}^V = 1$ and $D = D_3{}^V$, to be

$$1D_3{}^V = \frac{1}{EA}\sum_{i=1}^{7} pPL$$

Using the algebraic sum of the fifth column in Table 8-1 and the given values of A and E, we find the deflection to be

$$(1\text{ kN})D_3{}^V = \frac{(262.12\text{ kN}^2\text{-m})}{(200 \times 10^6\text{ kN/m}^2)(1.2 \times 10^{-3}\text{ m}^2)}$$

$$D_3{}^V = 0.001092\text{ m} = 1.092\text{ mm}$$

The deflection at any other point and in any direction can be found in the same manner by applying a unit virtual force to the truss in the

TABLE 8-1

Member	p, kN	P, kN	L, m	pPL
1	−0.625	−43.75	3.0	+ 82.03
2	+0.375	+26.25	3.6	+ 35.44
3	+0.625	− 6.25	3.0	− 11.72
4	−0.750	−22.50	3.6	+ 60.75
5	+0.625	+ 6.25	3.0	+ 11.72
6	+0.375	+18.75	3.6	+ 25.31
7	−0.625	−31.25	3.0	+ 58.59
				+262.12

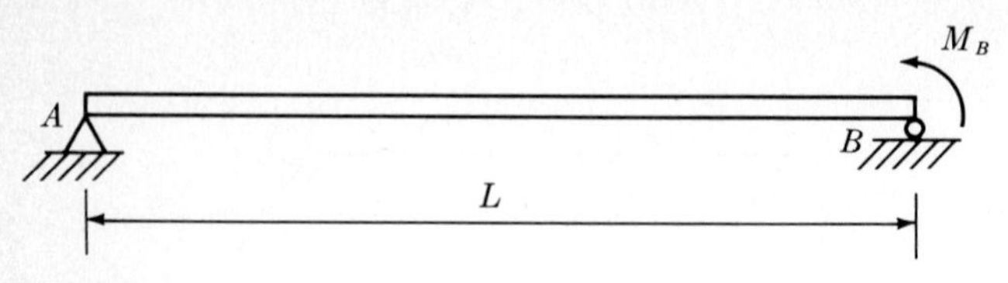

Fig. 8.19

desired direction. A negative result would indicate that the direction of the deflection is opposite to the direction of the unit virtual force.

EXAMPLE 8-5 The rotations at A and B due to an applied moment M_B are to be determined for the beam of Fig. 8-19 by the virtual-work method. Only bending strains are to be considered. EI for the beam is constant. We can use the energy expression given in Eq. (8-59),

$$\delta QD = \int_0^L \frac{mM}{EI}\,dx$$

To obtain the rotation at A we apply a unit virtual moment, $\delta Q_A = 1$ N-m to the beam as shown in Fig. 8-20a. The use of newton-meters as units of moment is convenient if we consider the units of E and I to be newtons per square meter and meters to the fourth power, respectively. Other units of moment can be used, but adjustments may be necessary to make the numerical results dimensionally correct. The expression for the internal virtual moment m corresponding to this load is

$$m = 1 - \frac{x}{L}$$

The expression for the real moment M due to the applied moment M_B

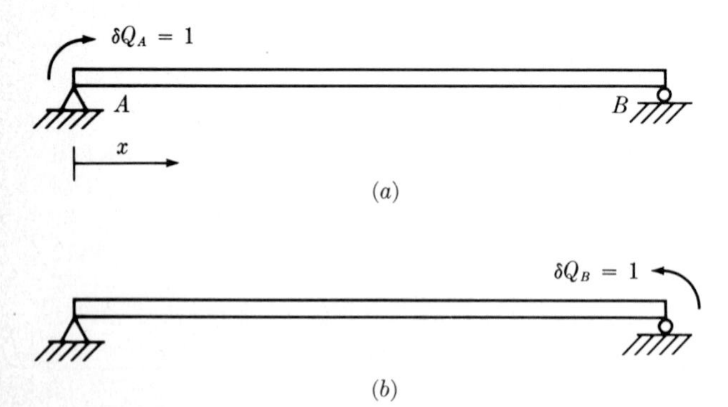

Fig. 8.20

in Fig. 8-19 is

$$M = \frac{x}{L} M_B$$

The value of the desired external displacement $D = \theta_A$ is found from Eq. (8-60) as

$$1\theta_A = \frac{1}{EI} \int_0^L \left(1 - \frac{x}{L}\right) \frac{x}{L} M_B \, dx$$

and evaluation results in

$$\theta_A = \frac{M_B L}{6EI}$$

The value of rotation at B due to M_B is found by applying a unit virtual moment to the beam at B, as shown in Fig. 8-20*b*. The expression for the internal virtual moment is

$$m = \frac{x}{L}$$

According to Eq. (8-60), the displacement $D = \theta_B$ is

$$1\theta_B = \int_0^L \frac{x}{L} \frac{x}{L} M_B \, dx$$

or $\qquad \theta_B = \dfrac{M_B L}{3EI}$

Note that the direction of the external virtual forces applied to the structure is arbitrary. For example, if δQ_A in Fig. 8-20*a* had been applied counterclockwise, the resulting value of θ_A would have had a negative sign. The negative sign would mean that the rotation was in a direction opposite that of the applied virtual moment δQ_A.

It is important to note that in using the virtual-work method, *the internal virtual quantity and the internal real quantity must be expressed according to the same sign convention.* In the truss of Example 8-4, tension in a member was considered to be positive. In Example 8-5, the moment expressions were written according to the beam sign convention for moment. Other conventions are possible, but the real and virtual quantities for each member must be expressed according to the same convention.

The evaluation of integrals when using the method of virtual work can often be simplified by using a technique commonly referred to as *visual integration.* Consider the evaluation of a deflection D using virtual work, and consider for illustrative purposes bending effects. The expression for evaluating the deflection is

$$1D = \int_0^L \frac{mM}{EI} \, dx \tag{8-63}$$

For statically determinate structures with straight members, the virtual moment m is a linear function and we can consider m in the general form

$$m = a + bx$$

where a and b are constants. If we substitute this expression for m into Eq. (8-63), we obtain

$$D = \int_0^L \frac{(a + bx)M}{EI}\,dx$$

$$= a\int_0^L \frac{M}{EI}\,dx + b\int_0^L \frac{Mx}{EI}\,dx \tag{8-64}$$

We note that the first integral quantity of Eq. (8-64) is the area of the M/EI diagram, and the second integral quantity is the moment of the area of the M/EI diagram. If we denote the area of the M/EI diagram by A and the location of its centroid by $\bar{x}$, we can write Eq. (8-64) as

$$D = aA + b\bar{x}A$$

or $$D = A(a + b\bar{x}) \tag{8-65}$$

In words, Eq. (8-65) states that the deflection can be obtained by multiplying the area of the M/EI diagram by the value of m at the centroid of the M/EI diagram. The use of Eq. (8-63) requires that the functions be continuous; therefore, the same is required for the use of Eq. (8-65). For problems that contain discontinuous functions, Eq. (8-65) is applied in steps to the continuous segments. This concept is demonstrated in the following example.

EXAMPLE 8-6 The vertical deflection at B is to be determined for the beam of Fig. 8-21 using virtual work and visual integration. For the beam, E = 200 GPa and $I = 50 \times 10^6$ mm^4.

A unit virtual force of 1 kN is applied upward at point B. The M/EI diagram and the m diagram are shown in Fig. 8-22. Because of the discontinuity of the m diagram at B, the M/EI diagram is divided at B so that each diagram is continuous over the length of each segment. The resulting trapezoidal area of the M/EI diagram between B and C is divided into two triangular areas for convenience in locating the cen-

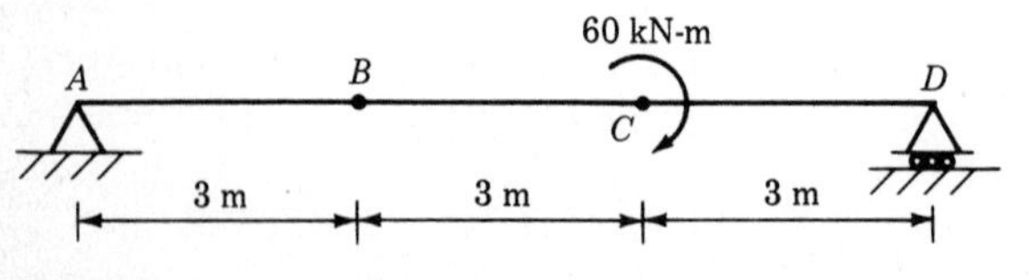

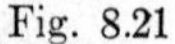
Fig. 8.21

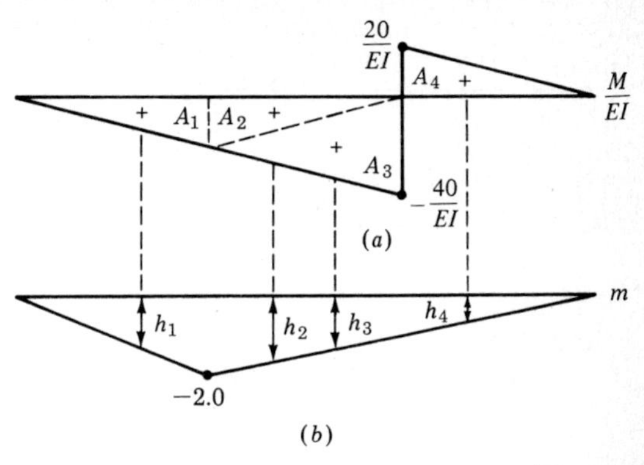

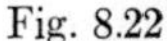
Fig. 8.22

troids. From Fig. 8-22 it is seen that

$$A_1 = (\tfrac{1}{2})(3)\left(-\frac{20}{EI}\right) = -\frac{30}{EI} \qquad h_1 = (\tfrac{2}{3})(-2.0) = -1.33$$

$$A_2 = (\tfrac{1}{2})(3)\left(-\frac{20}{EI}\right) = -\frac{30}{EI} \qquad h_2 = (\tfrac{5}{6})(-2.0) = -1.67$$

$$A_3 = (\tfrac{1}{2})(3)\left(-\frac{40}{EI}\right) = -\frac{60}{EI} \qquad h_3 = (\tfrac{4}{6})(-2.0) = -1.33$$

$$A_4 = (\tfrac{1}{2})(3)\left(\frac{20}{EI}\right) = \frac{30}{EI} \qquad h_4 = (\tfrac{2}{6})(-2.0) = -0.67$$

To obtain the vertical deflection at B, $D_B{}^V$, we evaluate the algebraic sum of the products A_ih_i:

$$D_B{}^V = \frac{1}{EI}(-30)(-1.33) + (-30)(-1.67) + (-60)(-1.33) + (30)(-0.67)$$

$$= \frac{149.7}{EI}$$

For the given values of E and I we obtain

$$(1\text{ kN})D_B{}^V = \frac{(149.7)(\text{kN-m}^2)(\text{kN-m})}{(200 \times 10^6\text{ kN/m}^2)(50 \times 10^{-6}\text{ m}^4)}$$

$$D_B{}^V = 0.0150\text{ m} = 15.0\text{ mm}$$

The visual-integration technique can also be used with the other forms of internal virtual work. Its application to frame problems is demonstrated in the following examples.

EXAMPLE 8-7 The horizontal displacement at the support denoted as ④ in Fig. 8-23*a* is to be determined by the method of virtual work. The

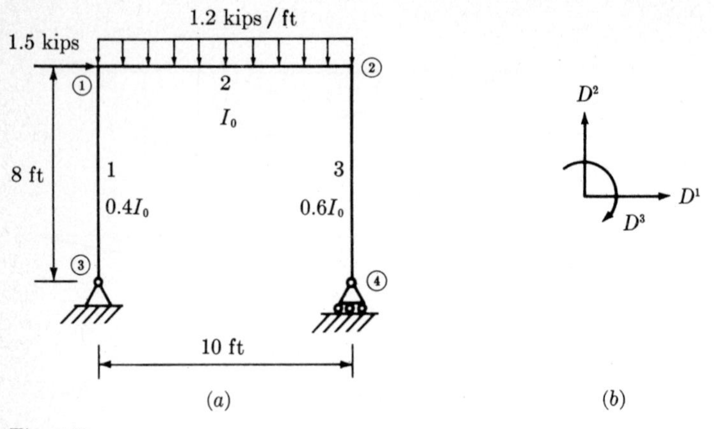

Fig. 8.23

moments of inertia of the members are given in terms of the reference quantity I_0. E for all members is constant. The units of E and I_0 are considered to be kips per square inch and inches to the fourth power, respectively.

For convenience, the deflections of points on such a frame can be described in terms of a framework coordinate system, as shown in Fig. 8-23*b*. Thus the deflection to be evaluated is denoted as $D_4{}^1$, where the superscript denotes the direction of the deflection and the subscript denotes the point under consideration. The deflection analysis is of the same form as in Examples 8-5 and 8-6, where the energy expression is given by Eq. (8-59) as

$$\delta QD = \int_0^L \frac{mM}{EI}\,dx$$

The total internal virtual work for the frame is the sum of the work done in each member.

With a unit virtual force $Q_4{}^1 = 1$ kip applied in the positive D^1 direction at ④, M and m diagrams for each member are obtained as shown in Fig. 8-24. The real and virtual moment diagrams for each member are constructed according to the same sign convention. As an example, for member 2, moment causing compression on the upper side of the member is considered positive. Note that applying the unit virtual force in the same direction as positive displacements in the framework coordinate system in Fig. 8-23*b* will result in a deflection of the proper sign.

The desired deflection is obtained using the technique of visual integration; i.e., the product of the area of each M/EI diagram and the

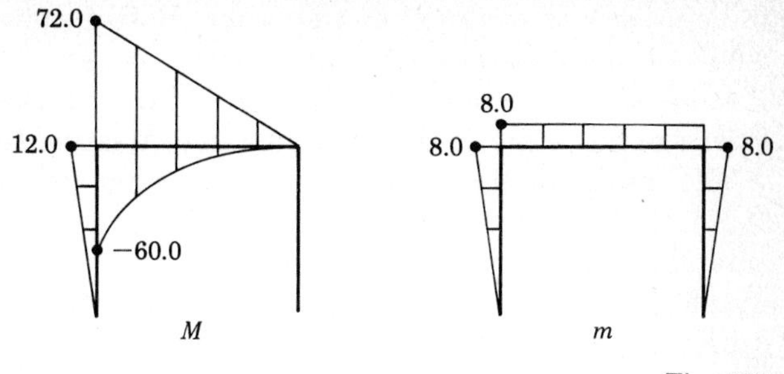

Fig. 8.24

value of m at its centroid:

$$D_4{}^1 = \frac{1728}{E}\left[\left(\frac{1}{0.4I_0}\right)(1/2)(12.0)(8)(2/3)(8) + \left(\frac{1}{I_0}\right)(1/2)(72.0)(10)(8.0) + \left(\frac{1}{I_0}\right)(1/3)(-60.0)(10)(8.00)\right]$$

$$= \frac{3.32 \times 10^6}{EI_0}\ \text{in.}$$

As anticipated, the deflection is in the positive direction of D^1, which is to the right.

The deflections at other points on the frame and in other directions can be determined in a similar manner. For example, if we wish to find the rotation of point ②, we apply a unit virtual moment $\delta Q_2{}^3 = 1$ ft-kip in a clockwise direction at ②. The evaluation of $D_2{}^3$ would follow the same procedure as that for $D_4{}^1$.

The virtual-work method can also be applied to deflection analyses of three-dimensional frames.

EXAMPLE 8-8 The deflection of point ① in the Y direction is to be determined for the frame in Fig. 8-25. The framework coordinate system

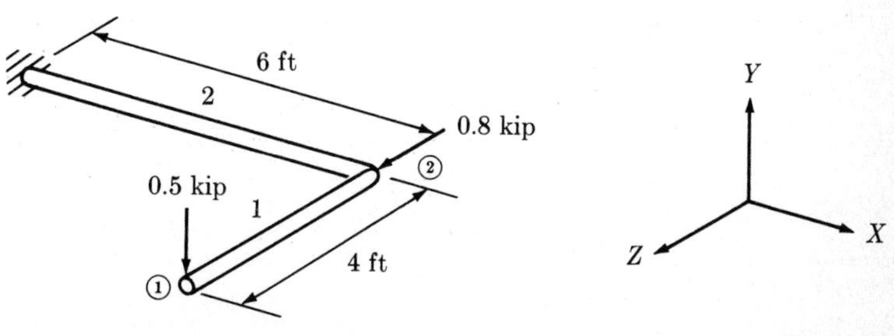

Fig. 8.25

to be used is shown next to the frame. Both members of the frame are made of extra-strong steel pipe with $I = 9.61$ in.^{4}, $J = 19.22$ in.^{4}, $E = 30 \times 10^3$ ksi, and $G = 12 \times 10^3$ ksi. The effects of bending and twisting strains are to be considered in evaluating the deflection.

The moments m and M in the expression for internal virtual work of Eq. (8-55) were developed with the moment considered to be acting only about the z axis. It can be shown that the expression for internal virtual work is the same for moment acting about the y axis. The general expression for equality of external and internal virtual work for the effects under consideration in this example is therefore

$$\delta QD = \int_0^L \frac{m^{yy}M^{yy}}{EI}\,dx + \int_0^L \frac{m^{zz}M^{zz}}{EI}\,dx + \int_0^L \frac{tT}{GJ}\,dx$$

The superscripts denote the axis about which the moment acts.

To evaluate the deflection in the Y direction at point ①, denoted as D_1^Y, we apply a unit force $\delta Q_1^Y = 1$ kip to the structure at point ① in the positive Y direction. The m, M, t, and T diagrams for the structure are as shown in Fig. 8-26. The M and m diagrams have been plotted on the side of the member that is in compression. Thereby, if M and m acting about an axis are on opposite sides of a member, the product of the area of the M/EI diagram and the value of m at its centroid is a negative quantity. Visual integration of twist is considered in the same manner as bending moment. T and t in Figs. 8-26c and e have been plotted with positive twist acting according to the right-hand screw rule referred

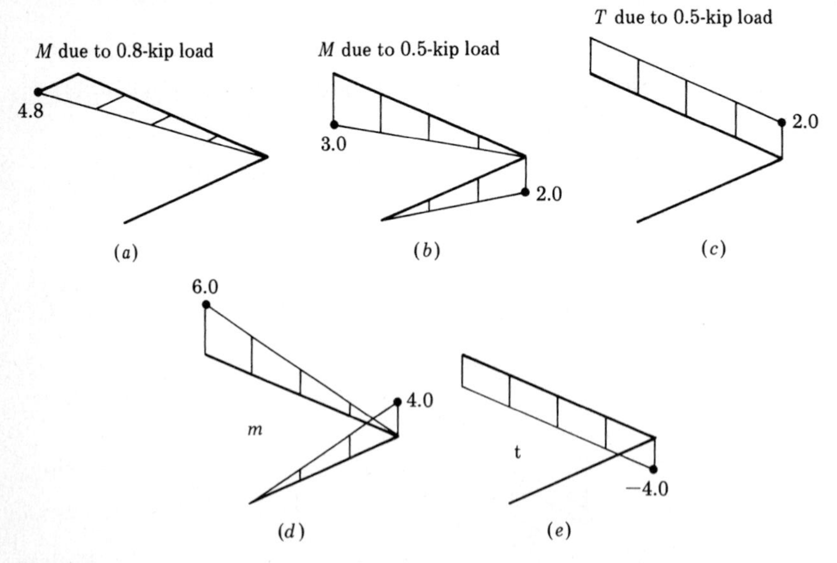

Fig. 8.26

to the X axis. The desired deflection is found from the diagrams of Fig. 8-26 to be

$$D_1{}^Y = \frac{1}{EI}[-(\tfrac{1}{2})(2.0)(4)(\tfrac{2}{3})(4.0) - (\tfrac{1}{2})(3.0)(6)(\tfrac{2}{3})(6.0)] + \frac{1}{GJ}[-(2.0)(6)(4.0)]$$

$$= -\frac{46.67}{EI} - \frac{48}{GJ}$$

For the given values of E, I, J, and G this becomes

$$D_1{}^Y = -\frac{(46.67)(1728)}{(30 \times 10^3)(9.61)} - \frac{(48)(1728)}{(19.22)(12 \times 10^3)}$$

$$= -0.64 \text{ in.}$$

The minus sign indicates that the deflection is in the negative Y direction, or downward.

The method of virtual work also lends itself well to the determination of deflections due to temperature changes. As an illustration of the approach to such problems, consider again the differential element of a member in Fig. 8-17. The element is considered to be acted upon by a single virtual stress $\delta\sigma_x$, and the element is then subjected to a set of real displacements in the form of strain due to temperature change. From strength-of-materials studies, the strain ϵ_x due to a change in temperature ΔT is given by

$$\epsilon_x = \alpha \, \Delta T$$

where α represents the coefficient of thermal expansion. The temperature change ΔT is the change in temperature that occurs at the location of the element. The virtual work done per volume of the element is therefore

$$\frac{\delta W_I}{dV} = \delta\sigma_x \epsilon_x = \delta\sigma_x \alpha \, \Delta T$$

The virtual stress $\delta\sigma_x$ is considered to be a function of the virtual axial force p and the virtual bending moment m, as in Eq. 8-50; thus the internal virtual work per volume of element becomes

$$\frac{\delta W_I}{dV} = \frac{p\alpha \, \Delta T}{A} - \frac{my\alpha \, \Delta T}{I}$$

Integrating over the area of the cross-section, we obtain the expression for the internal virtual work per differential length of the member,

$$\frac{\delta W_I}{dx} = \int^A \frac{p\alpha \, \Delta T}{A} \, dA - \int^A \frac{my\alpha \, \Delta T}{I} \, dA \tag{8-66}$$

If the change in temperature, ΔT, is considered to be constant throughout the member, and if we denote the magnitude of the temperature change by the constant C, then Eq. (8-66) becomes

$$\frac{\delta W_I}{dx} = \frac{p\alpha C}{A}\int^A dA - \frac{m\alpha C}{I}\int^A y\,dA$$

Since

$$\int^A dA = A \qquad \int^A y\,dA = 0$$

the virtual-work expression becomes

$$\frac{\delta W_I}{dx} = p\alpha C$$

For a member of length L the internal virtual work done on the member by a constant temperature change C is

$$\delta W_I = p\alpha CL \tag{8-67}$$

The virtual axial force is considered to be constant.

If the change in temperature varies linearly over the depth of the member, ΔT can be expressed as

$$\Delta T = C + Dy \tag{8-68}$$

where C and D are constants. The internal virtual work per differential length of the beam as given in Eq. (8-66) becomes

$$\frac{\delta W_I}{dx} = \frac{p\alpha}{A}\int^A (C + Dy)\,dA - \frac{m\alpha}{I}\int^A y(C + Dy)\,dA$$

With the zero values of certain integrals as above and the fact that $\int^A y^2 dA = I$, we obtain

$$\frac{\delta W_I}{dx} = p\alpha C - m\alpha D$$

The general expression for internal virtual work for a member of length L subjected to a linear change in temperature over its depth can thus be expressed as

$$\delta W_I = p\alpha CL - \alpha D\int_0^L m\,dx \tag{8-69}$$

In this expression it is assumed, as in Eq. (8-67), that the virtual axial force is constant. These developments can be extended to other variations of temperature change if desired.

The application of the virtual-work method to determining deflections caused by temperature changes is illustrated in the following examples.

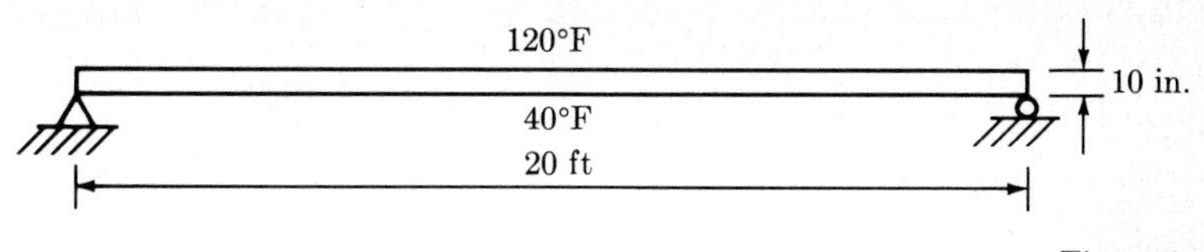

Fig. 8.27

EXAMPLE 8-9 The beam shown in Fig. 8-27 initially has a constant temperature of 70°F throughout. The vertical deflection at midspan resulting from a rise in the temperature of the upper surface to 120°F and a drop in the temperature of the lower surface to 40°F is to be determined. It is assumed that the temperature change between the upper and lower surface varies linearly. The coefficient of thermal expansion for the beam material is 6.5×10^{-6} per degree Fahrenheit.

Using the method of virtual work, we obtain the desired vertical deflection by applying a unit virtual force to the beam at midspan, as shown in Fig. 8-28. The expression for virtual moment in the left portion of the beam is therefore

$$m = -0.5x$$

The virtual work in the total beam can be obtained by using twice that of the left half of the beam. For the given temperature change the expression for ΔT of Eq. (8-68) is

$$\Delta T = 10 + 8y$$

where $C = 10$ and $D = 8$. From the general expression for internal virtual work of Eq. (8-69) and the fact that there is no virtual axial force p, the deflection at midspan is found to be

$$\begin{aligned}\Delta &= -(2)(6.5 \times 10^{-6})(8)(12) \int_0^{L/2} (-0.5x)\, dx \\ &= 0.0312 \text{ ft} \\ &= 0.374 \text{ in.}\end{aligned}$$

As anticipated, the deflection is upward. If desired, the resulting hori-

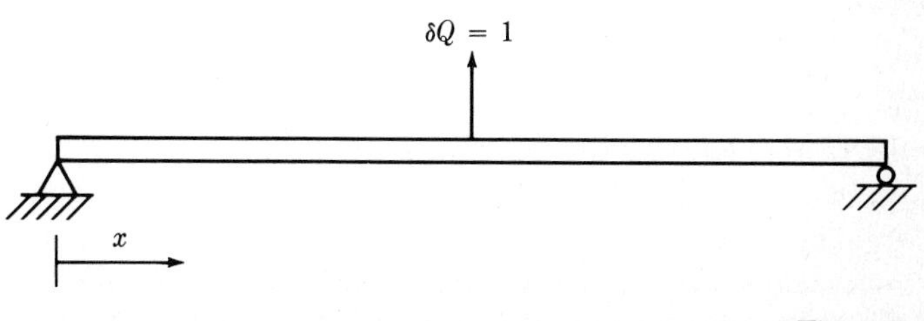

Fig. 8.28

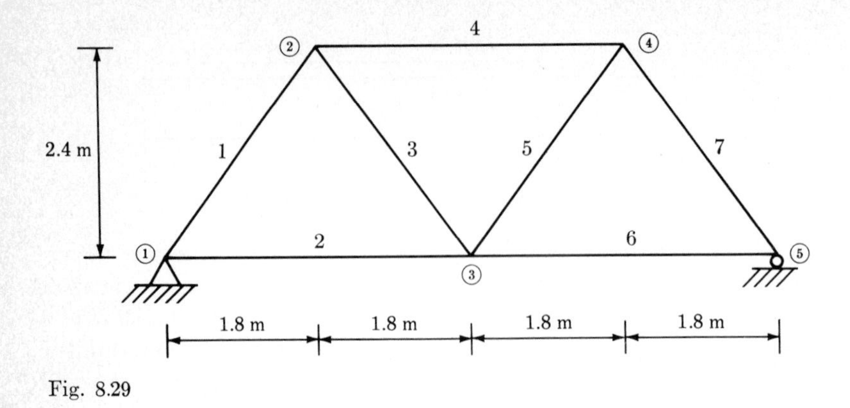

Fig. 8.29

zontal displacement of the right support could similarly be evaluated by applying a unit virtual force to the right end of the beam.

EXAMPLE 8-10 The upper members of the truss in Fig. 8-29, members 1, 4, and 7, are subjected to a temperature increase of 30°C. The resulting vertical deflection of joint ③ is to be determined. The coefficient of thermal expansion for the member materials is 11.7×10^{-6} per degree C.

To obtain the desired deflection we apply a unit virtual force to the truss in a downward direction at ③. The given truss was analyzed in Example 8-4 for the vertical deflection at ③ resulting from given loads, so we can use the values given in Table 8-1 for the virtual axial forces in the members. Using these values of virtual axial forces and Eq. (8-67), with $C = 30$, we find the value of the vertical deflection at ③ to be

$$D_3{}^V = [(-0.625)(3.0) + (-0.750)(3.6) + (-0.625)(3.0)](11.7 \times 10^{-6})(30)$$
$$= -2.26 \times 10^{-3} \text{ m} = -2.26 \text{ mm}$$

The minus sign indicates that the vertical deflection of joint ③ is upward.

8-8 CASTIGLIANO'S THEOREM

With the concepts of real work as developed in Secs. 8-2 and 8-3 it is possible to express the deflection at a point on a structure in terms of the derivative of the internal strain energy. To see this, consider a structure subjected simultaneously to a set of n loads, as in Fig. 8-30, where $n = 3$. Associated with the loading there will be internal work done on the structure, resulting in internal strain energy. Assuming that the reactions

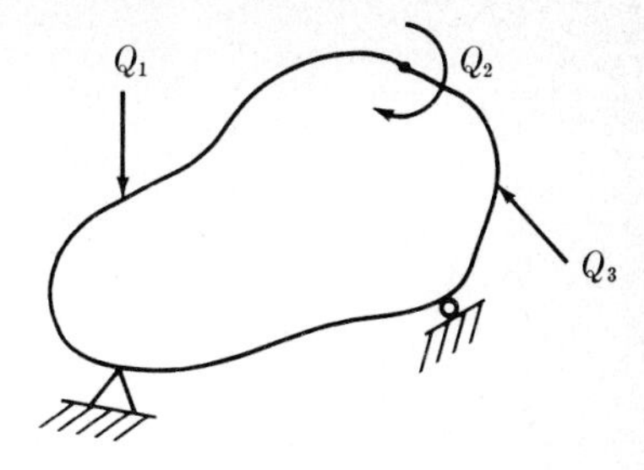

Fig. 8.30

do not displace during loading and therefore do no work, we know that

$$W_I = W_E$$

or, in terms of the external loads and their displacements,

$$W_I = \tfrac{1}{2} \sum_{j=1}^{n} Q_j D_j$$

For a clearer understanding of the developments to follow we can think of the internal work W_I as being a function of the external loads,

$$W_I = f(Q_1, Q_2, \ldots, Q_n)$$

If one of the n loads, let us say the ith load, is now increased by a differential amount, the change in strain energy can be expressed as

$$dW_I = \frac{\partial W_I}{\partial Q_i}\, dQ_i$$

The expression for the total strain energy in the structure is then

$$W_I + dW_I = \tfrac{1}{2} \sum_{j=1}^{n} Q_j D_j + \frac{\partial W_I}{\partial Q_i}\, dQ_i \tag{8-70}$$

Now suppose a differential amount of the ith load is applied to the structure *before* the set of n loads. The expression for strain energy due to this load is

$$dW_I = \tfrac{1}{2}\, dQ_i\, dD_i$$

If the structure is then subjected to the n simultaneous loads, the resulting expression for strain energy is

$$dW_I + W_I = \tfrac{1}{2}\, dQ_i\, dD_i + dQ_i\, D_i + \tfrac{1}{2} \sum_{j=1}^{n} Q_j D_j \tag{8-71}$$

The differential load dQ_i remains constant during the resulting displacements and performs work equal to the product of dQ_i and the displacement D_i.

The resulting strain energy in the structure will be the same for each of the above orders of applying the loads. Therefore, from Eqs. (8-70) and (8-71), we have

$$\tfrac{1}{2} \sum_{j=1}^{n} Q_j D_j + \frac{\partial W_I}{\partial Q_i} dQ_i = \tfrac{1}{2}\, dQ_i\, dD_i + dQ_i\, D_i + \tfrac{1}{2} \sum_{j=1}^{n} Q_j D_j$$

Eliminating like terms and neglecting the higher-order terms containing the product of dQ_i and dD_i, we obtain

$$\frac{\partial W_I}{\partial Q_i} dQ_i = dQ_i\, D_i$$

from which we have

$$D_i = \frac{\partial W_I}{\partial Q_i} \tag{8-72}$$

In other words, we have the following theorem:

CASTIGLIANO'S THEOREM: *If a linearly elastic structure is subjected to a set of loads, the displacement of any load is equal to the partial derivative of the strain energy with respect to that load.*

If the load is a moment, the displacement is the angle in radians through which the moment acts. If the load is a force, the displacement is the linear distance through which the load acts.

This method of evaluating deflections is known as Castigliano's *second* theorem.[1] Castigliano's *first* theorem can be derived in a similar manner; this theorem states that

$$Q_i = \frac{\partial W_I}{\partial D_i} \tag{8-73}$$

Thus the first theorem is an expression for a load in terms of the partial derivative of the strain energy with respect to the corresponding displacement. Application of the first theorem will not be considered here.

To apply Castigliano's second theorem as given in Eq. (8-72) we must express the internal strain energy of the structure in terms of the external loads. The expressions for internal work developed in Sec. 8-3 can be used for this purpose. For example, in evaluating a deflection due to bending

[1] Developed by the Italian engineer Alberto Castigliano (1847–1884), "Theorie de l'equilibre des systèmes elastiques et ses applications," Turin, 1879; translated into English by E. S. Andrews, "Elastic Stresses in Structures," Scott, Greenwood, and Son, London, 1919.

strains, the internal strain energy of bending as given in Eq. (8-32) is

$$W_I = \int_0^L \frac{M^2}{2EI}\,dx$$

The moment M must be expressed as a function of the external loads Q. From Eq. (8-72), the expression for the deflection can be written as

$$D_i = \int_0^L M \frac{\partial M}{\partial Q_i} \frac{dx}{EI} \tag{8-74}$$

Equation (8-74), where the partial derivative is taken before the integration is performed, is a simpler form to work with than Eq. (8-32).

Similar expressions can be written for the other types of strain. For example, in considering axial strain effects the expression for the deflection at a point is, from Eq. (8-31),

$$D_i = \int_0^L P \frac{\partial P}{\partial Q_i} \frac{dx}{EA} \tag{8-75}$$

The use of Eq. (8-75) for deflection calculations requires that the axial forces in the members be expressed in terms of the external loading.

The application of Castigliano's theorem to deflection calculations is illustrated in the following examples.

EXAMPLE 8-11 The deflection at the point of loading is to be determined for the beam of Fig. 8-31. EI is constant. The desired deflection, denoted as Δ, can be obtained from Eq. (8-74) as

$$\Delta = \int_0^L M \frac{\partial M}{\partial Q} \frac{dx}{EI}$$

From statics, the expression for moment is

$$M = \frac{Qx}{2} \qquad 0 \le x \le \frac{L}{2}$$

Therefore

$$\frac{\partial M}{\partial Q} = \frac{x}{2}$$

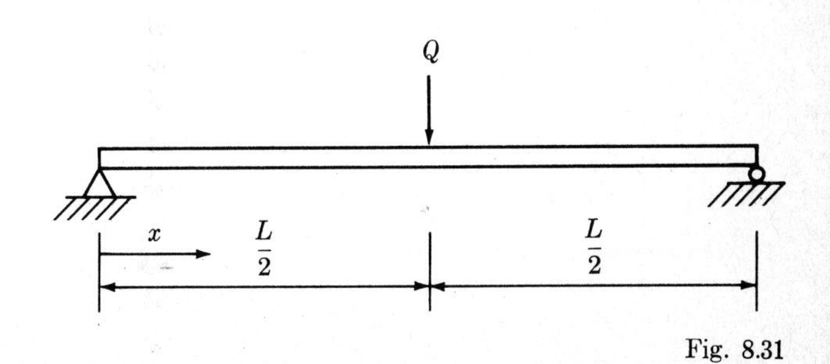

Fig. 8.31

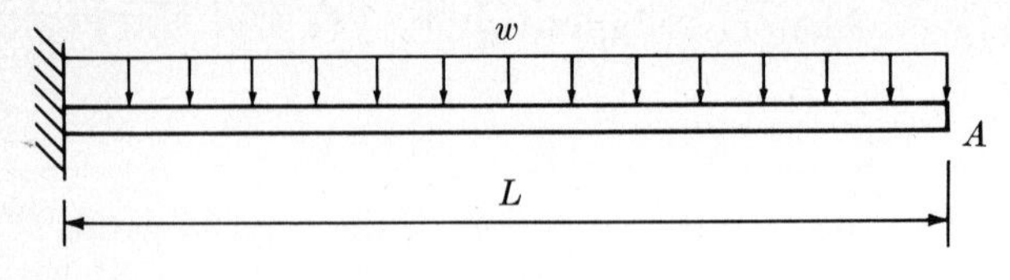

Fig. 8.32

Because of symmetry, the deflection can be obtained by using twice the value of the integral for the left half of the beam,

$$\Delta = \frac{2}{EI}\int_0^{L/2} \frac{Qx}{2}\frac{x}{2}\,dx$$

$$= \frac{QL^3}{48EI}$$

The positive sign indicates that the deflection is in the direction of the applied load.

With the form used in the above example it is possible to determine the deflection only at the point of given loading. However, there is a technique for obtaining deflections at points other than the given loading points. A fictitious force or moment is applied to the structure at the point and in the direction of the desired deflection. This fictitious quantity is set equal to zero *after* the integration has been performed, thus giving the desired deflection. The following example illustrates the procedure.

EXAMPLE 8-12 A cantilever beam is loaded with a uniform load w as shown in Fig. 8-32. The values of the vertical deflection and rotation at A are to be determined from Castigliano's theorem. The vertical deflection at A due to the uniform load, denoted as Δ_A, is obtained by applying a fictitious force Q to the beam as shown in Fig. 8-33. The expression for moment in the beam is therefore

$$M = -\frac{wx^2}{2} - Qx$$

and $$\frac{\partial M}{\partial Q} = -x$$

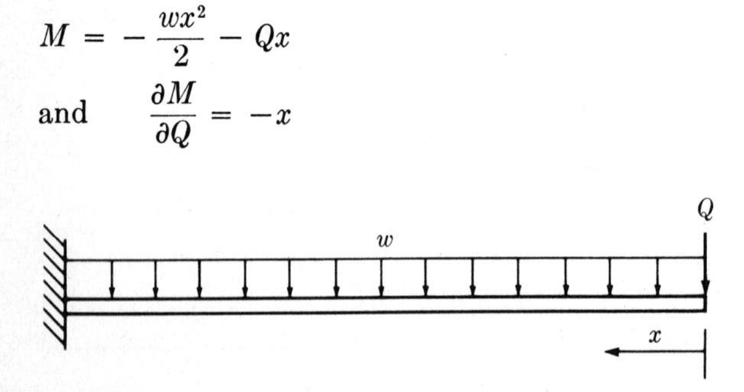

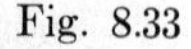

Fig. 8.33

The deflection is obtained from the integral expression of Eq. (8-74) as

$$\Delta_A = \frac{1}{EI}\int_0^L \left(-\frac{wx^2}{2} - Qx\right)(-x)\,dx$$

$$= \frac{wL^4}{8EI} + \frac{QL^3}{3EI}$$

We set the fictitious force Q equal to zero, and the desired deflection Δ_A is

$$\Delta_A = \frac{wL^4}{8EI}$$

The rotation at A due to the uniform load, denoted as θ_A, is obtained by applying a fictitious moment Q to the beam as shown in Fig. 8-34. Here the moment Q is applied in the direction opposite the anticipated direction of rotation to show that the procedure will yield the correct sign regardless of the assumed direction of the fictitious force.

The expression for moment in the beam is

$$M = -\frac{wx^2}{2} + Q \qquad \text{and} \qquad \frac{\partial M}{\partial Q} = 1$$

Therefore

$$\theta_A = \frac{1}{EI}\int_0^L \left(-\frac{wx^2}{2} + Q\right) 1\,dx$$

$$= \frac{1}{EI}\left(-\frac{wL^3}{6} + QL\right)$$

Setting Q equal to zero, we obtain

$$\theta_A = -\frac{wL^3}{6EI}$$

The minus sign indicates that the deflection is in a direction opposite that of the applied fictitious moment. If Q had been applied in a clockwise direction, the result would have been positive.

As we have seen from the above applications of Castigliano's theorem, the term involving the partial derivative corresponds to the virtual-force term in the method of virtual work. For example, the deflection due to bending by virtual work in Eq. (8-60) is

$$D = \int_0^L \frac{mM}{EI}\,dx$$

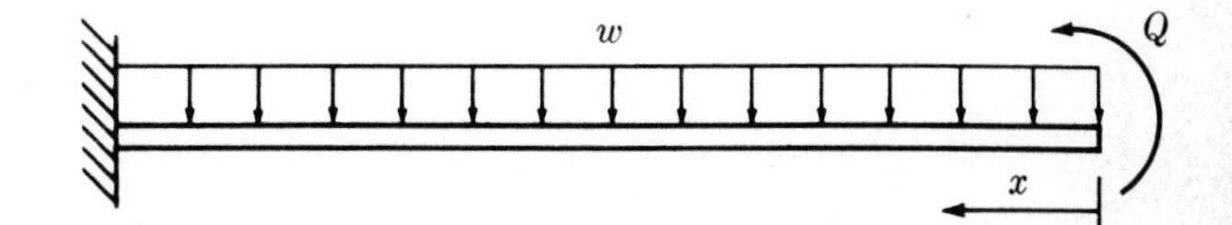

Fig. 8.34

From Castigliano's theorem, the deflection from Eq. (8-74) is

$$D = \int_G^L M \frac{\partial M}{\partial Q} \frac{dx}{EI}$$

The term m in Eq. (8-60) represents the moment due to a unit virtual (fictitious) force. The partial derivative $\partial M/\partial Q$ can be interpreted as the change in M per unit value of Q, which is the same as m. Similar comparisons hold for the terms representing deflections due to axial force, shear, and twist. The choice between the two methods, then, is simply a matter of which involves less work. It will be found that in most cases the method of virtual work is less involved.

Three basic energy methods for computing deflections have been discussed and illustrated in this chapter: real work, virtual work, and Castigliano's theorem. Although the choice of method for a particular problem is often a matter of personal preference, the versatility of the virtual-work method makes it a very attractive one for deflection analysis. The virtual-work principles will be employed in later chapters in developing the general matrix approach of analysis.

PROBLEMS

8-1. Use the method of real work to determine the horizontal deflection at A for the truss shown. For each member of the truss, E = 200 GPa and A = 1000 mm².

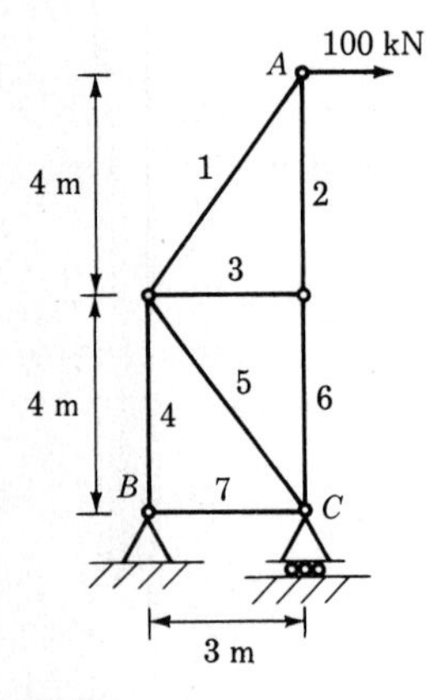

P8.1

8-2. Use the method of real work to determine the vertical deflection at ⑤ for the truss of Prob. 8-8.

8-3. Use the method of real work to determine the rotation at C for the beam of Prob. 7-3.

8-4. Use the method of real work to determine the horizontal deflection at D for the frame shown. For each member, $E = 30 \times 10^3$ ksi and $I = 300$ in.^{4}. Neglect axial deformations of the members.

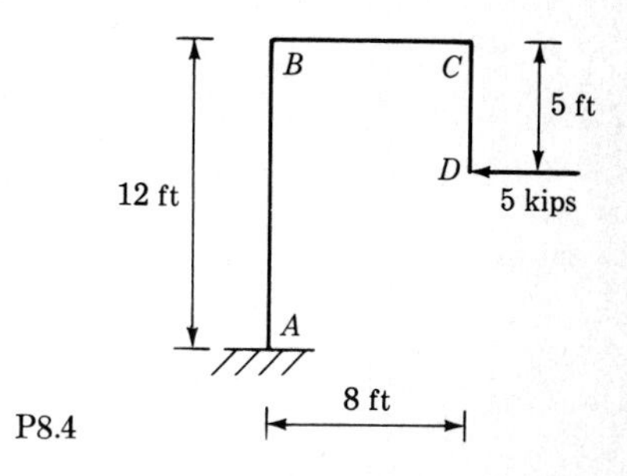

P8.4

8-5. Use the method of real work to determine the vertical deflection at D for the frame shown. For each member, $E = 200$ GPa, $G = 80$ GPa, $I = 15 \times 10^6$ mm^4, and $J = 30 \times 10^6$ mm^4.

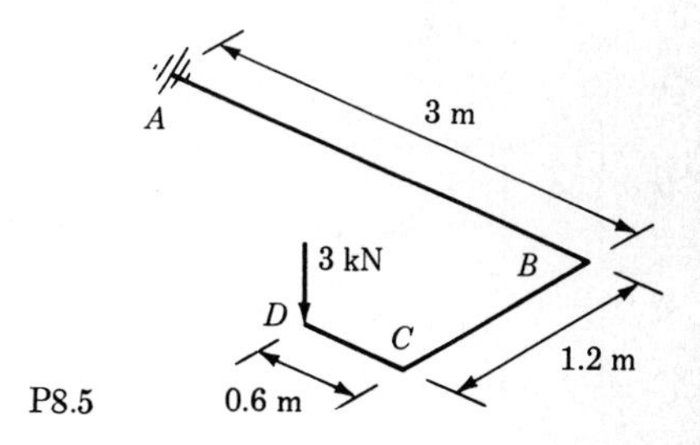

P8.5

8-6. Beam AB is simply supported at A and is supported by cable CD at C. $E = 200$ GPa for both the cable and the beam. Determine the deflection of the 40-kN load using the real-work method.

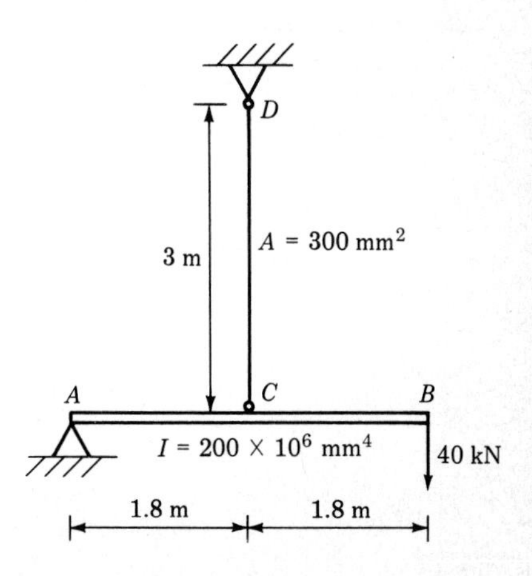

P8.6

8-7. For the truss shown, use the method of virtual work to determine (*a*) the vertical deflection at E; (*b*) the horizontal and vertical deflections at C. $E = 70$ GPa and $A = 1200$ mm^2 for each member.

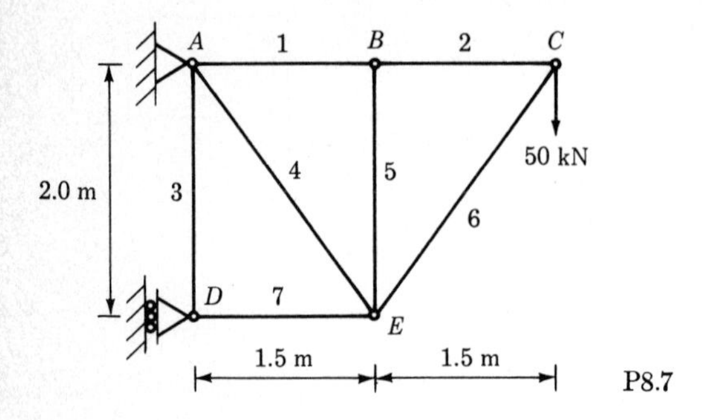

P8.7

8-8. For the truss shown, use the method of virtual work to determine (*a*) the horizontal deflection at ③; (*b*) the horizontal and vertical deflections at ⑤. For each member $A = 2.5$ in.^{2} and $E = 10 \times 10^3$ ksi.

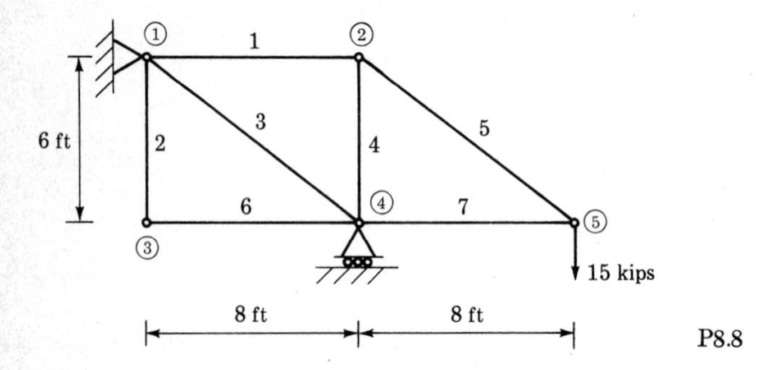

P8.8

8-9. For the truss shown, use the method of virtual work to determine (*a*) the vertical deflection at ⑥; (*b*) the horizontal deflection at ⑧. Express your answers in millimeters. For each member, $A = 1500$ mm^2 and $E = 200$ GPa.

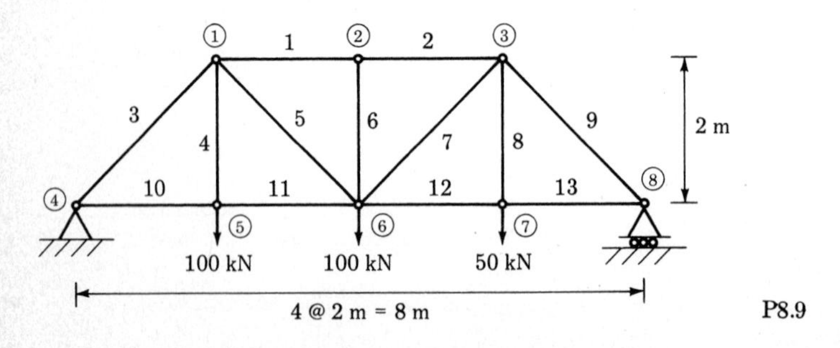

P8.9

8-10. For the truss of Prob. 3-6, use the method of virtual work to determine (*a*) the horizontal deflection at ⑤; (*b*) the vertical deflection at ⑤. $A = 5$ in.^{2} and $E = 30 \times 10^3$ ksi for each member.

8-11. Use the method of virtual work to solve Prob. 7-3.

8-12. Use the method of virtual work to solve Prob. 7-4.

8-13. Use the method of virtual work to solve Prob. 7-8.

8-14. Use the method of virtual work to solve Prob. 7-10.

8-15. Use the method of virtual work to determine the vertical deflection at D for the beam shown. $E = 200$ GPa and $I = 75 \times 10^6$ mm^4.

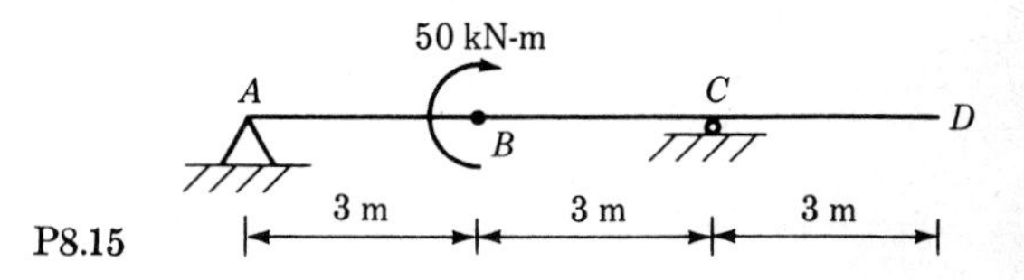

P8.15

8-16. For the structure shown, use the method of virtual work to determine (*a*) the vertical deflection at A; (*b*) the horizontal deflection at A; (*c*) the rotation at B. For each member $E = 30 \times 10^3$ ksi and $I = 400$ in.^{4}. Consider only bending deformations.

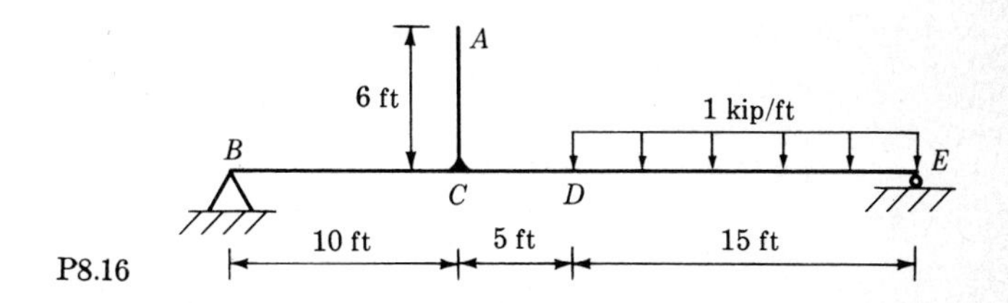

P8.16

8-17. Beam BD is simply supported at C and supported by cable BA at B. I for the beam is 200×10^6 mm^4, and the cross-sectional area of the cable is 350 mm^2. $E = 200$ GPa for each member. Determine the vertical deflection at D using the method of virtual work.

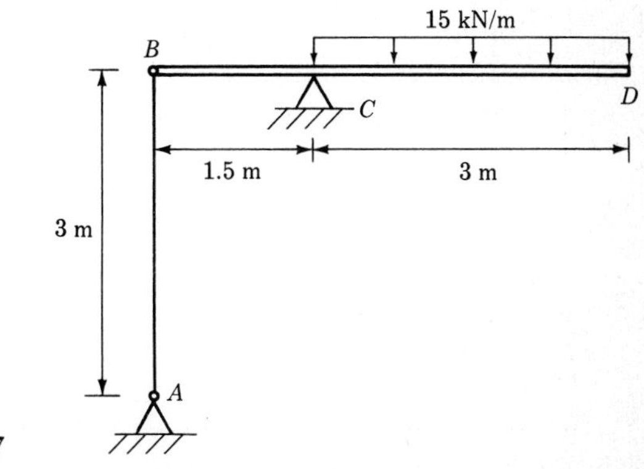

P8.17

8-18. For the frame shown, determine (*a*) the vertical deflection at A; (*b*) the vertical deflection at C. $E = 200$ GPa and $I = 150 \times 10^6$ mm^4 for each member.

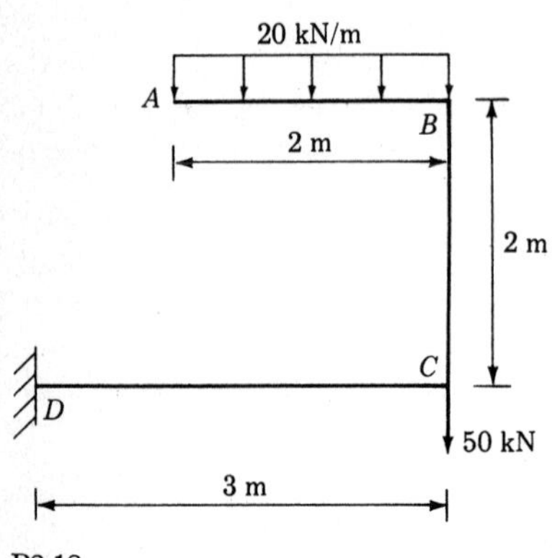

P8.18

8-19. Use the method of virtual work to solve Prob. 7-14. Determine also the rotation at D in radians.

8-20. Use the method of virtual work to solve Prob. 7-15. Determine also the rotation at D in radians.

8-21. For the truss shown, determine the direction and the distance that point E moves along the inclined support. For each member, $A = 1200$ mm^2 and $E = 200$ GPa.

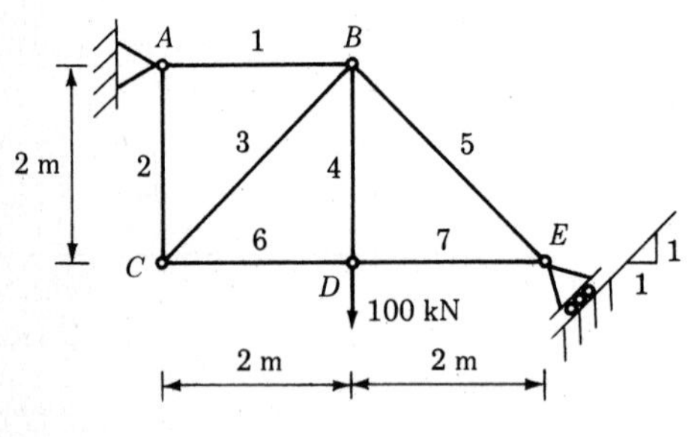

P8.21

8-22. Determine the direction and the distance that point D moves along the inclined support for the structure shown. $E = 30 \times 10^3$ ksi and $I = 250$ in.^{4} for each member. Consider only bending deformations.

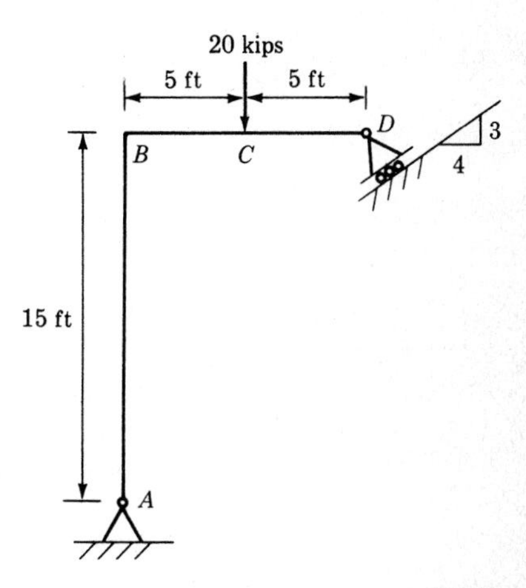

P8.22

8-23. For the three-dimensional frame shown, determine (*a*) the deflection in the Y direction at ④; (*b*) the deflection in the Z direction at ②; (*c*) the deflection in the X direction at ③. For each member, $E = 200$ GPa, $G = 80$ GPa, $I = 250 \times 10^6$ mm^4, and $J = 500 \times 10^6$ mm^4.

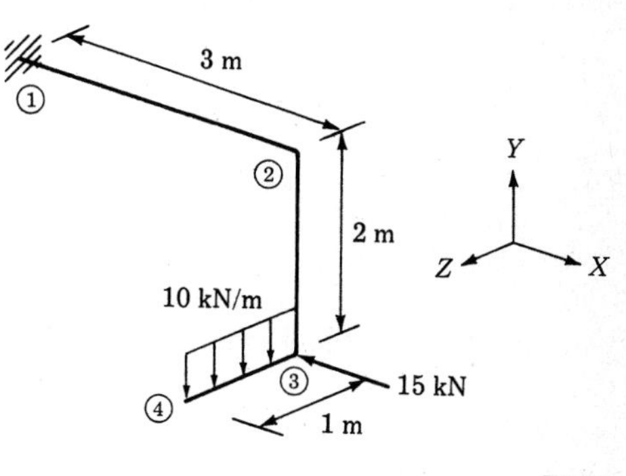

P8.23

8-24. For a steel W section, the value of K in Eqs. (8-33) and (8-56) is often assumed to be 1.0. The beam shown is a W24×55 for which $E = 30 \times 10^3$ ksi and $G = 12 \times 10^3$ ksi. Determine (a) the vertical deflection at C due to bending strains; (b) the vertical deflection at C due to shear strains. Neglect the dead weight of the beam in your calculations.

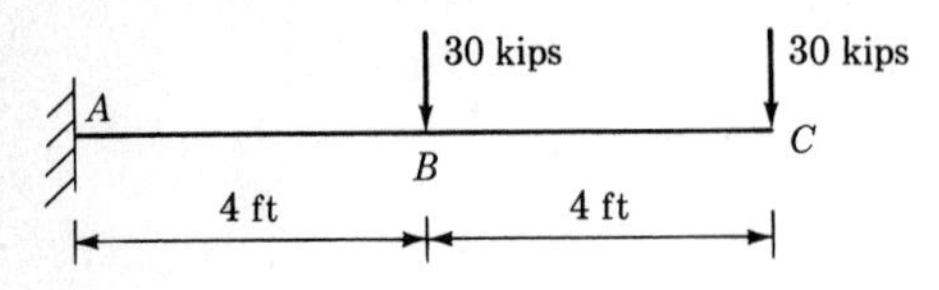

P8.24

8-25. For the frame shown, the outside surface temperature of AB is 32°C and the outside surface temperature of BC is 42°C. The inside surface of both members is 20°C. Determine the horizontal deflection at point C for the temperatures shown if the frame initially had a uniform temperature throughout of 26°C. The coefficient of thermal expansion for the member material is 11.7×10^{-6} per degree C. The depth of the members is 0.3 m. Assume the temperature varies linearly over the depth of the members.

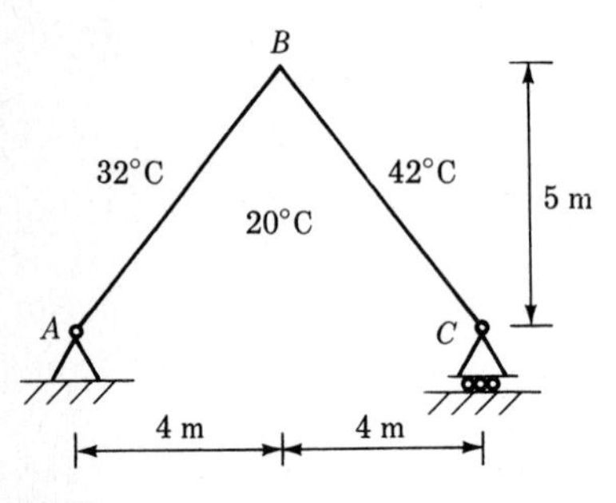

P8.25

8-26. Solve Prob. 8-4 by Castigliano's method.

8-27. For the truss of Prob. 8-8, determine the horizontal deflection at ③ by Castigliano's method.

8-28. Determine the vertical deflection at A for the frame in Prob. 8-18 using Castigliano's method.

CHAPTER

NINE

INFLUENCE COEFFICIENTS

9-1 INTRODUCTION

As we proceed into the analysis of more complicated structures it is helpful to have a firm understanding of how members and total structures respond to loads applied at particular points. In particular, it is helpful to understand how member displacements and member forces are related to each other. With an understanding of such relationships the task of analyzing the more complicated structures will be clearer.

These relationships can be expressed as *influence coefficients*. To clarify the basic idea of an influence coefficient let us consider the simple beam of Fig. 9-1. The beam is subjected to a single load Q, located a distance x from the support at A. From statics, the value of the reaction at A is given by

$$R_A = \left(1 - \frac{x}{L}\right) Q \tag{9-1}$$

For any value of x the quantity $1 - x/L$ describes the influence that the load Q has upon the value of reaction at A. The quantity $1 - x/L$, evaluated for a given value of x, therefore represents an influence coefficient. If desired, it can be referred to as a *force influence coefficient* in that when

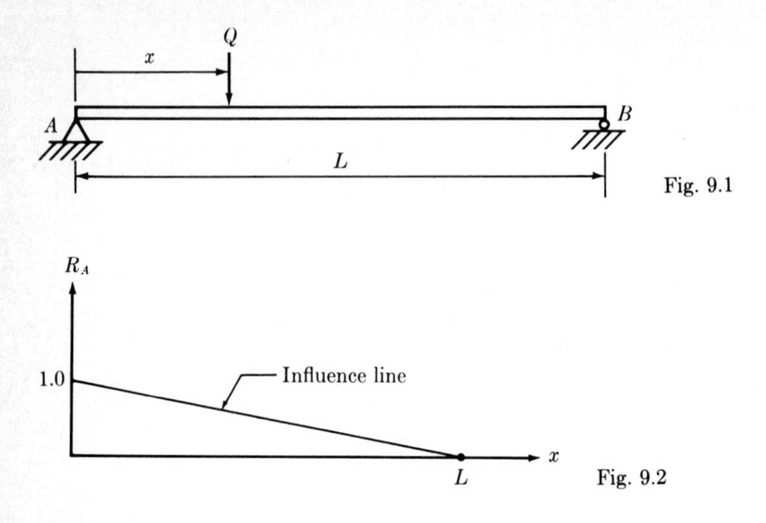

the force Q is multiplied by its value, the value of the force R_A is obtained. As we shall see later in the chapter, the displacements at a point on a structure due to applied loads can be expressed in terms of *flexibility influence coefficients*. Similarly, the force generated at a point on a structure due to applied deflections can be expressed in terms of *stiffness influence coefficients*.

If the value of Q in Fig. 9-1 is unity and the value of R_A in Eq. (9-1) is plotted against x, the result will be a straight-line relationship, as shown in Fig. 9-2. The resulting line is called an *influence line*. The value of the influence line for any value of x represents the value of the reaction at A due to a unit load placed on the beam at the position defined by x. The development and use of influence lines will be discussed in detail in the following chapter.

9-2 FLEXIBILITY INFLUENCE COEFFICIENTS

DEFINITION: *A flexibility influence coefficient for a structure, f_{ij}, is defined as the deflection at point i resulting from a unit force applied at j.*

Let us consider the simple beam of Fig. 9-3. A unit load is applied at point ① in Fig. 9-3*a*. The resulting deflection at ① is denoted as f_{11}. Correspondingly, the deflections at points ② and ③ are denoted as f_{21} and f_{31}, respectively. If a unit load is applied at point ②, as shown in Fig. 9-3*b*, the deflections at points ①, ②, and ③ are denoted as f_{12}, f_{22}, and f_{32}, respectively. Similarly, if a unit moment is applied at point ③, the values of deflection f_{13}, f_{23}, and f_{33} are obtained as shown in Fig. 9-3*c*.

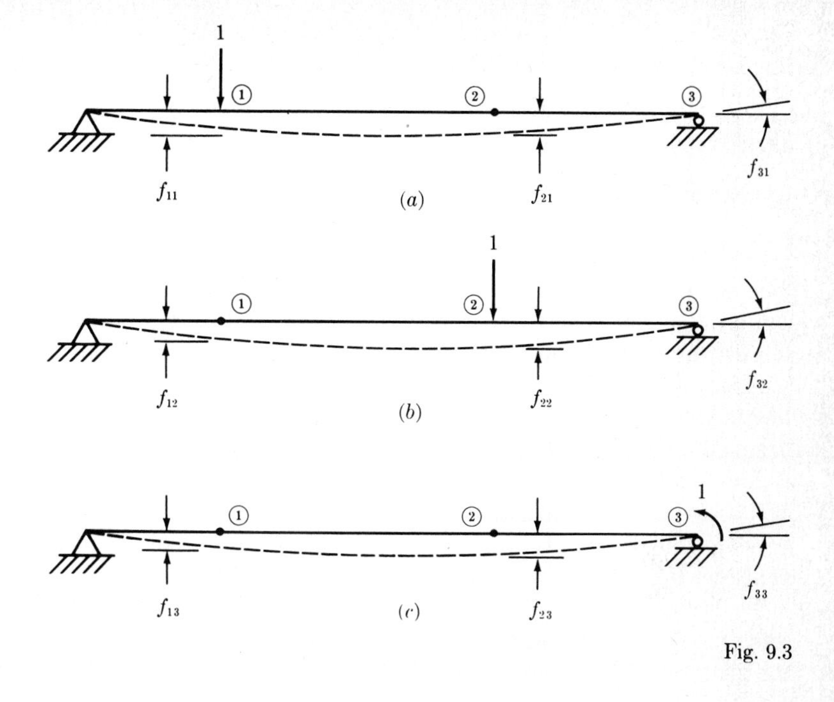

Fig. 9.3

The flexibility influence coefficient f_{ij} for $i \neq j$ can be either positive or negative. The sign depends on the direction of the applied unit force. The flexibility influence coefficient for $i = j$ is considered to be positive. The positive direction of deflection at a point is thus considered to be in the same direction as the positive unit force. The unit forces as applied in Fig. 9-3 result in all values of the flexibility coefficients being positive. As a variation, if the unit moment in Fig. 9-3*c* is applied in a clockwise direction at ③, rather than in a counterclockwise direction as shown, f_{13} and f_{23} are negative. Correspondingly, the values of f_{31} and f_{32} in Figs. 9-3*a* and *b* are negative quantities, since positive deflection at ③ is in the same direction as the applied unit moment.

For any value of loading Q_1, Q_2, and Q_3 applied to the beam of Fig. 9-3 at the points and in the directions in which the unit loads were applied, the total deflections at the points can be expressed as

$$\begin{aligned} D_1 &= f_{11}Q_1 + f_{12}Q_2 + f_{13}Q_3 \\ D_2 &= f_{21}Q_1 + f_{22}Q_2 + f_{23}Q_3 \\ D_3 &= f_{31}Q_1 + f_{32}Q_2 + f_{33}Q_3 \end{aligned} \tag{9-2}$$

To express the deflections as such requires that the behavior of the beam material be linearly elastic and that superposition be valid. The deflection expressions of Eq. (9-2) can be written in matrix form as

$$\begin{Bmatrix} D_1 \\ D_2 \\ D_3 \end{Bmatrix} = \begin{bmatrix} f_{11} & f_{12} & f_{13} \\ f_{21} & f_{22} & f_{23} \\ f_{31} & f_{32} & f_{33} \end{bmatrix} \begin{Bmatrix} Q_1 \\ Q_2 \\ Q_3 \end{Bmatrix} \tag{9-3}$$

or simply as

$$\mathbf{D} = \mathbf{FQ}$$

The square matrix **F**, whose elements are the flexibility influence coefficients f_{ij}, is referred to as a *flexibility matrix*. The flexibility matrix **F** relates the loads **Q** on the beam to the corresponding deflections **D**. We see from Eq. (9-3) or (9-2) that if $Q_1 = 1$ and $Q_2 = Q_3 = 0$, the displacements are $D_1 = f_{11}$, $D_2 = f_{21}$, and $D_3 = f_{31}$. The results correspond to the basic definition of a flexibility influence coefficient, as shown in Fig. 9-3*a*.

The flexibility matrix **F**, and thus the flexibility influence coefficients f_{ij}, possess a very important property: **F** is a symmetric matrix; that is, $f_{ij} = f_{ji}$. The proof of this is based on two important theorems of structural analysis, *Betti's theorem* and *Maxwell's reciprocal theorem*.

To develop Betti's theorem let us consider a structure subjected to two sets of loads, set A and set B, as shown in Fig. 9-4. If set A is applied to the structure first, then the expression for the external work done is

$$W_E = \tfrac{1}{2} \sum_{k=1}^{n^A} Q_k{}^A D_k{}^A$$

where the superscript A denotes quantities of set A. If set B is then applied, with set A still on the structure, the external work done will be

$$W_E = \tfrac{1}{2} \sum_{k=1}^{n^B} Q_k{}^B D_k{}^B + \sum_{k=1}^{n^A} Q_k{}^A D_k{}^B$$

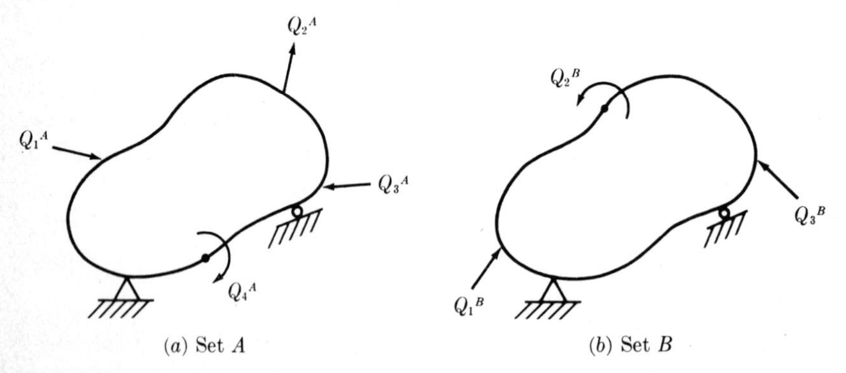

Fig. 9.4

The superscript B denotes quantities of set B. There is no factor of ½ in the second term because the loads of set A remain constant in value during the displacements due to set B. The total work done on the structure in applying the two sets is

$$W_E = \tfrac{1}{2} \sum_{k=1}^{n^A} Q_k{}^A D_k{}^A + \tfrac{1}{2} \sum_{k=1}^{n^B} Q_k{}^B D_k{}^B + \sum_{k=1}^{n^A} Q_k{}^A D_k{}^B \tag{9-4}$$

If the loads of set B are applied to the structure first, the external work done is

$$W_E = \tfrac{1}{2} \sum_{k=1}^{n^B} Q_k{}^B D_k{}^B$$

If set A is then applied to the structure with set B still on the structure, the external work done will be

$$W_E = \tfrac{1}{2} \sum_{k=1}^{n^A} Q_k{}^A D_k{}^A + \sum_{k=1}^{n^B} Q_k{}^B D_k{}^A$$

The total work done during this order of loading is represented by the expression

$$W_E = \tfrac{1}{2} \sum_{k=1}^{n^B} Q_k{}^B D_k{}^B + \tfrac{1}{2} \sum_{k=1}^{n^A} Q_k{}^A D_k{}^A + \sum_{k=1}^{n^B} Q_k{}^B D_k{}^A \tag{9-5}$$

The total work done by the two orders of loading is the same. Therefore, equating Eq. (9-4) to (9-5), we obtain

$$\sum_{k=1}^{n^A} Q_k{}^A D_k{}^B = \sum_{k=1}^{n^B} Q_k{}^B D_k{}^A \tag{9-6}$$

Equation (9-6) is a mathematical expression of Betti's theorem.

BETTI'S THEOREM: *For a linearly elastic structure, the work done by a set of external forces* Q^A, *acting through the displacements due to a set of external forces* Q^B, *is equal to the work done by the forces* Q^B, *acting through the displacements due to the set of forces* Q^A.

Maxwell's reciprocal theorem can be developed by applying Betti's theorem to a particular type of loading. Thus, if we consider the loads of set A to be a single unit load applied at point i and the loads of set B to be a single unit load applied at point j and recall that by definition f_{ij} is equal to the displacement at i due to a unit force applied at j, we have for Eq. (9-6)

$$1f_{ij} = 1f_{ji}$$

or

$$f_{ij} = f_{ji} \tag{9-7}$$

This is a mathematical expression of Maxwell's reciprocal theorem.

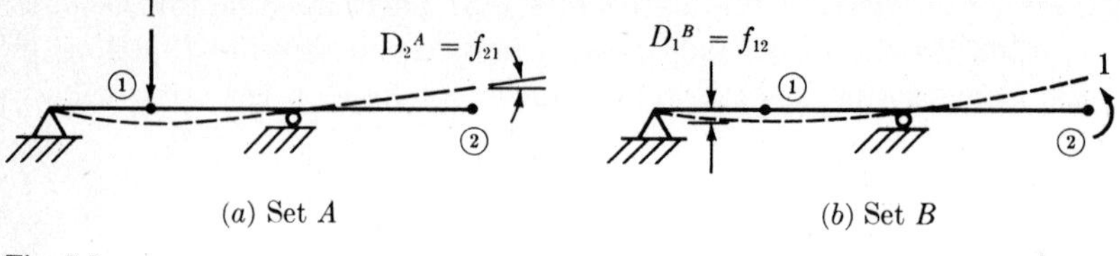

(a) Set A

(b) Set B

Fig. 9.5

MAXWELL'S RECIPROCAL THEOREM: *For a linearly elastic structure, the deflection component at any point i due to a component of force applied at any point j is equal to the component of deflection at j due to an equal force applied at i.*

By this theorem and Eq. (9-7) we have proved that the flexibility matrix **F** is symmetric.

The physical meaning of Betti's theorem and Maxwell's reciprocal theorem is demonstrated in Fig. 9-5. The structure considered is an overhanging beam. The loads of set A are considered to be a unit load applied at point ①, as shown in Fig. 9-5*a*. The resulting deflection at point ② is a rotation, denoted as $D_2{}^A = f_{21}$. The loads of set B are considered to be a unit moment applied to the beam at ② and in the direction of the deflection f_{21}, as shown in Fig. 9-5*b*. The resulting vertical deflection at ① is denoted as $D_1{}^B = f_{12}$. From Betti's theorem as given in Eq. (9-6),

$$1D_1{}^B = 1D_2{}^A$$

$$\text{or} \qquad f_{12} = f_{21}$$

It is very important in the application of these theorems that consistent components of deflection and force at a point be used.

In applying the concept of flexibility influence coefficients f_{ij} to structures it is often desirable to modify the subscript notation for clarity. For example, consider the flexibility coefficients for the loading on a structure shown in Fig. 9-6. The points under consideration are ① and ②. To avoid confusion and possible error, we can express the components of force at a point, and thus the resulting deflections, in terms of an XY coordinate system, as shown at point ②. It is helpful to include a sketch of the assumed positive directions of the axes. Corresponding to this notation for forces and deflections, the following form of subscripts for the flexibility coefficients can be used:

$$\begin{Bmatrix} D_1 \\ D_{2X} \\ D_{2Y} \end{Bmatrix} = \begin{bmatrix} f_{11} & f_{12X} & f_{12Y} \\ f_{2X1} & f_{2X2X} & f_{2X2Y} \\ f_{2Y1} & f_{2Y2X} & f_{2Y2Y} \end{bmatrix} \begin{Bmatrix} Q_1 \\ Q_{2X} \\ Q_{2Y} \end{Bmatrix}$$

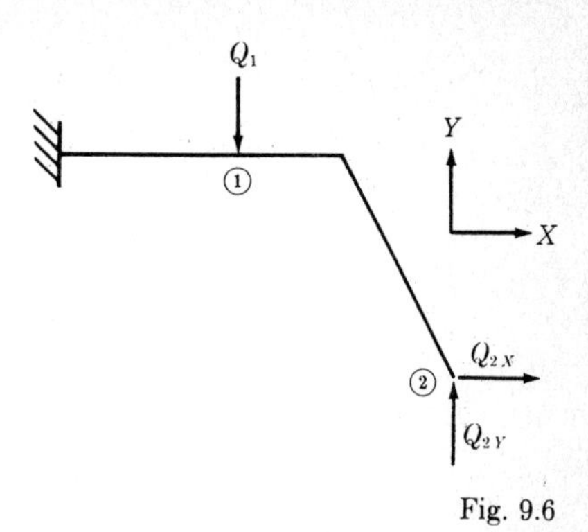

Fig. 9.6

Two subscripts are used when necessary to denote not only the point, but also the component at the point. Thus the term f_{2X2Y} is defined as the deflection at point ② in the X direction resulting from a unit force applied at point ② in the Y direction. Single-symbol subscripts were used at ① in the above expression. An additional symbol can be used if desired; f_{11} can be written as f_{1Y1Y} and D_1 and Q_1 can be written as D_{1Y} and Q_{1Y}.

The subscripts used in writing the flexibility influence coefficients of a structure in matrix form as the **F** matrix do not necessarily have the same meaning as those used in general references to an element of a matrix. An element of a matrix **A** is generally described as a_{ij}, where i refers to the row in which the element is located and j the column in which it is located. We can refer to an element of the above **F** matrix in the same manner, but the element must contain the subscripts appropriate to the meaning of a flexibility influence coefficient. Thus in the above expression the flexibility influence coefficient f_{2X2Y} can be referred to as element (2,3) of the **F** matrix.

EXAMPLE 9-1 The flexibility influence coefficients, and hence the elements of the flexibility matrix **F**, are to be determined for horizontal and vertical forces applied to point ① on the frame shown in Fig. 9-7. E and I for both members are the same. The forces and displacements at ① will be described in terms of the XY coordinate system shown. The flexibility influence coefficients f_{ij} to be determined relate the forces to the displacements at ① in the form

$$\begin{Bmatrix} D_{1X} \\ D_{1Y} \end{Bmatrix} = \begin{bmatrix} f_{1X1X} & f_{1X1Y} \\ f_{1Y1X} & f_{1Y1Y} \end{bmatrix} \begin{Bmatrix} Q_{1X} \\ Q_{1Y} \end{Bmatrix}$$

The value of f_{1X1X} is obtained by evaluating the deflection at ① in the X direction due to a unit force applied at ① in the X direction. For the virtual-work method the real and the virtual loads on the frame for

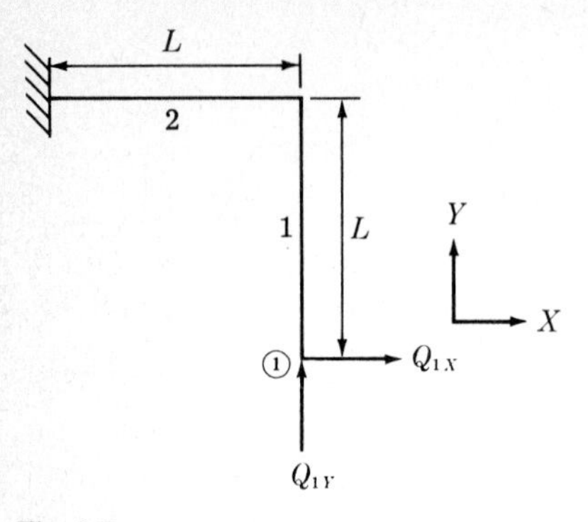

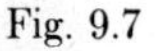

Fig. 9.7

determining f_{1X1X} are as shown in Fig. 9-8*a*. If we consider only the effects of bending, the expression for internal virtual work in each member is

$$\delta W_I = \int_0^L \frac{mM}{EI}\,dx$$

The real and virtual moment diagrams are the same for the evaluation of f_{1X1X} and are shown by dashed lines in Fig. 9-8*a*. Using the method of visual integration, we obtain

$$\begin{aligned} EIf_{1X1X} &= (\tfrac{1}{2})(L)(L)(\tfrac{2}{3})(L) + (L)(L)(L) \\ &= \frac{4L^3}{3} \end{aligned}$$

The value of f_{1X1Y}, which is equal to f_{1Y1X}, is obtained by applying a unit real force in the Y direction at ① and a unit virtual force in the X direction at ①, as shown in Fig. 9-8*b*. The resulting expression for f_{1X1Y} and

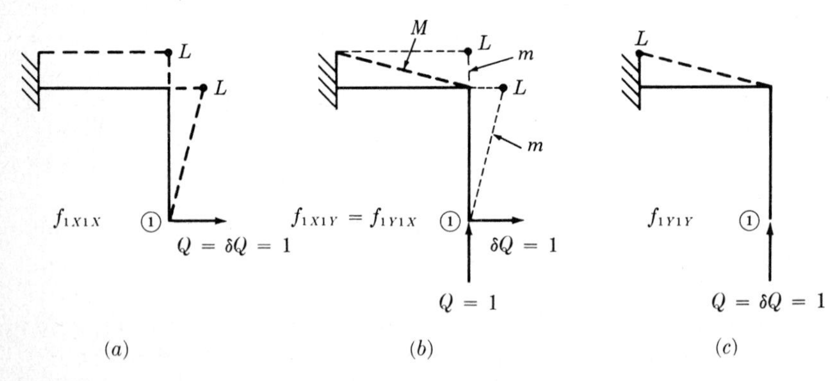

Fig. 9.8

f_{1Y1X} is

$$EIf_{1X1Y} = EIf_{1Y1X} = (1/2)(L)(L)(L) = \frac{L^3}{2}$$

The value of f_{1Y1Y} is obtained by applying a unit real and a unit virtual force in the Y direction at ①, as shown in Fig. 9-8*c*. Its value is found to be

$$EIf_{1Y1Y} = (1/2)(L)(L)(2/3)(L) = \frac{L^3}{3}$$

The flexibility relationship between the forces and the displacements can therefore be written as

$$\begin{Bmatrix} D_{1X} \\ D_{1Y} \end{Bmatrix} = \frac{L^3}{EI} \begin{bmatrix} 4/3 & 1/2 \\ 1/2 & 1/3 \end{bmatrix} \begin{Bmatrix} Q_{1X} \\ Q_{1Y} \end{Bmatrix}$$

9-3 STIFFNESS INFLUENCE COEFFICIENTS

DEFINITION: *A stiffness influence coefficient for a structure, k_{ij}, is defined as the force at point i resulting from a unit displacement imposed at point j.*

This definition implies that the displacements at all of the discrete points i under consideration other than the unit value at j are zero. As an illustration of the meaning of stiffness influence coefficients, consider the simple beam of Fig. 9-9, shown in Fig. 9-3 in terms of flexibility influence coefficients. In Fig. 9-9*a* a unit displacement is imposed upon the beam at point ①. The forces at points ①, ②, and ③ resulting from this displacement are denoted as k_{11}, k_{21}, and k_{31}, respectively. The force k_{11} is the force at ① associated with a unit displacement at ①. The force k_{21} is the force required at ② for zero displacement at ②. The force k_{31} is the moment required at ③ for zero rotation at point ③.

If a unit displacement is imposed only at point ②, the values of force at ①, ②, and ③ are as shown in Fig. 9-9*b* by k_{12}, k_{22}, and k_{32}, respectively. Similarly, the forces k_{13}, k_{23}, and k_{33} for a unit rotation at ③ are as shown in Fig. 9-9*c*.

The signs of the stiffness influence coefficients for $i = j$ are considered to be positive. Positive quantities of force and deflection can be defined as for flexibility influence coefficients; that is, the positive direction of deflection at a point is in the same direction assumed to be positive for the unit force. However, in developing stiffness influence coefficients unit displacements rather than unit forces are imposed on the structure; there-

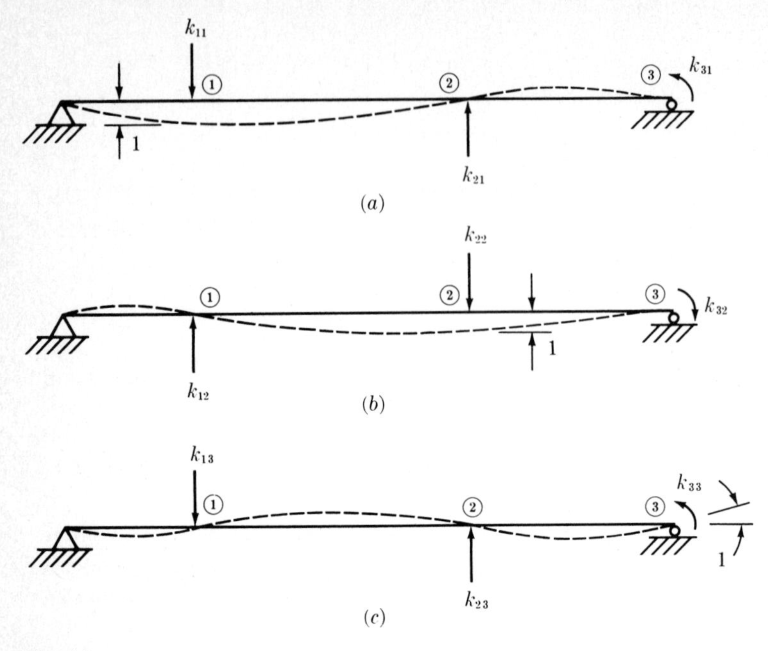

Fig. 9.9

fore it is often more convenient to think of positive quantities in a manner reversed from that used for flexibility coefficients. Thus the positive direction of a force at a point is considered to be in the same direction assumed positive for the unit displacement. The stiffness influence coefficients for $i \neq j$ can be either positive or negative. For the unit displacements shown in Fig. 9-9 the values of k_{21}, k_{12}, k_{32}, and k_{23} are negative.

For any values of displacement D_1, D_2, and D_3 imposed on the beam of Fig. 9-9 at the points and in the directions of the unit displacements the expressions for the resulting forces can be written as

$$\begin{aligned} Q_1 &= k_{11}D_1 + k_{12}D_2 + k_{13}D_3 \\ Q_2 &= k_{21}D_1 + k_{22}D_2 + k_{23}D_3 \\ Q_3 &= k_{31}D_1 + k_{32}D_2 + k_{33}D_3 \end{aligned} \tag{9-8}$$

These expressions can be written in matrix form as

$$\begin{Bmatrix} Q_1 \\ Q_2 \\ Q_3 \end{Bmatrix} = \begin{bmatrix} k_{11} & k_{12} & k_{13} \\ k_{21} & k_{22} & k_{23} \\ k_{31} & k_{32} & k_{33} \end{bmatrix} \begin{Bmatrix} D_1 \\ D_2 \\ D_3 \end{Bmatrix} \tag{9-9}$$

or simply as

$$\mathbf{Q} = \mathbf{KD}$$

The square matrix **K**, whose elements are the stiffness influence coefficients k_{ij}, is referred to as a *stiffness matrix*. The stiffness matrix **K** relates the displacements **D** imposed on the structure to the corresponding forces **Q**. The stiffness matrix, like the flexibility matrix **F**, is a symmetric matrix; that is, $k_{ij} = k_{ji}$. Its symmetry can be proved by means of Betti's theorem, developed in the previous section. If the loads of set A are considered to be the set of forces k_{ij} resulting from a unit displacement at point j and the loads of set B are considered to be the set of forces k_{ji} resulting from a unit displacement at point i, then from Eq. (9-6) we have

$$k_{ij}1 = k_{ji}1$$

or $\qquad k_{ij} = k_{ji}$

The values of the stiffness influence coefficients for a beam such as that of Fig. 9-9 are not as readily obtained as the flexibility coefficients. The values of the forces k_{ij} in Fig. 9-9 cannot be obtained from statics, because the beam is statically indeterminate. Statically indeterminate methods of analysis will be treated in later chapters. The values of the stiffness influence coefficients can, however, be obtained in another manner if we consider the relations of Eqs. (9-3) and (9-9):

$$\begin{Bmatrix} D_1 \\ D_2 \\ D_3 \end{Bmatrix} = \begin{bmatrix} f_{11} & f_{12} & f_{13} \\ f_{21} & f_{22} & f_{23} \\ f_{31} & f_{32} & f_{33} \end{bmatrix} \begin{Bmatrix} Q_1 \\ Q_2 \\ Q_3 \end{Bmatrix}$$

$$\begin{Bmatrix} Q_1 \\ Q_2 \\ Q_3 \end{Bmatrix} = \begin{bmatrix} k_{11} & k_{12} & k_{13} \\ k_{21} & k_{22} & k_{23} \\ k_{31} & k_{32} & k_{33} \end{bmatrix} \begin{Bmatrix} D_1 \\ D_2 \\ D_3 \end{Bmatrix}$$

Once we have obtained the values of the flexibility influence coefficients f_{ij} in Eq. (9-3) by deflection methods of analysis, we can express the values of loading **Q** in terms of the displacement **D** by using the inverse of the flexibility matrix. Thus

$$\begin{Bmatrix} Q_1 \\ Q_2 \\ Q_3 \end{Bmatrix} = \begin{bmatrix} f_{11} & f_{12} & f_{13} \\ f_{21} & f_{22} & f_{23} \\ f_{31} & f_{32} & f_{33} \end{bmatrix}^{-1} \begin{Bmatrix} D_1 \\ D_2 \\ D_3 \end{Bmatrix}$$

This is the relationship between loads and displacements desired in Eq. (9-9). Therefore the stiffness influence coefficients can be obtained by inverting the flexibility matrix; that is,

$$\begin{bmatrix} k_{11} & k_{12} & k_{13} \\ k_{21} & k_{22} & k_{23} \\ k_{31} & k_{32} & k_{33} \end{bmatrix} = \begin{bmatrix} f_{11} & f_{12} & f_{13} \\ f_{21} & f_{22} & f_{23} \\ f_{31} & f_{32} & f_{33} \end{bmatrix}^{-1} \qquad (9\text{-}10)$$

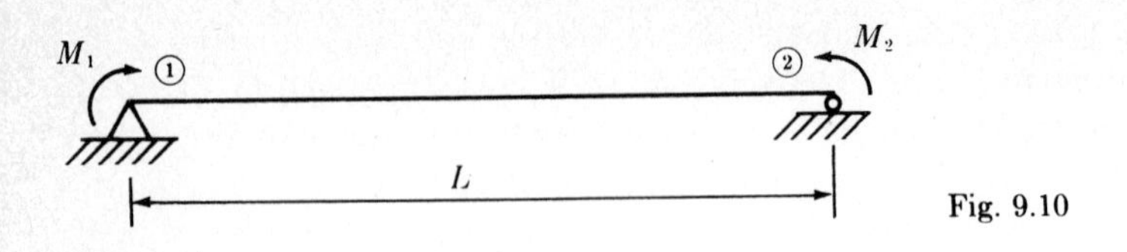

Fig. 9.10

or in general,

$$\mathbf{K} = \mathbf{F}^{-1} \tag{9-11}$$

EXAMPLE 9-2 The stiffness influence coefficients k_{ij} are to be determined for the loads on a simple beam as shown in Fig. 9-10. The stiffness coefficients to be evaluated relate the rotational displacements at points ① and ② to the moments M_1 and M_2 in the form

$$\begin{Bmatrix} M_1 \\ M_2 \end{Bmatrix} = \begin{bmatrix} k_{11} & k_{12} \\ k_{21} & k_{22} \end{bmatrix} \begin{Bmatrix} D_1 \\ D_2 \end{Bmatrix}$$

The values of k_{11} and k_{21} correspond to the condition shown in Fig. 9-11*a*. A unit value of rotation is imposed in a clockwise direction at ①. The moment at ① associated with this unit rotation is, by definition, k_{11}. The value of moment at point ② required for zero rotation at ② is k_{21}. Similarly, the values of k_{22} and k_{12} are shown in Fig. 9-11*b*. As proved above, $k_{12} = k_{21}$, and from Fig. 9-11 we see that they are negative quantities.

Both conditions shown in Fig. 9-11 are statically indeterminate. In particular, they are statically indeterminate to the first degree.

From the results obtained in Example 8-5 we observe that the flexibility matrix **F** for the beam is

$$\mathbf{F} = \frac{L}{6EI}\begin{bmatrix} 2 & 1 \\ 1 & 2 \end{bmatrix}$$

The value of the stiffness matrix **K**, and hence the values of the stiffness influence coefficients k_{ij}, can be found by inverting the **F** matrix. Using the method of cofactors discussed in Appendix A to invert **F**, we find the

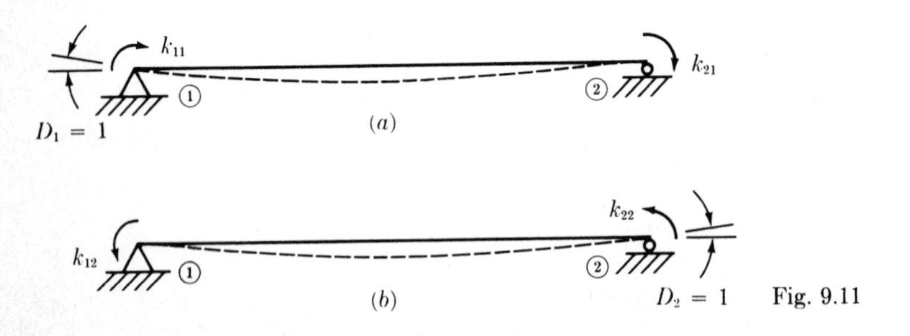

Fig. 9.11

value of the **K** matrix to be

$$\mathbf{K} = \begin{bmatrix} k_{11} & k_{12} \\ k_{21} & k_{22} \end{bmatrix} = \frac{6EI}{L}\begin{bmatrix} 2/3 & -1/3 \\ -1/3 & 2/3 \end{bmatrix} = \frac{2EI}{L}\begin{bmatrix} 2 & -1 \\ -1 & 2 \end{bmatrix}$$

The stiffness relationship between displacements and moments is therefore

$$\begin{Bmatrix} M_1 \\ M_2 \end{Bmatrix} = \frac{2EI}{L}\begin{bmatrix} 2 & -1 \\ -1 & 2 \end{bmatrix}\begin{Bmatrix} D_1 \\ D_2 \end{Bmatrix}$$

As observed earlier, k_{12} and k_{21} are negative quantities.

Note that the simply supported beam of Examples 8-5 and 9-2 was statically determinate, and we were therefore able to determine the flexibility influence coefficients f_{ij} by previously developed deflection methods and then obtain the stiffness influence coefficients k_{ij} by inverting the **F** matrix. If the given structure is statically indeterminate, the values of f_{ij} cannot be determined by the methods discussed thus far. Their evaluation becomes more involved. The procedure of determining stiffness and flexibility matrices for more involved structures will be treated in greater detail in a later chapter on general matrix analysis.

9-4 MEMBER-STIFFNESS MATRICES

It is essential to have an understanding of member stiffness relationships before we develop the general matrix stiffness (displacement) method in Chap. 14. Knowledge of member stiffnesses is also helpful in developing other methods of analysis in chapters prior to that. The types of member stiffnesses commonly encountered in the analysis of structures made up of individual members are those due to the effects of axial force, bending moment, twist, and sometimes shear.

Let us consider first the stiffness of a member subjected to axial forces and displacements at its ends as shown in Fig. 9-12. The forces and displacements as shown are considered to be positive quantities. From the basic definition of a stiffness influence coefficient given in Sec. 9-3, we

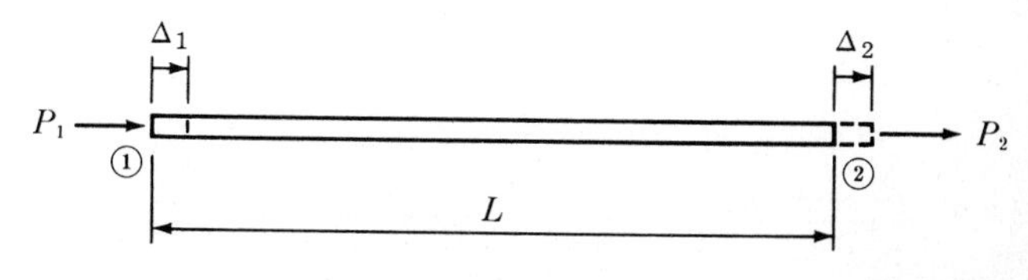

Fig. 9.12

can write the stiffness relationship for the member as

$$\begin{Bmatrix} P_1 \\ P_2 \end{Bmatrix} = \begin{bmatrix} k_{11} & k_{12} \\ k_{21} & k_{22} \end{bmatrix} \begin{bmatrix} \Delta_1 \\ \Delta_2 \end{bmatrix} \tag{9-12}$$

where the values of the stiffness influence coefficients k_{ij} correspond to the conditions shown in Fig. 9-13. From previous studies we know that the general expression for axial deformation in a member due to a force P is

$$e = \frac{PL}{EA} \tag{9-13}$$

or $$P = \frac{EA}{L} e \tag{9-14}$$

where A = cross-sectional area of the member
L = length of the member
E = modulus of elasticity of the material

For a displacement of $\Delta_1 = 1$ as shown in Fig. 9-13a, we obtain from Eq. (9-14)

$$k_{11} = \frac{EA}{L}(1)$$

For equilibrium of the member in Fig. 9-13a, $k_{21} = -k_{11}$. Proceeding in the same manner with a unit displacement at ② as shown in Fig. 9-13b, we see that

$$k_{22} = \frac{EA}{L}(1)$$

and $k_{12} = -k_{22}$. For any values of displacements Δ_1 and Δ_2 at the member ends, Eq. (9-12) can be written as

$$\begin{Bmatrix} P_1 \\ P_2 \end{Bmatrix} = \frac{EA}{L} \begin{bmatrix} 1 & -1 \\ -1 & 1 \end{bmatrix} \begin{Bmatrix} \Delta_1 \\ \Delta_2 \end{Bmatrix} \tag{9-15}$$

Equation (9-15) represents the stiffness of a member due to axial loading in terms of displacements Δ_1 and Δ_2 at the ends of the member.

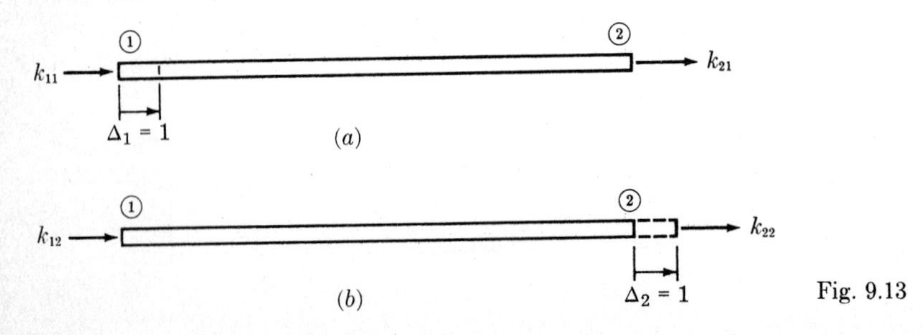

Fig. 9.13

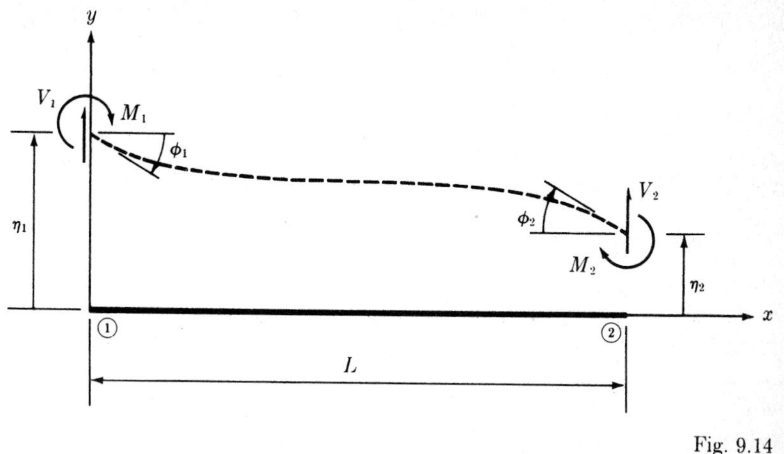

Fig. 9.14

Now let us develop the stiffness matrix of a member of constant cross-section subjected to bending moment. The member is subjected to moments at its end points. A member subjected to such loading is represented in the undeformed and the deformed states in Fig. 9-14. Quantities of force and displacement as shown are considered to be positive.

The stiffness relationship to be developed is of the form

$$\begin{Bmatrix} M_1 \\ V_1 \\ M_2 \\ V_2 \end{Bmatrix} = \begin{bmatrix} k_{11} & k_{12} & k_{13} & k_{14} \\ k_{21} & k_{22} & k_{23} & k_{24} \\ k_{31} & k_{32} & k_{33} & k_{34} \\ k_{41} & k_{42} & k_{43} & k_{44} \end{bmatrix} \begin{Bmatrix} \phi_1 \\ \eta_1 \\ \phi_2 \\ \eta_2 \end{Bmatrix} \tag{9-16}$$

To determine the values of the influence coefficients k_{ij} we can consider a unit value of each of the member-end displacements in separate steps as shown in Fig. 9-15. Only the moments at the ends of the member are shown in Fig. 9-15. As we shall see, the values of shear at the ends can

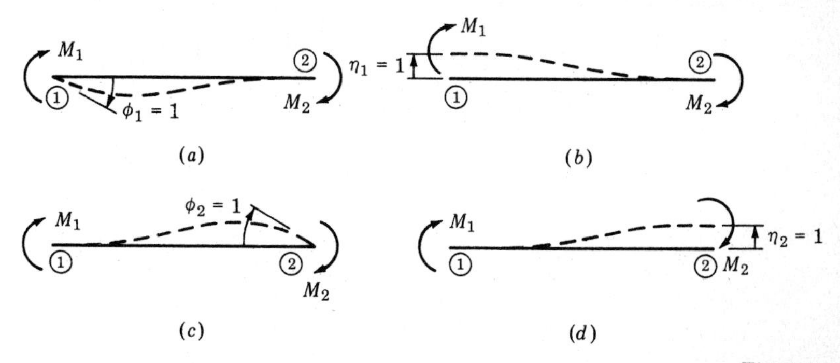

Fig. 9.15

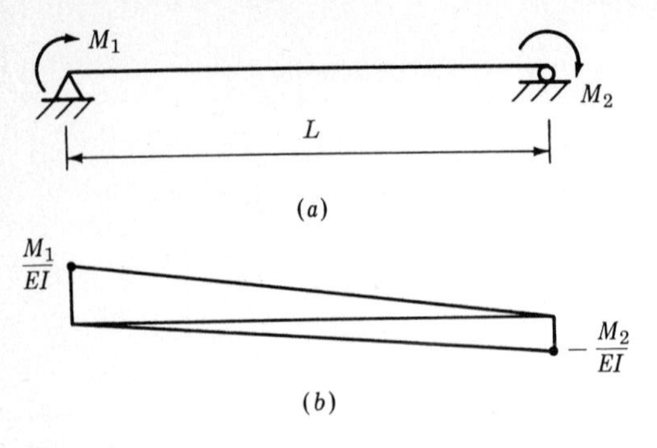

Fig. 9.16

be obtained if the member-end moments are known. The values of k_{i1}, the first column of the stiffness matrix in Eq. (9-16), are represented in Fig. 9-15*a* and can be evaluated using the moment-area method as shown in Fig. 9-16. The moments M_1 and M_2 can be considered to act on a simply supported beam as shown in Fig. 9-16*a* under the condition that they result in the necessary end displacements of Fig. 9-15*a*; that is, $\phi_1 = 1$, $\phi_2 = \eta_1 = \eta_2 = 0$. The M/EI diagram is shown in Fig. 9-16*b*. Note that the moment M_2 is a negative quantity according to the beam sign convention. Recalling from Sec. 7-2 that the change in slope between ① and ②, $\phi_2 - \phi_1$, is the area under the M/EI diagram, and that $\phi_2 = 0$ and $\phi_1 = -1$, we obtain from Fig. 9-16*b*

$$0 - (-1) = \left(\frac{1}{2}\right)\left(\frac{M_1}{EI}\right)(L) + \left(\frac{1}{2}\right)\left(-\frac{M_2}{EI}\right)(L)$$

$$= \frac{L}{2EI}(M_1 - M_2) \tag{9-17}$$

The end displacement conditions also require that the moment of the M/EI diagram about ① be zero,

$$t_{12} = 0 = \left(\frac{1}{2}\right)\left(\frac{M_1}{EI}\right)(L)\left(\frac{L}{3}\right) + \left(\frac{1}{2}\right)\left(-\frac{M_2}{EI}\right)(L)\left(\frac{2}{3}\right)(L)$$

$$= \frac{L^2}{6EI}(M_1 - 2M_2)$$

from which

$$M_2 = \tfrac{1}{2}M_1$$

Substituting this result into Eq. (9-17), we obtain

$$M_1 = \frac{4EI}{L}$$

and therefore

$$M_2 = \frac{2EI}{L}$$

The end shears V_1 and V_2 in Fig. 9-14 can be obtained from statics:

$$V_1 = -\frac{M_1 + M_2}{L} = -\frac{6EI}{L^2}$$

$$V_2 = -V_1 = \frac{6EI}{L^2}$$

From these results we can state that

$$k_{11} = \frac{4EI}{L} \qquad k_{21} = -\frac{6EI}{L^2} \qquad k_{31} = \frac{2EI}{L} \qquad k_{41} = \frac{6EI}{L^2}$$

The values of k_{i2}, the second column of the stiffness matrix in Eq. (9-16), are obtained from the conditions of Fig. 9-15*b*. The conditions of the moment-area method to be met are $t_{12} = 1$ and the change in slope between ① and ② is zero. It should be recalled from Sec. 7-2 that t_{12} for this condition is a positive quantity. The M/EI diagram is the same as that used before in Fig. 9-16*b*. Proceeding as above, we obtain

$$k_{12} = -\frac{6EI}{L^2} \qquad k_{22} = \frac{12EI}{L^3} \qquad k_{32} = -\frac{6EI}{L^2} \qquad k_{42} = -\frac{12EI}{L^3}$$

The values of k_{ij} for the third and fourth columns of Eq. (9-16) can be obtained in a similar manner using the conditions shown in Figs. 9-15*c* and *d*, respectively. The resulting stiffness matrix for the member due to bending is

$$\begin{Bmatrix} M_1 \\ V_1 \\ M_2 \\ V_2 \end{Bmatrix} = \frac{2EI}{L^3} \begin{bmatrix} 2L^2 & -3L & L^2 & 3L \\ -3L & 6 & -3L & -6 \\ L^2 & -3L & 2L^2 & 3L \\ 3L & -6 & 3L & 6 \end{bmatrix} \begin{Bmatrix} \phi_1 \\ \eta_1 \\ \phi_2 \\ \eta_2 \end{Bmatrix} \tag{9-18}$$

The bending stiffness matrix of Eq. (9-18) is an often used matrix in the general matrix stiffness method of analysis for framed structures. It is also very useful in developing the classical methods of statically indeterminate analysis.

The stiffness matrix of Eq. (9-18) can also be developed by using the basic differential equation of bending. Such an approach can also be applied to more advanced stability and dynamic analyses. Consider again the deflected member as shown in Fig. 9-14. We begin the development by considering the general fourth-order differential equation for bending,

$$y^{iv} = \frac{w}{EI} \tag{9-19}$$

where w is the distributed load on the beam. Because we are concerned here with loads applied only at the member ends, we set w equal to zero, and Eq. (9-19) becomes

$$y^{iv} = 0 \tag{9-20}$$

Successive integration of Eq. (9-20) results in the general solution

$$y = Ax^3 + Bx^2 + Cx + D \tag{9-21}$$

From the geometric boundary conditions in Fig. 9-14 we have

$$\begin{aligned} \eta_1 &= y(0) & \eta_2 &= y(L) \\ \phi_1 &= -y'(0) & \phi_2 &= -y'(L) \end{aligned}$$

Positive deflection and slope are in accordance with the directions of the x and y axes in Fig. 9-14. Substitution of these boundary conditions into Eq. (9-21) results in the expressions

$$\begin{aligned} \phi_1 &= -y'(0) = -0 - 0 - C - 0 \\ \eta_1 &= y(0) = 0 + 0 + 0 + D \\ \phi_2 &= -y'(L) = -3AL^2 - 2BL - C - 0 \\ \eta_2 &= y(L) = AL^3 + BL^2 + CL + D \end{aligned}$$

These equations can be written in matrix form as

$$\begin{Bmatrix} \phi_1 \\ \eta_1 \\ \phi_2 \\ \eta_2 \end{Bmatrix} = \begin{bmatrix} 0 & 0 & -1 & 0 \\ 0 & 0 & 0 & 1 \\ -3L^2 & -2L & -1 & 0 \\ L^3 & L^2 & L & 1 \end{bmatrix} \begin{Bmatrix} A \\ B \\ C \\ D \end{Bmatrix}$$

or $\quad \mathbf{D} = \mathbf{MA}$ (9-22)

where $\mathbf{D}$ represents the member displacements and $\mathbf{M}$ represents the 4×4 matrix of coefficients of the undetermined constants $\mathbf{A}$.

Using the second-order expression for moment

$$y'' = \frac{M}{EI}$$

and the expression for shear

$$y''' = \frac{V}{EI}$$

at the boundaries, we obtain the equations:

$$\begin{aligned} M_1 &= EIy''(0) = EI(0 + 2B + 0 + 0) \\ V_1 &= EIy'''(0) = EI(6A + 0 + 0 + 0) \\ M_2 &= -EIy''(\mathrm{L}) = -EI(6AL + 2B + 0 + 0) \\ V_2 &= -EIy'''(\mathrm{L}) = -EI(6A + 0 + 0 + 0) \end{aligned}$$

Positive moment and shear are in accordance with the beam sign convention. These equations can be written in matrix form as

$$\begin{Bmatrix} M_1 \\ V_1 \\ M_2 \\ V_2 \end{Bmatrix} = EI \begin{bmatrix} 0 & 2 & 0 & 0 \\ 6 & 0 & 0 & 0 \\ -6L & -2 & 0 & 0 \\ -6 & 0 & 0 & 0 \end{bmatrix} \begin{Bmatrix} A \\ B \\ C \\ D \end{Bmatrix}$$

or $\quad \mathbf{Q} = \mathbf{NA} \qquad (9\text{-}23)$

where $\mathbf{Q}$ represents the member forces and $\mathbf{N}$ represents the 4×4 matrix of coefficients of the column matrix $\mathbf{A}$. Equation (9-22) can be written as

$$\mathbf{A} = \mathbf{M}^{-1}\mathbf{D}$$

which, when substituted into Eq. (9-23), leads to a matrix relation between member displacements $\mathbf{D}$ and member forces $\mathbf{Q}$,

$$\mathbf{Q} = \mathbf{NM}^{-1}\mathbf{D} \qquad (9\text{-}24)$$

The inverse of $\mathbf{M}$ can be obtained by using the method of cofactors in Appendix A. Performing the necessary steps of inversion, we find the inverse of $\mathbf{M}$ to be

$$\mathbf{M}^{-1} = \frac{1}{L^4} \begin{bmatrix} -L^2 & 2L & -L^2 & -2L \\ 2L^3 & -3L^2 & L^3 & 3L^2 \\ -L^4 & 0 & 0 & 0 \\ 0 & L^4 & 0 & 0 \end{bmatrix}$$

Therefore

$$\mathbf{NM}^{-1} = \frac{EI}{L^4} \begin{bmatrix} 0 & 2 & 0 & 0 \\ 6 & 0 & 0 & 0 \\ -6L & -2 & 0 & 0 \\ -6 & 0 & 0 & 0 \end{bmatrix} \begin{bmatrix} -L^2 & 2L & -L^2 & -2L \\ 2L^3 & -3L^2 & L^3 & 3L^2 \\ -L^4 & 0 & 0 & 0 \\ 0 & L^4 & 0 & 0 \end{bmatrix}$$

The result is the same as that obtained earlier in Eq. (9-18).

It is desirable at times to work with a reduced size of the stiffness matrix in Eq. (9-18). The stiffness relation can be expressed in terms of a 2×2 matrix. The reduced size of the stiffness matrix can be obtained by considering the equilibrium equations for the member in Fig. 9-14. We see that

$$V_1 = -V_2 \qquad (9\text{-}25)$$

and $\quad M_2 = -M_1 - V_1 L \qquad (9\text{-}26)$

From Eq. (9-25) and the fact that the shear is constant over the length of the member, it is sufficient to write Eq. (9-18) in the form

$$\begin{Bmatrix} M_1 \\ V \\ M_2 \end{Bmatrix} = \frac{2EI}{L^3} \begin{bmatrix} 2L^2 & -3L & L^2 & 3L \\ -3L & 6 & -3L & -6 \\ L^2 & -3L & 2L^2 & 3L \end{bmatrix} \begin{Bmatrix} \phi_1 \\ \eta_1 \\ \phi_2 \\ \eta_2 \end{Bmatrix} \tag{9-27}$$

where $\quad V = V_1 = -V_2$

Since Eq. (9-26) can serve as a supplemental equation for determining the value of shear in terms of the end moments, it is sufficient to write Eq. (9-27) in the form

$$\begin{Bmatrix} M_1 \\ M_2 \end{Bmatrix} = \frac{2EI}{L^3} \begin{bmatrix} 2L^2 & -3L & L^2 & 3L \\ L^2 & -3L & 2L^2 & 3L \end{bmatrix} \begin{Bmatrix} \phi_1 \\ \eta_1 \\ \phi_2 \\ \eta_2 \end{Bmatrix} \tag{9-28}$$

The deflection of a member due to bending can be sufficiently described by the angle θ as shown in Fig. 9-17, where *deflections are referenced to the chord connecting the end points of the member* rather than to the member coordinate axes, as in Fig. 9-14. From Figs. 9-14 and 9-17 we see that

$$\phi_1 = \theta_1 + \frac{\eta_1 - \eta_2}{L} \qquad \text{and} \qquad \phi_2 = \theta_2 + \frac{\eta_1 - \eta_2}{L}$$

Substituting these values of ϕ_1 and ϕ_2 into Eq. (9-28), for M_1 we obtain

$$M_1 = \frac{2EI}{L^3}\left[2L^2\left(\theta_1 + \frac{\eta_1 - \eta_2}{L}\right) - 3L\eta_1 + L^2\left(\theta_2 + \frac{\eta_1 - \eta_2}{L}\right) + 3L\eta_2\right]$$

which reduces to

$$M_1 = \frac{2EI}{L^3}(2L^2\theta_1 + L^2\theta_2)$$

Proceeding in the same manner for M_2, we obtain

$$\begin{Bmatrix} M_1 \\ M_2 \end{Bmatrix} = \frac{2EI}{L^3} \begin{bmatrix} 2L^2 & L^2 \\ L^2 & 2L^2 \end{bmatrix} \begin{Bmatrix} \theta_1 \\ \theta_2 \end{Bmatrix}$$

or

$$\begin{Bmatrix} M_1 \\ M_2 \end{Bmatrix} = \frac{2EI}{L} \begin{bmatrix} 2 & 1 \\ 1 & 2 \end{bmatrix} \begin{Bmatrix} \theta_1 \\ \theta_2 \end{Bmatrix} \tag{9-29}$$

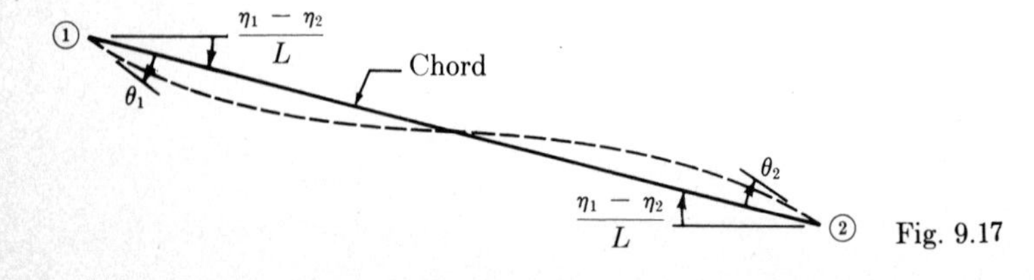

Fig. 9.17

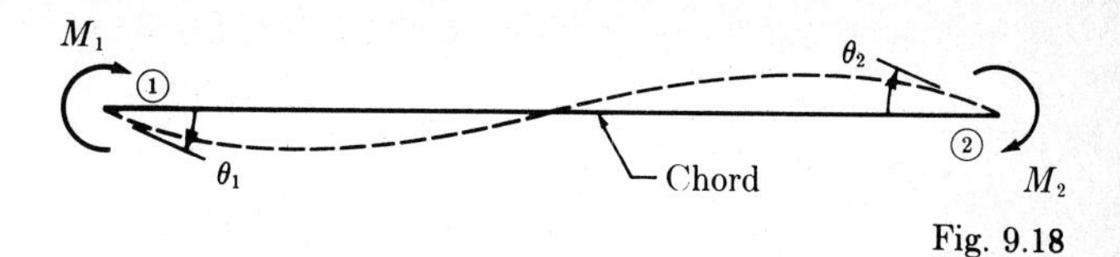

Fig. 9.18

In this form the stiffness of the member due to bending is represented by the 2×2 matrix

$$\mathbf{K} = \frac{2EI}{L}\begin{bmatrix} 2 & 1 \\ 1 & 2 \end{bmatrix} \tag{9-30}$$

The flexibility matrix for the member can be obtained by inverting the stiffness matrix of Eq. (9-30). Thus

$$\mathbf{F} = \frac{L}{6EI}\begin{bmatrix} 2 & -1 \\ -1 & 2 \end{bmatrix} \tag{9-31}$$

Accordingly, moments are related to displacements in the form

$$\begin{Bmatrix} \theta_1 \\ \theta_2 \end{Bmatrix} = \frac{L}{6EI}\begin{bmatrix} 2 & -1 \\ -1 & 2 \end{bmatrix}\begin{Bmatrix} M_1 \\ M_2 \end{Bmatrix} \tag{9-32}$$

It must be emphasized that moments and rotations for the forms of **K** and **F** in Eqs. (9-30) and (9-31) are considered positive as shown in Fig. 9-18.

The stiffness of a member due to twisting effects can be developed in a manner similar to that used for axial loading. For a member of circular cross-section the deformation resulting from twist is as shown in Fig. 9-19. The deformation in a length L associated with twisting moment is denoted by the angle ψ. The twist vectors T_1 and T_2 are drawn in accordance with the right-hand screw rule. The quantities shown are considered positive.

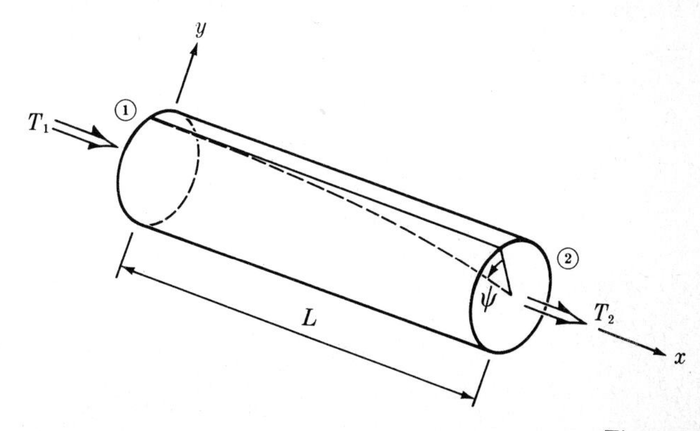

Fig. 9.19

The general stiffness relation for the member can be written as

$$\begin{Bmatrix} T_1 \\ T_2 \end{Bmatrix} = \begin{bmatrix} k_{11} & k_{12} \\ k_{21} & k_{22} \end{bmatrix} \begin{Bmatrix} \psi_1 \\ \psi_2 \end{Bmatrix}$$

From mechanics of materials, the general relation between twist and deformation is

$$T = \frac{GJ}{L}\psi \tag{9-33}$$

where J is the polar moment of inertia of the cross-section and G is the modulus of elasticity in shear. Following the same approach as in the case of axial force and axial displacement, whereby a unit value of ψ_1 and a unit value of ψ_2 are applied to the member, the stiffness relationship due to twist is found by the use of Eq. (9-33) to be

$$\begin{Bmatrix} T_1 \\ T_2 \end{Bmatrix} = \frac{GJ}{L} \begin{bmatrix} 1 & -1 \\ -1 & 1 \end{bmatrix} \begin{Bmatrix} \psi_1 \\ \psi_2 \end{Bmatrix} \tag{9-34}$$

The effects of shearing strain in a member are generally small in comparison to bending effects. They are, therefore, usually neglected. The development of the member stiffness matrix including shearing effects can be found in the text by Przemieniecki.[1]

The developments of this section can be summarized in general forms of the member stiffness matrix. A general three-dimensional stiffness expression for a member, considering axial, bending, and twisting effects, can be obtained from previously developed expressions. To do so, consider a member of length L as shown in Fig. 9-20. The member is subjected at each end to an axial force q^1, two components of shearing force q^2 and q^3, two components of moment q^5 and q^6, and twist q^4. As before, the forces are assumed to act at the centroid of the cross-section, and the moments are assumed to act about the two principal axes which are the y and z axes. Moment and twist are shown in accordance with the right-hand screw rule. Quantities as shown are considered positive. The displacements associated with each of the force quantities are denoted by d^i and are assumed positive when they act in the same direction as the force. The resulting stiffness matrix is a 12×12 matrix. Assembling the results of previous developments, we can express the stiffness relation for

[1] J. S. Przemieniecki, "Theory of Matrix Structural Analysis," McGraw-Hill Book Company, New York, 1968.

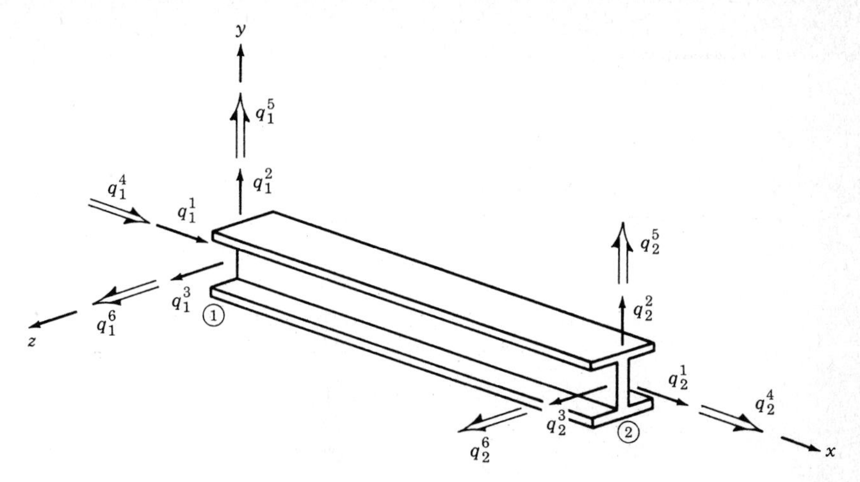

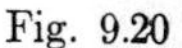
Fig. 9.20

the three-dimensional member of Fig. 9-20 as

$$
\begin{Bmatrix} q_1^1 \\ q_1^2 \\ q_1^3 \\ q_1^4 \\ q_1^5 \\ q_1^6 \\ q_2^1 \\ q_2^2 \\ q_2^3 \\ q_2^4 \\ q_2^5 \\ q_2^6 \end{Bmatrix}
=
\begin{bmatrix}
\frac{EA}{L} & & & & & & & & & & & \\
0 & \frac{12EI_z}{L^3} & & & & & & & & & & \\
0 & 0 & \frac{12EI_y}{L^3} & & & & & \text{Symmetric} & & & & \\
0 & 0 & 0 & \frac{GJ}{L} & & & & & & & & \\
0 & 0 & -\frac{6EI_y}{L^2} & 0 & \frac{4EI_y}{L} & & & & & & & \\
0 & \frac{6EI_z}{L^2} & 0 & 0 & 0 & \frac{4EI_z}{L} & & & & & & \\
-\frac{EA}{L} & 0 & 0 & 0 & 0 & 0 & \frac{EA}{L} & & & & & \\
0 & -\frac{12EI_z}{L^3} & 0 & 0 & 0 & -\frac{6EI_z}{L^2} & 0 & \frac{12EI_z}{L^3} & & & & \\
0 & 0 & -\frac{12EI_y}{L^3} & 0 & \frac{6EI_y}{L^2} & 0 & 0 & 0 & \frac{12EI_y}{L^3} & & & \\
0 & 0 & 0 & -\frac{GJ}{L} & 0 & 0 & 0 & 0 & 0 & \frac{GJ}{L} & & \\
0 & 0 & -\frac{6EI_y}{L^2} & 0 & \frac{2EI_y}{L} & 0 & 0 & 0 & \frac{6EI_y}{L^2} & 0 & \frac{4EI_y}{L} & \\
0 & \frac{6EI_z}{L^2} & 0 & 0 & 0 & \frac{2EI_z}{L} & 0 & -\frac{6EI_z}{L^2} & 0 & 0 & 0 & \frac{4EI_z}{L}
\end{bmatrix}
\begin{Bmatrix} d_1^1 \\ d_1^2 \\ d_1^3 \\ d_1^4 \\ d_1^5 \\ d_1^6 \\ d_2^1 \\ d_2^2 \\ d_2^3 \\ d_2^4 \\ d_2^5 \\ d_2^6 \end{Bmatrix}
\qquad (9\text{-}35)
$$

It should be noted that some quantities such as rotations d_1^6 and d_2^6 in Fig. 9-20 are considered positive in the direction opposite to that assumed in Fig. 9-14. This necessitates a change in the sign of the associated stiffness influence coefficients for shear as they appear in Eq. (9-18). Likewise, for displacements such as d_1^2 and d_2^2, the sign of the influence coefficients for moments is changed. Subscripts are used with the moment-of-inertia terms in Eq. (9-35) to distinguish between the two axes. The constant J

in the influence coefficient for twist can be referred to in general as the *torsional constant* for the member. Its value depends on the shape of the cross-section.[1] We have already considered the special case of a circular cross-section where J is the polar moment of inertia of the cross-section.

An often used type of analysis is the consideration of axial and bending effects in the analysis of two-dimensional (plane) frames. For such two-dimensional analyses, Eq. (9-35) reduces to

$$\begin{Bmatrix} P_1 \\ V_1 \\ M_1 \\ P_2 \\ V_2 \\ M_2 \end{Bmatrix} = E \begin{bmatrix} \frac{A}{L} & & & & & \\ 0 & \frac{12I}{L^3} & & \text{Symmetric} & & \\ 0 & \frac{6I}{L^2} & \frac{4I}{L} & & & \\ -\frac{A}{L} & 0 & 0 & \frac{A}{L} & & \\ 0 & -\frac{12I}{L^3} & -\frac{6I}{L^2} & 0 & \frac{12I}{L^3} & \\ 0 & \frac{6I}{L^2} & \frac{2I}{L} & 0 & -\frac{6I}{L^2} & \frac{4I}{L} \end{bmatrix} \begin{Bmatrix} \Delta_1 \\ \eta_1 \\ \phi_1 \\ \Delta_2 \\ \eta_2 \\ \phi_2 \end{Bmatrix} \tag{9-36}$$

In this expression, force and displacement quantities q and d of Fig. 9-20 are replaced by the notation used previously in this chapter. Positive quantities are as indicated in Fig. 9-20.

As mentioned earlier, the member stiffness matrices developed in this section will be used in later chapters in the development of methods for the analysis of complex structures.

PROBLEMS

9-1. For the beam shown, numerically show that $f_{4R2Y} = f_{2Y4R}$; that is, apply a unit load in the Y direction at ② and compute the resulting rotation at ④. Then apply a unit moment at ④ and compute the resulting vertical deflection at ②. For the beam, $E = 200$ GPa and $I = 50 \times 10^6$ mm^4.

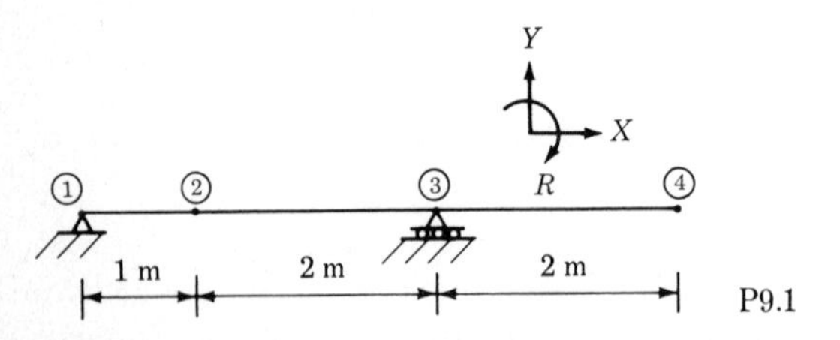

P9.1

[1] J. T. Oden, "Mechanics of Elastic Structures," McGraw-Hill Book Company, New York, 1967.

9-2. For the truss shown, show that $f_{3X2Y} = f_{2Y3X}$. EA is the same for each member.

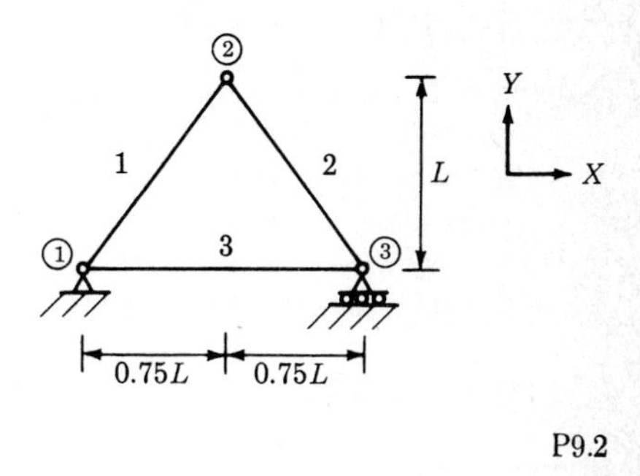

P9.2

9-3. For the frame shown, numerically show that $f_{1Y3R} = f_{3R1Y}$. Consider only bending deformations in the members. For each member, $I = 100$ in.^{4} and $E = 30 \times 10^3$ ksi.

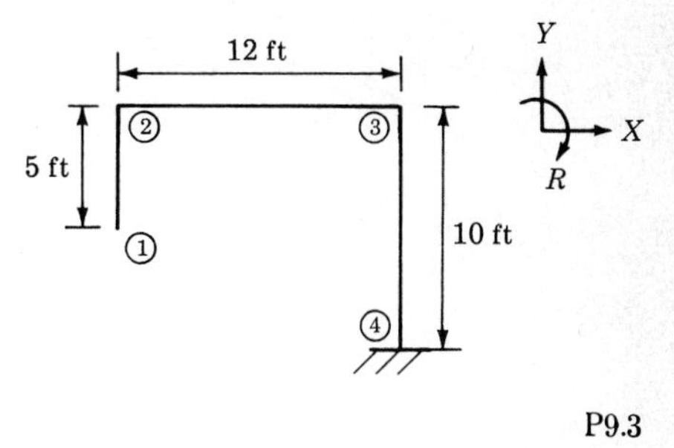

P9.3

9-4. For the beam shown, determine (*a*) the flexibility relation between

$$\begin{Bmatrix} Q_{2Y} \\ Q_{3Y} \end{Bmatrix} \quad \text{and} \quad \begin{Bmatrix} D_{2Y} \\ D_{3Y} \end{Bmatrix}$$

and (*b*) the corresponding stiffness matrix using inversion of the flexibility matrix. EI for the beam is constant.

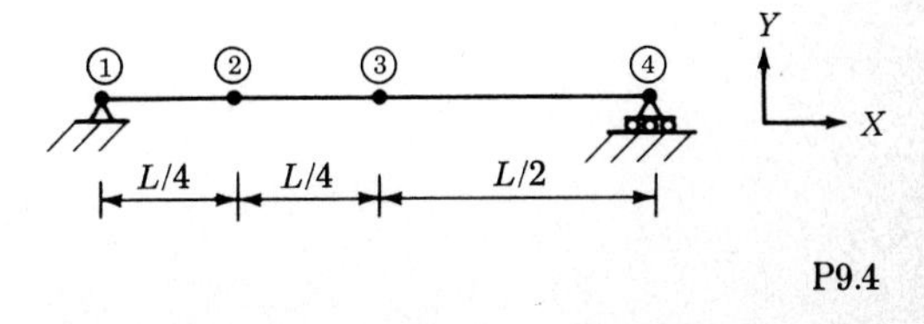

P9.4

9-5. For the pin-connected truss of Prob. 9-2, determine the flexibility and stiffness matrices for the relations between

$$\begin{Bmatrix} Q_{2X} \\ Q_{2Y} \end{Bmatrix} \quad \text{and} \quad \begin{Bmatrix} D_{2X} \\ D_{2Y} \end{Bmatrix}$$

9-6 and 9-7. The pinned-pinned member shown can be displaced at ① in the direction indicated by D. Draw the deflected configuration of the member for a displacement D and (*a*) determine the expression for the resulting axial force in the member in terms of the angle ω and the member stiffness properties E, A, and L; (*b*) using equilibrium conditions at ①, determine the expression for the force Q that would be associated with D.

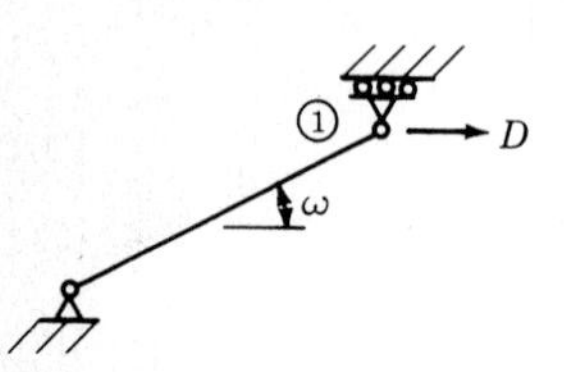

P9.6

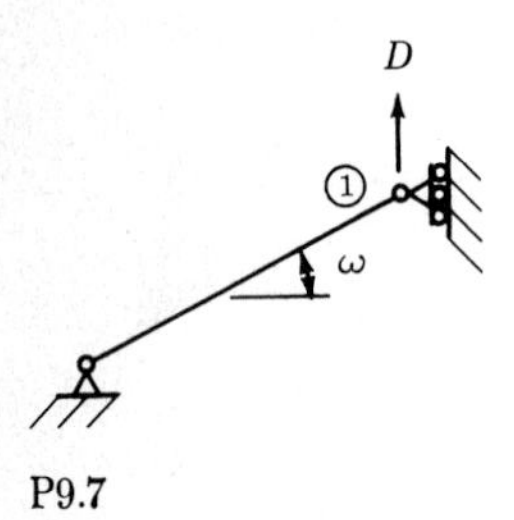

P9.7

CHAPTER
TEN

INFLUENCE LINES

10-1 DEVELOPMENT OF INFLUENCE LINES

In the last chapter we were concerned with the meaning and the development of influence coefficients. We saw that influence coefficients represent the influence of a unit force, or a unit deflection, at a *particular point* on a structure upon the deflection, or force, at *another point* on the structure. As discussed in Sec. 9-1, it is possible to express the effects of a force placed at various positions on a structure by means of influence lines.

DEFINITION: *An influence line is defined as a function whose value at a point represents the value of some structural quantity due to a unit force placed at the point.*

The structural quantities most often considered are reactions—shear, moment, or deflection at a point—and for trusses, member forces.

In previous chapters we determined a reaction or the shear, moment, or deflection due to a given load at a given point on the structure. We have thus far given little attention to the variation in the structural quantities as loads are moved about on the structure. For structures in which the positions of the loads can vary it is essential to understand

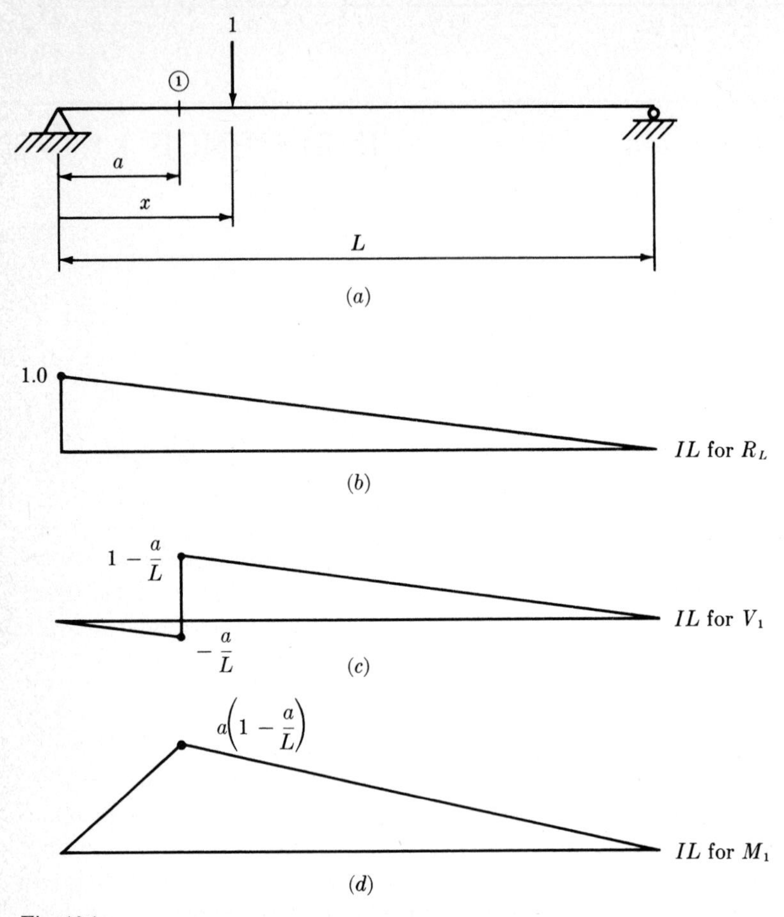

Fig. 10.1

how the structural quantities are affected. The effects can be described by influence lines.

For the basic idea of influence lines let us consider the simple beam of Fig. 10-1*a*. A unit load is applied to the beam at some position described as x. The units of the applied force are arbitrary. From statics, the value of the reaction at the left end is

$$R_L = \left(1 - \frac{x}{L}\right)(1) \tag{10-1}$$

The reaction is considered positive when it acts in an upward direction. Also from statics, the values of shear and moment at point ①, which is located a distance a from the left support, are

$$V_1 = \begin{cases} -R_R = -\dfrac{x}{L}(1) & 0 \le x \le a \\ R_L = \left(1 - \dfrac{x}{L}\right)(1) & a \le x \le L \end{cases} \tag{10-2}$$

and

$$M_1 = \begin{cases} R_R(L-a) = \dfrac{x}{L}(L-a)(1) & 0 \le x \le a \\ R_L a = \left(1 - \dfrac{x}{L}\right)a(1) & a \le x \le L \end{cases} \tag{10-3}$$

Shear and moment are considered positive according to the usual beam convention, as shown in Figs. 2-9*d* and *a*, respectively. Plotting Eqs. (10-1) to (10-3) results in the influence lines, as shown in Figs. 10-1*b*, *c*, and *d*. The coordinate axes R_L, V_1, M_1, and x are not generally included on the diagrams. The diagrams are simply labeled as shown on the right. It may be helpful, however, to retain the axes until their significance has been thoroughly grasped.

The influence lines in Fig. 10-1 were constructed from general equations for the structural quantity under consideration. Influence lines can also be constructed by placing the unit load at an appropriate number of positions, evaluating for each position the desired quantity under consideration, plotting the results, and drawing the influence line. The general expressions above help to clarify the meaning of an influence line by showing that the desired structural quantity is a function of x which reflects the position of the unit load.

Another approach to obtaining the influence line for shear at point ① in the beam of Fig. 10-1*a* is to observe that the shear at ① is equal to the value of the left reaction when the unit load is located anywhere to the right of ① and that it is equal to minus the value of the right reaction when the unit load is placed anywhere to the left of ①. Thus the influence line for V_1 can be obtained from the influence lines for R_L and $-R_R$, as shown in Fig. 10-2*a*, with those portions of the reaction influence lines that are not applicable disregarded. The values of the influence line at ① are obtained from the ratios of the distance of ① from the left and right reactions.

The influence line for moment at ① in the same beam can be obtained in a similar manner. Note that the value of M_1 is equal to $R_L a$ when the unit load is placed anywhere to the right of ① and is equal to $R_R(L - a)$ when the unit load is placed anywhere to the left of ①. The influence lines for the reactions can thus be used to determine the desired influence line. For the portion of the beam to the left of ① we can use the influence line for R_R and multiply its value by $L - a$. For the portion of the beam to the

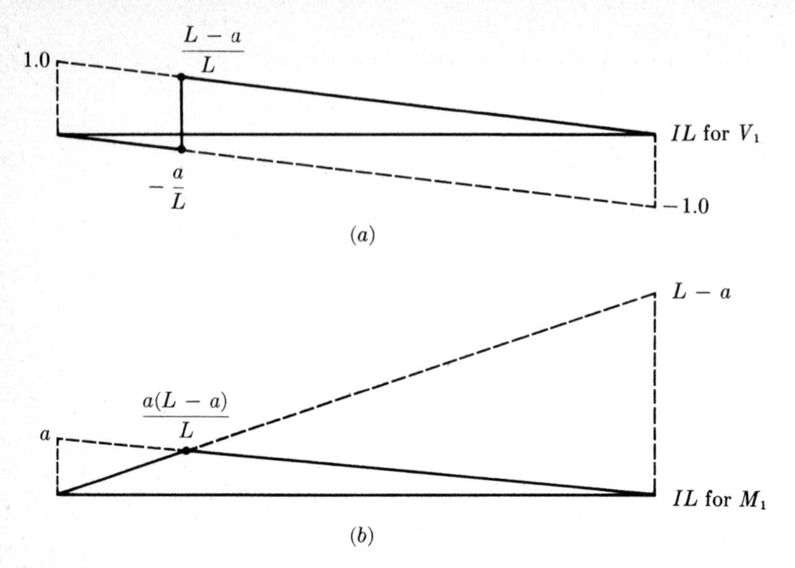

Fig. 10.2

right of ① we can use the influence line for R_L and multiply its value by a. The resulting construction is shown in Fig. 10-2*b*. The value of the influence line at ① can be obtained from the ratio of the distance of ① from the left or right support.

As an illustration of the use of influence lines for frames, consider the frame of Fig. 10-3*a*. Influence lines for shear and moment will be constructed for point P on member DB. P is 6 ft from the reaction at D. The unit load will be moved across the horizontal member from A to C. The position of the unit load is described by the distance x, which is measured from point A as shown in Fig. 10-3*b*.

The coordinate axes adjacent to member DB in Fig. 10-3*b* are used to define positive shear and moment for the member. Positive shear and moment are in accordance with the beam sign convention. We see from Fig. 10-3*b* that we can determine the shear and moment at P if we know the value of the horizontal reaction at D, R_{DH}. The reaction components at D and E are considered positive, as shown. Taking the summation of moments about point D and using the relation between R_{EH} and R_{EV}, we find

$$R_{EH} = \frac{(x-4)(1)}{26}$$

For equilibrium of forces in the horizontal direction on the entire frame,

$$R_{DH} = R_{EH}$$

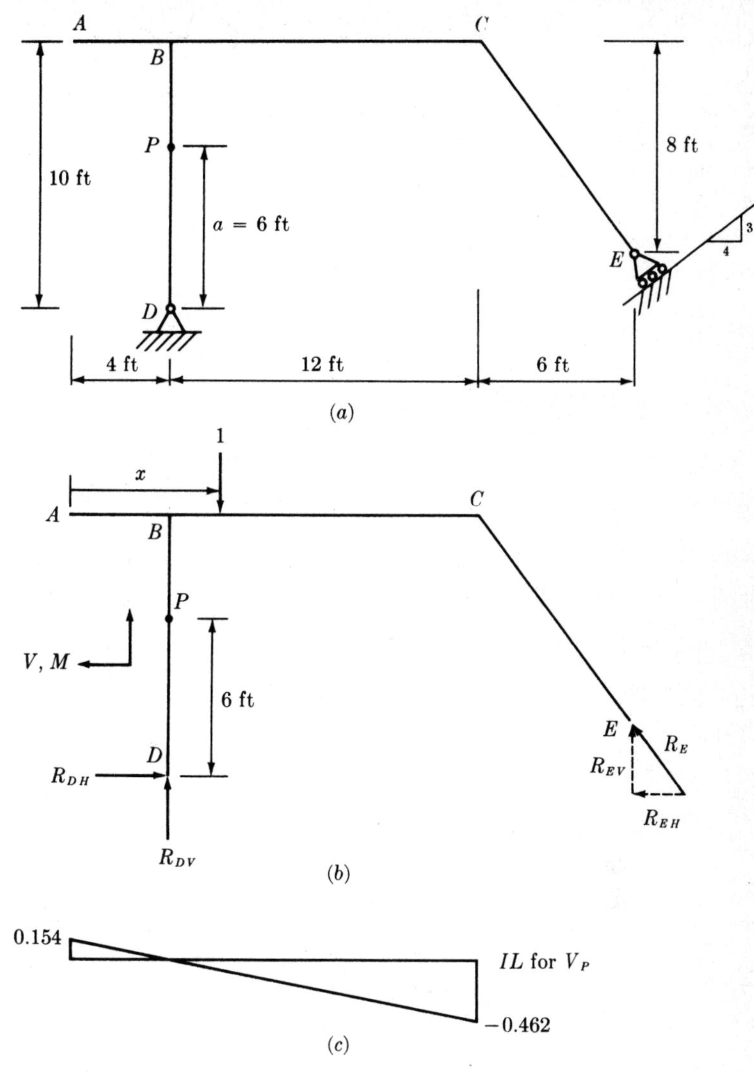
A
C
B
P
10 ft
a = 6 ft
8 ft
3
4
E
D
4 ft
12 ft
6 ft
(a)
1
x
A
C
B
P
V, M
6 ft
E
R_E
R_{EV}
D
R_{DH}
R_{EH}
R_{DV}
(b)
0.154
IL for V_P
−0.462
(c)

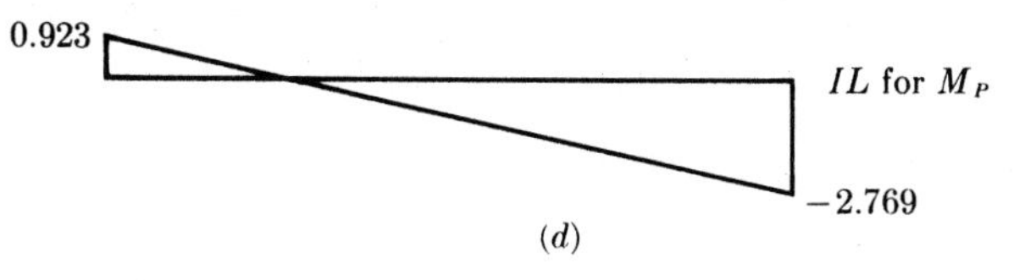
0.923
IL for M_P
−2.769
(d)

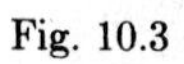
Fig. 10.3

The value of shear at the point under consideration is therefore

$$V_P = -R_{DH} = -R_{EH} = -\frac{(x-4)(1)}{26} \qquad 0 \leq x \leq 16$$

The resulting influence line for shear at point P is shown in Fig. 10-3*c*.

The influence line for moment at the same point is shown in Fig. 10-3*d*. Its development needs little explanation; the moment at P is equal to minus the product of the 6-ft length and the previously determined expression for R_{DH}. The resulting expression for moment is

$$M_P = -6R_{DH} = -\frac{(3)(x-4)(1)}{13}$$

As previously indicated, the sign of the moment is indicated by the coordinate axes next to member BD in Fig. 10-3*b*.

These concepts can be applied to the construction of influence lines for beams used in floor systems. The plan view of a typical building framing system is shown in Fig. 10-4. Similar systems are used in bridge construction. In such systems the live load is carried by the floor slab, which is supported by the joists. The joists are supported by the floor beams, which are in turn supported by the girders. The girders are supported by the building columns.

Influence lines for shear and moment will be constructed for the floor beam denoted by M-M in Fig. 10-4. An isolated sketch of this beam is shown in Fig. 10-5*a*. The supports of the beam are considered to be represented by the pinned-roller supports, as shown. To help in visualizing the interactions between the structural elements we assume the joists to rest on top of the beam and the floor slabs to rest on top of the joists. The schematic sketch of Fig. 10-5*a* shows that, regardless of the type of load-

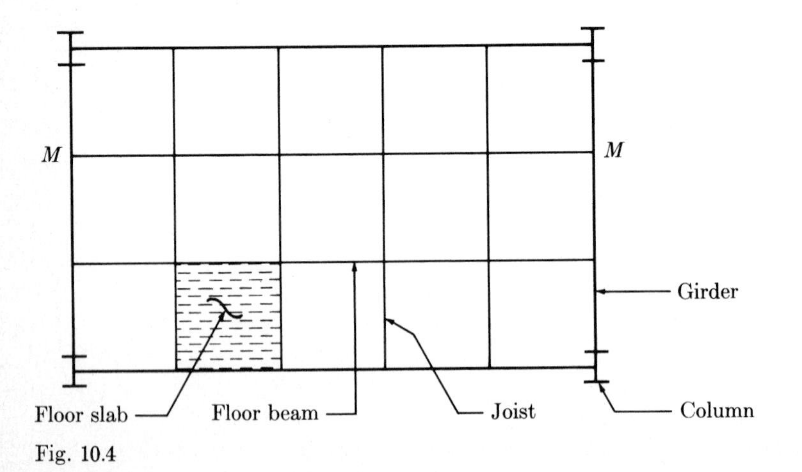

Fig. 10.4

ing on the floor slab, the beam is subjected to concentrated loads brought in by the joists. The portion of the floor beam between two joists is referred to as a *panel.* The points at which the floor beam supports the joists are referred to as *panel points,* indicated by $A, B, \ldots, F$ in Fig. 10-5*a*.

The influence line for shear in panel DE of the floor beam is constructed first. If the unit load is moved across the floor from point A to point F, the resulting influence line is as shown in Fig. 10-5*b*. The portion of the influence line from A to D is obtained in the same manner as for a simple beam. Even though the unit load is applied through the joists, the shear in panel DE is not dependent on the unit-load distribution to any two joists from A to D. A free-body diagram of the beam to the left of point E will clarify this point. However, when the unit load is located between D and E, the shear in panel DE is dependent on the ratio of the joist loads at D and E. If x describes the location of the unit load, measured from the left support, the force on the beam from the joist at D is $[(8 - x)1]/2$, where $6 \leq x \leq 8$. The shear in panel DE is equal to the reaction at the left support minus this force at D. Thus

$$V_{DE} = R_L - \left(4 - \frac{x}{2}\right)(1)$$

$$= \left[\frac{10 - x}{10} - \left(4 - \frac{x}{2}\right)\right](1)$$

$$= \left(-3 + \frac{2x}{5}\right)(1) \qquad 6 \leq x \leq 8$$

When $x = 6$ m, $V_{DE} = -0.6$, and when $x = 8$ m, $V_{DE} = 0.2$. It is also apparent that the shear in the panel is equal to zero when $x = 7.5$ m. The construction of the influence line from E to F follows the same procedures as from A to D.

The influence line for moment at midspan of the beam is shown in Fig. 10 5*c*. Construction of the influence lines from A to C and from D to F is again the same as for a simple beam. When the unit load is located between C and D, the expression for moment at midspan is found to be a constant value of 2.0.

The influence lines for shear in panel DE and for moment at midspan can also be developed by the technique described in Fig. 10-2 for a simple beam. The developments are shown in Fig. 10-6.

Influence lines are constructed for trusses in much the same manner as for beams. From the basic definition of an influence line, an influence line for a member in the truss is a function which indicates how the force in the member is influenced by a unit load moving across the truss. For many of the simple trusses the member force can be expressed in general

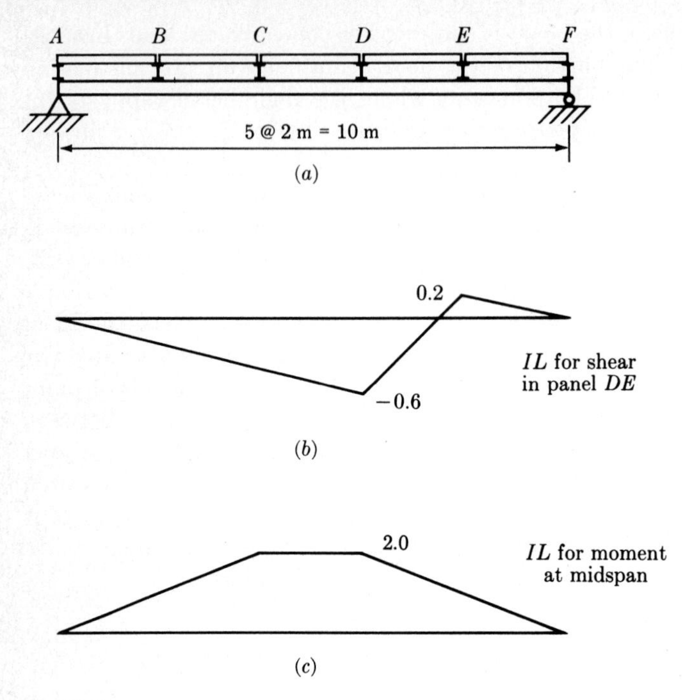

Fig. 10.5

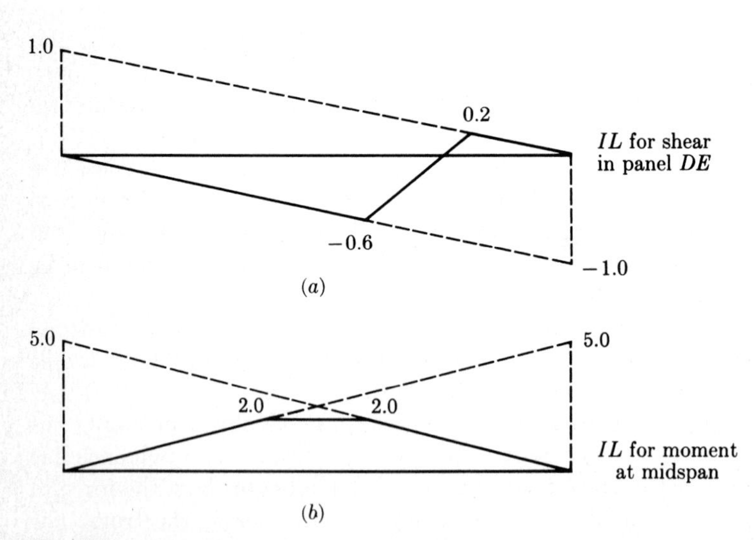

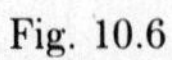

Fig. 10.6

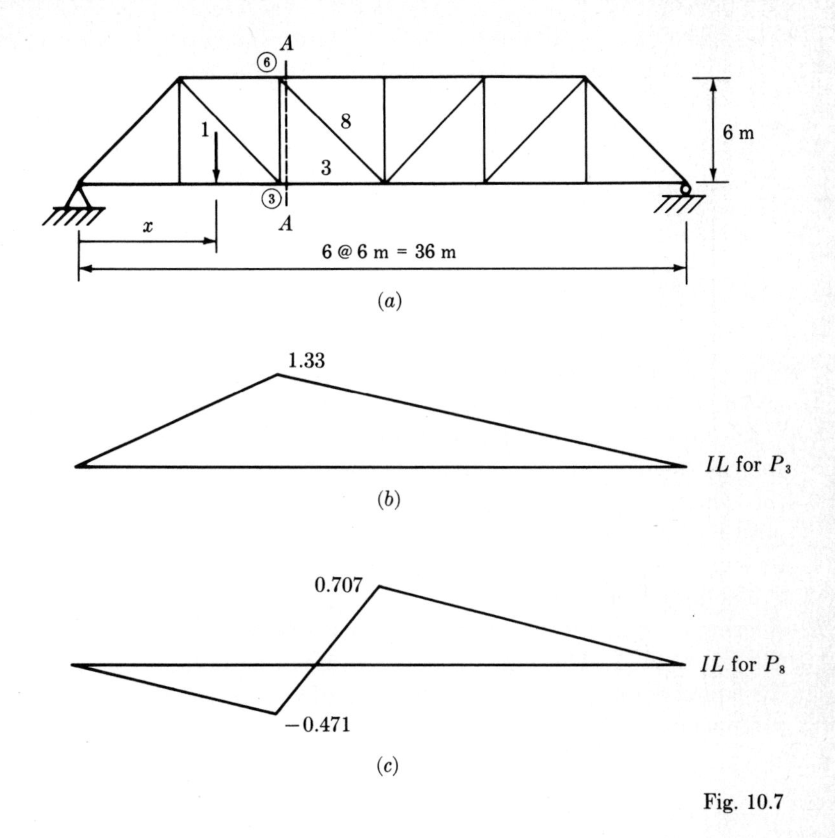

Fig. 10.7

form as a function of the parameter x, which describes the position of the unit load, as for beams.

Let us develop influence lines for the forces in members 3 and 8 of the truss in Fig. 10-7a, where a unit load is considered to be moved across the truss at the lower chord level. The position of the unit load is described by the distance x, measured from the left support. Loads applied to such a truss are brought into the truss at the lower joints by floor beams spanning between the trusses. The floor beams support the floor slab, and thus the loading, as for the building frame of Fig. 10-4.

The force in member 3 can be obtained by the method of sections. Cutting the structure at section A-A and considering a free-body diagram of the left section, we find the force in member 3 by taking the summation of moments about joint ⑥. The resulting expression for P_3 is

$$P_3 = \begin{cases} \dfrac{12R_L - (12 - x)(1)}{6} & 0 \le x \le 12 \\ 2R_L & 12 \le x \le 36 \end{cases}$$

The value of R_L is obtained by considering the summation of moments for the entire truss about the right support. The result is

$$R_L = \frac{(36 - x)(1)}{36}$$

The expressions for the force in member 3 therefore are

$$P_3 = \begin{cases} \dfrac{x}{9}(1) & 0 \leq x \leq 12 \\ \dfrac{(36 - x)(1)}{18} & 12 \leq x \leq 36 \end{cases}$$

The resulting influence line for the force in member 3 is shown in Fig. 10-7*b*. Tension, as before, is considered positive.

As stated earlier, the influence line can also be constructed by applying the unit load to the truss at a number of points on the lower chord—at the joints, for example—and evaluating the force in P_3 for each of these positions. The values of P_3 can then be plotted and the points connected by straight lines.

The influence line for the force in member 8 can be obtained by considering the same section of the truss used for P_3. Consider the section to the left of *A-A*. The force in member 8 can be obtained by considering the equilibrium of forces in the vertical direction. In terms of the vertical component of P_8, we obtain

$$P_{8V} = R_L - 1 \qquad 0 \leq x \leq 12$$

As the unit load is moved to the right of joint ③, with $12 \leq x \leq 18$, a portion of the load is transmitted to joint ③ by the floor system. The expression for the vertical component of force in member 8 is therefore

$$P_{8V} = R_L - \frac{(18 - x)(1)}{6} \qquad 12 \leq x \leq 18$$

With the unit load in the range $18 \leq x \leq 36$, we have

$$P_{8V} = R_L$$

Substituting the expression for R_L into these expressions and noting that $P_8 = 1.414P_{8V}$, we obtain

$$P_8 = \begin{cases} -\dfrac{1.414x}{36}(1) & 0 \leq x \leq 12 \\ \dfrac{1.414(5x - 72)}{36}(1) & 12 \leq x \leq 18 \\ \dfrac{1.414(36 - x)}{36}(1) & 18 \leq x \leq 36 \end{cases}$$

The resulting influence line for the force in member 8 is shown in Fig. 10-7*c*.

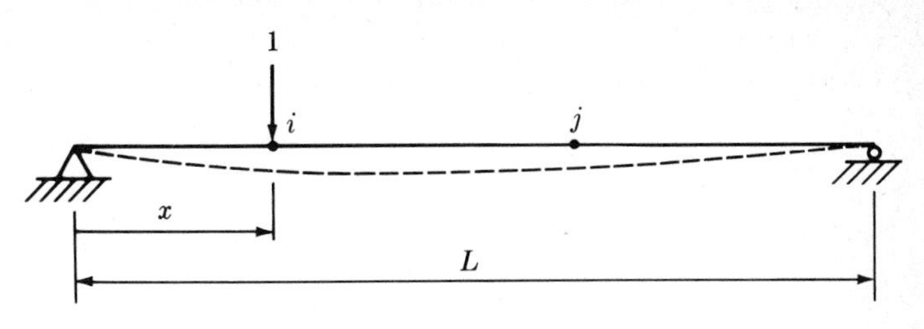

Fig. 10.8

The procedure for developing influence lines for deflection in a structure can be seen by considering the simple beam of Fig. 10-8, where the influence line for vertical deflection at point j is to be developed. The point at which the unit force acts, located by the distance x, is denoted as i; point i can be at any position along the length of the beam. From the basic definition of a flexibility influence coefficient in Sec. 9-2, the deflection at point j due to a unit force at i is equal to f_{ji}. However, from Maxwell's reciprocal theorem we have seen that $f_{ji} = f_{ij}$, provided, as we are assuming, that the structure is linearly elastic. Therefore we can conclude that the deflected shape of the beam resulting from placing a unit force at j is the influence line for the vertical deflection at j.

As this development shows, the influence line for deflection at any point on a structure is the deflected shape of the structure resulting from the application of a unit force at the point. Influence lines can thus be developed for either linear or rotational deflections. The applied unit force is either a force or a moment, depending on the type of deflection being considered.

10-2 INFLUENCE LINES FROM DEFLECTED SHAPES

Influence lines for force quantities in a structure can be developed from deflected shapes of the structure. As we shall see, this method of constructing influence lines is particularly useful for more complicated structures. The concept of constructing influence lines from deflected shapes is based on the Müller-Breslau principle.[1]

MÜLLER-BRESLAU PRINCIPLE: *The influence line for any force quantity in a structure is represented to some scale by the deflected shape of the structure resulting from moving the force quantity under consideration through a small displacement.*

[1] Developed by Prof. Heinrich Müller-Breslau (1851–1925), *Z. Archit. Ing. Ver.*, Hannover, 1885.

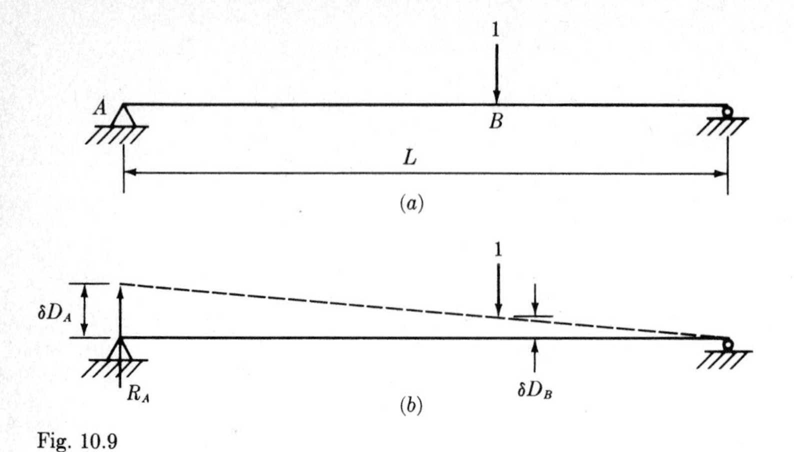

Fig. 10.9

It must be emphasized that the Müller-Breslau technique applies to the construction of influence lines for *force* quantities.

In order to verify this principle let us consider the simple beam of Fig. 10-9*a*, for which influence lines were previously developed in Fig. 10-1. The Müller-Breslau principle will be applied first to the development of the influence line for the reaction at A. A unit load is placed on the beam at some arbitrary point B, as shown. If the reaction at A is moved through a small displacement, the deflected configuration of the beam is as shown in Fig. 10-9*b*. If the small displacement at A is assumed to be a virtual displacement, the expression for the virtual work done is

$$R_A \delta D_A - 1\delta D_B = 0$$

$$\text{or} \qquad R_A = \frac{\delta D_B}{\delta D_A} \quad (1)$$

Because the magnitude of the imposed deflection δD_A is arbitrary, we can for convenience assume a value of unity. Thus the influence line for R_A is represented directly by the deflected shape. The resulting influence line is the same as that obtained earlier in Fig. 10-1*b*.

Let us consider next the application of the Müller-Breslau principle to the development of influence lines for shear. The influence line for shear at point C on the beam of Fig. 10-10*a* is to be developed. Point C is located a distance a from the left support. The unit load is again applied to the beam at some arbitrary point B. The displacement to be given to the beam is a displacement in the direction of the force quantity under consideration, which is shear. The desired type of displacement for shear is obtained if a pinned-bar connection is introduced in the beam at C, as shown in Fig. 10-10*b*. Such a connection permits the desired type of dis-

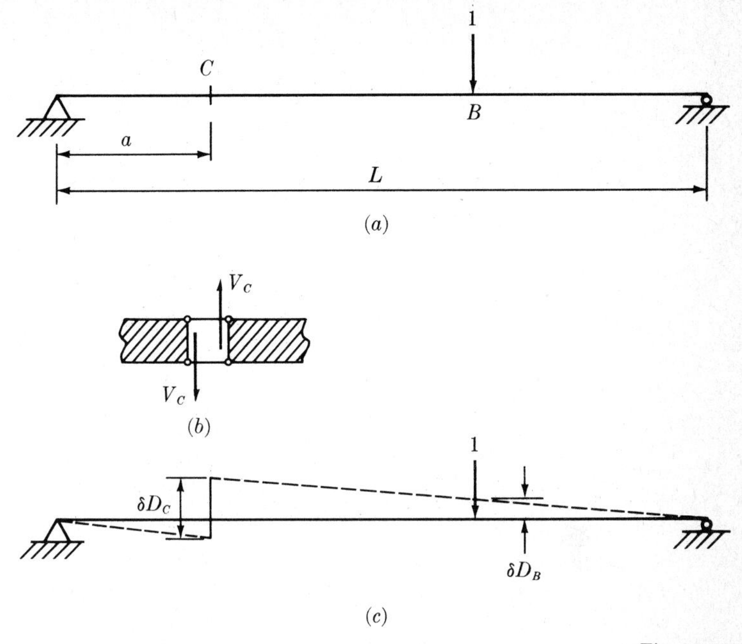

Fig. 10.10

placement but prevents relative angular displacement between the two portions of the beam. The direction of the displacement to be imposed is in the same direction as positive quantities of shear. The deflected shape of the beam is therefore as shown in Fig. 10-10c. The two portions of the beam remain parallel during the displacement. With the displacement considered to be a virtual displacement, the expression for virtual work is

$$V_C \delta D_C - 1 \delta D_B = 0$$

$$\text{or} \qquad V_C = \frac{\delta D_B}{\delta D_C} \quad (1)$$

If, as before, we assume the value of the relative displacement δD_C to be unity, the influence line for shear at C is represented directly by the deflected shape.

Influence lines for moment can be obtained by introducing a pinned connection in the structure at the point under consideration. Thus, if a pinned connection is introduced at C in Fig. 10-11 and the beam is deflected through an angle of δD_C at C, the virtual work done is

$$M_C \delta D_C - 1 \delta D_B = 0$$

$$\text{or} \qquad M_C = \frac{\delta D_B}{\delta D_C} \quad (1)$$

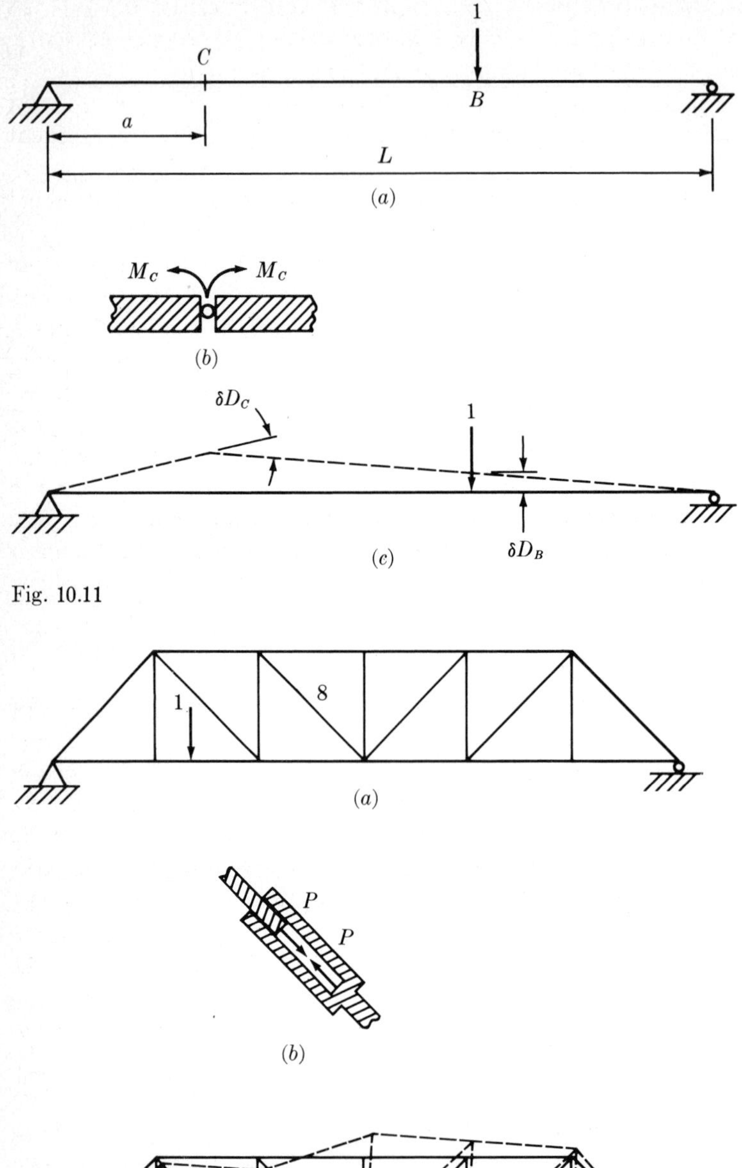

Fig. 10.11

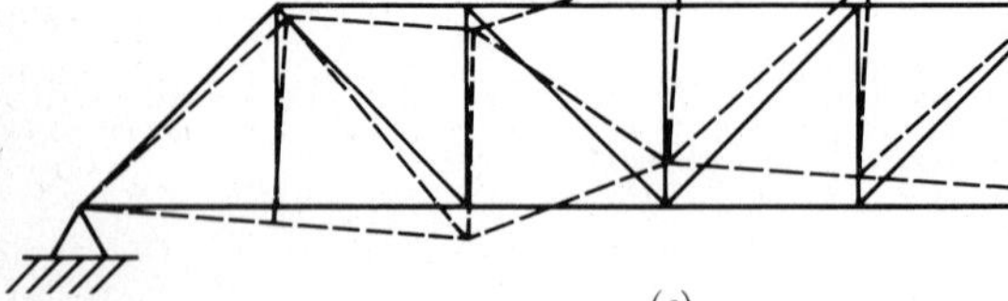

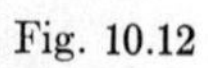
Fig. 10.12

The rotational displacement at C is in the same direction as positive moment at C. Assuming a unit value of virtual rotation at C, $\delta D_C = 1$, we see that the influence line for moment is represented directly by the deflected shape. The equivalence of the results obtained with the deflected shape and the influence line previously developed in Fig. 10-1 are evident from Fig. 10-11*c*, where the magnitude of the vertical displacement at C is $(a/L)(L - a)\delta D_C$. For $\delta D_C = 1$ the value is the same as that obtained previously in Fig. 10-1*d*.

Influence lines for member forces of trusses can also be obtained from deflected shapes. Consider the truss of Fig. 10-7, shown again in Fig. 10-12*a*. The influence line for the force in member 8 is to be determined. To generate the deflected shape we introduce a piston-and-cylinder connection at some point on member 8. Such a connection is shown in cross-section in Fig. 10-12*b*. If a tensile force in member 8 is moved through a small displacement, with the displacement represented by a contraction of the piston-and-cylinder connection, the deflected shape of the truss is as shown in Fig. 10-12*c*. Consideration of the virtual work involved would show that for a unit value of displacement in member 8 the influence line for the force in member 8 is represented by the deflected shape of the lower chord in Fig. 10-12*c*.

Deflected shapes can also be used for constructing influence lines for frames such as the one discussed in Fig. 10-3.

The use of deflected shapes to determine influence lines becomes even more helpful in complicated structures. For example, consider the structure of Fig. 10-13*a*. Influence lines are to be determined for a unit load placed anywhere along the length of the structure. Applying the Müller-Breslau principle, we obtain the influence line for the reaction at A by permitting the reaction at A to displace a unit distance. The resulting deflected shape, representing the influence line for R_A, is shown in Fig. 10-13*b*. The influence line for shear at B is obtained by permitting positive shear at B to relatively displace a unit value. Because the portion of the beam to the left of B cannot move, the unit displacement is imposed on the portion of the beam to the right of B. The displaced portion of the beam must remain parallel to the stationary portion at B. The result is shown in Fig. 10-13*c*. The influence line for moment at C is obtained by inserting a pin at C and imposing a unit rotation in the direction of positive moment. The deflected shape, and hence the influence line for moment at C, is as shown in Fig. 10-13*d*. Because the deflected shapes obtained for statically determinate structures are composed of straight-line segments, the value of an influence line can be determined at any point on the beam provided the lengths of the members are given.

Up to this point we have been concerned with influence lines only for statically determinate structures. Although methods for analyzing stati-

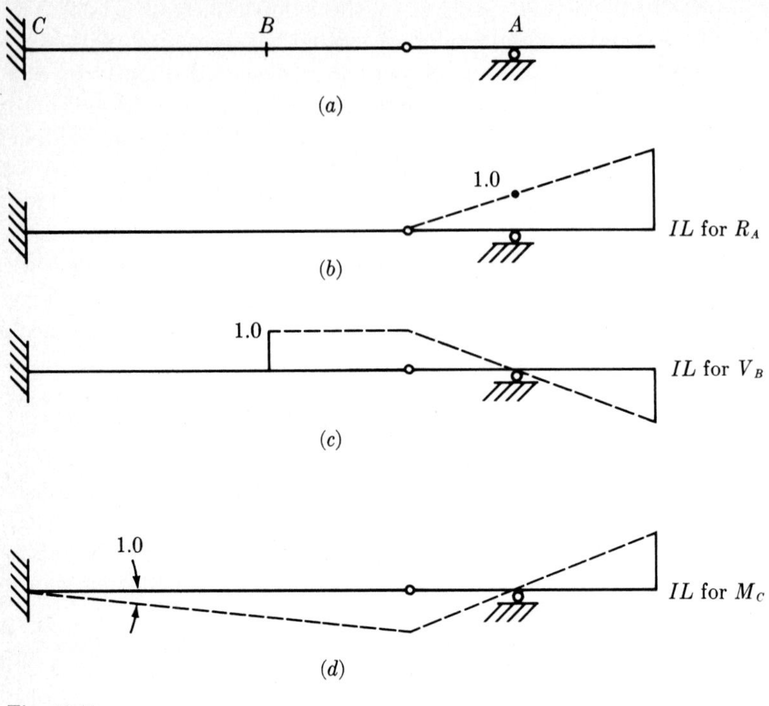

Fig. 10.13

cally indeterminate structures are not treated until later in the text, it is appropriate to point out here the applicability of deflected shapes in constructing influence lines for statically indeterminate structures. Often it is desirable to develop influence lines only in a qualitative sense, that is, to determine only the shape and not the values of the influence lines. Such qualitative influence lines are very useful in studying loading patterns on a structure. Evidence of this will be given in the following section, where the uses of influence lines are discussed.

The development of qualitative influence lines for statically indeterminate structures follows the same general procedure as demonstrated above and as stated in the Müller-Breslau principle. Consider the continuous beam of Fig. 10-14*a*. To develop a particular influence line, we impose a unit displacement in the direction of the positive force quantity under consideration. The resulting deflected shape must be such that it is consistent with the given restraints of the structure. Instead of straight-line segments, as for statically determinate structures, the deflected shapes for statically indeterminate structures are curved segments. Achieving a good deflected shape for a statically indeterminate structure often re-

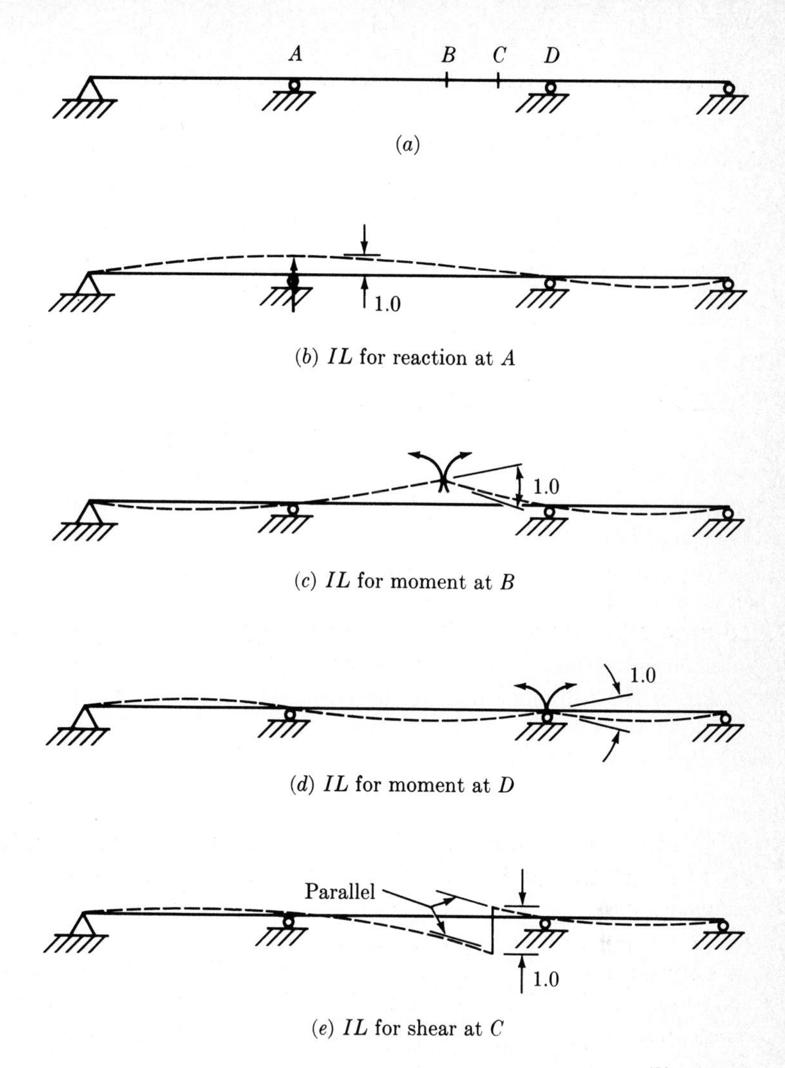

(a)

(b) IL for reaction at A

(c) IL for moment at B

(d) IL for moment at D

(e) IL for shear at C

Fig. 10.14

quires more thought and consideration than for a statically determinate structure.

The deflected shape, and hence the influence line, for the reaction at *A* is shown in Fig. 10-14*b*. The deflected shape is obtained by moving the reaction at *A* through a unit value of displacement. Influence lines for moment at *B* and *D* are obtained in Figs. 10-14*c* and *d*, respectively, by inserting pins at the points and imposing unit rotations. The influence line for shear at *C* is obtained in Fig. 10-14*e* by imposing a unit value of relative displacement transverse to the axis of the beam. As noted, the segments of the deflected shape must be parallel at C.

For statically indeterminate structures such as the continuous beam of Fig. 10-14 it is sometimes desirable to determine the values of an influence line at various points on the structure; that is, quantitative influence lines are desired. The determination of quantitative influence lines for statically indeterminate structures is treated in a later chapter. In addition to using analytic methods, we can also evaluate influence lines from model analysis of the structure.[1]

10-3 USES OF INFLUENCE LINES

Once we have developed an influence line for a structure, we can use it to determine the value of the structural quantity for various types of loading. Let us begin with the use of influence lines for concentrated loads. Consider again the simple beam of Fig. 10-1, with $a = 1.5$ m and $L = 6$ m. The influence lines for shear and moment at ① are shown in Figs. 10-15*b* and *c*. The value of shear or moment at ① due to a concentrated load placed on the beam is equal to the product of the value of the influence line at the point where the load is placed and the value of the load. For example, if a concentrated load of 50 kN is placed on the beam 2 m from the left support, the value of V_1 from Fig. 10-15*b* is $(0.667)(50) = 33.33$ kN. The value of the influence line at this point, 0.667, can also be obtained from Eq. 10-2, with $x = 2$ and $L = 6$. Similarly, the value of M_1 from the influence line of Fig. 10-15*c* or Eq. 10-3 is found to be $(1.0)(50) = 50.0$ kN-m.

Now let us consider the case where the beam is subjected to a series of concentrated loads, as shown in Fig. 10-15*d*. The resulting shear and moment will be obtained for the same point on the beam as above, with the influence lines of Figs. 10-15*b* and *c*.

The shear and moment are obtained by summing the influences of the individual concentrated loads, with the values of the influence line used where the loads are applied. Thus for shear

$$\begin{aligned} V_1 &= (-0.167)(50) + (0.417)(25) + (0.25)(25) \\ &= 8.325 \text{ kN} \end{aligned}$$

Note that the value of the influence line for the 50-kN load is negative. The shear at the point under consideration due to this load is therefore negative. The total value of shear due to all of the loads, however, is positive.

[1] T. M. Charlton, "Model Analysis of Structures," John Wiley & Sons, Inc., New York, 1954.

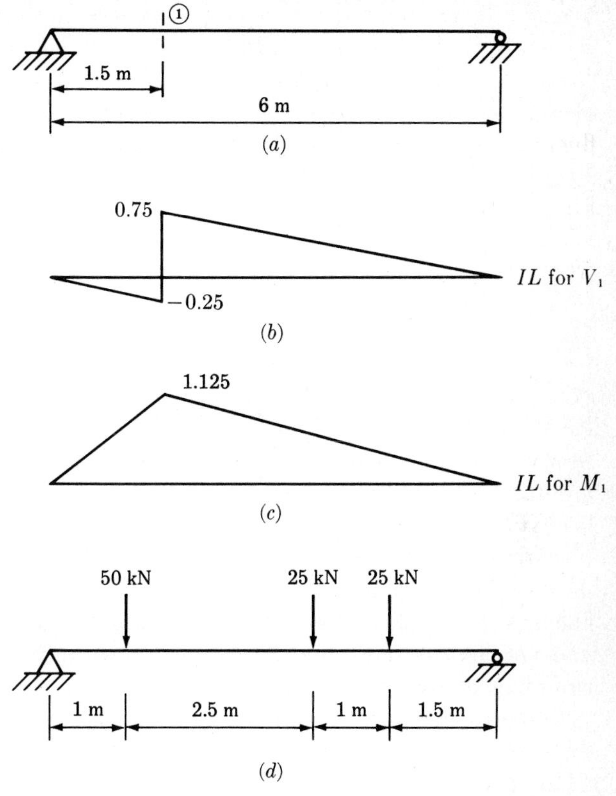

Fig. 10.15

In a similar manner the moment is found to be

$$
\begin{aligned}
M_1 &= (0.75)(50) + (0.625)(25) + (0.375)(25) \\
&= 62.5 \text{ kN-m}
\end{aligned}
$$

All loads for this beam contribute positively to the total moment.

Influence lines can also be used to determine the value of a structural quantity due to distributed loads. Consider a segment of a member loaded from point b to point c with a distributed load w_x, as shown in Fig. 10-16*a*. The subscript x indicates that the magnitude of the distributed load is a function of the location on the beam. In many cases the distributed load has a constant value. To develop general results, however, we shall consider the distributed load to be a function of x. When used in connection with influence lines, distributed loads are considered positive when they act in the same direction as the unit load used to develop the influence line.

A segment of an influence line for the member is shown in Fig. 10-16*b*.

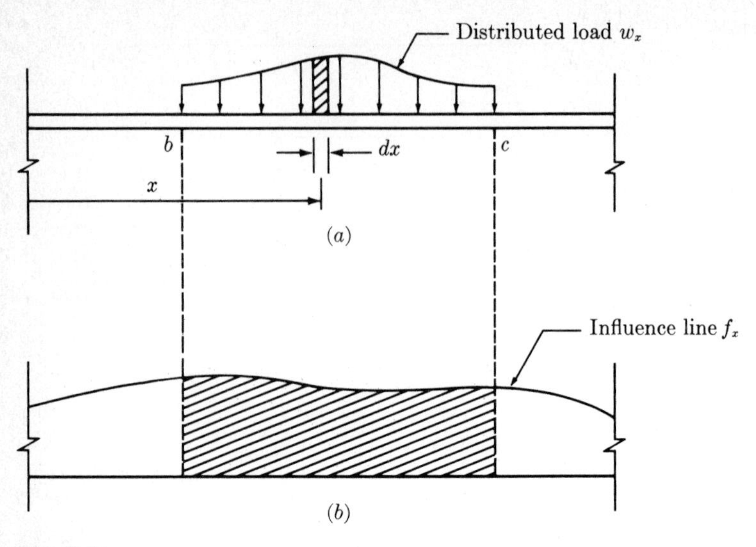

Fig. 10.16

The ordinate of this influence line is denoted by the function f_x. For convenience, in the following discussion the structural quantity for which the influence line has been constructed will be denoted by the letter F.

We see from Fig. 10-16*a* that the distributed load w_x acting on a differential length of the beam dx is equivalent to a concentrated load of magnitude $w_x\,dx$. The influence of this concentrated load on some structural quantity is therefore

$$dF = w_x\,dx\,f_x \tag{10-4}$$

That is, the value of the structural quantity is equal to the product of the concentrated load and the value of the influence line at that point. For the total effect Eq. (10-4) is integrated over the length of the beam on which the distributed load acts. Thus

$$F = \int_b^c w_x f_x\,dx \tag{10-5}$$

If the distributed load is uniform, w_x is constant and Eq. (10-5) can be written as

$$F = w \int_b^c f_x\,dx \tag{10-6}$$

From Eq. (10-6) we can state that the value of a structural quantity due to a uniform load is equal to the product of the magnitude of the uniform load and the area under the corresponding portion of the influence line. For example, if the beam in Fig. 10-15*a* is loaded with a uniform load of

20 kN/m over its entire length, the value of moment at a section 1.5 m from the left support is found from Fig. 10-15*c* to be

$$F = M_1 = \frac{(6)(1.125)(20)}{2}$$
$$= 67.5 \text{ kN-m}$$

The value of shear at ① for the same load is found from Fig. 10-15*b* to be

$$F = V_1 = \left[\frac{(1.5)(-0.25)}{2} + \frac{(4.5)(0.75)}{2}\right](20)$$
$$= 30 \text{ kN}$$

If the uniform load were placed only over the right 4.5 m of the beam, from Fig. 10-15*b* we see that the value of shear at the section under consideration would be

$$F = V_1 = \frac{(4.5)(0.75)(20)}{2}$$
$$= 33.75 \text{ kN}$$

Influence lines can be used advantageously in this manner to determine the placement of the uniform load for maximum effects.

When a structure is subjected to the wheel loads of a moving vehicle, it is often desirable to determine the location of these wheel loads which results in the maximum shear or moment at a point on the structure. Influence lines are a valuable aid in determining the location. For some types of structures the location of the loads for maximum effects can be determined by inspection or a trial-and-error procedure, with the shear and moment evaluated as in the previous paragraphs. For a simple beam a direct technique can be used. Suppose the beam of Fig. 10-17*a* is to be subjected to a series of concentrated moving loads. The maximum shear and moment are to be determined at point B, 4 m from the left support. The influence lines for shear and moment are shown in Figs. 10-17*b* and *c*, and the series of moving loads to be considered are shown in Fig. 10-17*d*. The problem is essentially to determine the position of these loads for which the shear and moment at point B on the beam are maximum. A systematic procedure of moving the loads across the beam is followed. The loads in this discussion will be moved from right to left.

Let us first determine the position of the loads for maximum shear. This is accomplished by moving the loads in successive steps across the beam and observing the manner in which the shear at B changes. From Fig. 10-17*b* we see that if a load P is initially located over the right support and then moved 1 m to the left, the value of shear at B increases from 0 to $0.1P$. The slope of the influence line can be used to evaluate the change in the quantity under consideration. It is seen in Fig. 10-17*b*

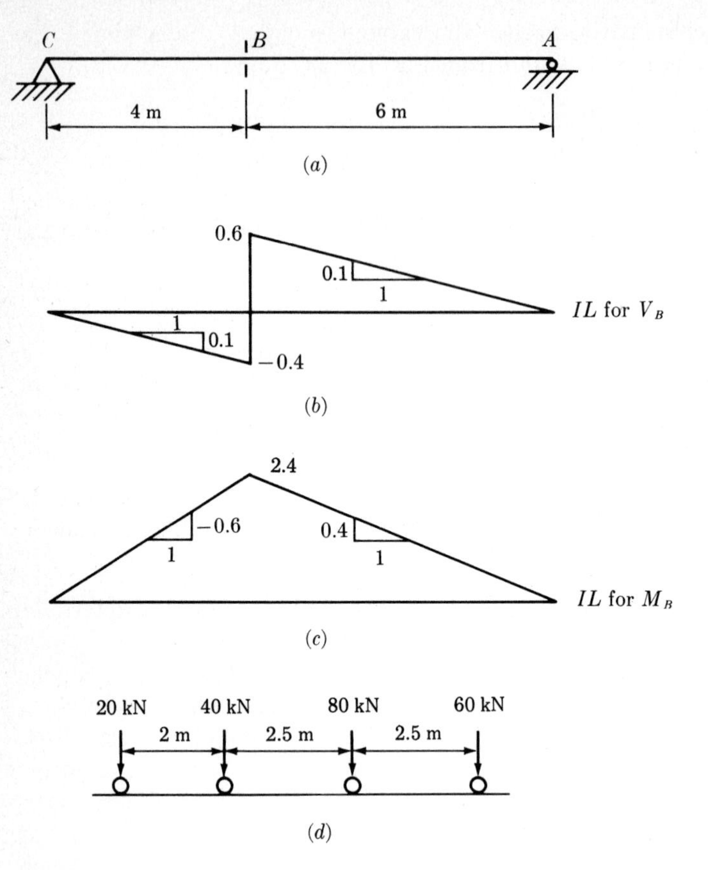

Fig. 10.17

that the slope of the influence line from A to B is $0.6/6 = 0.1/1$. The slope from B to C is $0.4/4 = 0.1/1$. Note that the slope from B to C is considered positive because the movement of a concentrated load from right to left in this section of the beam causes the shear at B to become less negative. Thus we can conclude that the change in the value of a particular quantity during the movement of a load is equal to the product of the slope of the influence line, the magnitude of the load, and the distance the load is moved.

If the loads of Fig. 10-17d are moved onto the beam from the right, the value of shear at B increases continuously until the 20-kN load reaches B. As the 20-kN load moves past B, there is an abrupt decrease in the shear at B. However, as the loads continue to move left, the value of shear at B again increases, until the 40-kN load reaches B. Point B

experiences a series of increases and abrupt decreases in shear value during the movement of the load across the beam. The maximum shear value is obtained with one of the loads just to the right of B. The maximum value of shear can be obtained by examining the change in shear at B during each movement. If the change in shear is positive for a movement, the maximum shear value has not yet been obtained. When the change becomes negative for a movement, the location of the loads for maximum shear is that location prior to the movement.

The 20-kN load is considered initially to be just to the right of B. The loads are moved to the left until the 40-kN load is just to the right of B. During this movement the change in shear at B is

$$\begin{aligned} \Delta V_B &= (-1.0)(20) + (0.1)(140)(2) + (0.1)(60)(1) \\ &= 14.0 \text{ kN} \end{aligned}$$

Note that the 60-kN load comes onto the beam during this movement, but it moves only 1 m along the beam. The results show that the shear at B increases by 14.0 kN during this movement.

The procedure is repeated, with the 40-kN load moved past B and the 80-kN load moved up to B. The resulting change in shear is

$$\begin{aligned} \Delta V_B &= (-1.0)(40) + (0.1)(180)(2.5) + (0.1)(20)(2) \\ &= 9.0 \text{ kN} \end{aligned}$$

The 20-kN load leaves the beam during this movement, and the shear again increases.

Moving the 60-kN load over to B results in

$$\begin{aligned} \Delta V_B &= (-1.0)(80) + (0.1)(140)(2.5) + (0.1)(40)(1.5) \\ &= -39.0 \text{ kN} \end{aligned}$$

The shear at B for this movement decreases. The maximum value of shear is therefore obtained with the 80-kN load just to the right of B. With the loads in this position, the maximum value of V_B is found from the influence line of Fig. 10-17b to be

$$\begin{aligned} V_{B\,\max} &= (-0.15)(40) + (0.6)(80) + (0.35)(60) \\ &= 63.0 \text{ kN} \end{aligned}$$

The maximum moment at B can similarly be obtained from the influence line of Fig. 10-17c. As loads are moved from A to B, the moment at B increases, so the slope of the influence line in this section of the beam is positive. Loads moving from B to C cause a decrease in the moment at B, so the slope of this segment of the influence line is negative. The maximum moment at B will occur with one of the loads at B.

There is a continuous increase in the moment at B as the loads are moved onto the beam from the right until the 20-kN load is at B. As the 20-kN load is moved past B, its contribution to the moment at B becomes negative. The position with the 20-kN load at B will be used as the initial position in calculating the change in moment during successive movements. If we move the loads to the left from this position until the 40-kN load is at B, the change in moment at B is

$$\begin{aligned}\Delta M_B &= (-0.6)(20)(2) + (0.4)(120)(2) + (0.4)(60)(1) \\ &= 96.0 \text{ kN-m}\end{aligned}$$

This results in an increase in the moment at B. The loads are again moved to the left until the 80-kN load is at B. The change in moment at B for this movement is

$$\begin{aligned}\Delta M_B &= (-0.6)(20)(2) + (-0.6)(40)(2.5) + (0.4)(140)(2.5) \\ &= 56.0 \text{ kN-m}\end{aligned}$$

The moment again increases. Moving the 60-kN load to B results in

$$\begin{aligned}\Delta M_B &= (-0.6)(40)(1.5) + (-0.6)(80)(2.5) + (0.4)(60)(2.5) \\ &= -96.0 \text{ kN-m}\end{aligned}$$

The moment at B decreases for this movement; therefore the maximum moment at B is obtained with the 80-kN load at B. With the loads in this position, the maximum moment is found from the influence line of Fig. 10-17c to be

$$\begin{aligned}M_{B\,\max} &= (0.9)(40) + (2.4)(80) + (1.4)(60) \\ &= 312.0 \text{ kN-m}\end{aligned}$$

The following example illustrates another aspect of moving loads. This aspect has to do with determining the direction in which loads should be facing to produce the maximum effects.

EXAMPLE 10-1 The maximum moment at point A is to be determined for the bridge girder of Fig. 10-18a. The loading is a standard truck loading of Fig. 10-18b, which is 4 kips on the front wheel, 16 kips on the rear wheel, and 16 kips on the trailer wheel. The spacing between wheels is 14 ft. The problem is to determine which wheel is to be placed over point A and in which direction the truck is to be facing to produce the maximum moment at A.

The influence line for the moment at point A is shown in Fig. 10-19. The slopes of the influence line are as shown for loads moving from right to left. The truck will first be considered to be facing to the right; that is,

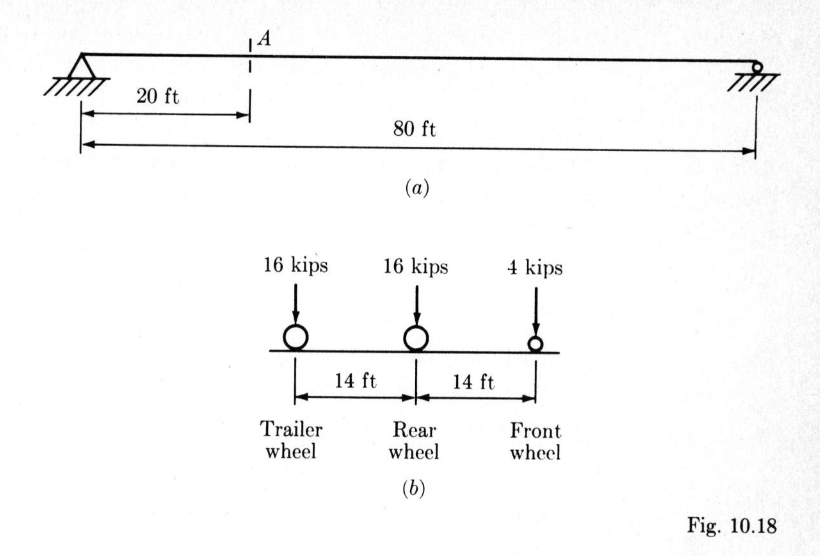

Fig. 10.18

the 4-kip load will be on the right. The 16-kip trailer wheel will be placed at A for the initial position. If the loads are moved to the left until the rear wheel is at A, the change in moment at A is

$$\begin{aligned} \Delta M_A &= (-0.75)(16)(14) + (0.25)(20)(14) \\ &= -98.0 \text{ ft-kips} \end{aligned}$$

The initial position therefore produces the maximum moment at A. The value of maximum moment is

$$\begin{aligned} M_{A\,\max} &= (15)(16) + (11.5)(16) + (8)(4) \\ &= 456 \text{ ft-kips} \end{aligned}$$

The truck is next considered to be facing to the left with the front wheel initially at A. Moving the loads to the left until the rear wheel is at A

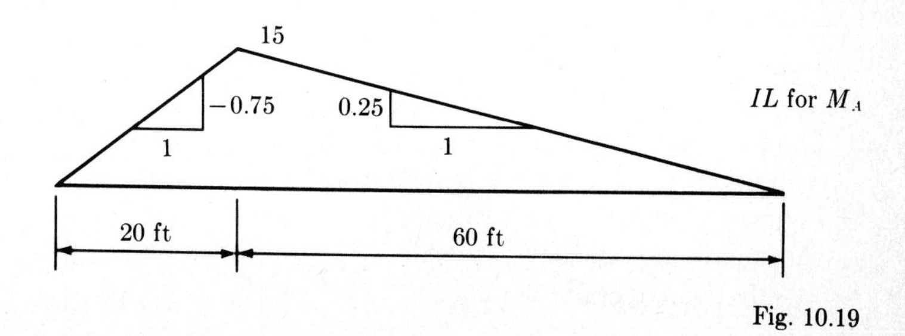

Fig. 10.19

results in

$$\Delta M_A = (-0.75)(4)(14) + (0.25)(32)(14)$$
$$= 70 \text{ ft-kips}$$

which indicates that the moment increases during the movement. With the rear wheel at A, the truck is moved to the left until the trailer wheel is at A. This results in a change in moment of

$$\Delta M_A = (-0.75)(4)(6) + (-0.75)(16)(14) + (0.25)(16)(14)$$
$$= -130 \text{ ft-kips}$$

The decrease in moment indicates that the maximum moment is obtained with the rear wheel at A. The maximum moment is found to be

$$M_{A\,\max} = (4.5)(4) + (15)(16) + (11.5)(16)$$
$$= 442 \text{ ft-kips}$$

The previously determined maximum value is larger. The maximum value of moment at A is therefore 456 ft-kips, with the truck facing to the right with the trailer wheel at A. It should be noted that the positioning of the wheels is not necessarily the same as this for determining the maximum moment at other points on the girder.

This method of determining the increase or decrease in a structural quantity is also applicable to other types of influence lines. In some cases it may not be necessary to move the loads through all possible positions. Often by inspecting the magnitude of the loads the procedure can be reduced to examining only a few loads at the position under consideration. In some cases where there are only a few loads, the location can be determined simply by inspection. When in doubt, however, the general procedure should be used.

The influence lines for trusses can be used to determine the maximum force in a member due to moving loads just as was done for beams. The following example illustrates their application.

EXAMPLE 10-2 The maximum force in member 8 of the truss in Fig. 10-7 is to be determined for simultaneous uniform and concentrated loadings. The uniform load is 20 kN/m which can be placed anywhere along the length of the truss and the concentrated loads are 50 kN each, spaced 3 m apart. The previously constructed influence line of Fig. 10-7c can be used to determine the maximum force in member 8.

The loads are placed in the position shown in Fig. 10-20 to develop the maximum force in member 8. Only the lower chord line of the truss is shown in the figure. The influence line indicates that the maximum force in the member is positive (tensile). The uniform load is placed over the length of the truss for which the value of the influence line is positive.

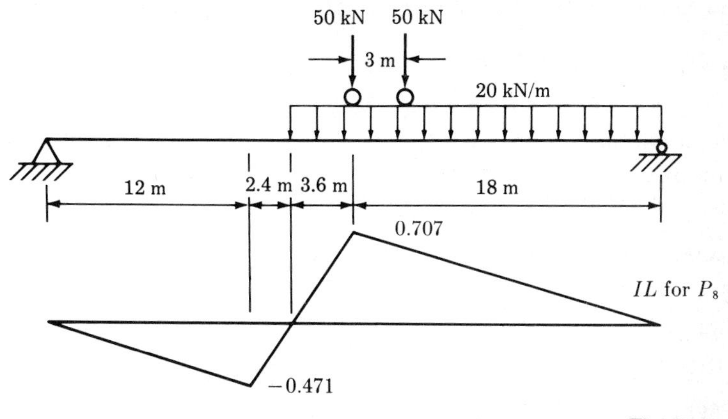

Fig. 10.20

This length extends from the right support to the point at which the value of the influence line is zero. The influence line is found to have zero value at a point 14.4 m to the right of the left support. The concentrated loads are placed so that one is at the point of maximum value on the influence line and the other is 3 m to the right.

From the previously developed methods for determining the magnitude of structural quantities from influence lines, the maximum tensile force in member 8 is found to be

$$P_{8\,\max} = (20)\left[\frac{(21.6)(0.707)}{2}\right] + (50)(0.707) + (50)(0.589)$$
$$= 217.5 \text{ kN}$$

If desired, the maximum compressive force in the member could be obtained by appropriately placing the loads on the left 14.4 m of the truss.

We have seen how influence lines can be used to determine the value of a particular structural quantity for various types of loading; it should be apparent how qualitative influence lines can be used to study load placements for maximum effects. As an illustration, consider the continuous beam previously presented in Fig. 10-14*a*. The beam and its influence line for moment at *B* are shown in Fig. 10-21. Let us assume that the continuous beam is to be subjected to uniform loading, the positioning of which can vary over the length of the beam. From the influence line in Fig. 10-21*b* we see that the maximum value of positive moment at *B* is obtained by placing the uniform load from *A* to *D*, with no load on the remaining portions of the beam. Similarly, the maximum negative moment at *B* is obtained by placing the uniform load from *E* to *A* and from *D* to *F*, with no load from *A* to *D*. The same procedure can be followed

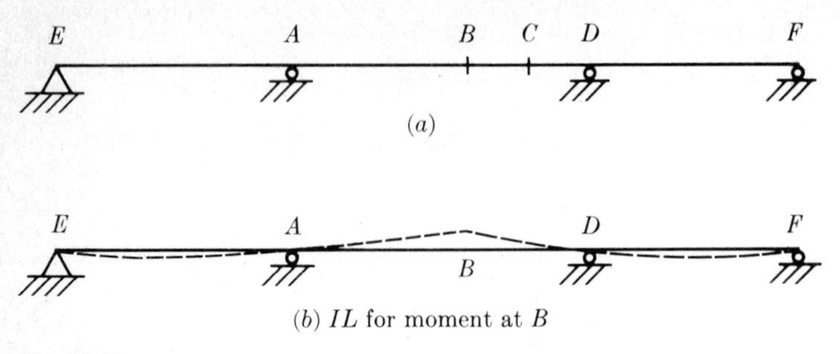

(b) IL for moment at B

Fig. 10.21

in studying the other structural quantities for which influence lines were constructed in Fig. 10-14.

Although a continuous beam was used in the discussion of the previous paragraph, the same concepts apply to frames and other types of structures. Qualitative influence lines are particularly useful in positioning floor loads in building frames.

In many instances we wish to determine the position of loading which will result in the maximum value of a particular structural quantity. Qualitative influence lines, developed from deflected shapes, can serve as very effective means for such studies. After the position of the loading is determined, an analysis can be performed to determine the magnitude of the structural quantity under consideration.

10-4 ABSOLUTE MAXIMUM SHEAR AND MOMENT

In addition to determining the maximum shear and moment at a certain point in a beam, it is often desirable to determine the magnitude and location of the absolute maximum shear and moment in the beam. Although influence lines are not involved in the following discussion, it is well to discuss the maximums at this point in association with the concentrated loads of the previous sections. The concentrated loads we shall consider are those caused by moving vehicles such as trucks, trains, or cranes. The beams on which these loads act are considered to be simply supported. The evaluation of the maximums for loads and beams other than these can become considerably more complicated.

The absolute maximum shear in a simply supported beam subjected to moving concentrated loads requires little discussion. It will occur at a point next to one of the reactions and therefore merely entails positioning the loads such that the maximum value of the reaction is obtained.

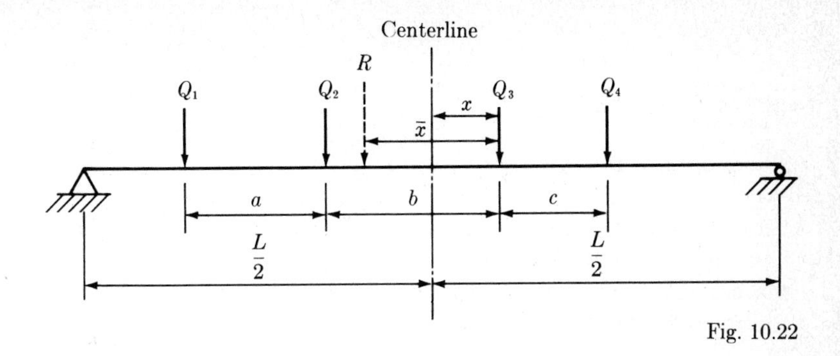

Fig. 10.22

We shall determine the location of the absolute maximum moment with reference to the beam of Fig. 10-22. The beam is subjected to a series of concentrated loads Q. The absolute maximum moment will occur under one of the concentrated loads. We want to know the load under which it will occur and where this load is positioned on the beam. The load under which the maximum moment will occur is found by trial and error, although it is sometimes obvious from inspection. The location of the load for absolute maximum moment can be determined by considering the variable x in Fig. 10-22, which describes the distance from the centerline of the beam to a load at which the maximum moment is being considered.

Let us consider moment at load Q_3. To develop an expression for x which will yield the desired position of the loads, we first obtain the resultant R of the loads. Its distance from Q_3 is denoted as $\bar{x}$. The value of the reaction at the right end of the beam, R_R, can be obtained in terms of R by taking the summation of moments about the left end of the beam. Thus

$$R_R = R\left(\frac{1}{2} - \frac{\bar{x}}{L} + \frac{x}{L}\right)$$

The moment under load Q_3, denoted as M_3, is therefore

$$\begin{aligned} M_3 &= R_R\left(\frac{L}{2} - x\right) - Q_4c \\ &= R\left(\frac{1}{2} - \frac{\bar{x}}{L} + \frac{x}{L}\right)\left(\frac{L}{2} - x\right) - Q_4c \\ &= R\left(\frac{L}{4} - \frac{\bar{x}}{2} + \frac{x\bar{x}}{L} - \frac{x^2}{L}\right) - Q_4c \end{aligned}$$

To obtain the value of x for which the moment is maximum under Q_3 we differentiate the expression for M_3 with respect to x and set it equal to zero. Thus

$$\frac{dM_3}{dx} = \frac{R\bar{x}}{L} - \frac{2Rx}{L} = 0$$

Therefore $\quad x = \frac{\bar{x}}{2}$

From this result the general statement can be made that the maximum moment occurs under a load when the loads are positioned such that the centerline of the beam is located midway between the load under consideration and the resultant of the loads.

EXAMPLE 10-3 The absolute maximum shear and moment are to be determined for the beam of Fig. 10-23. The beam is loaded with the standard truck load considered in Fig. 10-18. To obtain the absolute maximum shear we position the loads such that the 16-kip trailer wheel is next to the left support. The value of the absolute maximum shear in the beam, which is equal to the value of the left reaction, is

$$V = R_L = \frac{(16)(80) + (16)(66) + (4)(52)}{80}$$
$$= 31.8 \text{ kips}$$

It is obvious from inspection that the absolute maximum moment will occur under the 16-kip rear wheel. The location of the resultant of the loads is found to be 4.67 ft from the rear wheel, as shown in Fig. 10-24. The loads are positioned according to the previously developed procedure such that the centerline of the beam is midway between the resultant and the rear wheel. The resulting position is shown in Fig. 10-24, where $x = 2.33$ ft. The value of the absolute maximum moment under the rear wheel is found to be

$$M = (40 - 2.33)R_R - (4)(14)$$
$$= (37.67)(36)\left(\frac{37.67}{80}\right) - 56$$
$$= 582.6 \text{ ft-kips}$$

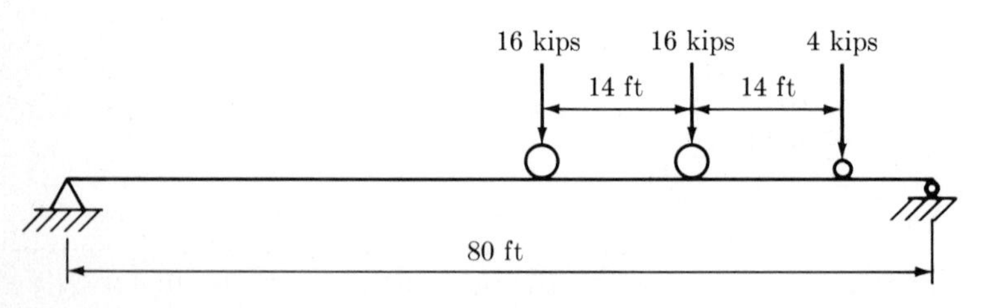

Fig. 10.23

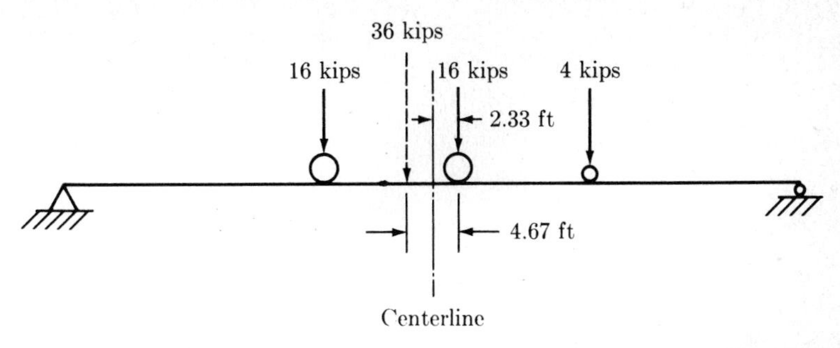

Fig. 10.24

It was stated earlier that the evaluation of absolute maximum shear and moment for structures other than simple beams and for other types of loading can become quite involved. Certain structures, such as cantilever beams, are exceptions to this; the absolute maximum shear and moment can be obtained for a cantilever beam without difficulty. In general, though, the analysis becomes quite involved for such structures as continuous beams and frames. It is possible for the more involved cases to plot the maximum values at various points, obtained from influence lines for these points, and then to determine from such a plot the absolute maximum. Such a procedure would become tedious if performed manually, but it would be feasible to formulate the problem for machine computation.

PROBLEMS

10-1. For the beam shown, construct influence lines for (*a*) the reaction at D; (*b*) the shear just to the left of B; (*c*) the shear at C; (*d*) the moment at C; (*e*) the moment at B.

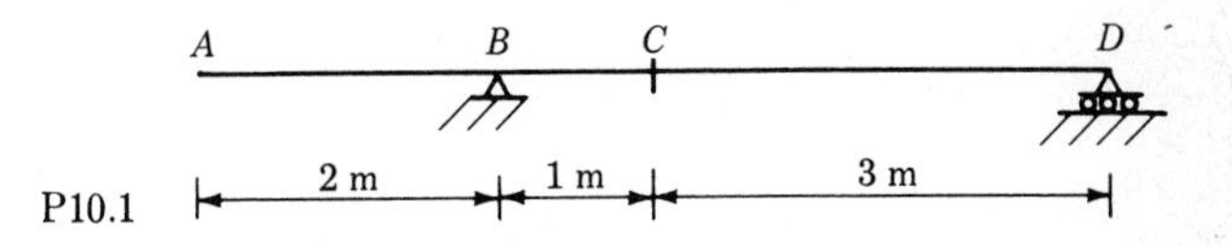

P10.1

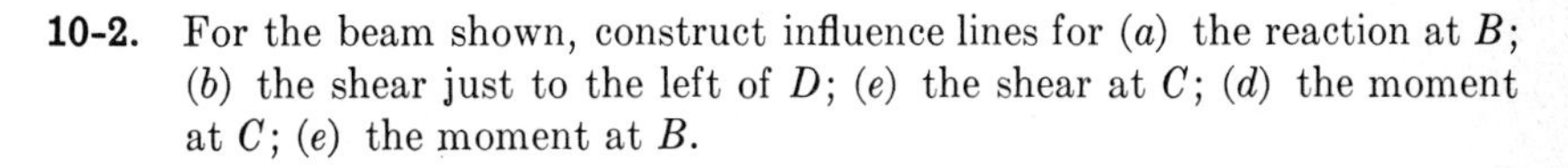

10-2. For the beam shown, construct influence lines for (*a*) the reaction at B; (*b*) the shear just to the left of D; (*e*) the shear at C; (*d*) the moment at C; (*e*) the moment at B.

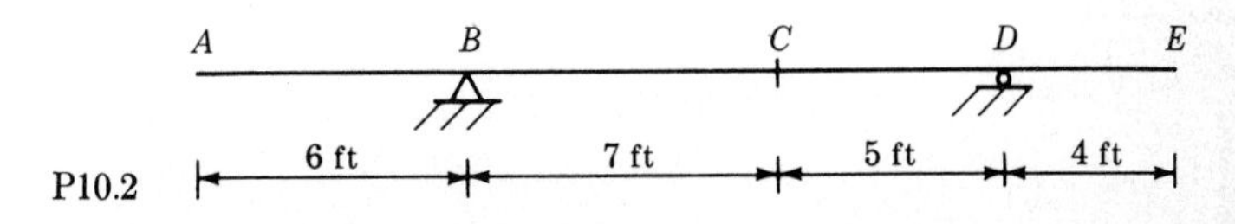

P10.2

10-3. For the beam shown, construct influence lines for (*a*) the reaction at C; (*b*) the shear at B; (*c*) the shear just to the left of E; (*d*) the moment at B; (*e*) the moment at C.

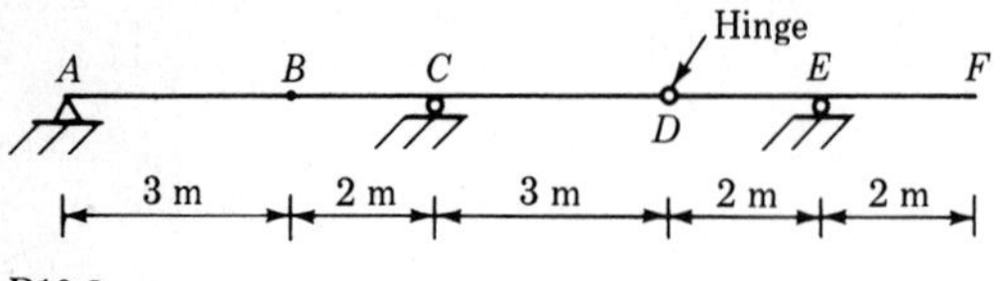

P10.3

10-4. For the beam shown, construct influence lines for (*a*) the reaction at A; (*b*) the reaction at C; (*c*) the shear just to the right of C; (*d*) the moment at C.

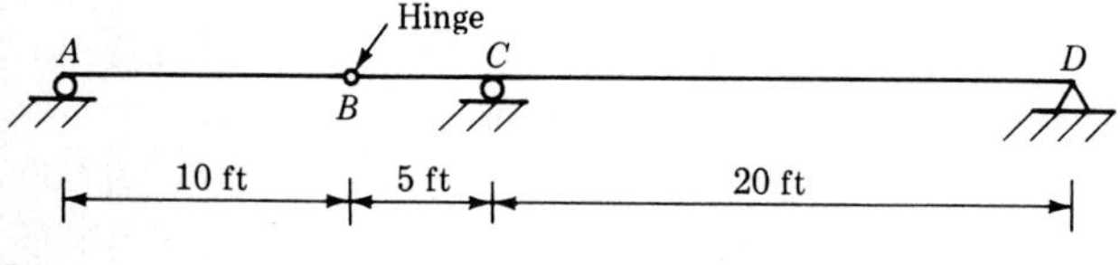

P10.4

10-5. For the beam shown, construct influence lines for (*a*) the reaction at C; (*b*) the vertical reaction at A; (*c*) the shear just to the left of C; (*d*) the moment at A; (*e*) the moment at C.

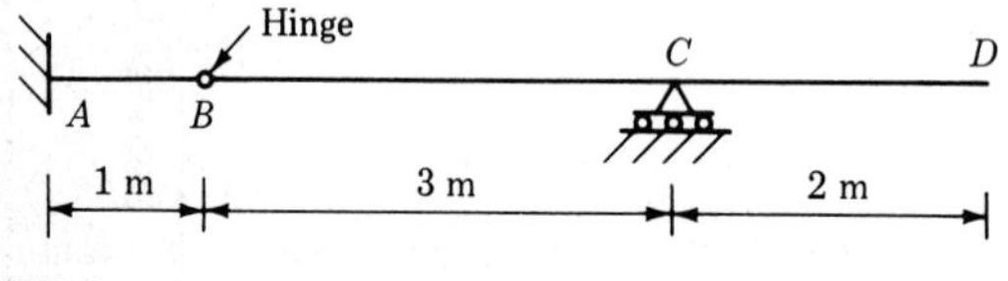

P10.5

10-6. For a unit load moving from C to G on the truss shown, construct influence lines for the numbered members.

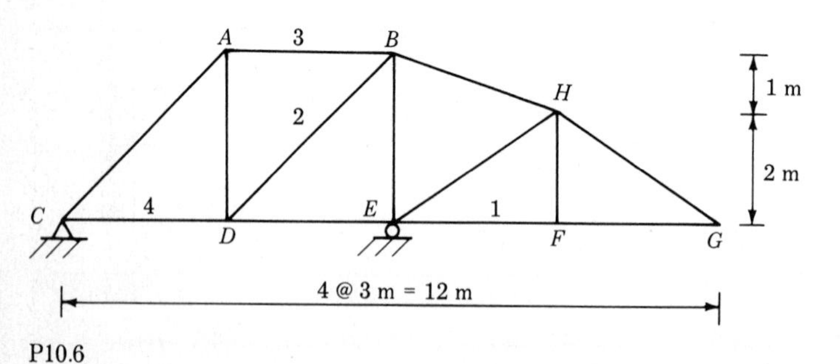

P10.6

10-7. For a unit load moving from D to H on the truss shown, construct influence lines for the numbered members.

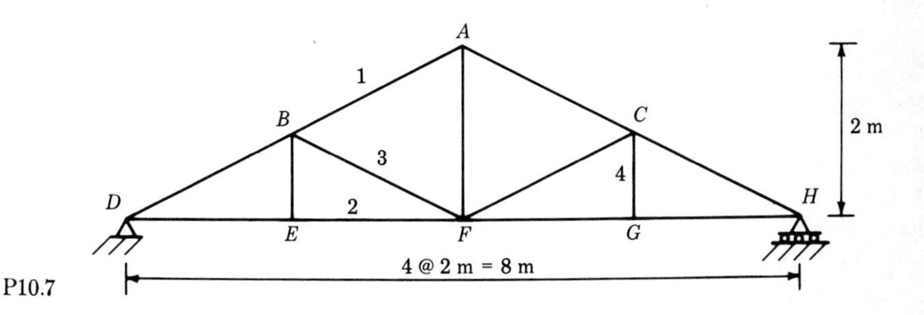

P10.7

10-8. For a unit load moving from ④ to ⑧ on the truss shown, construct influence lines for the numbered members.

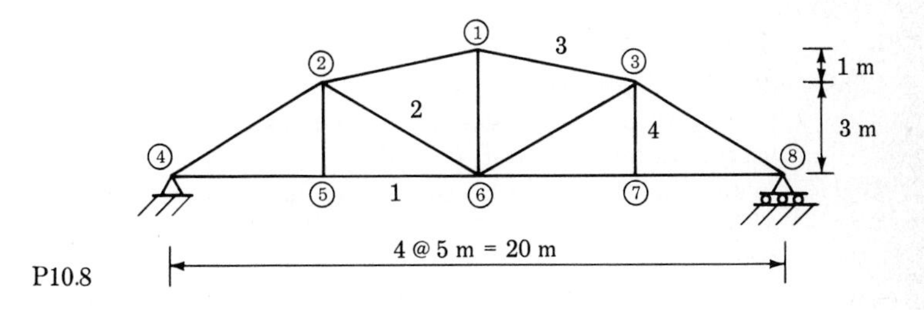

P10.8

10-9. For a unit load moving from A to D on the truss shown, construct influence lines for the numbered members.

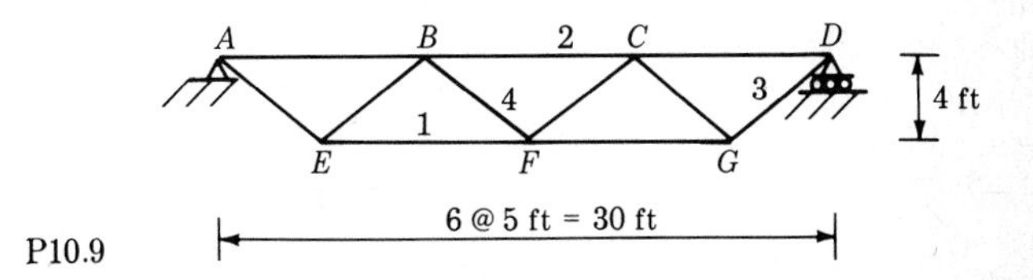

P10.9

10-10. For a unit load moving from A to E on the floor system shown, construct influence lines for (*a*) the reaction at H; (*b*) the shear just to the right of G in beam GH; (*c*) the moment at B in beam FG.

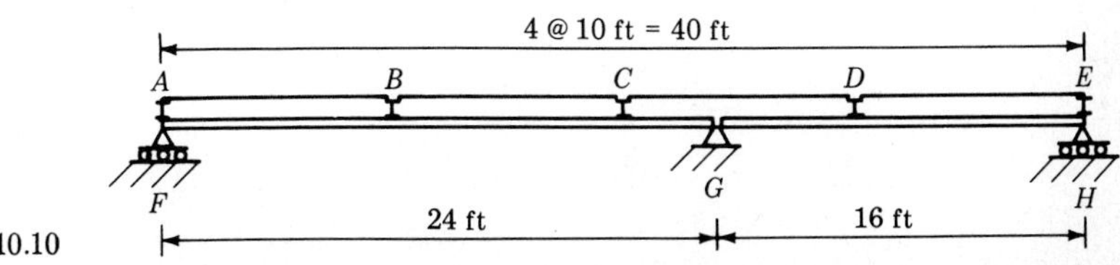

P10.10

10-11. For a unit load moving from A to D on the truss-and-beam structure shown, construct influence lines for (*a*) the reaction at A; (*b*) the reaction at D; (*c*) the moment at C; (*d*) the numbered members of the truss.

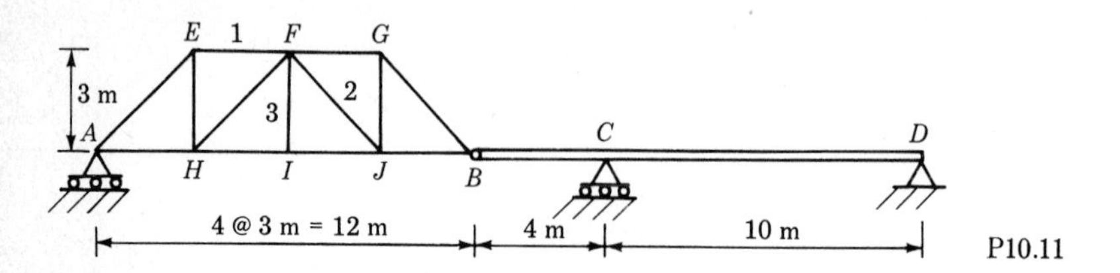

P10.11

10-12. A uniform load of 20 kN/m can be placed along any part of the beam in Prob. 10-1. Use the influence lines developed in Prob. 10-1 to determine the positions of the loading and the resulting values of (*a*) the maximum downward reaction at D; (*b*) the maximum positive bending moment at C.

10-13. A uniform load of 800 lb/ft and two 1.5-kip loads, spaced at 5 ft, can be placed along any part of the floor of Prob. 10-10. Use the influence lines developed in Prob. 10-10 to determine the positions of the loading and the resulting values of (*a*) the maximum reaction at H; (*b*) the maximum moment at B in beam FG.

10-14. A uniform load of 15 kN/m can be placed along any part of the truss-and-beam structure in Prob. 10-11. Use the influence lines developed in Prob. 10-11 to determine the positions of the loading and the resulting values of (*a*) the maximum downward reaction at D; (*b*) the maximum compression in member 2 of the truss.

10-15. For the beam shown, sketch qualitative influence lines for (*a*) the reaction at C; (*b*) shear at A; (*c*) moment at B; (*d*) moment at A. Indicate the position of uniform loading to obtain the maximum values of the above quantities.

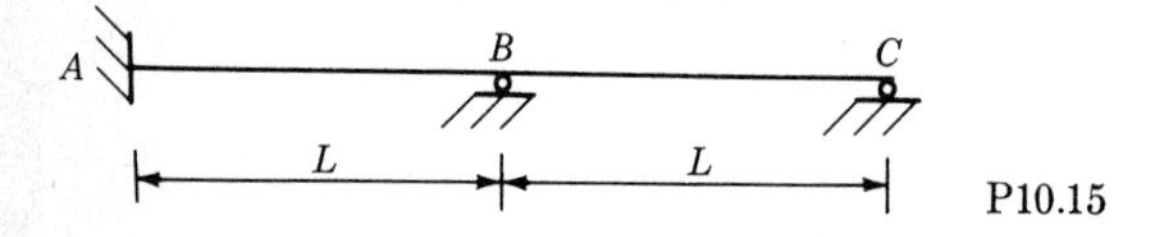

P10.15

10-16. For the beam shown, sketch qualitative influence lines for (*a*) the reaction at B; (*b*) shear just to the right of B; (*c*) moment at D; (*d*) moment at C. Indicate the position of uniform loading to obtain the maximum values of the above quantities.

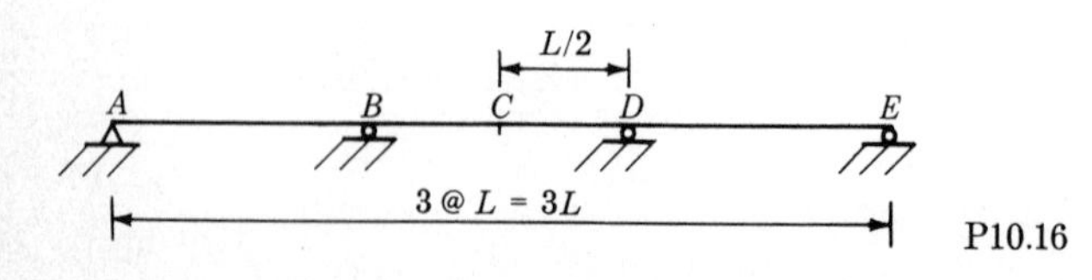

P10.16

10-17. Use qualitative influence lines to show the positions of uniform loads on the beam shown which result in (*a*) the maximum upward reaction at C; (*b*) the maximum negative moment at B; (*c*) the maximum positive moment at A.

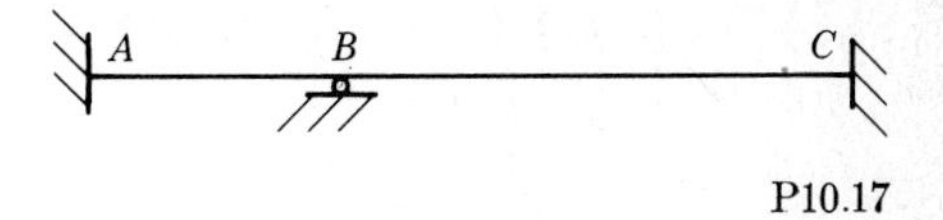

P10.17

10-18. Use qualitative influence lines to show the positions of uniform loads on the floor and roof of the frame shown which result in (*a*) the maximum compressive axial force in column AB; (*b*) the maximum moment in the column at C; (*c*) the maximum positive moment at D.

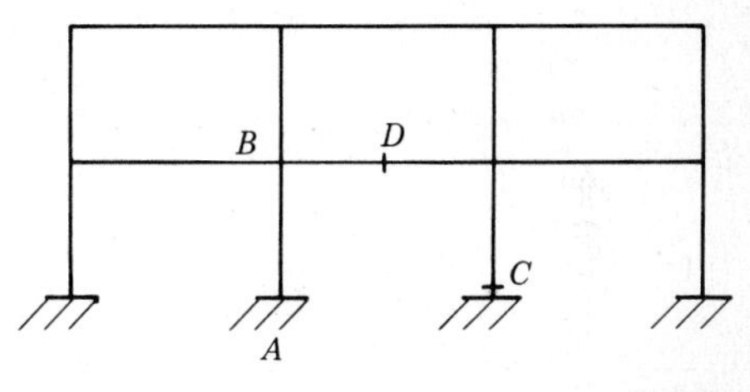

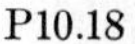
P10.18

10-19. The simply supported beam is subjected to a set of four concentrated loads as shown. Determine the positions of the loads and the resulting values of (*a*) the absolute maximum shear in the beam; (*b*) the absolute maximum moment in the beam.

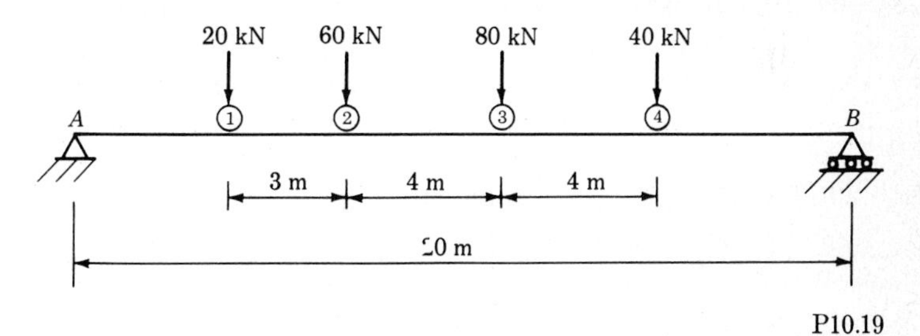

P10.19

10-20. If the four concentrated loads in Fig. P10-19 are placed in the truss-and-beam structure of Fig. P10-11, use the influence line developed in Prob. 10-11 for the moment at C and determine the position of the four loads that will result in the maximum negative moment at C.

10-21. If the four concentrated loads in Fig. P10-19 are placed on the truss of Fig. P10-8, use the influence lines developed in Prob. 10-8 to determine the position of the loads that will result in the maximum value of (*a*) P_1; (*b*) P_2; (*c*) P_3. Determine these maximum values.

CHAPTER

ELEVEN

METHOD OF CONSISTENT DISPLACEMENTS

11-1 BASIC CONCEPTS OF THE METHOD

The method of consistent displacements, sometimes referred to as the *general method*, is the first method of analyzing statically indeterminate structures that we shall consider. The slope-deflection method, the moment-distribution method, and the general matrix stiffness method will be treated in succeeding chapters.

Before discussing the analysis of any particular type of structure or numerical examples, let us concentrate on the basic concepts involved in an analysis by the method of consistent displacements. First it is necessary to understand to what degree a structure is statically indeterminate. Recall from previous discussions that a structure is statically indeterminate when there are unknown force quantities in excess of the number of equations of equilibrium available. When such a condition does exist, a solution can be obtained by considering deflection conditions. Thus conditions for consistent displacements furnish the additional equations required for an analysis.

To develop an understanding of the method of consistent displacements let us first consider as an example the statically indeterminate

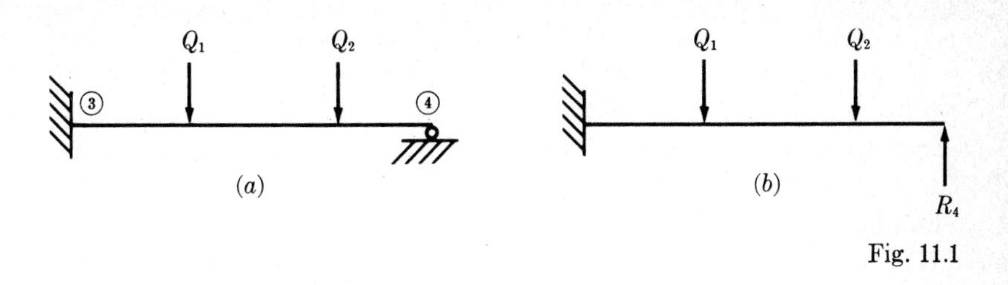

Fig. 11.1

beam of Fig. 11-1*a*. For the support conditions shown the structure is statically indeterminate to the first degree; that is, a free-body diagram of the beam would show that there is one more unknown force quantity than there are equations of equilibrium. An additional equation can be obtained from deflection considerations. We consider that one of the supporting forces is a *redundant force* and the restraining action of this redundant is temporarily removed. The redundant force is considered to be an active force on the resulting structure. The choice of the redundant force is arbitrary. We could select any one of the reaction components, or we could consider as redundant an internal force such as moment and introduce a hinge somewhere along the length of the beam. If we consider the reaction at ④ to be the redundant force and remove its restraining effect, we obtain a beam loaded as in Fig. 11-1*b*.

In selecting redundants it must be remembered that the structure remaining after the restraining effects of the redundant forces are removed must be stable. This remaining structure is commonly referred to as the *primary structure*. Thus the resulting cantilever beam in Fig. 11-1*b* is the primary structure. The primary structure supports not only the given loads, but also the forces representing the redundants.

Proceeding with the analysis, we concern ourselves next with the deflections of the primary structure. The deflections of the primary structure due to the given loads and to the redundant forces are considered separately. For the beam under consideration the given loads will result in a displacement at point ④ of $D_4{}^Q$ as shown in Fig. 11-2*a*. The superscript Q implies that the deflection is due to the given loads. The use of such a superscript makes it possible to distinguish between the deflections due

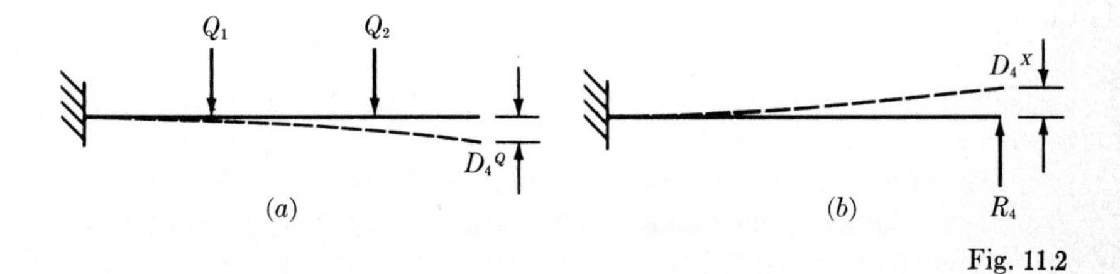

Fig. 11.2

to the given loads and the deflections due to the redundant forces. The reason for making a clear distinction between the two types of deflections will become apparent in the following paragraphs. The deflection at point ④ resulting from the redundant force R_4 is denoted in Fig. 11-2*b* as D_4^X. The superscript X implies that the displacement is associated with the redundant force whose value as yet is unknown.

The deflection equation required for the analysis of the given beam is obtained by considering the net deflection of point ④ due to the two types of loading. The net deflection at ④ must be zero. Therefore, from Fig. 11-2 we can state that

$$D_4^Q + D_4^X = 0 \tag{11-1}$$

In writing the condition for consistent displacements at ④ in this manner we assume that D_4^Q and D_4^X are expressed in accordance with some positive direction of displacement. Thus, if an upward deflection is considered positive, D_4^Q is a negative quantity. Obtaining the correct signs for deflections and for forces is essential. The use of deflection and force coordinate systems can be helpful in more complicated analyses.

Recall from Chap. 9 that a flexibility influence coefficient f_{ij} is defined as the deflection at point i resulting from a unit force applied at point j. From this definition of a flexibility influence coefficient we see that D_4^X is equal to the flexibility coefficient f_{44} multiplied by R_4; hence we can write Eq. (11-1) as

$$D_4^Q + f_{44}R_4 = 0 \tag{11-2}$$

from which

$$R_4 = -\frac{D_4^Q}{f_{44}} \tag{11-3}$$

D_4^Q and f_{44} are deflection quantities of the statically determinate primary structure. Their values can be obtained by the deflection methods discussed in previous chapters.

To extend the method further let us consider a more involved form of the beam as shown in Fig. 11-3*a*. With the additional span as shown and the simple support at ⑤, the beam is now statically indeterminate to the second degree. Two additional equations of deflection must therefore be obtained for analysis of the beam. The reactions at ④ and ⑤ are selected as the redundant forces. The deflected shape of the primary structure due to the given loads is shown in Fig. 11-3*b*. The deflected shape of the primary structure due to the redundant forces R_4 and R_5 is shown in Fig. 11-3*c*. For the conditions of zero displacements at ④ and ⑤ in the

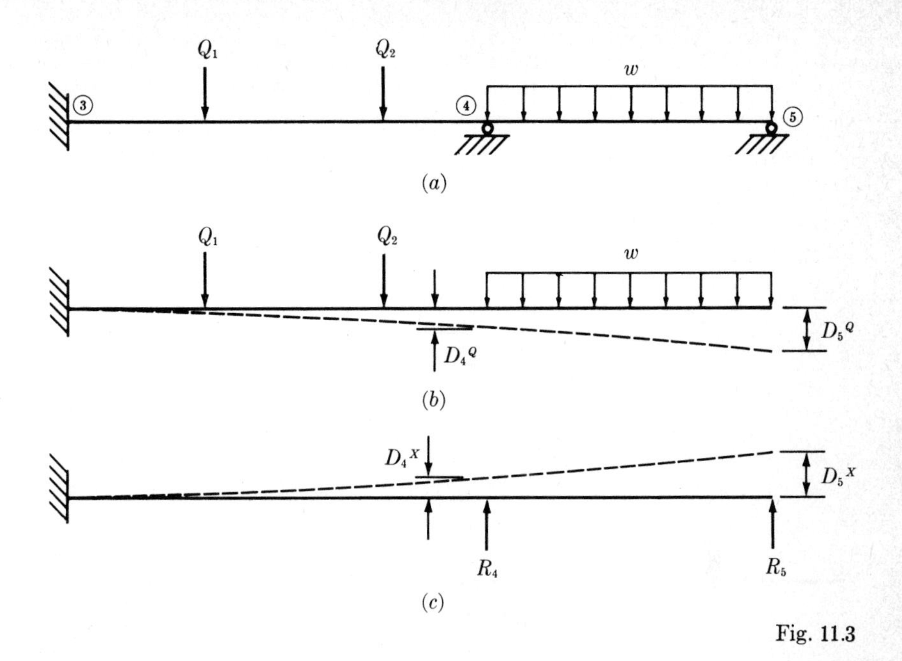

Fig. 11.3

given beam we can write

$$D_4^Q + D_4^X = 0 \qquad D_5^Q + D_5^X = 0 \tag{11-4}$$

D_4^X and D_5^X can be expressed in terms of flexibility coefficients and the redundant forces R_4 and R_5. Equation (11-4) can therefore be written as

$$D_4^Q + f_{44}R_4 + f_{45}R_5 = 0 \qquad D_5^Q + f_{54}R_4 + f_{55}R_5 = 0 \tag{11-5}$$

In matrix form this becomes

$$\begin{Bmatrix} D_4^Q \\ D_5^Q \end{Bmatrix} + \begin{bmatrix} f_{44} & f_{45} \\ f_{54} & f_{55} \end{bmatrix} \begin{Bmatrix} R_4 \\ R_5 \end{Bmatrix} = \begin{Bmatrix} 0 \\ 0 \end{Bmatrix} \tag{11-6}$$

The values of the redundant forces R_4 and R_5 are obtained from a solution of Eq. (11-6); that is,

$$\begin{Bmatrix} R_4 \\ R_5 \end{Bmatrix} = - \begin{bmatrix} f_{44} & f_{45} \\ f_{54} & f_{55} \end{bmatrix}^{-1} \begin{Bmatrix} D_4^Q \\ D_5^Q \end{Bmatrix} \tag{11-7}$$

All quantities appearing on the right side of Eq. (11-7) can be obtained from deflection calculations for the statically determinate primary structure.

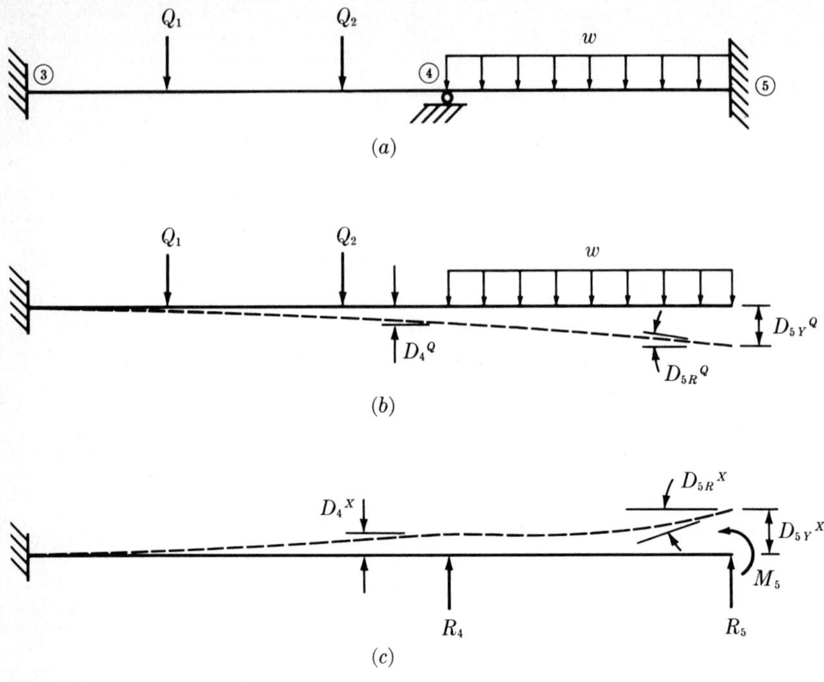

Fig. 11.4

As one further extension of the above beam problem, let us consider the support at ⑤ to be fixed, as shown in Fig. 11-4*a*. Choosing the reaction at ④ and the reaction components at ⑤ to be the redundants, the deflected primary structures for the two types of loading are as shown in Figs. 11-4*b* and *c*. Additional subscripts have been used in Fig. 11-4 with the deflection quantities at ⑤ in order to distinguish between the components. The subscript Y denotes a deflection in the vertical direction and a subscript R denotes a rotation.

Following the previous procedure, the condition for zero vertical displacement at ④ and zero vertical displacement and rotation at ⑤ can be written in matrix form as

$$\begin{Bmatrix} D_4{}^Q \\ D_{5Y}{}^Q \\ D_{5R}{}^Q \end{Bmatrix} + \begin{bmatrix} f_{44} & f_{45Y} & f_{45R} \\ f_{5Y4} & f_{5Y5Y} & f_{5Y5R} \\ f_{5R4} & f_{5R5Y} & f_{5R5R} \end{bmatrix} \begin{Bmatrix} R_4 \\ R_5 \\ M_5 \end{Bmatrix} = \begin{Bmatrix} 0 \\ 0 \\ 0 \end{Bmatrix} \tag{11-8}$$

Additional subscripts, as discussed in Sec. 9-2, are used with the flexibility influence coefficients to clarify their meaning.

The values of the unknown redundants are obtained by inverting the flexibility matrix of Eq. (11-8) and premultiplying the negative of the matrix $\mathbf{D}^Q$ by this inverse. Thus

$$\begin{Bmatrix} R_4 \\ R_5 \\ M_5 \end{Bmatrix} = - \begin{bmatrix} f_{44} & f_{45Y} & f_{45R} \\ f_{5Y4} & f_{5Y5Y} & f_{5Y5R} \\ f_{5R4} & f_{5R5Y} & f_{5R5R} \end{bmatrix}^{-1} \begin{Bmatrix} D_4{}^Q \\ D_{5Y}{}^Q \\ D_{5R}{}^Q \end{Bmatrix} \tag{11-9}$$

This procedure is applicable to various types of structures, as we shall see in the next section. The general procedure can be summarized as follows. Depending on the degree of indeterminacy of the structure under consideration, certain forces are selected as redundant quantities. These unknown force quantities can be represented by a column matrix **X**. The number of elements in **X** is therefore equal to the degree of indeterminacy. As demonstrated above, the values of the unknown redundants are obtained from known conditions for displacement at the action points of the redundants. For the condition of zero displacement at the action points of the redundants, we can write the matrix expression

$$\mathbf{D}^Q + \mathbf{F}\mathbf{X} = \mathbf{0} \tag{11-10}$$

from which

$$\mathbf{X} = -\mathbf{F}^{-1}\mathbf{D}^Q \tag{11-11}$$

F is a flexibility matrix for the primary structure. The values of the flexibility influence coefficients making up **F** are obtained by evaluating the deflections for unit values of the redundant forces applied to the primary structure. The procedure for developing flexibility influence coefficients was discussed in Sec. 9-2. The column matrix $\mathbf{D}^Q$ in Eqs. (11-10) and (11-11) represents the deflections of the primary structure due to given loads.

This method also lends itself well to the consideration of support settlements and other types of initial displacements. For example, let us assume that the beam support at point ④ in Fig. 11-3 settles 0.1 in. The expression for consistent displacements given in Eq. (11-6) now becomes

$$\begin{Bmatrix} D_4{}^Q \\ D_5{}^Q \end{Bmatrix} + \begin{bmatrix} f_{44} & f_{45} \\ f_{54} & f_{55} \end{bmatrix} \begin{Bmatrix} R_4 \\ R_5 \end{Bmatrix} = \begin{Bmatrix} -0.1 \\ 0 \end{Bmatrix}$$

With positive displacements assumed in the upward direction, the settlement is a negative quantity. The resulting expression for the redundants is

$$\begin{Bmatrix} R_4 \\ R_5 \end{Bmatrix} = - \begin{bmatrix} f_{44} & f_{45} \\ f_{54} & f_{55} \end{bmatrix}^{-1} \begin{Bmatrix} 0.1 + D_4{}^Q \\ D_5{}^Q \end{Bmatrix}$$

Care must be exercised in such analyses to have the given settlements and the computed deflections in the same dimensions.

11-2 APPLICATIONS OF THE METHOD

The method of consistent displacements can be used in the analysis of various types of statically indeterminate structures. This procedure entails the computation of deflection quantities. Various methods for computing deflections were presented in Chaps. 7 and 8. The computation method used for a given structure is often a matter of personal choice, but in some cases certain methods have distinct advantages over others. In the examples that follow the deflections are computed by an appropriate method. It may well be that some other method of evaluating the deflections seems preferable; however, the emphasis in these examples is on the indeterminate aspect of analysis, and not so much on comparisons between deflection methods.

EXAMPLE 11-1 The fixed-end moment at point ① is to be determined for the beam in Fig. 11-5. The beam is subjected to a uniform load w. EI for the beam is constant. The given beam is statically indeterminate to the first degree. The choice of the redundant quantity is arbitrary, but because we are to determine the moment at ① we shall select this moment as the redundant quantity. When the restraint of the moment at ① is removed, the resulting primary structure is a simply supported beam. The primary structure loaded with the given uniform load is shown in Fig. 11-6*a*, and the primary structure loaded with a unit value of the redundant M_1 is shown in Fig. 11-6*b*. The condition for consistent displacements at ① is written as

$$D_1^Q + f_{11}M_1 = 0$$

Deflections and moment as shown in Fig. 11-6 are considered positive.

The deflection quantities D_1^Q and f_{11} are evaluated by the moment-area method. From the M/EI diagram for the uniform load of Fig. 11-6*a*, the expression for D_1^Q is

$$D_1^Q = \frac{t_{21}}{L} = \frac{2}{3}\frac{wL^2}{8EI}L\frac{L}{2}\frac{1}{L} = \frac{wL^3}{24EI}$$

From the M/EI diagram for the unit moment in Fig. 11-6*b*, the expression

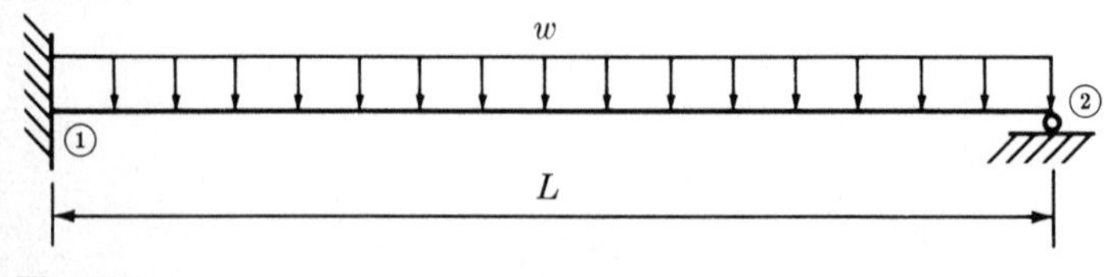

Fig. 11.5

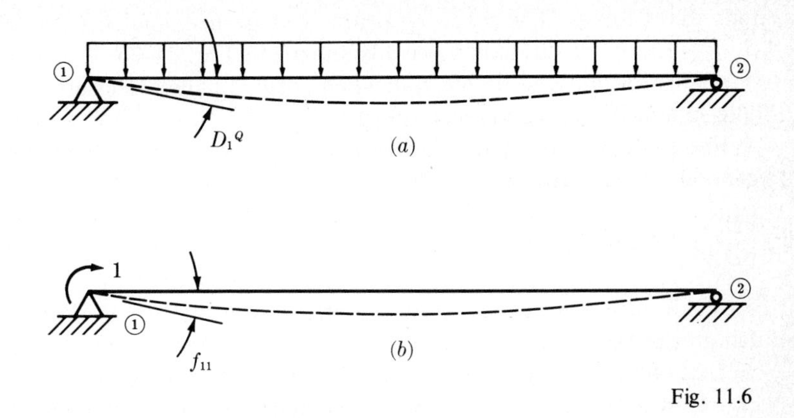

Fig. 11.6

for the flexibility influence coefficient f_{11} is

$$f_{11} = \frac{t_{21}}{L} = \frac{1}{2}\frac{1}{EI}L\frac{2L}{3}\frac{1}{L} = \frac{L}{3EI}$$

The value of the redundant moment is therefore

$$M_1 = -\frac{D_1^Q}{f_{11}} = -\frac{wL^3}{24EI}\frac{3EI}{L} = -\frac{wL^2}{8}$$

The minus sign indicates that the moment acts on the beam in the direction opposite to that in which the unit load is acting in Fig. 11-6*b*.

EXAMPLE 11-2 The values of the reaction components are to be determined for the frame of Fig. 11-7*a*. *EI* for each member of the frame is the same. The frame is statically indeterminate to the second degree. The horizontal and vertical components of reaction at ② are selected as the

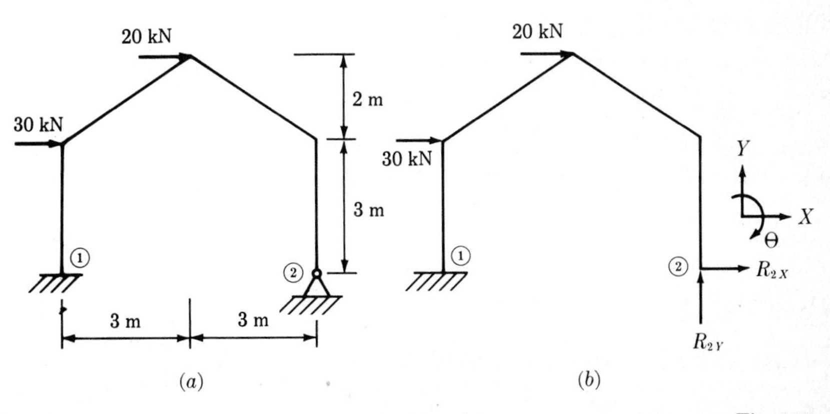

Fig. 11.7

redundant forces. The resulting primary structure, loaded with the given loads and the redundant forces, is shown in Fig. 11-7*b*. The redundant forces are assumed to be positive when acting in the direction of the positive X and Y axes, as shown.

The condition for consistent displacements at point ②, as given in the general matrix expression of Eq. (11-10), is

$$\begin{Bmatrix} D_{2X}^Q \\ D_{2Y}^Q \end{Bmatrix} + \begin{bmatrix} f_{2X2X} & f_{2X2Y} \\ f_{2Y2X} & f_{2Y2Y} \end{bmatrix} \begin{Bmatrix} R_{2X} \\ R_{2Y} \end{Bmatrix} = \begin{Bmatrix} 0 \\ 0 \end{Bmatrix}$$

We shall evaluate the deflection quantities by the method of virtual work using visual integration. The loadings on the primary structure to obtain each of the required deflection quantities by virtual work are shown in Fig. 11-8. Forces are applied to the structure at point ② in the positive direction of the X and Y axes. The internal virtual work for each member is obtained from the expression.

$$\delta W_I = \int_0^L \frac{mM}{EI}\,dx$$

The real moment M and the virtual moment m diagrams are constructed on the primary structure in Fig. 11-8; the M diagrams are indicated by solid lines, and the m diagrams by dashed lines.

Equating external virtual work to internal virtual work and using the visual-integration technique, we obtain from Fig. 11-8*a*

$$EID_{2X}^Q = -(\tfrac{1}{2})(3.61)(40)(3.67) - (3)(40)(1.5) - (\tfrac{1}{2})(3)(150)(1) = -669.97$$

From Fig. 11-8*b* we have

$$EID_{2Y}^Q = -(\tfrac{1}{2})(3.61)(40)(5) - (3)(40)(6) - (\tfrac{1}{2})(3)(150)(6) = -2431.00$$

From Fig. 11-8*c*,

$$EIf_{2X2X} = 2[(\tfrac{1}{2})(3)(3)(2) + (3.61)(3)(4) + (\tfrac{1}{2})(3.61)(2)(4.33)] = 135.90$$

From Fig. 11-8*d*,

$$EIf_{2X2Y} = (\tfrac{1}{2})(3.61)(3)(4.33) + (3.61)(3)(4) + (\tfrac{1}{2})(3.61)(3)(3.67) + (3)(6)(1.5) = 113.64$$

From Fig. 11-8*e*,

$$EIf_{2Y2Y} = (\tfrac{1}{2})(3.61)(3)(2) + (3.61)(3)(4.5) + (\tfrac{1}{2})(3.61)(3)(5) + (3)(6)(6) = 194.64$$

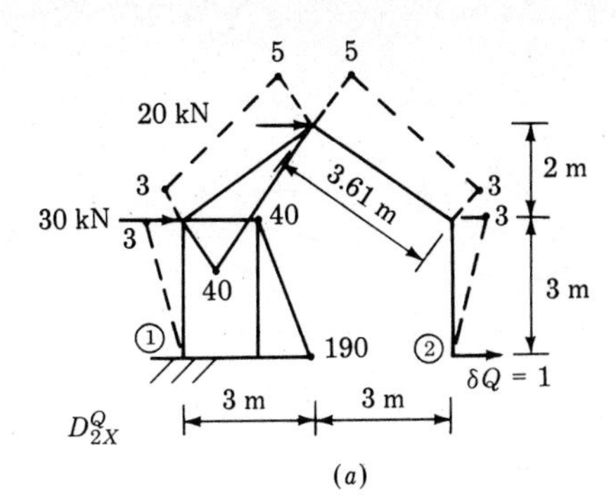

(a)

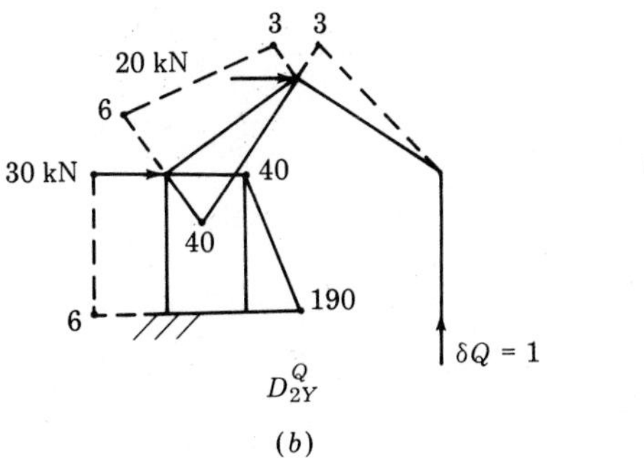

(b)

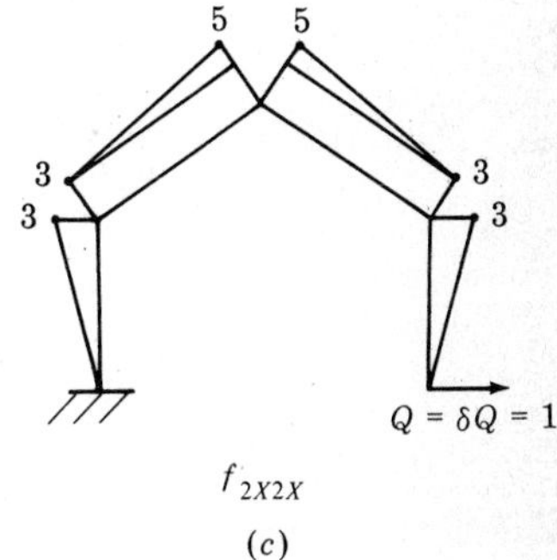

(c)

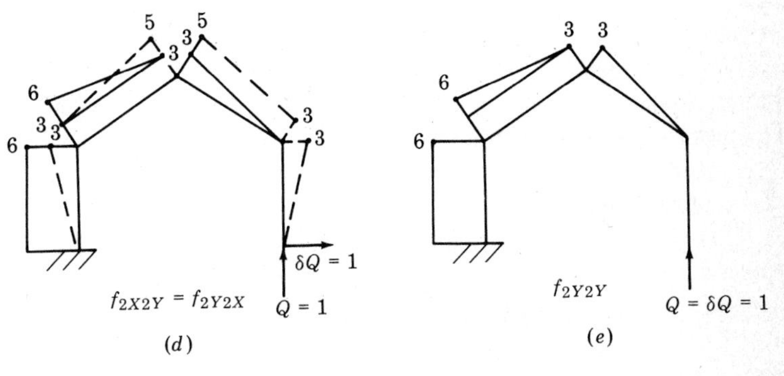

(d)

(e)

Fig. 11.8

The expression for consistent displacements at point ②, written in matrix form, is

$$\frac{1}{EI}\begin{bmatrix} 135.90 & 113.64 \\ 113.64 & 194.64 \end{bmatrix} \begin{Bmatrix} R_{2X} \\ R_{2Y} \end{Bmatrix} = \frac{1}{EI}\begin{Bmatrix} 669.97 \\ 2431.00 \end{Bmatrix}$$

Solving for the redundants, we obtain

$$\begin{Bmatrix} R_{2X} \\ R_{2Y} \end{Bmatrix} = \begin{Bmatrix} -10.77 \\ 18.78 \end{Bmatrix} \text{ kN}$$

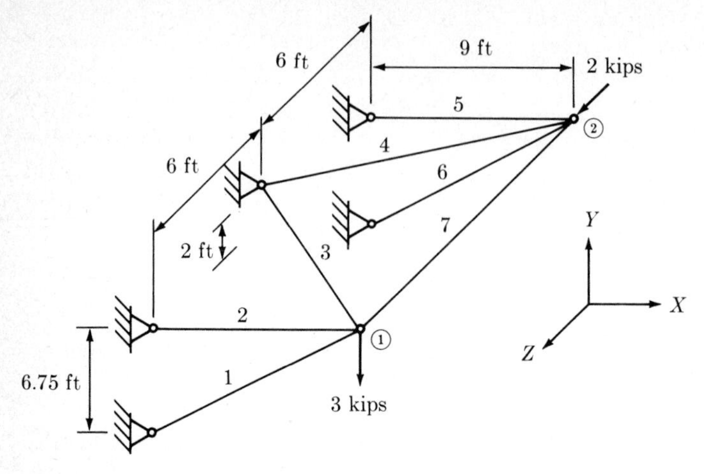

Fig. 11.9

The negative sign of R_{2X} indicates that the horizontal reaction at ② acts to the left.

The reaction components at point ① are obtained from statics as

$$R_{1X} = -39.23 \text{ kN} \qquad R_{1Y} = -18.78 \text{ kN} \qquad M_1 = -77.32 \text{ kN-m}$$

The signs of the reactions at ① are in accordance with the $XY\Theta$ coordinate system of Fig. 11-7*b*.

EXAMPLE 11-3 The bar forces associated with the given external loads are to be determined for the ball-and-socket-connected truss of Fig. 11-9. The cross-sectional areas of members 3 and 4 are given as A and all other member areas are given as $2A$. E for all members is the same. The three-dimensional coordinate system shown is for convenience in referring to loads and deflections.

The given truss is statically indeterminate to the first degree. The axial force in member 7 is selected as the redundant force. If we cut member 7 at some point i along its length, the resulting primary structure is as

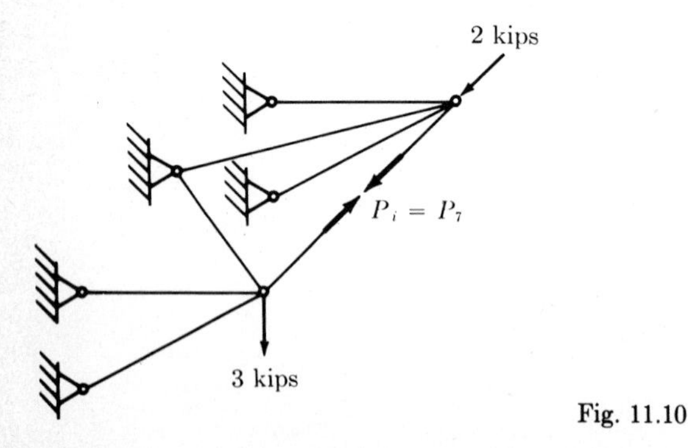

Fig. 11.10

shown in Fig. 11-10. The redundant force at i is denoted as P_i, which is equal to the force in member 7. Cutting member 7 will result in a small relative displacement at the cut ends, although in Fig. 11-10 it appears that member 7 has a segment removed from its length. It is necessary to draw the cut member in this manner so that the member force can be shown. The redundant force P_i is assumed positive when it causes tension in member 7. The condition for consistent displacement at the cut is given by

$$D_i^Q + f_{ii}P_i = 0$$

where D_i^Q is the relative displacement between the cut ends of the member resulting from the given loads and f_{ii} is the flexibility influence coefficient for unit tensile forces applied at each cut end of the bar at i.

The method of virtual work will be used to evaluate the deflection quantities. The expression for internal virtual work in the truss is

$$\delta W_I = \sum \frac{pPL}{EA}$$

The loadings on the truss for determining the deflection quantities are shown in Figs. 11-11a and b. The values of the bar forces resulting from each condition of loading are obtained from statics. The results of such an analysis are shown in Table 11-1. As before, tensile forces in the members are considered to be positive. The value of D_i^Q is found from the data of Table 11-1 to be

$$\begin{aligned} D_i^Q &= \frac{1}{EA}(-1.83)(-3.67)(11.0) + \frac{1}{2EA}[(-0.56)(-5.0)(11.25) \\ &\quad + (1.94)(4.0)(9.0) + (1.94)(3.89)(9.0) + (-0.56)(-1.11)(11.25)] \\ &= \frac{162.0}{EA} \end{aligned}$$

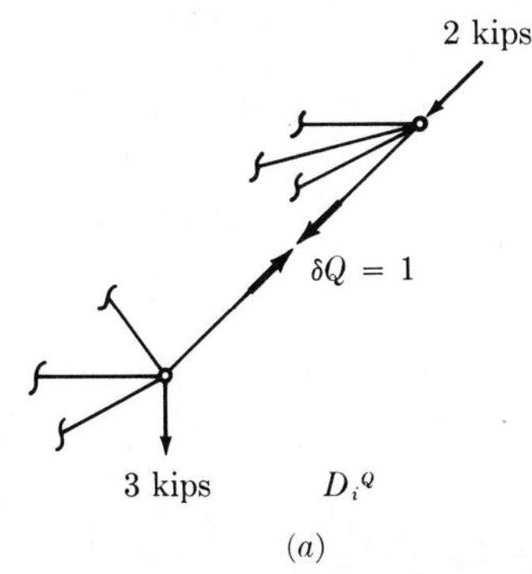

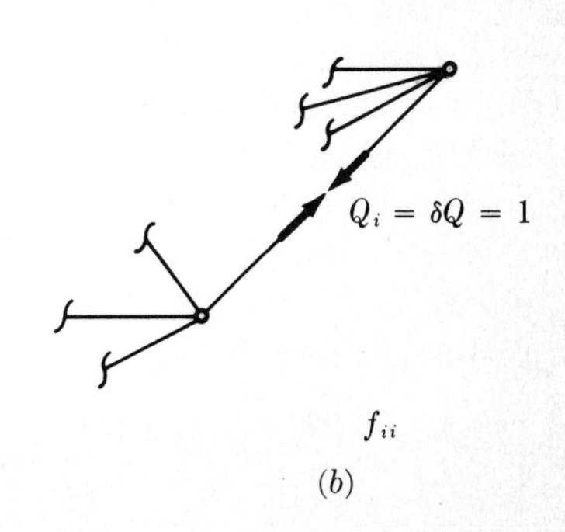

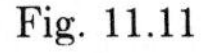

Fig. 11.11 (a) (b)

TABLE 11-1

Member	Length, ft	P for $Q_{1Y} = -3$, $Q_{2Z} = 2$	p and P for $Q_i = 1$
1	11.25	−5.0	−0.56
2	9.0	4.0	1.94
3	11.0	0	−1.83
4	11.0	−3.67	−1.83
5	9.0	3.89	1.94
6	11.25	−1.11	−0.56
7	12.0	0	1.0

The value of f_{ii} is found to be

$$f_{ii} = \frac{2}{EA}(-1.83)^2(11.0) + \frac{2}{2EA}[(-0.56)^2(11.25) + (1.94)^2(9.0)] + \frac{1}{2EA}(1.0)^2(12.0)$$

$$= \frac{117.6}{EA}$$

Therefore the value of force in member 7 which is equal to P_i is

$$P_7 = P_i = -\frac{D_i^Q}{f_{ii}} = -\frac{162.0}{EA}\frac{EA}{117.6} = -1.38 \text{ kips}$$

The remaining bar forces can be obtained by multiplying the values in the fourth column of Table 11-1 by −1.38 and adding the results to the values in the third column.

EXAMPLE 11-4 A fixed-end beam is subjected to a temperature change that varies linearly through the depth of the beam.

We are to determine the expression for the resulting fixed-end moments in terms of the coefficient of thermal expansion α for the beam material, the constant D of Eq. (8-68), and EI, and then show that the resulting vertical deflection at midspan on the beam is zero.

A fixed-end beam as shown in Fig. 11-12a is statically indeterminate to the second degree. The end moments M_1 and M_2 are selected as the two redundants. The primary structure, loaded with the redundants, is shown in Fig. 11-12b. The values of the redundants are obtained from the consistent-displacement expression

$$\begin{Bmatrix} D_{1\theta}^Q \\ D_{2\theta}^Q \end{Bmatrix} + \begin{bmatrix} f_{11} & f_{12} \\ f_{21} & f_{22} \end{bmatrix} \begin{Bmatrix} M_1 \\ M_2 \end{Bmatrix} = \begin{Bmatrix} 0 \\ 0 \end{Bmatrix}$$

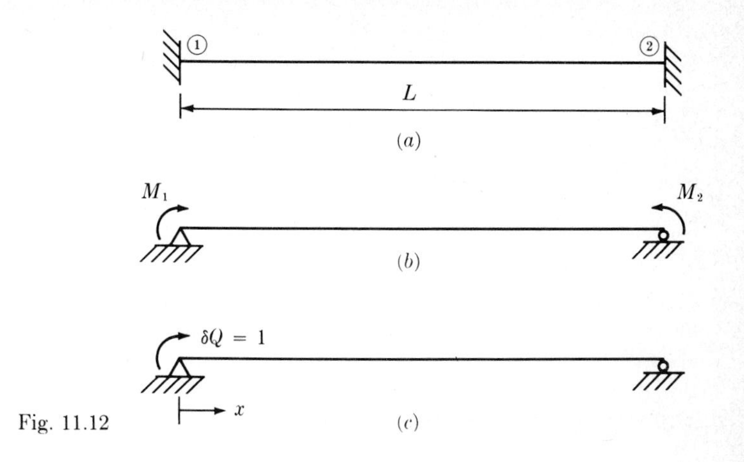

Fig. 11.12

The end rotations of the primary structure are assumed to be positive in the directions shown for M_1 and M_2 in Fig. 11-12*b*. The value of $D_{1\theta}{}^Q$ is obtained by applying a unit virtual moment to the beam as shown in Fig. 11-12*c*. The internal virtual work due to the temperature change is, according to Eq. (8-69),

$$\delta W_I = -\alpha D \int_0^L m\, dx$$

From Fig. 11-12*c* we see that $m = 1 - x/L$. Equating external virtual work to internal virtual work, we find the rotation at ① to be

$$D_{1\theta}{}^Q = -\alpha D \int_0^L \left(1 - \frac{x}{L}\right) dx = -\frac{\alpha DL}{2}$$

From inspection, $D_{2\theta}{}^Q = D_{1\theta}{}^Q$.

The values of the flexibility influence coefficients f_{ij} can be obtained directly from Eq. (9-31) with appropriate changes in sign. The matrix expression for consistent displacements thus becomes

$$-\frac{1}{2}\begin{Bmatrix}\alpha DL \\ \alpha DL\end{Bmatrix} + \frac{L}{6EI}\begin{bmatrix}2 & 1 \\ 1 & 2\end{bmatrix}\begin{Bmatrix}M_1 \\ M_2\end{Bmatrix} = \begin{Bmatrix}0 \\ 0\end{Bmatrix}$$

In solving the expression for M_1 and M_2, the inverse of the flexibility matrix is given directly by the stiffness matrix of Eq. (9-30) with appropriate changes in sign. The resulting values of the fixed-end moments are found to be

$$M_1 = M_2 = \alpha DEI$$

A positive value of D, which according to Eq. (8-68) means that the temperature on the top of the beam is higher than on the bottom, will

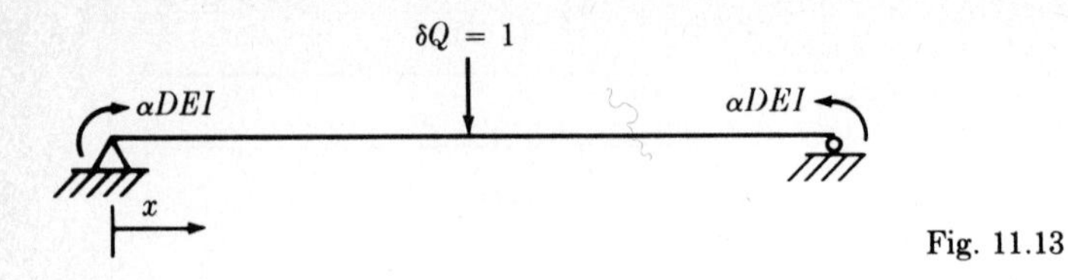

Fig. 11.13

therefore result in end moments causing compression on the upper fibers of the beam.

The vertical deflection at midspan on the beam is determined by subjecting the primary structure to the temperature change and to the values of end moment as shown in Fig. 11-13. A unit virtual force is applied in the downward direction at midspan. The value of deflection at midspan is obtained from the expression

$$D_V = -\alpha D \int_0^L m\, dx + \int_0^L \frac{mM}{EI}\, dx$$

where, from Fig. 11-13,

$$m = 0.5x \qquad M = \alpha DEI \qquad 0 \leq x \leq \frac{L}{2}$$

The value of D_V is therefore found to be

$$D_V = -2\alpha D \int_0^{L/2} 0.5x\, dx + \frac{2}{EI} \int_0^{L/2} 0.5x\, \alpha DEI\, dx = 0$$

It can be shown that the vertical deflection at any point on a fixed-end beam is zero for a linear temperature change. This is not the case for other types of end support.

11-3 THREE-MOMENT EQUATION

A general equation can be developed for analysis of continuous beams with no support movement. The equation relates the moments in the beam at three consecutive support points to the loadings on the intermediate spans and is therefore referred to as the *three-moment equation.* The result of applying the three-moment equation to the analysis of a beam is a set of simultaneous equations with the moments at the supports as the unknowns.

The three-moment equation can be developed by the method of consistent displacements. Consider the continuous beam of Fig. 11-14. The support points are denoted from left to right as i, j, and k. The properties of the span to the left of point j are denoted with the subscript i and the

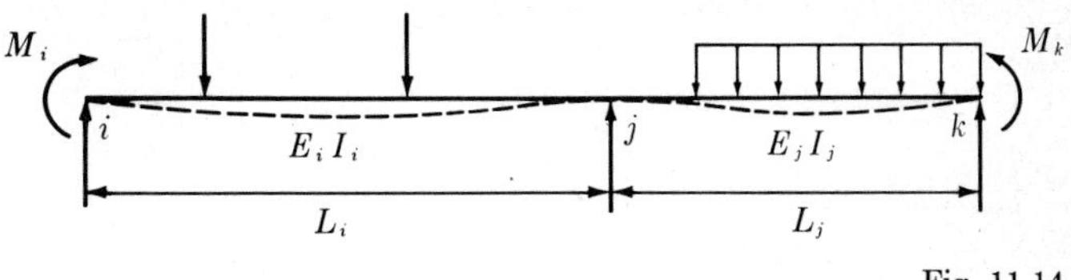

Fig. 11.14

properties of the span to the right of point j are denoted by the subscript j. The beam is subjected to arbitrary transverse loading on each span and to possible moments at its end points of M_i and M_k. The deflected shape of the beam is shown by the dashed curve in Fig. 11-14.

The beam of Fig. 11-14 is statically indeterminate to the first degree. Because we are concerned with the moment in the beam at j, M_j is selected as the redundant quantity. Removing the restraint of M_j, which is equivalent to inserting a pin in the beam at j, results in a deflected shape of the primary structure as shown in Fig. 11-15a. The relative rotation between the beam segments at j is represented by $D_j{}^Q$. The value of the redundant moment M_j is found from the deflection condition

$$D_j{}^Q + f_{jj}M_j = 0 \tag{11-12}$$

The value of the flexibility influence coefficient f_{jj} is obtained by subjecting the primary structure to unit moments at j, as shown in Fig. 11-15b.

The rotation quantity $D_j{}^Q$ is determined by considering the two segments of the beam as shown in Figs. 11-16a and c. From Figs. 11-16a and c and Fig. 11-15a we see that

$$D_j{}^Q = D^Q_{j\text{-left}} + D^Q_{j\text{-right}}$$

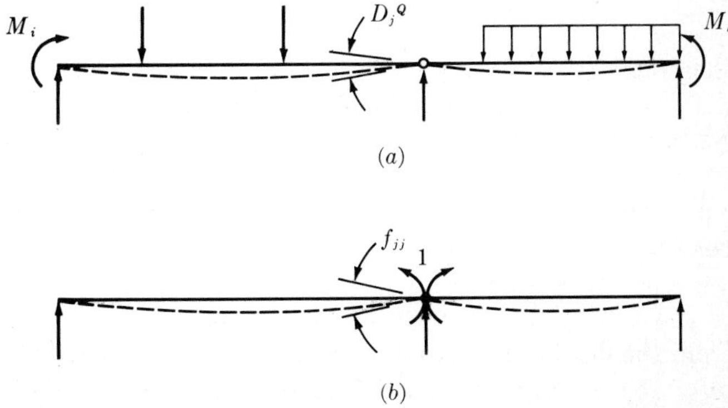

(a)

(b)

Fig. 11.15

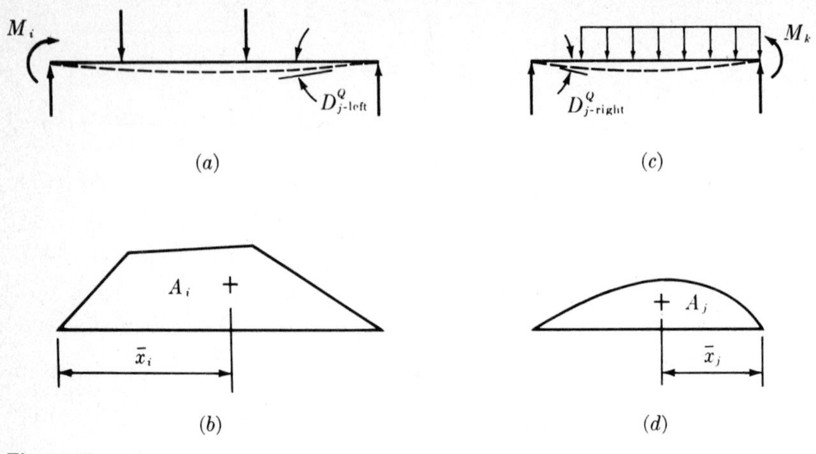

Fig. 11.16

The rotation $D^Q_{j\text{-left}}$ is a function of the transverse loading on the left span and the moment M_i. The amount of rotation due to the transverse loading can be expressed in general terms by use of the moment-area method. The area of the moment diagram due to the transverse loading is denoted as A_i in Fig. 11-16*b*. The location of the centroid of the area from the left support is denoted as the distance $\bar{x}_i$. Using the flexibility relation of Eq. (9-32) to obtain the rotation at j due to the moment at i, we obtain the total rotation of the left segment at j,

$$D^Q_{j\text{-left}} = \frac{A_i \bar{x}_i}{E_i I_i L_i} + \frac{M_i L_i}{6E_i I_i}$$

Similarly, from Figs. 11-16*c* and *d*,

$$D^Q_{j\text{-right}} = \frac{A_j \bar{x}_j}{E_j I_j L_j} + \frac{M_k L_j}{6E_j I_j}$$

Therefore

$$D_j{}^Q = \frac{M_i L_i}{6E_i I_i} + \frac{M_k L_j}{6E_j I_j} + \frac{A_i \bar{x}_i}{E_i I_i L_i} + \frac{A_j \bar{x}_j}{E_j I_j L_j} \tag{11-13}$$

The value of f_{jj} can be obtained in a similar manner by observing in Fig. 11-15*b* that

$$f_{jj} = f_{jj\text{-left}} + f_{jj\text{-right}}$$

From the flexibility relation of Eq. (9-32), this becomes

$$f_{jj} = \frac{L_i}{3E_i I_i} + \frac{L_j}{3E_j I_j} \tag{11-14}$$

Substituting Eqs. (11-13) and (11-14) into Eq. (11-12) results in the expression

$$\frac{M_iL_i}{6E_iI_i} + \frac{M_kL_j}{6E_jI_j} + \frac{A_i\bar{x}_i}{E_iI_iL_i} + \frac{A_j\bar{x}_j}{E_jI_jL_j} + \left(\frac{L_i}{3E_iI_i} + \frac{L_j}{3E_jI_j}\right)M_j = 0$$

which can be written in the form

$$\frac{M_iL_i}{E_iI_i} + 2M_j\left(\frac{L_i}{E_iI_i} + \frac{L_j}{E_jI_j}\right) + \frac{M_kL_j}{E_jI_j} = -\frac{6A_i\bar{x}_i}{E_iI_iL_i} - \frac{6A_j\bar{x}_j}{E_jI_jL_j} \tag{11-15}$$

This expression is the general form of the three-moment equation. In this form the moments at i, j, and k are expressed in terms of the span properties and in terms of the transverse loading by means of the moment-area terms. The moment quantities in Eq. (11-15) are positive according to the beam sign convention; that is, positive moment causes compression in the upper fibers.

If EI for the beam is constant, then Eq. (11-15) becomes

$$M_iL_i + 2M_j(L_i + L_j) + M_kL_j = -\frac{6A_i\bar{x}_i}{L_i} - \frac{6A_j\bar{x}_j}{L_j} \tag{11-16}$$

For the condition of uniform loading on each span, where

$$A_i = \frac{2L_i}{3}\frac{w_iL_i^2}{8} = \frac{w_iL_i^3}{12} \qquad A_j = \frac{w_jL_j^3}{12}$$

$$\bar{x}_i = \frac{L_i}{2} \qquad \bar{x}_j = \frac{L_j}{2}$$

Eq. (11-16) becomes

$$M_iL_i + 2M_j(L_i + L_j) + M_kL_j = -\frac{w_iL_i^3}{4} - \frac{w_jL_j^3}{4} \tag{11-17}$$

The three-moment equations developed above are expressed in terms of moments to the right and to the left of some point j. In applying the three-moment equation to the analysis of a particular beam we locate the j point successively at the interior support points, which results in as many equations as there are unknown moments. Applications of the method are given in the following examples.

EXAMPLE 11-5 The moments in the beam at the reaction points and the values of the reactions are to be determined for the continuous beam shown in Fig. 11-17 by the three-moment equation. EI for the beam is constant. Because the loading on each span is uniform and EI is constant, we use the form of the three-moment equation given in Eq. (11-17). For the given beam there are three unknown values of moment: M_B, M_C, and M_D. The moments at A and E are zero, so three equations must be obtained to determine the unknowns.

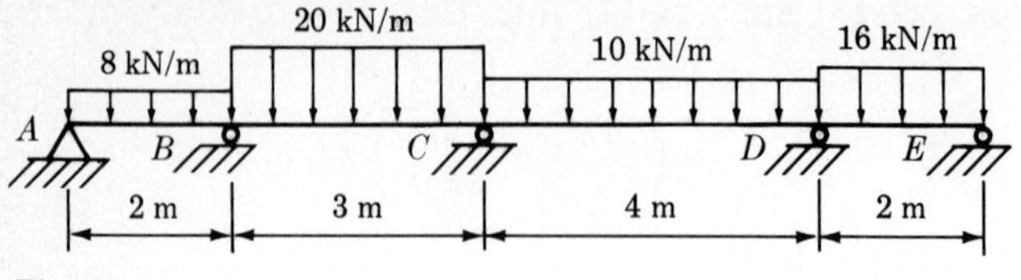

Fig. 11.17

Considering the j point in the three-moment equation to be located first at support point B, we obtain

$$0 + 2M_B(2 + 3) + M_C(3) = -\frac{(8)(2)^3}{4} - \frac{(20)(3)^3}{4}$$

or $\quad 10M_B + 3M_C = -151.0$

With the j point at C,

$$M_B(3) + 2M_C(3 + 4) + M_D(4) = -\frac{(20)(3)^3}{4} - \frac{(10)(4)^3}{4}$$

$$3M_B + 14M_C + 4M_D = -295.0$$

With the j point at D,

$$M_C(4) + 2M_D(4 + 2) + 0 = -\frac{(10)(4)^3}{4} - \frac{(16)(2)^3}{4}$$

$$4M_C + 12M_D = -192.0$$

The three desired simultaneous equations are therefore

$$\begin{aligned} 10M_B + 3M_C &= -151.0 \\ 3M_B + 14M_C + 4M_D &= -295.0 \\ 4M_C + 12M_D &= -192.0 \end{aligned}$$

or, in matrix form,

$$\begin{bmatrix} 10 & 3 & 0 \\ 3 & 14 & 4 \\ 0 & 4 & 12 \end{bmatrix} \begin{Bmatrix} M_B \\ M_C \\ M_D \end{Bmatrix} = \begin{Bmatrix} -151.0 \\ -295.0 \\ -192.0 \end{Bmatrix}$$

A solution of this matrix equation such as by the computer program in Appendix B results in

$$\begin{Bmatrix} M_B \\ M_C \\ M_D \end{Bmatrix} = \begin{Bmatrix} -10.37 \\ -15.78 \\ -10.74 \end{Bmatrix} \text{kN-m}$$

All moments at the supports are negative, causing compression in the lower fibers of the beam.

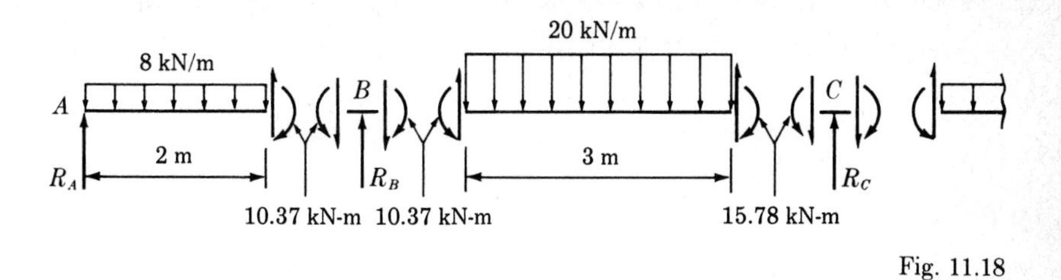

Fig. 11.18

The values of the reactions are obtained by considering free-body diagrams of the beam spans. The free-body diagrams for the two left spans are shown in Fig. 11-18. The reaction at A can be obtained by taking the summation of moments about B,

$$2R_A - (8)(2)(1) + 10.37 = 0$$
$$R_A = 2.82 \text{ kN}$$

The value of an interior reaction such as R_B is obtained by determining first the values of shear in the beam to the right and to the left of the reaction. The value of shear to the left of R_B is obtained by taking the summation of vertical forces on the left-span free body. The value of the shear to the right of R_B is obtained by taking the summation of moments about C on the free-body diagram between B and C. From Fig. 11-18, the values of shear are

$$V_{B\text{-left}} = (8)(2) - 2.82 = 13.18 \text{ kN}$$
$$V_{B\text{-right}} = \frac{10.37 - 15.78 + (20)(3)(1.5)}{3} = 28.20 \text{ kN}$$

Therefore

$$R_B = 13.18 + 28.20 = 41.38 \text{ kN}$$

Similarly, the values of the remaining reactions are found to be

$$R_C = 53.06 \text{ kN}$$
$$R_D = 40.11 \text{ kN}$$
$$R_E = 10.63 \text{ kN}$$

In the above example, if either or both of the ends of the continuous beam had been fixed, the number of unknown moments would have been greater than the number of equations obtainable from the general approach. To obtain additional equations for fixed-end conditions we add an imaginary span to the beam in place of the fixed support. In applying the three-moment equation to the modified beam, the length of the imaginary span is set equal to zero. The justification for this and a demonstration of the procedure are given in the following section.

11-4 EVALUATION OF FIXED-END MOMENTS

A necessary prerequisite to the methods of analysis that are presented in the following chapters is the evaluation of fixed-end moments for individual members. Fixed-end moments, denoted as FEM, are defined as the moments at the ends of a member when the ends of the member are fixed against rotation. Fixed-end moments are usually due to transverse loading on a member but, as we shall see in the moment-distribution method of Chap. 13, fixed-end moments can also be developed by translation of the member ends. In this section we shall see how previously developed techniques can be used to determine expressions for fixed-end moments due to various loading conditions.

EXAMPLE 11-6 The fixed-end moments are to be determined for the uniformly loaded member shown in Fig. 11-19. The member is assumed to have a constant EI value and as such is commonly referred to as a *prismatic* member.

The fixed-end moments are denoted by FEM with subscripts as shown in Fig. 11-19. The first subscript denotes the point under consideration, and the second subscript denotes the other end of the member.

The beam is statically indeterminate to the second degree. In arriving at this conclusion we note that there are three possible reaction components at each end, giving a total of six unknown reactions. There are three equations of equilibrium. However, there is no axial force, and we are therefore left with a total of four unknown reaction components, moment and shear at each end, and two equations of equilibrium.

As mentioned earlier in this chapter, the choice of redundants is arbitrary. We shall select the two fixed-end moments as the redundant quantities. The primary structure is therefore a simply supported beam. The redundant moments and rotations at A and B will be considered positive when they act in the directions of FEM_{AB} and FEM_{BA} shown in Fig. 11-19. The consistent displacement equations for zero rotation at A and B, written in matrix form, are

$$\begin{Bmatrix} D_A{}^Q \\ D_B{}^Q \end{Bmatrix} + \begin{bmatrix} f_{AA} & f_{AB} \\ f_{BA} & f_{BB} \end{bmatrix} \begin{Bmatrix} \text{FEM}_{AB} \\ \text{FEM}_{BA} \end{Bmatrix} = \begin{Bmatrix} 0 \\ 0 \end{Bmatrix}$$

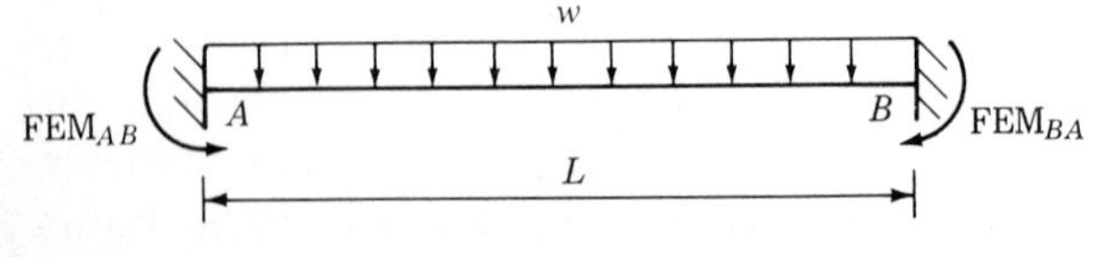

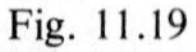
Fig. 11.19

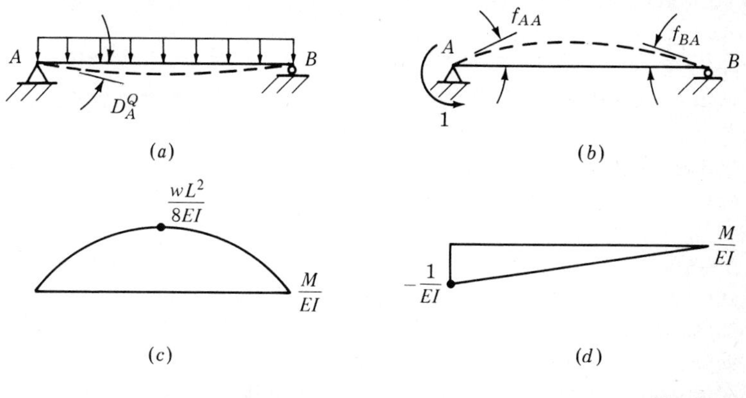

Fig. 11.20

We can note that, because of symmetry, $\text{FEM}_{AB} = \text{FEM}_{BA}$, $D_A{}^Q = D_B{}^Q$, and $f_{AA} = f_{BB}$. As shown earlier, $f_{AB} = f_{BA}$. The above consistent displacement equations can therefore be reduced to the single equation

$$D_A{}^Q + (f_{AA} + f_{BA})\text{FEM}_{AB} = 0$$

The deflection quantities can conveniently be evaluated using the moment-area method. The loading and the deflection quantities, as well as the M/EI diagrams, are shown in Fig. 11-20. In the calculations, the signs of the deflection quantities as evaluated by moment-area theorems are adjusted to make the results consistent with the convention selected above for positive rotations. From Figs. 11-20*a* and *c* we obtain

$$D_A{}^Q = -\frac{t_{BA}}{L} = -\left(\frac{2}{3}\right)\left(\frac{wL^2}{8EI}\right)(L)\left(\frac{L}{2}\right)\left(\frac{1}{L}\right) = -\frac{wL^3}{24EI}$$

From Figs. 11-20*b* and *d*,

$$f_{AA} = -\frac{t_{BA}}{L} = -\left(\frac{1}{2}\right)\left(-\frac{1}{EI}\right)(L)\left(\frac{2L}{3}\right)\left(\frac{1}{L}\right) = \frac{L}{3EI}$$

$$f_{BA} = -\frac{t_{AB}}{L} = -\left(\frac{1}{2}\right)\left(-\frac{1}{EI}\right)(L)\left(\frac{L}{3}\right)\left(\frac{1}{L}\right) = \frac{L}{6EI}$$

Therefore,

$$-\frac{wL^3}{24EI} + \left(\frac{L}{3EI} + \frac{L}{6EI}\right)\text{FEM}_{AB} = 0$$

from which we obtain

$$\text{FEM}_{AB} = \frac{wL^2}{12}$$

EXAMPLE 11-7 The fixed-end moments are to be determined for the member shown in Fig. 11-21 by the three-moment equation. *EI* for the

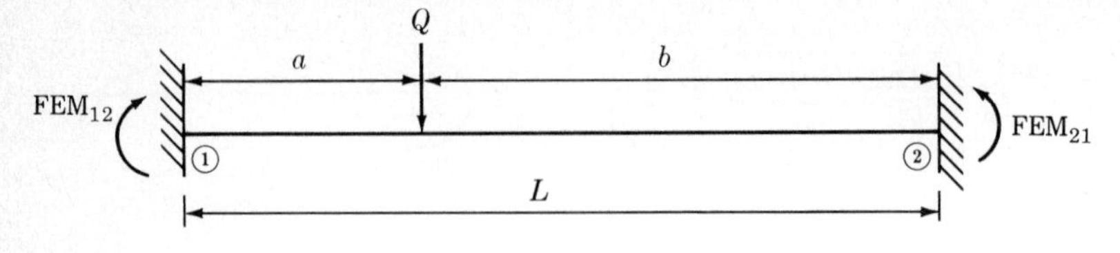

Fig. 11.21

member is constant. Fixed-end moments will be considered positive as shown in Fig. 11-21 to correspond to positive moment as used in the three-moment equation. For the given loading, the form of the three-moment equation as given in Eq. (11-16) is used. In order to obtain two equations for the unknown moments we add imaginary spans to each end of the beam, as shown in Fig. 11-22. In setting the lengths of the imaginary spans to zero we can consider a term such as $-6A_i\bar{x}_i/L_i$ of Eq. (11-16) to be of the form $-6CL_i$, where C represents a constant by which L_i^2 is multiplied to obtain the quantity $A_i\bar{x}_i$. Thus we avoid dividing by zero, and the term is zero when L_i is zero. With j at point ① in Fig. 11-22, we obtain

$$0 + 2M_1(0 + L) + M_2L = -0 - \frac{6A_j\bar{x}_j}{L}$$

The reason for using an imaginary span of zero length is apparent from the second theorem of the moment-area method. A tangent drawn to the elastic curve at ① for the given beam of Fig. 11-21 will pass through point ②. Therefore t_{21} is equal to zero. If we were to express t_{21} in terms of M_1, M_2, and the loading-moment diagram and equate the resulting expression for t_{21} to zero, we would obtain the same expression as above.

For the given loading A_j and $\bar{x}_j$, as defined in Fig. 11-16, are

$$A_j = \frac{abQ}{2} \qquad \bar{x}_j = \frac{a + 2b}{3}$$

The first three-moment equation is therefore

$$2M_1L + M_2L = -\frac{ab(a + 2b)Q}{L}$$

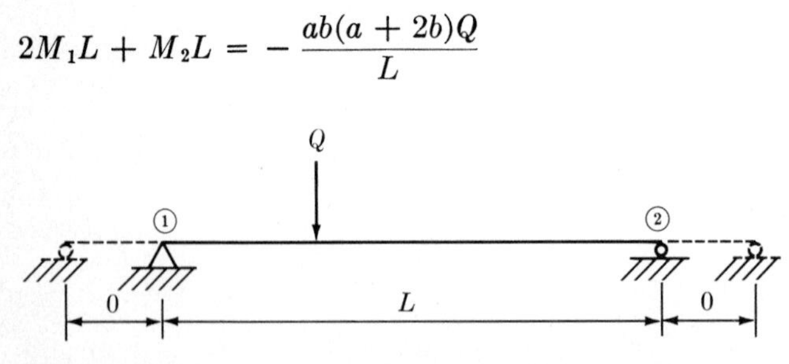

Fig. 11.22

With j at point ② in Fig. 11-22, we have

$$M_1L + 2M_2(L + 0) + 0 = -\frac{6A_i\bar{x}_i}{L_i} - 0$$

The values of A_i and $\bar{x}_i$ are

$$A_i = \frac{abQ}{2} \qquad \bar{x}_i = \frac{2a + b}{3}$$

The second three-moment equation is therefore

$$M_1L + 2M_2L = -\frac{ab(2a + b)Q}{L}$$

The two equations for M_1 and M_2, written in matrix form, are

$$L\begin{bmatrix} 2 & 1 \\ 1 & 2 \end{bmatrix}\begin{Bmatrix} M_1 \\ M_2 \end{Bmatrix} = -\begin{Bmatrix} \dfrac{ab(a + 2b)Q}{L} \\ \dfrac{ab(2a + b)Q}{L} \end{Bmatrix}$$

Solving this matrix expression yields the fixed-end moments

$$M_1 = \text{FEM}_{12} = -\frac{ab^2Q}{L^2}$$

$$M_2 = \text{FEM}_{21} = -\frac{a^2bQ}{L^2}$$

The results obtained in the above example can be used to obtain the fixed-end moments for various forms of distributed loading. As an illustration, let us consider the loading on a member as shown in Fig. 11-23. The load varies linearly from zero at ② to an intensity of w per unit length at ①. To obtain the fixed-end moments using the results of the above example, we consider the result of the distributed load over a length dx at position x on the member:

$$Q = \left(\frac{x}{L}\right)(w)\,dx$$

We also note that $a = L - x$ and $b = x$. The fixed-end moments can be obtained by substituting these values into the fixed-end-moment expres-

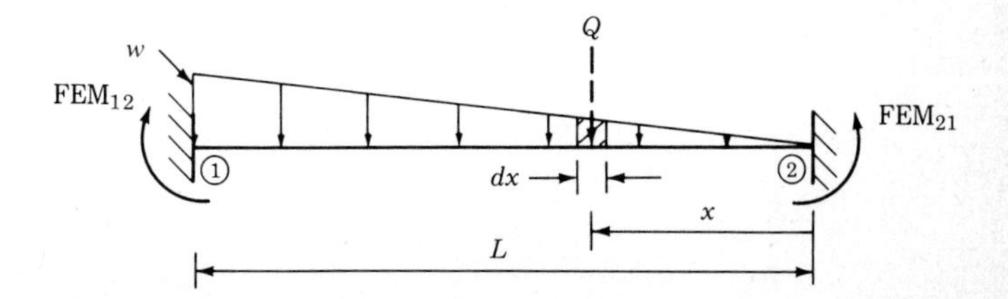

Fig. 11.23

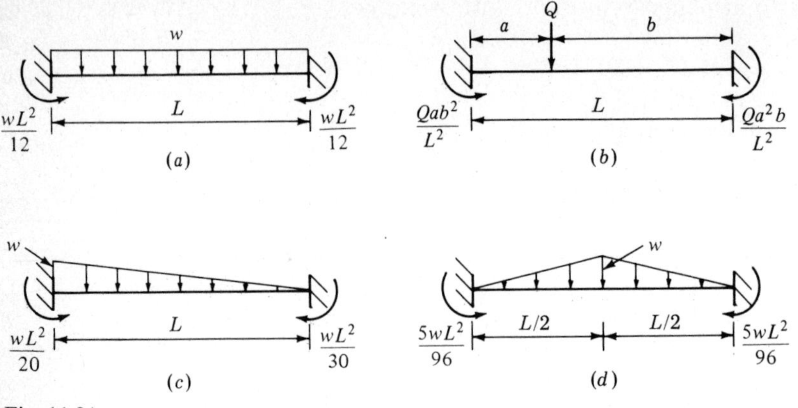

Fig. 11.24

sions for a concentrated load and integrating over the length of the member:

$$\mathrm{FEM}_{12} = -\int_0^L \frac{(L-x)(x^2)}{L^2}\left(\frac{x}{L}\right)(w)\,dx = -\frac{wL^2}{20}$$

$$\mathrm{FEM}_{21} = -\int_0^L \frac{(L-x)^2(x)}{L^2}\left(\frac{x}{L}\right)(w)\,dx = -\frac{wL^2}{30}$$

Expressions for some of the commonly encountered forms of loading are summarized in Fig. 11-24. It should be noted that the fixed-end moments for a member loaded with a series of concentrated loads can be obtained by superimposing the results of individual loads as obtained from Fig. 11-24*b*.

PROBLEMS

11-1 through 11-4. Determine the axial forces in all members and the reactions for the pin-connected trusses shown. The modulus of elasticity E and the cross-sectional area A are the same for all members.

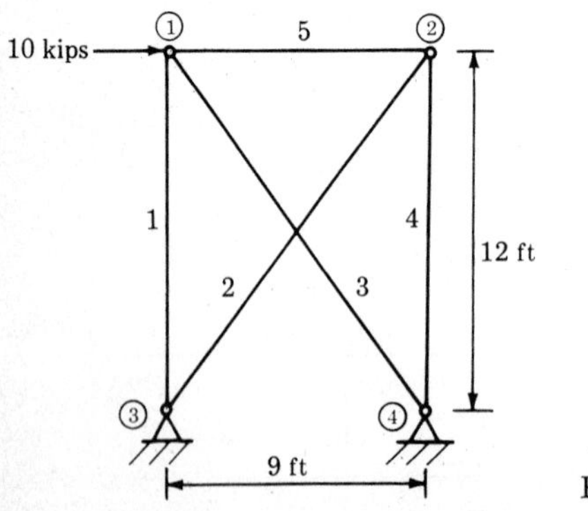

P11.1

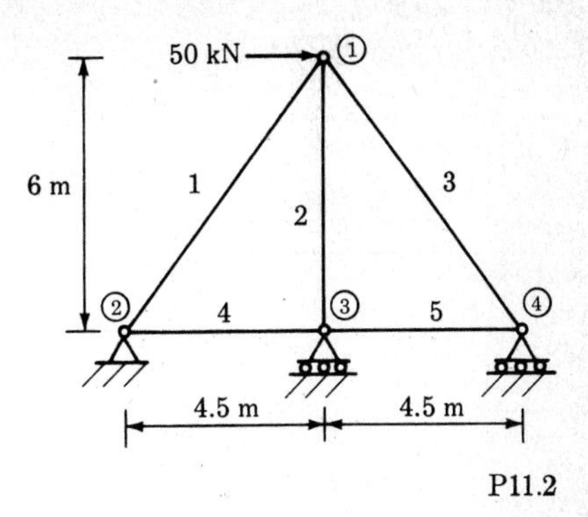

P11.2

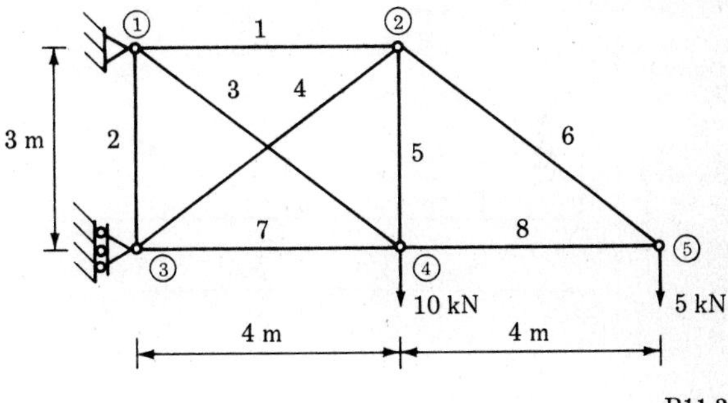

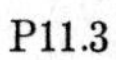
P11.3

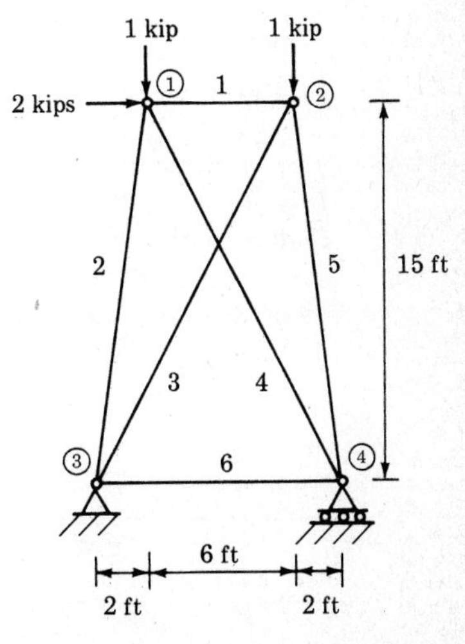

P11.4

11-5 through 11-8. Determine the values of the reaction components for the beams shown. EI is constant.

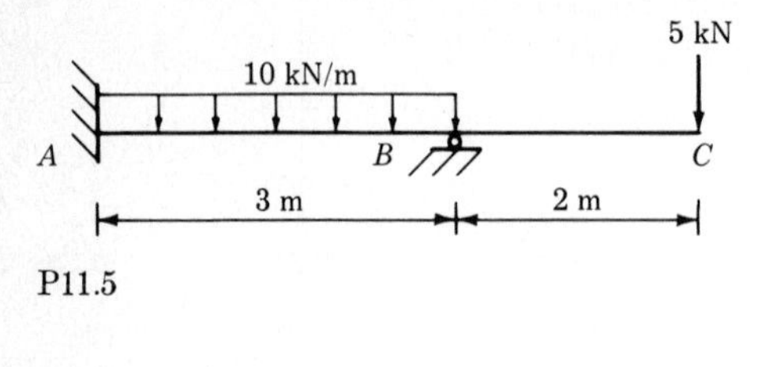

P11.5

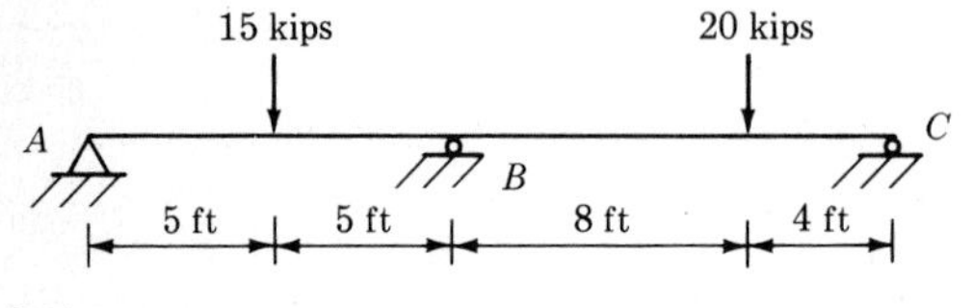

P11.6

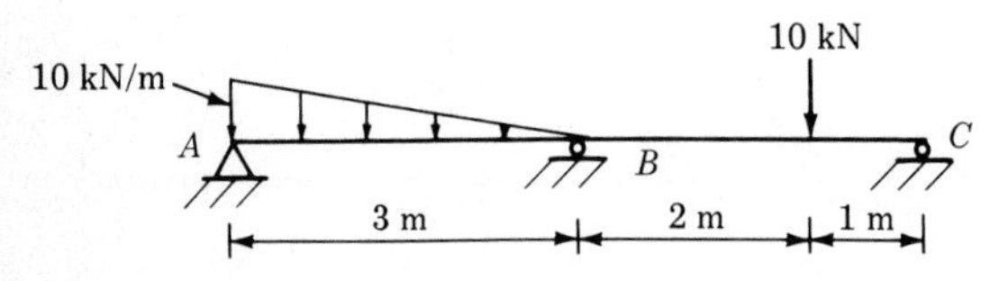

P11.7

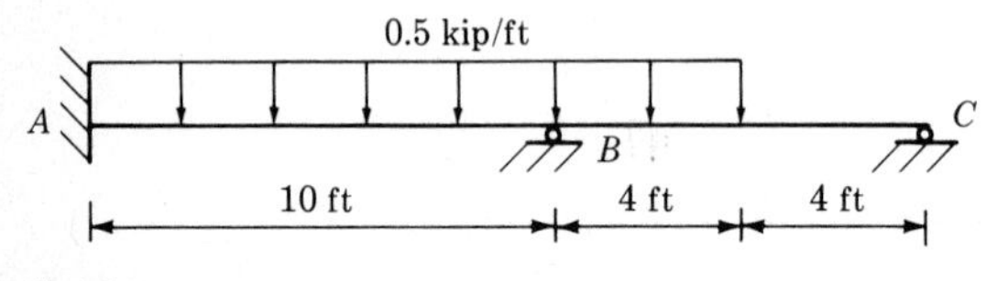

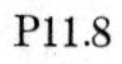

P11.8

11-9. The continuous beam shown has a constant I value of 500 in.^{4}, and $E = 30 \times 10^3$ ksi. Determine the values of the reactions at A, B, and C due to the given uniform load if (*a*) there is no settlement at the support points; (*b*) support point C settles 0.10 in.; (*c*) support point B settles 0.25 in. and support point C settles 0.15 in. Construct the moment diagram for the beam for each condition on the tension side of the member.

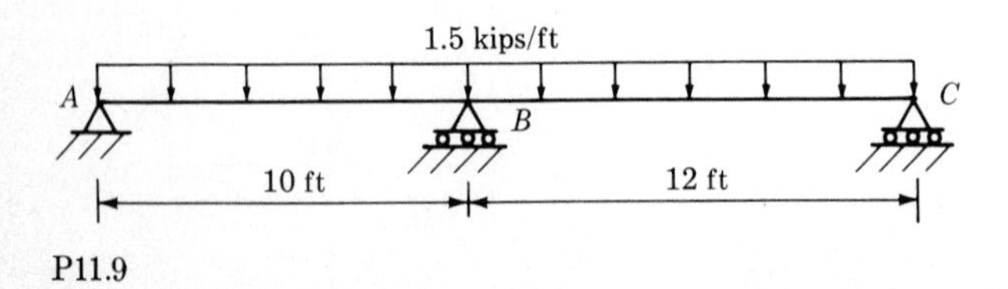

P11.9

11-10. For the frame shown, (*a*) determine the components of reaction at A and C; (*b*) construct the moment diagram for the frame on the tension side of the members. EI is the same for each member.

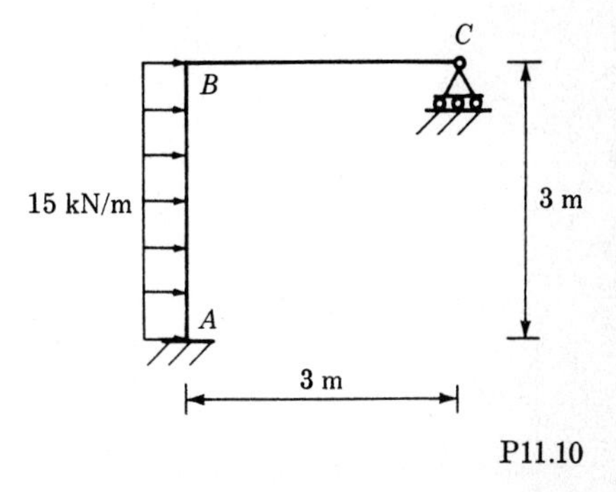

P11.10

11-11 and 11-12. Determine the values of the reaction components for the frames shown. EI is the same for each member.

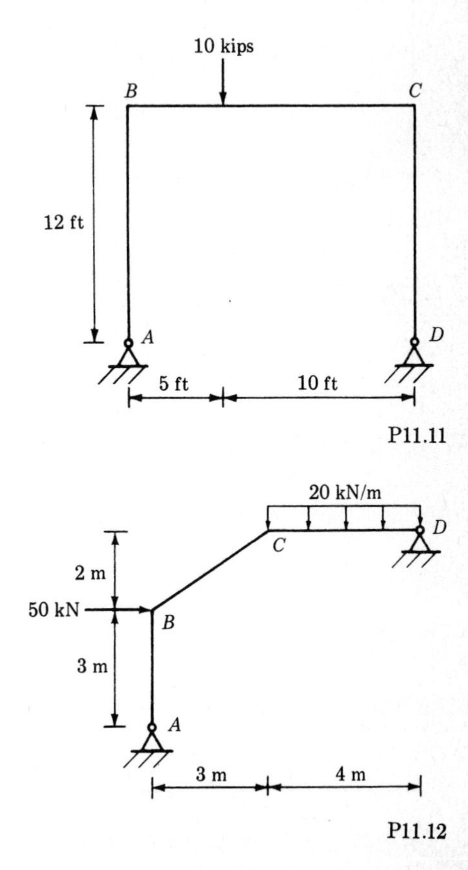

P11.12

11-13. Determine the values of the reaction components for the frame shown in Fig. P12-10 using the method of consistent displacements. EI is constant.

11-14. Determine the values of the reaction components for the frame shown in Fig. P13-11 using the method of consistent displacements. E is constant.

11-15. For the frame shown, (*a*) determine the reaction components at ① and ④; (*b*) construct the moment diagram for the frame on the compression side of the members.

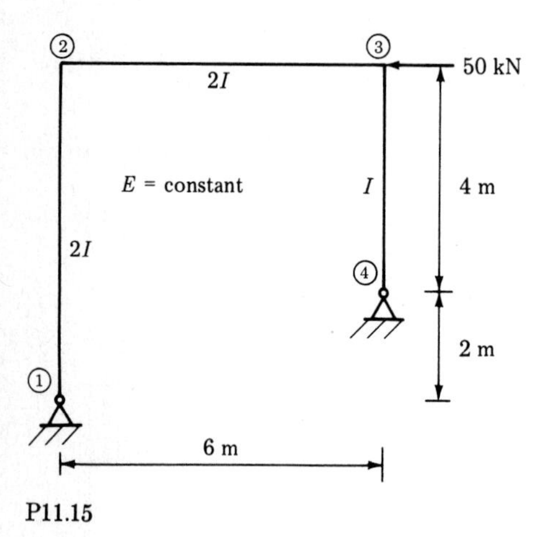

P11.15

11-16. Beam AC is fixed at A and is supported by cable BD at B. Determine the moment in the beam at A and the force in cable BD. $E = 30 \times 10^3$ ksi for both the cable and the beam.

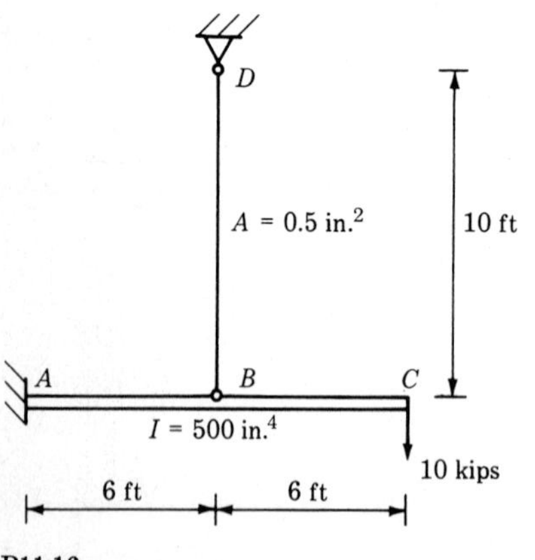

P11.16

11-17. For the frame shown, (*a*) determine the reaction components at A and D; (*b*) construct the moment diagram for the frame on the tension side of the members.

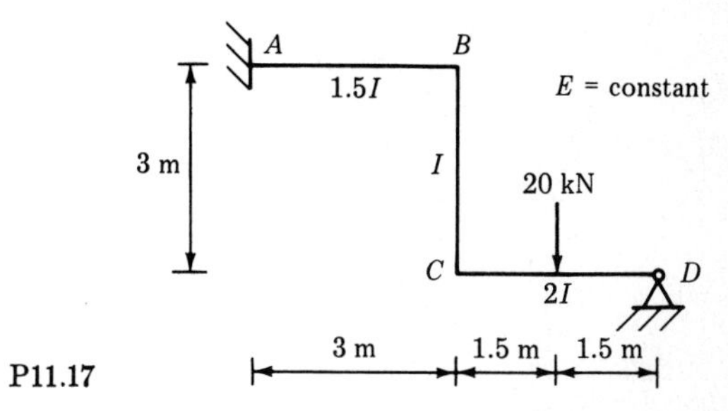

P11.17

11-18. Determine the axial force in each member of the space truss shown. E and A are the same for all members.

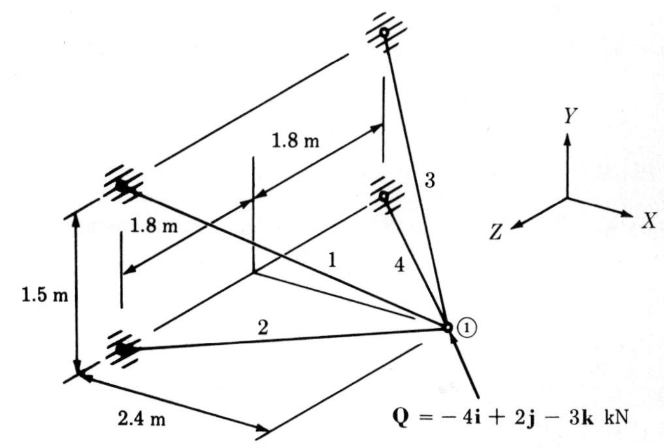

P11.18

11-19. Frame ABC is fixed at A and has an elastic support at C. For members AB and BC, $E = 30 \times 10^3$ ksi, $G = 12 \times 10^3$ ksi, $I = 50$ in.^{4}, and $J = 100$ in.^{4}. Spring CD has a stiffness of 5 kips/in. Determine the axial force in spring CD and the moment at A.

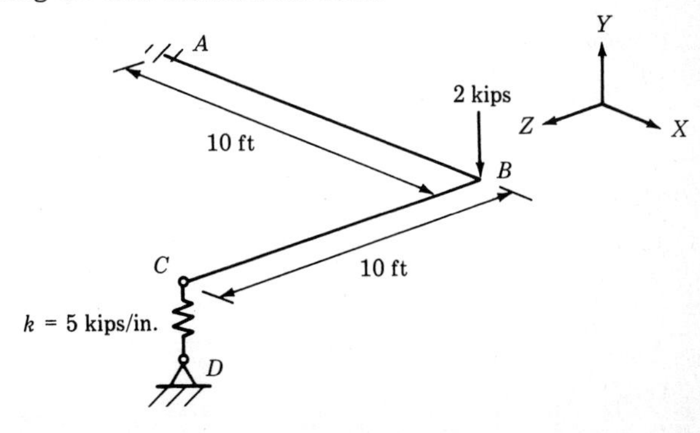

P11.19

11-20 and 11-21. For the continuous beams shown, use the three-moment equation and determine (*a*) the values of moment at A, B, C, and D; (*b*) the values of vertical reaction at A, B, C, and D. Construct the shear and moment diagrams using the beam sign convention. E is constant.

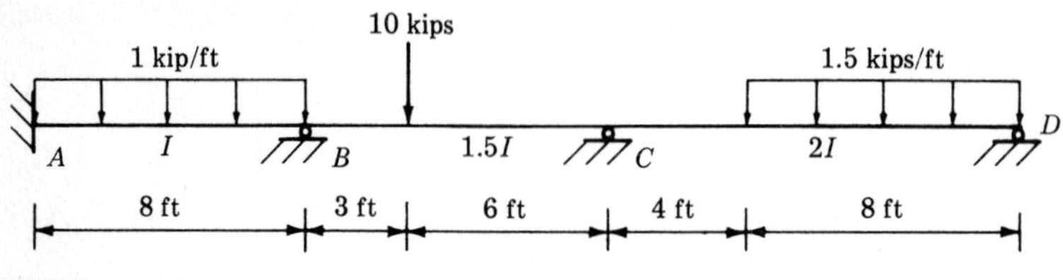

P11.20

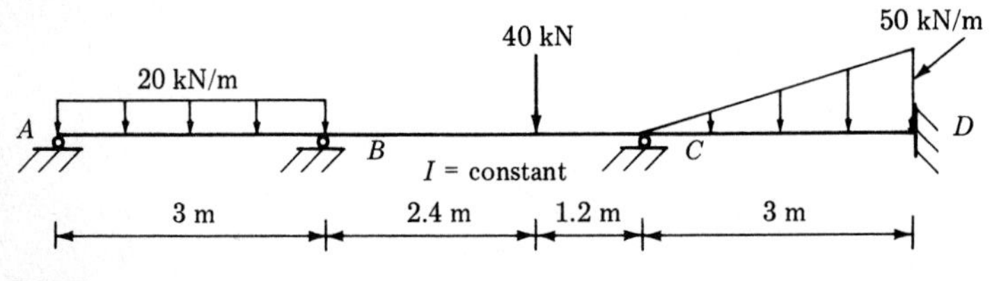

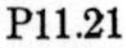
P11.21

11-22. Determine the values of the fixed-end moments at A and B for the members in Figs. P11-22*a* and *c*, and the expressions for the fixed-end moments for the members in Figs. P11-22*b* and *d*. EI is constant.

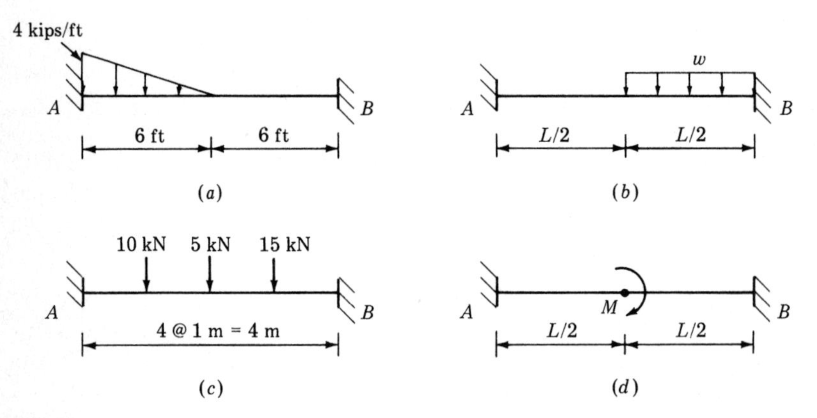

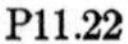
P11.22

CHAPTER

TWELVE

SLOPE-DEFLECTION METHOD

12-1 BASIC CONCEPTS OF THE METHOD

The slope-deflection method can be used to analyze statically indeterminate structures composed of moment-resisting members such as continuous beams and frames. The method is based upon two general equations which express the moments at the ends of a member in terms of the fixed-end moments and the displacements of the ends of the member. In the analysis of a structure the slope-deflection equations are applied to each member of the structure. Using appropriate equations of equilibrium along with the equations for each member, we obtain a set of simultaneous equations with displacements as the unknowns. Substitution of the resulting displacement values back into the slope-deflection equations leads to the desired values of moment in the members.

Let us consider the member shown in Fig. 12-1 to be a typical member of a statically indeterminate structure, with end points denoted as i and j. The member is subjected to possible transverse loading and is connected at its ends to other members of the structure, or possibly to a support. The moments in the ends of the members can be considered a combination of moments due to a fixed-end condition and moments due to displacements of the member ends. These two forms of moment are shown in

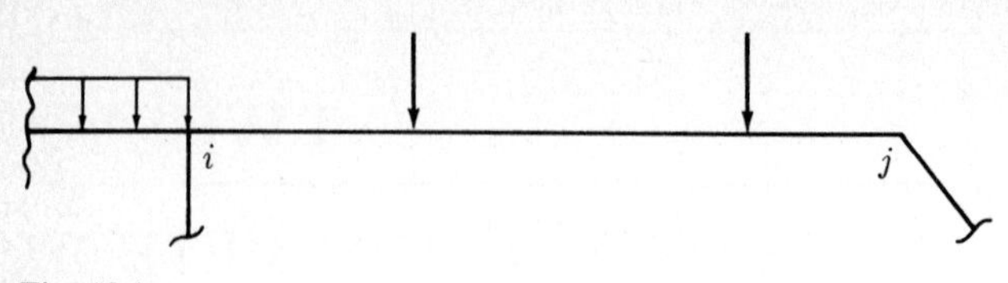

Fig. 12.1

Figs. 12-2a and b, respectively. Figure 12-2a illustrates the condition in which the member ends are fixed. Figure 12-2b illustrates a generally deflected member, where the deflection is described in terms of the rotation and translation of the member ends. Associated with these displacements are moments M_i' and M_j' generated at the member ends. The expressions for M_i' and M_j' in terms of the member-end displacements can readily be obtained from the member-stiffness relation of Eq. (9-18).

If we denote the total moment in the member at the i end as M_i and that at the j end as M_j, the combination of the effects in Figs. 12-2a and b and the use of the stiffness relation of Eq. (9-18) results in the expressions

$$\begin{aligned} M_i &= \text{FEM}_{ij} + \frac{4EI}{L}\phi_i + \frac{2EI}{L}\phi_j - \frac{6EI}{L}\beta \\ M_j &= \text{FEM}_{ji} + \frac{2EI}{L}\phi_i + \frac{4EI}{L}\phi_j - \frac{6EI}{L}\beta \end{aligned} \tag{12-1}$$

where $\beta = (\eta_i - \eta_j)/L$. Equations (12-1) are the slope-deflection equations. Moments are considered positive when they act on a member end

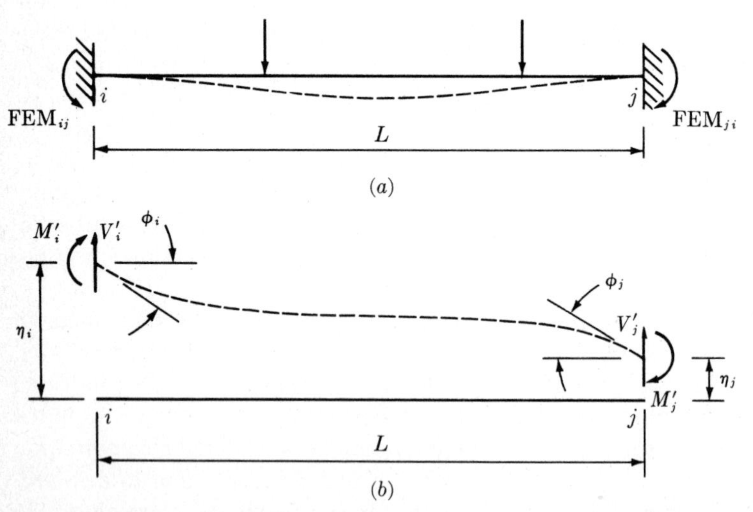

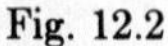

Fig. 12.2

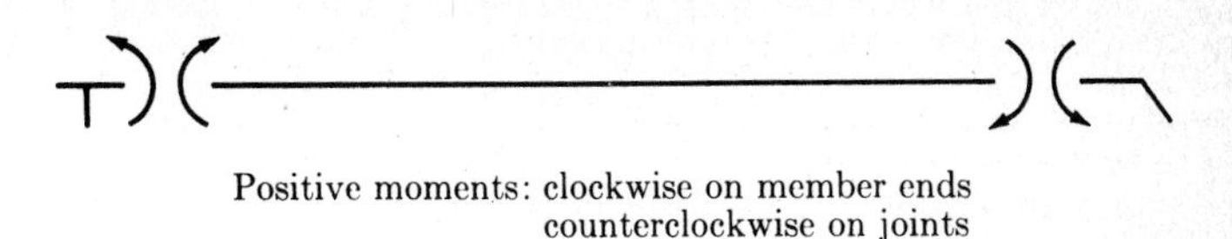

Positive moments: clockwise on member ends
counterclockwise on joints

Fig. 12.3

or on a joint in the directions as shown in Fig. 12-3. Thus the value of FEM_{ij} shown in Fig. 12-2*a* is a negative quantity and FEM_{ji} is a positive quantity.

Member displacements are considered positive in the directions shown in Fig. 12-2*b*. From the above definition $\beta = (\eta_i - \eta_j)/L$ we see that β is the angle that the member chord makes with the original direction of the member. It can be said that β is positive when the chord of the member rotates in a clockwise direction. Sometimes, however, it may be preferable to determine the sign of β directly from η_i and η_j.

It is convenient under certain conditions to use a simplified form of the slope-deflection equations (12-1). If the j end of the member in Fig. 12-2 is pinned, then M_j in Eqs. (12-1) is zero. With $M_j = 0$ we can multiply the second of Eqs. (12-1) by $-\frac{1}{2}$ and add the result to the first equation. The result can be expressed as

$$M_i = \text{FEM}_{ij} - \frac{\text{FEM}_{ji}}{2} + \frac{3EI}{L}(\phi_i - \beta) \tag{12-2}$$

Equation (12-2) is referred to as the *simplified slope-deflection equation*. The simplified equation can be used for members that have zero moment at either the i or the j end. Equation (12-2) replaces both of Eqs. (12-1) for such members. It should be noted that the simplified equation contains one fewer unknown rotation and, as we shall see in the next section, by its use the work in the analysis of a total structure can be reduced.

12-2 ANALYSIS OF CONTINUOUS BEAMS

The analysis of continuous beams by the slope-deflection method is a fairly straightforward procedure. Provided there are no movements of the supports, the possible displacements of the beam are rotations of the beam at the supports, represented by the angle ϕ in the general slope-deflection equations of Eq. (12-1). If the supports are permitted to displace, this type of displacement is represented by the parameter β.

By writing the slope-deflection equations for each span of a continuous beam, we obtain a set of equations in terms of the unknown displacements. Generally these equations cannot be solved directly for the unknowns. By using moment-equilibrium equations at the support points and satisfying geometric boundary conditions, we obtain a set of simultaneous equations for the unknown displacements. Solving the simultaneous equations for the displacements and substituting the values of displacements into the original slope-deflection equations results in the values of moment in the beam. The following example illustrates the procedure.

EXAMPLE 12-1 The moments at the support points in the continuous beam of Fig. 12-4 are to be determined by the slope-deflection method. The relative moments of inertia for each span are as shown. E is constant. To apply the slope-deflection equations we must first determine the fixed-end moments for each span. From Fig. 11-24 we obtain

$$\mathrm{FEM}_{12} = -\frac{wL^2}{12} = -\frac{(18)(6)^2}{12} = -54.0 \text{ kN-m}$$

$$\mathrm{FEM}_{21} = 54.0 \text{ kN-m}$$

$$\mathrm{FEM}_{23} = -\frac{QL}{8} = -\frac{(36)(3)}{8} = -13.5 \text{ kN-m}$$

$$\mathrm{FEM}_{32} = 13.5 \text{ kN-m}$$

We recognize that there is zero moment at ③. Therefore, the simplified slope-deflection equation (12-2) can be used for member ②-③. Because no support displacements are permitted, $\beta = 0$ for both spans. Furthermore, for the fixed support at ①, $\phi_1 = 0$. The slope-deflection equations for the two spans are therefore

$$M_{12} = -54.0 + \frac{2E(3I)}{6}\phi_2 = -54.0 + 1.0EI\phi_2$$

$$M_{21} = 54.0 + \frac{4E(3I)}{6}\phi_2 = 54.0 + 2.0EI\phi_2$$

$$M_{23} = -13.5 - \left(\frac{13.5}{2}\right) + \frac{3EI}{3}\phi_2 = -20.25 + 1.0EI\phi_2$$

Considering a free-body diagram of a segment of the beam at joint ② as shown in Fig. 12-5, we see that

$$M_{21} + M_{23} = 0$$

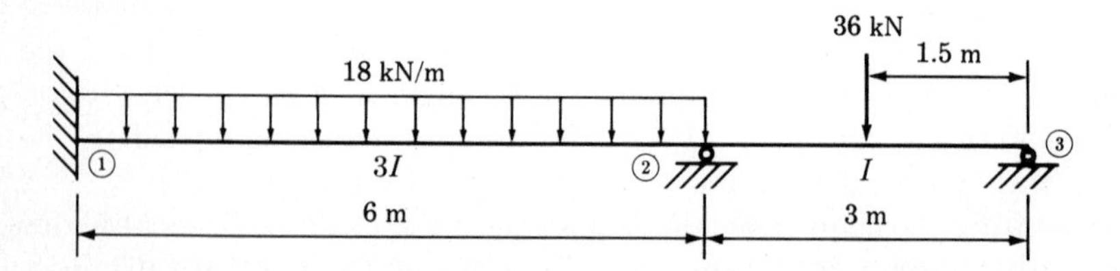

Fig. 12.4

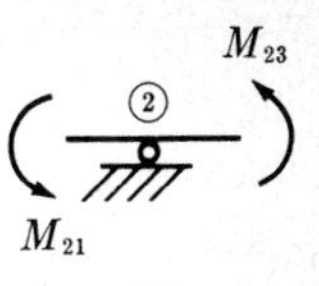

Fig. 12.5

Moments are considered positive in the directions shown in Fig. 12-3. Using this moment relationship, we add together the second and third of the above equations, which results in

$$0 = 33.75 + 3.0EI\phi_2$$

which results in the value

$$\phi_2 = -\frac{11.25}{EI}$$

Substituting this value of deflection into the original slope-deflection expressions, we obtain

$$M_{12} = -65.25 \text{ kN-m} \qquad M_{21} = -M_{23} = 31.50 \text{ kN-m}$$

According to the moment sign convention of Fig. 12-3, these moments cause compression in the bottom fibers of the beam at both ① and ②.

The effects of support displacements can readily be included in the slope-deflection analysis of continuous beams. For example, suppose that the support at point ② in the above example is permitted to displace a given amount. We would therefore have a known value of β for each span; in writing the slope-deflection equations for each member we would have an additional constant term in the equations due to the values of β. The solution of the problem would then proceed in the same manner as above.

Another condition sometimes encountered in continuous beams is an overhanging span. Such problems can be solved by replacing the overhanging span by an equivalent concentrated moment at the support point. The procedure for the analysis is then the same as before, except that there is a known value of moment at the end of the member adjacent to the overhanging span.

12-3 ANALYSIS OF FRAMES

The analysis of frames in which joint translation is prevented follows the same general procedure as for continuous beams. In writing the slope-deflection equations for the three members of the frame of Fig. 12-6, for

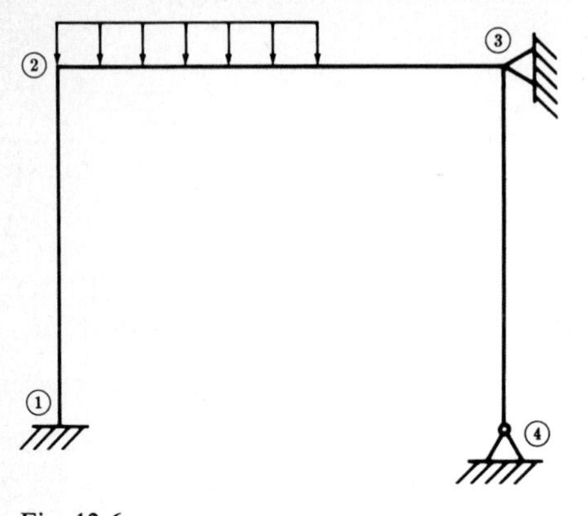

Fig. 12.6

example, we see that the unknown deflections are the rotations at points ② and ③. With zero moment at ④, the simplified slope-deflection equation (12-2) can be used for the right member. Because translations of the joints are prevented, there are no terms containing β. Two simultaneous equations for the unknowns are obtained by summing the moments at ② and ③.

For frames in which translations of joints are permitted it is necessary to consider additional factors in order to obtain a solution. The following example will illustrate the procedure.

EXAMPLE 12-2 The moments in the members of the frame of Fig. 12-7 are to be determined by the slope-deflection method. The relative moments of inertia for the members are as shown. E is constant. The fixed-end moments for the members are obtained from the general expressions

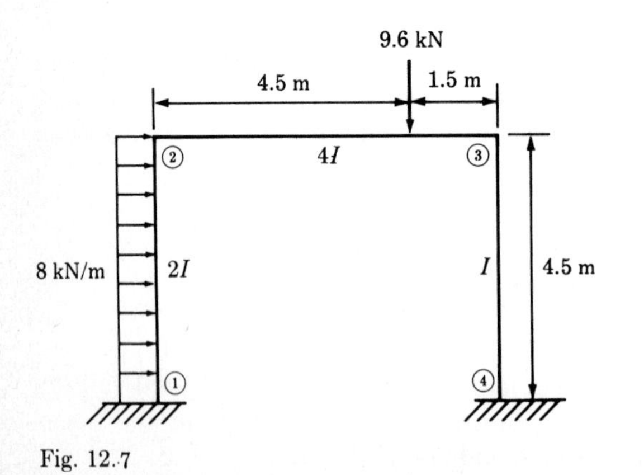

Fig. 12.7

of Fig. 11-24 as

$$\text{FEM}_{12} = -\frac{(8)(4.5)^2}{12} = -13.5 \text{ kN-m}$$
$$\text{FEM}_{21} = 13.5 \text{ kN-m}$$
$$\text{FEM}_{23} = -\frac{(9.6)(4.5)(1.5)^2}{6^2} = -2.7 \text{ kN-m}$$
$$\text{FEM}_{32} = \frac{(9.6)(4.5)^2(1.5)}{6^2} = 8.1 \text{ kN-m}$$

Because axial deformations are being neglected, the lateral translation at ② is equal to that at ③. Therefore we can let $\beta_{12} = \beta_{43} = \beta$. Noting also that $\phi_1 = \phi_4 = 0$, we obtain the following set of slope-deflection equations for the three members:

$$M_{12} = -13.5 + EI(0.889\phi_2 - 2.667\beta)$$
$$M_{21} = 13.5 + EI(1.778\phi_2 - 2.667\beta)$$
$$M_{23} = -2.7 + EI(2.667\phi_2 + 1.333\phi_3)$$
$$M_{32} = 8.1 + EI(1.333\phi_2 + 2.667\phi_3)$$
$$M_{34} = EI(0.889\phi_3 - 1.333\beta)$$
$$M_{43} = EI(0.444\phi_3 - 1.333\beta)$$

Considering the summation of moments for segments of the frame at joint ② and at joint ③, we obtain the expressions

$$M_{21} + M_{23} = 0 = 10.8 + EI(4.445\phi_2 + 1.333\phi_3 - 2.667\beta)$$
$$M_{32} + M_{34} = 0 = 8.1 + EI(1.333\phi_2 + 3.556\phi_3 - 1.333\beta)$$

A third expression is obtained by considering the summation of horizontal forces acting on the entire structure for which

$$\Sigma F_H = 0$$

The forces involved in this equilibrium equation are the uniform horizontal load and the shear in the columns at points ① and ④. Therefore we can write

$$\Sigma F_H = 0 = (8)(4.5) - V_{12} - V_{43}$$

Forces acting to the right on the structure are considered positive. Expressions for the shears V_{12} and V_{43} can be obtained in terms of the deflections by considering the free-body diagrams of the columns as shown in Fig. 12-8. Taking the summation of moments about ② on the left column in Fig. 12-8a, we obtain

$$V_{12} = -\frac{M_{12} + M_{21}}{4.5} + 18.0$$

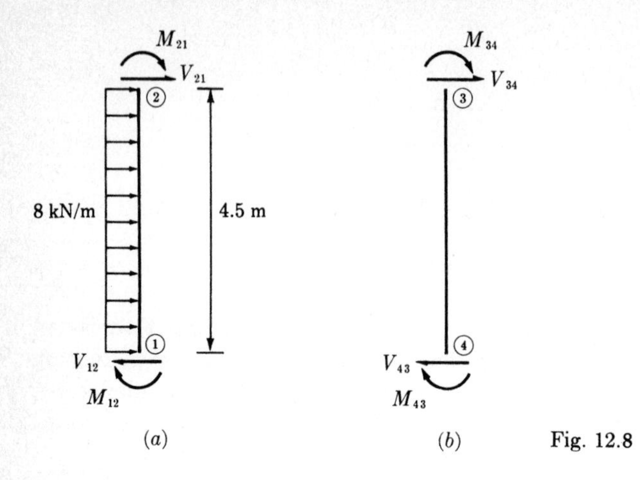

Fig. 12.8

Using the above expressions for M_{12} and M_{21} in terms of ϕ_2 and β, wc obtain

$$V_{12} = -\frac{EI}{4.5}(2.667\phi_2 - 5.333\beta) + 18.0$$

Taking the summation of moments about point ③ on the right column in Fig. 12-8*b*, we obtain

$$V_{43} = -\frac{M_{34} + M_{43}}{4.5}$$

Using the above expressions for M_{34} and M_{43}, we obtain

$$V_{43} = -\frac{EI}{4.5}(1.333\phi_3 - 2.667\beta)$$

The expression for the summation of horizontal forces is then

$$0 = 81.0 + EI(2.667\phi_2 + 1.333\phi_3 - 8.0\beta)$$

The resulting simultaneous equations, written in the general matrix form for simultaneous equations, are

$$EI\begin{bmatrix} 4.445 & 1.333 & -2.667 \\ 1.333 & 3.556 & -1.333 \\ 2.667 & 1.333 & -8.0 \end{bmatrix}\begin{Bmatrix} \phi_2 \\ \phi_3 \\ \beta \end{Bmatrix} = \begin{Bmatrix} -10.8 \\ -8.1 \\ -81.0 \end{Bmatrix}$$

Solving this matrix equation by a computer program such as that of Appendix B gives us

$$\begin{Bmatrix} \phi_2 \\ \phi_3 \\ \beta \end{Bmatrix} = \frac{1}{EI}\begin{Bmatrix} 4.449 \\ 0.433 \\ 11.680 \end{Bmatrix}$$

Substituting these values of displacement into the original slope-deflection equations, we obtain the following values of moments in the frame:

$M_{12} = -40.70$ kN-m
$M_{21} = -M_{23} = -9.74$ kN-m
$M_{32} = -M_{34} = 15.19$ kN-m
$M_{43} = -15.38$ kN-m

This slope-deflection procedure can also be applied to frames with more than one bay and with more than one story. For multistory frames the additional equations are obtained by considering the summation of horizontal forces above each floor level. As the number of bays and stories increases, so does the number of simultaneous equations to be solved. However, with the aid of a digital computer, the solution of a large number of simultaneous equations is feasible.

Another type of frame sometimes encountered is one with sloping, rather than vertical, members. The analysis of such a frame by the slope-deflection method is illustrated in the following example.

EXAMPLE 12-3 The frame of Fig. 12-9 is to be analyzed by the slope-deflection method. The moment of inertia I and the modulus of elasticity E are the same for all members. The given frame has three degrees of freedom, rotation at points ② and ③ and translation of the horizontal member. The rotations at ② and ③ are denoted by the usual slope-deflection notation as ϕ_2 and ϕ_3. Before writing the slope-deflection equations for the members, we should investigate the expression for the third degree of freedom. To obtain a general expression for this type of displacement let us consider the slope of the two side members to be described by the angles θ_1 and θ_2, as shown in Fig. 12-10. The horizontal translation of the frame can be described in terms of the horizontal dis-

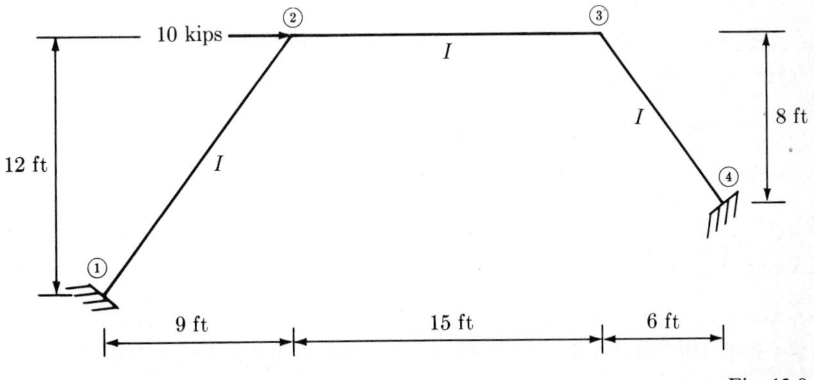

Fig. 12.9

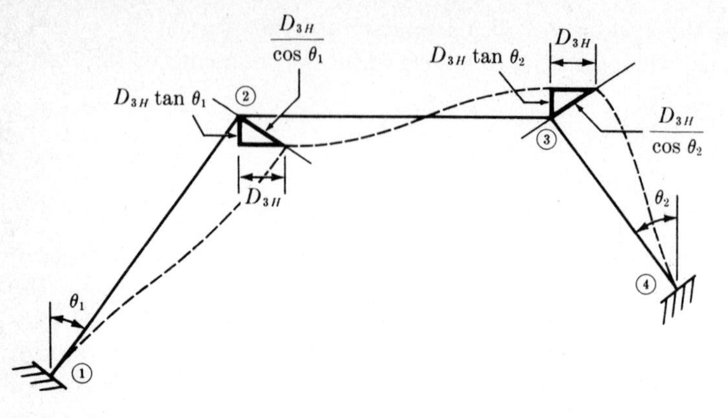

Fig. 12.10

placement of joint ③. This displacement is denoted as D_{3H} in Fig. 12-10. For a given value of D_{3H} the frame will deflect as shown in Fig. 12-10. We are interested in determining the resulting values of member displacements transverse to the axes of the members, that is, η_i and η_j, in terms of the displacement D_{3H}. The horizontal displacement of point ② is equal to the horizontal displacement of ③ because axial deformations are being neglected. Therefore, from geometric considerations of the deflections at ② and ③ in Fig. 12-10,

$$\beta_{12} = \frac{1}{L_{12}}\frac{D_{3H}}{\cos\theta_1} \qquad \beta_{34} = \frac{1}{L_{34}}\frac{D_{3H}}{\cos\theta_2}$$

$$\beta_{23} = -\frac{1}{L_{23}}D_{3H}(\tan\theta_1 + \tan\theta_2)$$

For the member lengths and slopes of the given structure in Fig. 12-9, we obtain the following values of β in terms of D_{3H}:

$$\beta_{12} = \frac{D_{3H}}{(15)(0.8)} = 0.0833D_{3H}$$

$$\beta_{23} = -\frac{D_{3H}}{15}(0.75 + 0.75) = -0.1D_{3H}$$

$$\beta_{34} = \frac{D_{3H}}{(10)(0.8)} = 0.125D_{3H}$$

The fixed-end moments for all members of the given structure are zero. Noting that $\phi_1 = \phi_4 = 0$, we find the following slope-deflection equations for the members:

$$M_{12} = EI(0.1333\phi_2 - 0.4\beta_{12})$$
$$M_{21} = EI(0.2667\phi_2 - 0.4\beta_{12})$$
$$M_{23} = EI(0.2667\phi_2 + 0.1333\phi_3 - 0.4\beta_{23})$$
$$M_{32} = EI(0.1333\phi_2 + 0.2667\phi_3 - 0.4\beta_{23})$$
$$M_{34} = EI(0.4\phi_3 - 0.6\beta_{34})$$
$$M_{43} = EI(0.2\phi_3 - 0.6\beta_{34})$$

Then, taking the summation of moments for segments of the frame at points ② and ③, we obtain the expressions

$$M_{21} + M_{23} = 0 = EI(0.5333\phi_2 + 0.1333\phi_3 - 0.4\beta_{12} - 0.4\beta_{23})$$
$$M_{32} + M_{34} = 0 = EI(0.1333\phi_2 + 0.6667\phi_3 - 0.4\beta_{23} - 0.6\beta_{34})$$

A third equilibrium expression is obtained by considering the summation of moments about point O in Fig. 12-11. Point O is located at the intersection point of lines extended from the two sloping members. The member forces at points ① and ④ that are involved in the summation of moments about O are shown on the diagram in Fig. 12-11. Taking the summation of moments about O, we obtain

$$27.5V_{12} + M_{12} + 22.5V_{43} + M_{43} - (10)(10) = 0$$

The shear V_{12} can be expressed in terms of the moments M_{12} and M_{21}, and the shear V_{43} can be expressed in terms of the moments M_{34} and M_{43} by considering free-body diagrams of the two sloping members. The moments in the resulting expressions can be expressed in terms of the

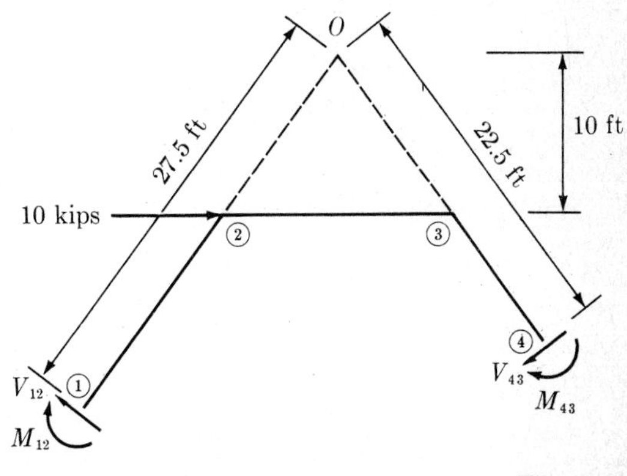

Fig. 12.11

deflections using the above slope-deflection equations. The expression for the summation of moments about O thus becomes

$$0 = EI(-0.6\phi_2 - 1.15\phi_3 + 1.0667\beta_{12} + 2.1\beta_{34}) - 100$$

With the previously developed expressions for β in terms of D_{3H}, the three simultaneous equations for the structure are

$$\begin{aligned} 0 &= EI(0.5333\phi_2 + 0.1333\phi_3 + 0.0067D_{3H}) \\ 0 &= EI(0.1333\phi_2 + 0.6667\phi_3 - 0.0350D_{3H}) \\ 100 &= EI(-0.6\phi_2 - 1.15\phi_3 + 0.3514D_{3H}) \end{aligned}$$

The matrix form of these equations is

$$EI\begin{bmatrix} 0.5333 & 0.1333 & 0.0067 \\ 0.1333 & 0.6667 & -0.0350 \\ -0.6000 & -1.1500 & 0.3514 \end{bmatrix}\begin{Bmatrix} \phi_2 \\ \phi_3 \\ D_{3H} \end{Bmatrix} = \begin{Bmatrix} 0 \\ 0 \\ 100 \end{Bmatrix}$$

Solving this matrix equation with a computer program such as that of Appendix B, we obtain

$$\begin{Bmatrix} \phi_2 \\ \phi_3 \\ D_{3H} \end{Bmatrix} = \frac{1}{EI}\begin{Bmatrix} -8.982 \\ 19.235 \\ 332.189 \end{Bmatrix}$$

With this value of D_{3H}, the values of β are therefore

$$\beta_{12} = 0.0833D_{3H} = \frac{27.671}{EI}$$

$$\beta_{23} = -0.1D_{3H} = -\frac{33.219}{EI}$$

$$\beta_{34} = 0.125D_{3H} = \frac{41.523}{EI}$$

Substituting these values of β and the above values of ϕ_2 and ϕ_3 into the original slope-deflection equations, we obtain for the moments in the frame

$$\begin{aligned} M_{12} &= -12.26 \text{ ft-kips} \\ M_{21} &= -M_{23} = -13.46 \text{ ft-kips} \\ M_{32} &= -M_{34} = 17.22 \text{ ft-kips} \\ M_{43} &= -21.07 \text{ ft-kips} \end{aligned}$$

The values of the remaining reaction components can be obtained from free-body considerations.

PROBLEMS

12-1 and 12-2. For the beams shown, determine the values of moment in the beams at the support points and the values of the reactions. Construct the shear and moment diagrams using the beam sign convention. EI is constant.

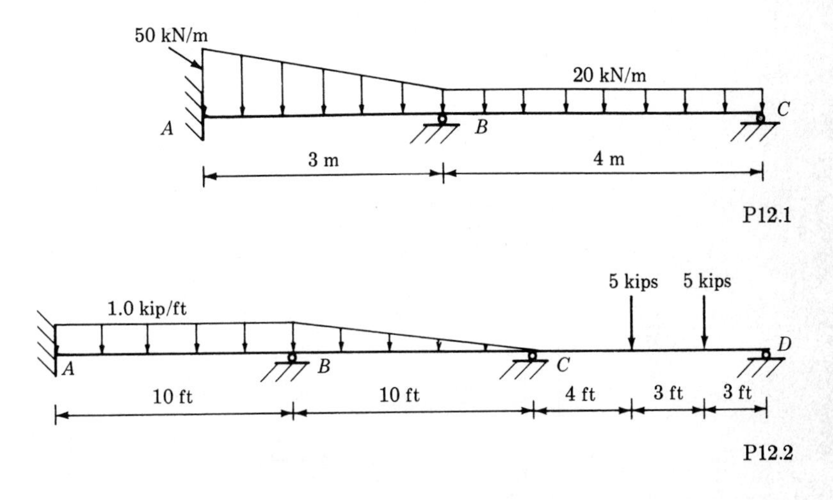

12-3. Use the slope-deflection method to solve Prob. 11-20.
12-4. Use the slope-deflection method to solve Prob. 11-21.
12-5 through 12-7. For the frames shown, (*a*) determine the moments in the members; (*b*) construct the moment diagrams on the tension side of the members.

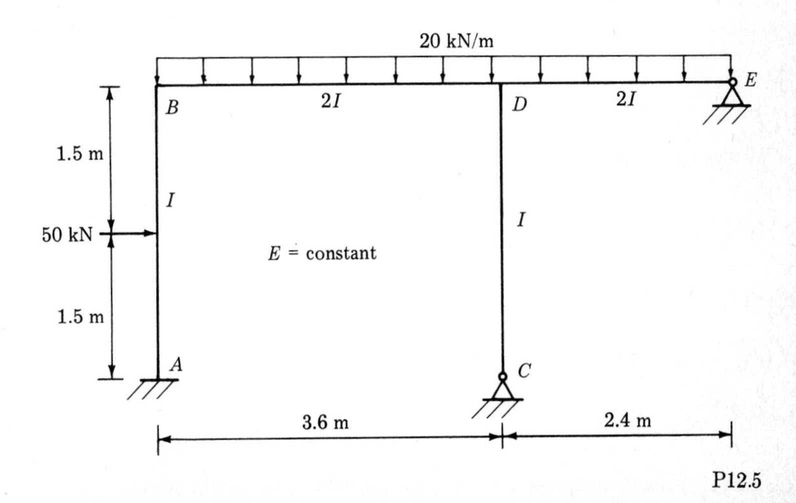

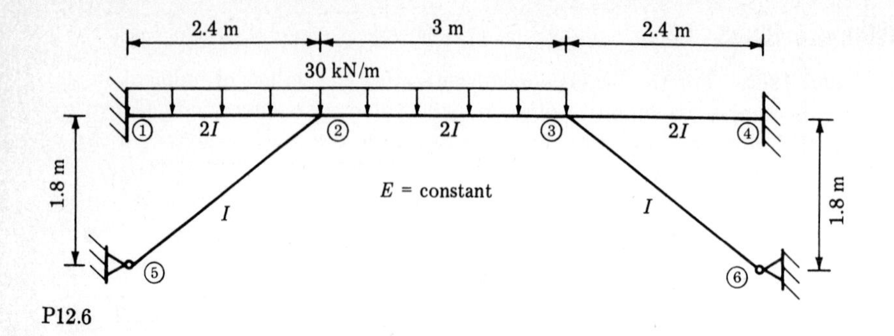

P12.6

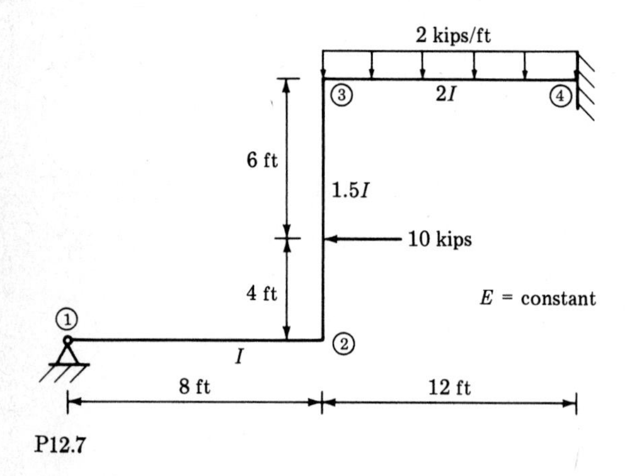

P12.7

12-8. Determine the moments in the members of the frame shown. E is constant.

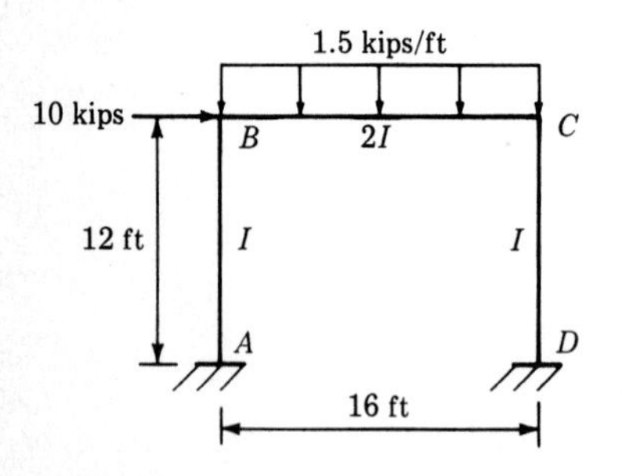

P12.8

12-9. Use the slope-deflection method to solve Prob. 11-11.

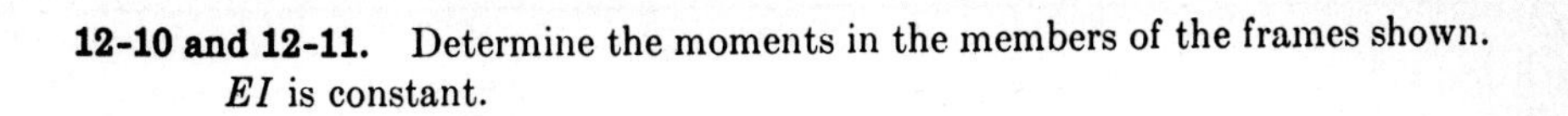

12-10 and 12-11. Determine the moments in the members of the frames shown. EI is constant.

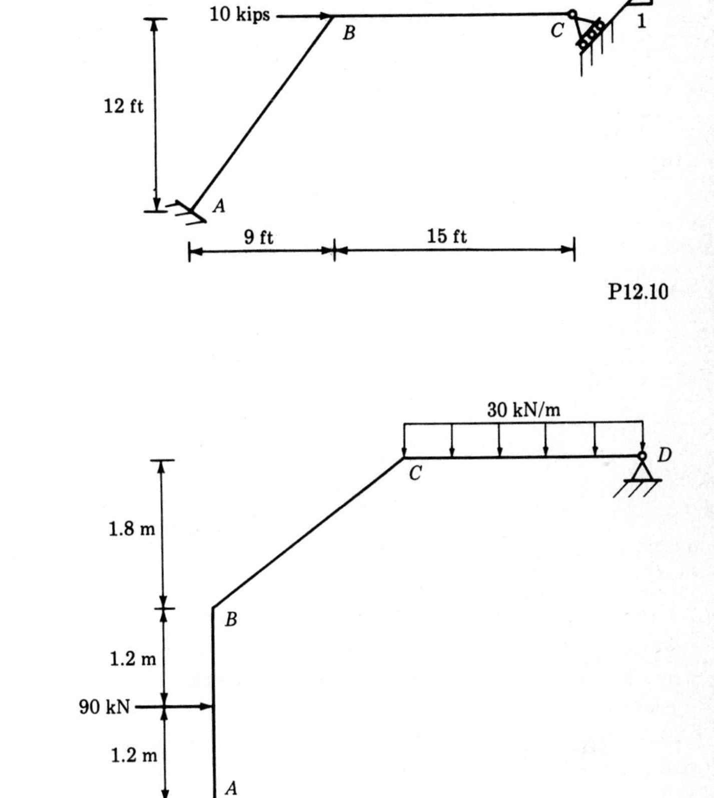

P12.11

CHAPTER

THIRTEEN

MOMENT-DISTRIBUTION METHOD

13-1 TERMINOLOGY

The moment-distribution method of analysis provides a means of analyzing statically indeterminate beams and frames by relatively simple hand calculations. It is basically an iteration procedure. We begin the iteration procedure by assuming all joints of the structure to be temporarily restrained against displacement and evaluating the moments in the members corresponding to this condition. After a number of cycles of releasing the joints and distributing the member moments, we obtain the true values of moments in the members. The method is sometimes referred to as the *Hardy Cross method*, in tribute to the person who conceived its basic concepts.[1]

Before developing the procedures of the method, let us define some terms.

Carryover factor

If we apply a moment to the simply supported end of the beam in Fig. 13-1, a specific amount of moment is generated at the fixed end. The

[1] Hardy Cross, Analysis of Continuous Frames by Distributed Fixed-end Moments, *Proc. ASCE*, May, 1930.

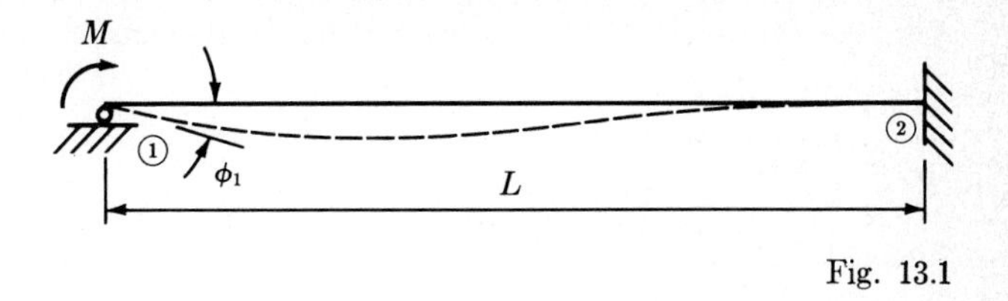

Fig. 13.1

carryover factor is defined as the factor by which the moment in the beam at ①, M_1, is multiplied to give the value of moment which is carried over to the opposite end,

$$M_2 = C_{12}M_1$$

where C_{12} represents the carryover factor from point ① to point ②. M_1 in Fig. 13-1 is equal to the applied moment M. Considering EI for the beam to be constant, we can obtain the amount of moment developed at point ② from the stiffness relation of Eq. (9-18). For the given conditions in Fig. 13-1, $\eta_1 = \phi_2 = \eta_2 = 0$. Therefore from Eq. (9-18) we obtain

$$M_1 = \frac{4EI}{L}\phi_1 \qquad M_2 = \frac{2EI}{L}\phi_1$$

Moments are considered positive when they act in the directions shown in Fig. 13-2. This is the same sign convention used in the slope-deflection method of Chap. 12. From these expressions for M_1 and M_2 we see that

$$M_2 = \tfrac{1}{2}M_1$$

The carryover factor is therefore

$$C_{12} = \tfrac{1}{2} \tag{13-1}$$

Stiffness factor

Let us again consider the beam of Fig. 13-1. The *absolute stiffness factor* is defined as the stiffness of the beam at ① against rotation. The absolute

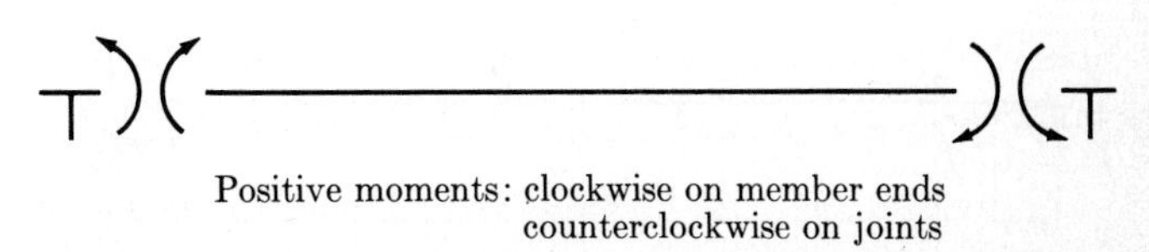

Positive moments: clockwise on member ends
counterclockwise on joints

Fig. 13.2

stiffness factor is therefore represented by element k_{11} in the stiffness matrix of Eq. (9-18),

$$k_{11} = \frac{M_1}{\phi_1} = \frac{4EI}{L} \tag{13-2}$$

The ratio I/L in Eq. (13-2) is often referred to simply as the *stiffness factor* of the beam and is denoted by the letter K. The stiffness factor of the beam at ① is thus written as

$$K_{12} = \frac{I}{L} \tag{13-3}$$

If the opposite end of the member is pinned rather than fixed, as shown in Fig. 13-3, we obtain a stiffness at ① referred to as the *modified stiffness factor*. From Eq. (9-18), the relation between rotations and moments is

$$\begin{Bmatrix} M_1 \\ M_2 \end{Bmatrix} = \frac{2EI}{L} \begin{bmatrix} 2 & 1 \\ 1 & 2 \end{bmatrix} \begin{Bmatrix} \phi_1 \\ \phi_2 \end{Bmatrix}$$

Using the fact that M_2 is zero, we obtain the expression

$$M_2 = 0 = \frac{2EI}{L} (\phi_1 + 2\phi_2)$$

from which

$$\phi_2 = -\frac{\phi_1}{2}$$

Therefore we find that

$$M_1 = \frac{2EI}{L} \left[2\phi_1 + 1 \left(-\frac{\phi_1}{2} \right) \right] = \frac{3EI}{L} \phi_1$$

or $$\frac{M_1}{\phi_1} = \frac{3EI}{L}$$

We see that the value of stiffness is three-fourths that for the condition of the opposite end being fixed. The modified stiffness factor, denoted as K', is therefore expressed as

$$K' = \frac{3K}{4} \tag{13-4}$$

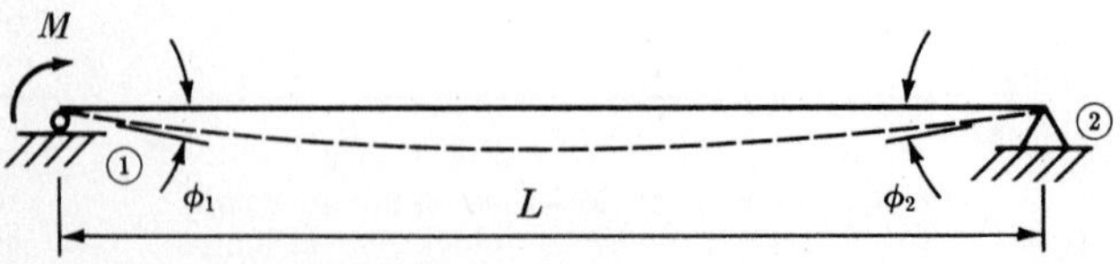

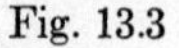

Fig. 13.3

Distribution factor

If an external moment is applied to the joint of a structure where two or more members meet, we can determine in what proportions the members resist the moment by means of distribution factors. Consider the structure of Fig. 13-4. Three members meet at the joint denoted as ①. Joint ① is a rigid joint, and the ends of all the members opposite ① are fixed. An external moment M is applied to the structure at ①, producing a rotation of θ. For equilibrium at joint ①

$$M_{12} + M_{13} + M_{14} - M = 0 \tag{13-5}$$

As shown in Fig. 13-2, member moments are considered positive when they act in a counterclockwise direction on the joint. From the moment-rotation relation of Eq. (13-2) and the definition of a stiffness factor in Eq. (13-3) we can state that

$$M_{12} = 4EK_{12}\theta$$
$$M_{13} = 4EK_{13}\theta$$
$$M_{14} = 4EK_{14}\theta$$

Equation (13-5) can therefore be written as

$$4E\theta(K_{12} + K_{13} + K_{14}) = M$$

or $\quad M = 4E\theta\Sigma K$

from which

$$\theta = \frac{M}{4E\Sigma K}$$

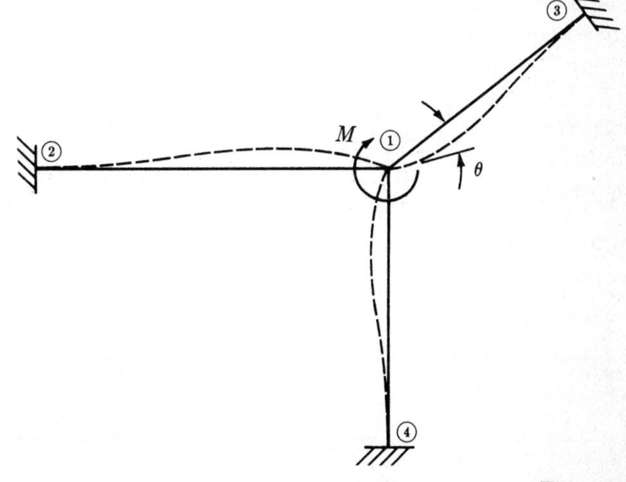

Fig. 13.4

The above expression for M_{12} can now be written as

$$M_{12} = 4EK_{12}\frac{M}{4E\Sigma K}$$
$$= \frac{K_{12}}{\Sigma K}M$$

The ratio $K_{12}/\Sigma K$ indicates the portion of the moment M that is resisted by member 1-2. The ratio is referred to as the *distribution factor* for the member. From the definition of a stiffness factor K in Eq. (13-3), the distribution factor (DF) for the ith member is defined in general as

$$\text{DF} = \frac{K_i}{\Sigma K} = \frac{I_i/L_i}{\Sigma(I/L)} \tag{13-6}$$

The distribution factor for a member is thus equal to the stiffness of the member divided by the sum of the stiffness of all members meeting at the joint.

13-2 DEVELOPMENT OF THE METHOD

The moment-distribution method of analysis will be introduced in terms of the continuous beam of Fig. 13-5. The beam is fixed at its ends but is free to rotate at support point ②. For a moment-distribution analysis we consider initially that all joints are temporarily fixed. Thus we begin with a beam supported as shown in Fig. 13-6a, where joint ② is temporarily restrained against rotation. The fixed-end moments for each member due to the given loads are evaluated. Using the expressions for fixed-end moments in Fig. 11-24 and the sign convention for moments of Fig. 13-2, we obtain

$$\text{FEM}_{12} = -\frac{(1.2)(20)^2}{12} = -40.0 \text{ ft-kips} \qquad \text{FEM}_{21} = 40.0 \text{ ft-kips}$$

$$\text{FEM}_{23} = -\frac{(8)(10)}{8} = -10.0 \text{ ft-kips} \qquad \text{FEM}_{32} = 10.0 \text{ ft-kips}$$

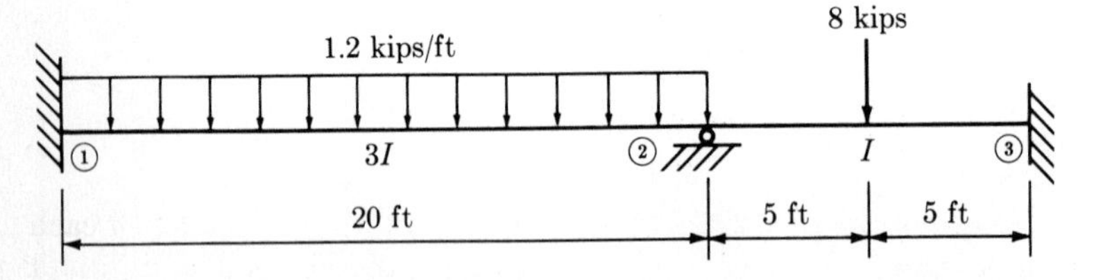

Fig. 13.5

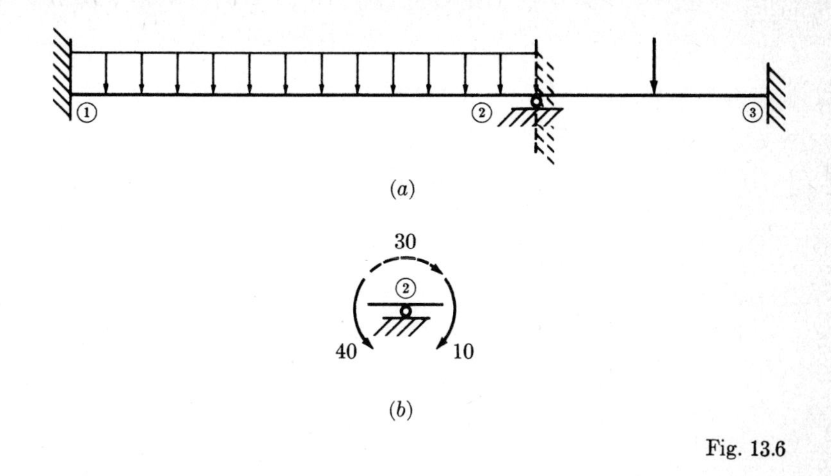

Fig. 13.6

Considering the free-body diagram of joint ② in Fig. 13-6*b*, we see that the temporary restraint is resisting rotation by 30 ft-kips. The resisting moment is a negative quantity, according to the sign convention for joints in Fig. 13-2. To obtain the true condition at ② we remove the temporary rotational restraint at ②. The member will therefore rotate at ② until it reaches a position of equilibrium, generating a moment of −30 ft-kips in the two members at ②. The moment is distributed between the two members in accordance with their relative stiffnesses. From Eq. (13-6) we find that

$$\mathrm{DF}_{21} = \frac{3I/20}{(3I/20) + (I/10)} = 0.6$$

$$\mathrm{DF}_{23} = \frac{I/10}{(3I/20) + (I/10)} = 0.4$$

Therefore the moments developed in the ends of each member are

$$M_{21} = (0.6)(-30) = -18.0$$
$$M_{23} = (0.4)(-30) = -12.0$$

Half these moments are carried over to the opposite ends of the members. The values of the moments carried over are

$$M_{12} = -\frac{18.0}{2} = -9.0$$

$$M_{32} = -\frac{12.0}{2} = -6.0$$

The beam is now in its true position. The total moment at the end of each member is obtained by algebraically summing the fixed-end values and the values due to releasing the restraint at ②. For example, the total

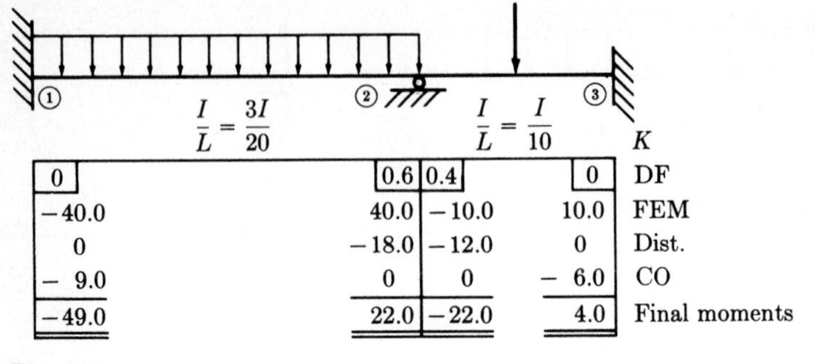

①	②	②	③	K
0	0.6	0.4	0	DF
−40.0	40.0	−10.0	10.0	FEM
0	−18.0	−12.0	0	Dist.
− 9.0	0	0	− 6.0	CO
−49.0	22.0	−22.0	4.0	Final moments

Fig. 13.7

moment at ① is

$$M_{12} = -40.0 - 9.0 = -49.0 \text{ ft-kips}$$

The above procedure can be recorded in tabular form as shown in Fig. 13-7. It should be pointed out that the final values of moment at ② satisfy the condition for equilibrium of moments. Although they were not directly used, the values of the distribution factors at points ① and ③ are zero. This can be verified by considering the expression for distribution factors in Eq. (13-6). The fixed support results in a value of infinity in the denominator, and therefore DF = 0 for the beam at these points.

To extend the developments further let us consider a variation in the support of the above beam as shown in Fig. 13-8. Instead of being fixed, point ③ is now simply supported. It is possible for joint ③ as well as joint ② to rotate.

The analysis is begun by considering both joints to be temporarily restrained against rotation as shown in Fig. 13-9*a*. The two points are then released one at a time. Releasing joint ② first, we obtain a condition that is identical to the final condition of the beam in Fig. 13-5. Joint ② is now considered to be temporarily restrained in the deflected position, as shown in Fig. 13-9*b*. The temporary restraint at ③ is then removed. We see from Fig. 13-9*b* that the moment at ③ is balanced to zero, but

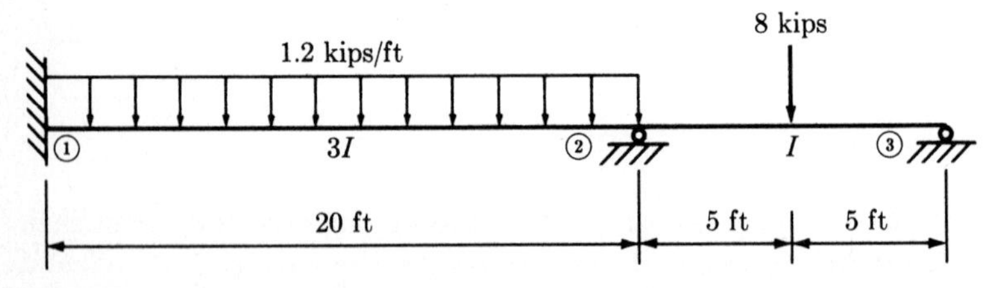

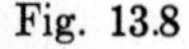
Fig. 13.8

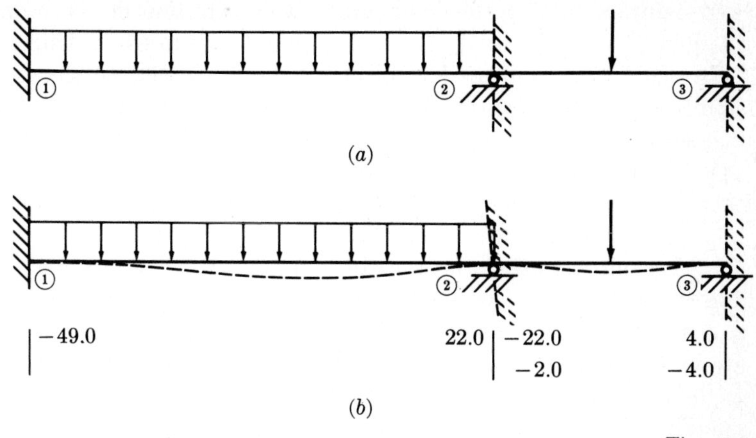

Fig. 13.9

half the −4.0 ft-kips of moment is carried over to ②. The moment at ② therefore becomes unbalanced. Note, however, that the imbalance at ② of 2 ft-kips is considerably less than the original imbalance of 30.0 ft-kips. Joint ③ is next restrained in its deflected position, and joint ② is released. The cycle is repeated until the imbalance of moments at joints ② and ③ is within the desired degree of accuracy.

This procedure can be condensed and performed in tabular form, where the operations of each cycle are simultaneous. The procedure is shown in the table of Fig. 13-10. The analysis is complete when the imbalance of moments becomes small enough that it can be neglected. The final

①	②	②	③	
$\frac{I}{L} = \frac{3I}{20}$		$\frac{I}{L} = \frac{I}{10}$		K
0	0.6	0.4	1	DF
−40.0	40.0	−10.0	10.0	FEM
0	−18.0	−12.0	−10.0	Dist. 1
− 9.0	0	− 5.0	− 6.0	CO 1
0	3.0	2.0	6.0	Dist. 2
1.5	0	3.0	1.0	CO 2
0	− 1.8	− 1.2	− 1.0	Dist. 3
− 0.9	0	− 0.5	− 0.6	CO 3
0	0.3	0.2	0.6	Dist. 4
−48.4	23.5	−23.5	0	Final moments

Fig. 13.10

step before summing the moments is a distribution of the moments at the joints. Thus the final values of moments are in equilibrium.

In using the moment-distribution method of analysis many analysts simply follow the orderly pattern of computations shown in Fig. 13-10 instead of trying to rationalize each step. In such an approach the stiffness factor K is first computed for each member. Often the stiffnesses of each member are given in relative terms. The distribution factors at each joint are then evaluated from the values of K. It should be noted for checking purposes that the distribution factors for an interior joint must add up to 1.0. The fixed-end moments for each span are evaluated and entered in the table as shown in Fig. 13-10. The signs of the moments are in accordance with the sign convention for moment acting on a joint, as defined in Fig. 13-2. The next step in the analysis is to determine the amount of moment imbalance at each joint and then distribute minus this amount to the members in accordance with their distribution factors. For example, at joint ② in Fig. 13-10 the initial imbalance is $40.0 - 10.0 = 30.0$ ft-kips. An amount of -30.0 ft-kips is thus distributed to the members in accordance with their distribution factors, $(-30.0)(0.6) = -18.0$ to the left and $(-30.0)(0.4) = -12.0$ to the right. It is convenient to draw a line after the distributed values in each column to set off the various cycles of analysis. Half the moments distributed into the member ends is then carried over to the opposite ends. The result is that the joints again become imbalanced, but by a smaller amount than in previous cycles. This procedure is continued until the imbalance at each joint is within the desired degree of accuracy.

In structures where one end of a member is simply supported, such as the right span in the beam above, a considerable amount of work can be saved by using the modified stiffness factor of Eq. (13-4). Figure 13-11 shows the moment-distribution computations for the above beam with the modified stiffness factor of the right span. According to Eq. (13-4),

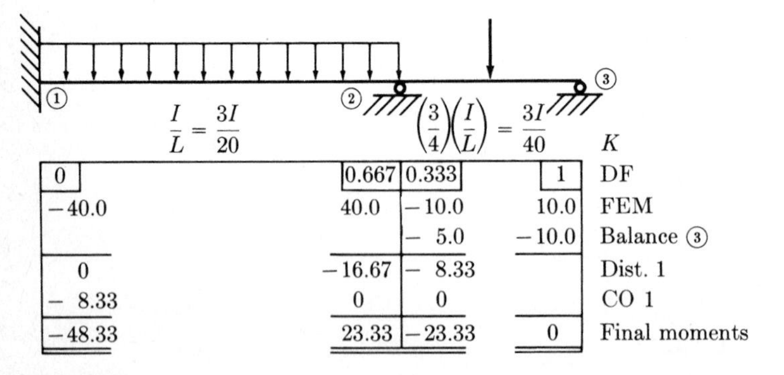

①	② left	② right	③	
0	0.667	0.333	1	DF
−40.0	40.0	−10.0	10.0	FEM
		− 5.0	−10.0	Balance ③
0	−16.67	− 8.33		Dist. 1
− 8.33	0	0		CO 1
−48.33	23.33	−23.33	0	Final moments

Fig. 13.11

the stiffness factor for a member end with the far end simply supported is three-fourths the stiffness factor for the far end fixed. We therefore first evaluate the modified stiffness factor as shown in Fig. 13-11. The distribution factors at joint ② thus differ from the values used previously. If we now balance the simply supported joint and carry over the appropriate amount of moment to ②, the remaining steps of analysis are performed as usual on the portion of the structure excluding the simply supported member. It is evident from Fig. 13-11 that a considerable amount of computation is avoided. Note that after the step of balancing joint ③ no moments are carried over from ② to ③.

In the discussions of moment distribution above none of the joints in the beams was permitted to displace in a direction transverse to the axis of a member. Where such joint translation is possible, the analysis of a structure by moment distribution is more involved. Such analyses will be discussed in the following section with regard to frames.

Structures with overhanging members can be analyzed by replacing the overhanging span with an equivalent concentrated moment at the support point. The following example illustrates the manner in which overhanging spans can be treated and will also serve to summarize the general steps of moment distribution discussed thus far.

EXAMPLE 13-1 The moments in the continuous beam of Fig. 13-12 are to be determined by the moment-distribution method. *EI* for the beam is constant. First we replace the overhanging span at the right end with an equivalent moment at ④ of -6.0 kN-m. The sign of the moment is negative because the joint is tending to be rotated clockwise by the 3-kN load. The -6.0 kN-m moment is entered in the moment-distribution table of Fig. 13-13 on the row of fixed-end moments. The fixed-end moments on the other spans of the beam are

$$\mathrm{FEM}_{12} = -\frac{(10)(2)}{8} = -2.5 \text{ kN-m} \qquad \mathrm{FEM}_{21} = 2.5 \text{ kN-m}$$

$$\mathrm{FEM}_{23} = -\frac{(15)(3)^2}{12} = -11.25 \text{ kN-m} \qquad \mathrm{FEM}_{32} = 11.25 \text{ kN-m}$$

$$\mathrm{FEM}_{34} = -\frac{(9)(4)^2}{12} = -12.0 \text{ kN-m} \qquad \mathrm{FEM}_{43} = 12.0 \text{ kN-m}$$

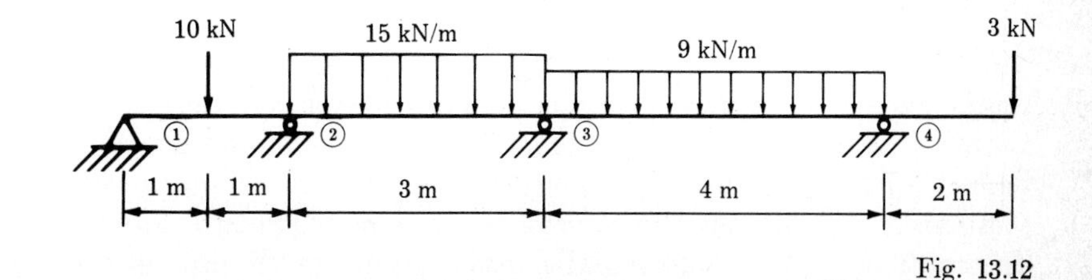

Fig. 13.12

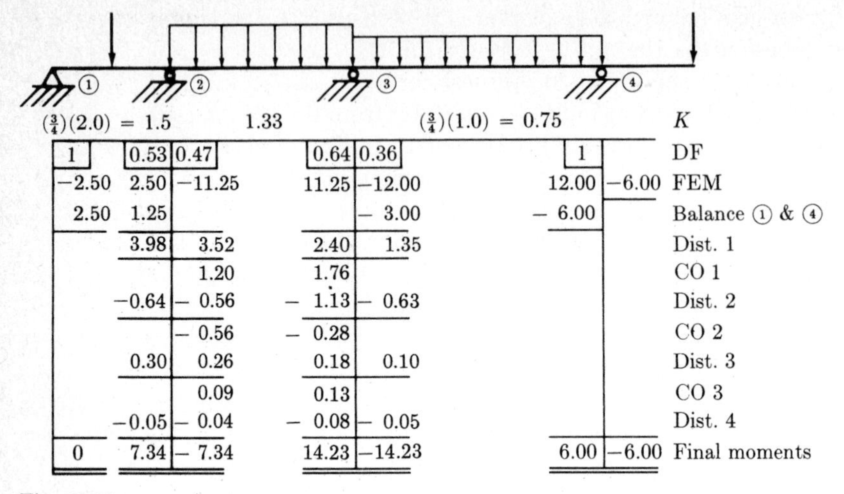

$(\frac{3}{4})(2.0) = 1.5$		1.33		$(\frac{3}{4})(1.0) = 0.75$			K
1	0.53	0.47	0.64	0.36	1		DF
−2.50	2.50	−11.25	11.25	−12.00	12.00	−6.00	FEM
2.50	1.25			− 3.00	− 6.00		Balance ① & ④
	3.98	3.52	2.40	1.35			Dist. 1
		1.20	1.76				CO 1
	−0.64	− 0.56	− 1.13	− 0.63			Dist. 2
		− 0.56	− 0.28				CO 2
	0.30	0.26	0.18	0.10			Dist. 3
		0.09	0.13				CO 3
	−0.05	− 0.04	− 0.08	− 0.05			Dist. 4
0	7.34	− 7.34	14.23	−14.23	6.00	−6.00	Final moments

Fig. 13.13

The stiffness factors for each span are evaluated on a relative basis. Because the moment of inertia for the beam is constant, we can assume a convenient value of I for purposes of determining the stiffness factors. A value of 4 is assumed for I, yielding a value of $I/L = 1.0$ for the span between ③ and ④. The corresponding values of I/L for the other spans are shown in Fig. 13-13. Because joints ① and ④ are simply supported, modified stiffness factors are used in the end spans. Thus the I/L values for these spans are multiplied by ¾. The distribution factors corresponding to the stiffness values K are then evaluated.

As shown in Fig. 13-13, the first step in determining moment distribution is to balance the simply supported joints ① and ④. The equivalent moment of the overhanging span is included in the balancing of joint ④. After carrying over half the balancing moments to ② and ③, the moment-distribution analysis follows the usual procedure to completion.

13-3 ANALYSIS OF FRAMES

Moment distribution for the analysis of frames in which none of the joints translates follows the same general procedure as for continuous beams. It is not unusual in frame analyses to have more than two members meeting at a point. Care must be taken in such cases to consider the stiffness of all the members at the meeting point when evaluating the distribution factors. Additional considerations must also be given to

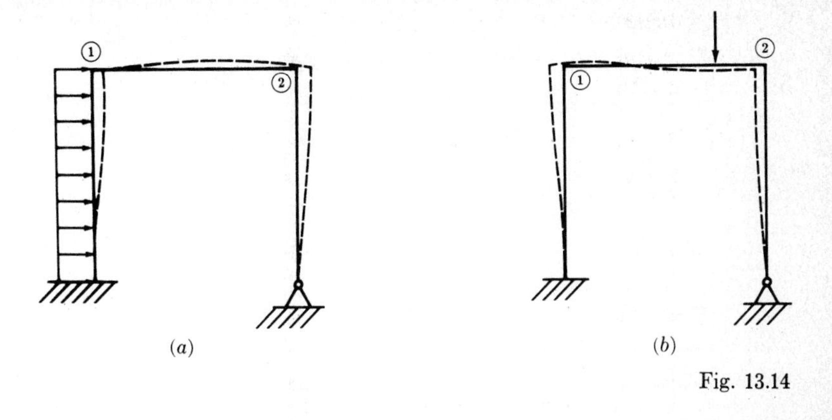

Fig. 13.14

the form in which the calculations are kept. One method of recording the calculations is on a sketch of the structure. The use of such a form for calculations is discussed below.

A typical frame in which translation of some joints is permitted is shown in Fig. 13-14. For the uniform horizontal loading in Fig. 13-14*a* the structure will deflect to the right. With axial deformations in the members disregarded, points ① and ② will deflect to the right by the same amount. In Fig. 13-14*b* points ① and ② will deflect to the left by the same amount. Translation of the joints results in a distribution of moments in the frame that is different from that for the frame restrained against joint translation.

Before demonstrating the procedure for analyzing structures in which translation of joints is possible, we need some relationships between member moments and member translations. If the ends of a fixed-fixed member deflect transverse to the axis of the member by a relative amount Δ, as shown in Fig. 13-15*a*, then from Eq. (9-18) we find that

$$M_1 = M_2 = \frac{6EI\Delta}{L^2} \qquad (13\text{-}7)$$

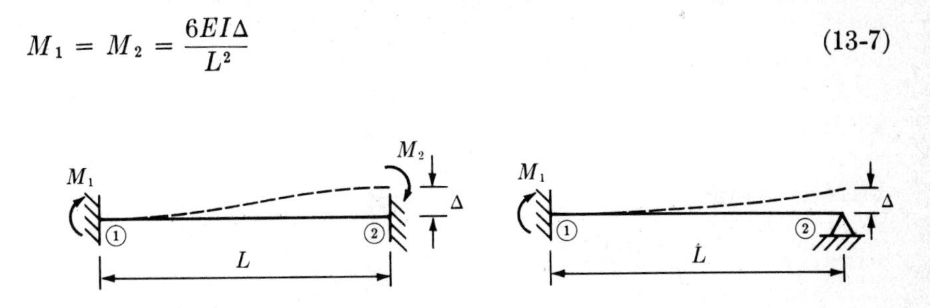

Fig. 13.15

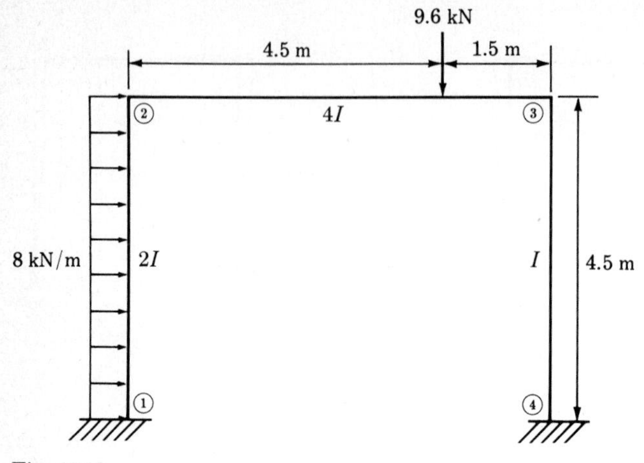

Fig. 13.16

where $\Delta = \eta_1 - \eta_2$. If joint ② is simply supported rather than fixed, as shown in Fig. 13-15*b*, it can be found from Eq. (9-18) that

$$M_1 = \frac{3EI\Delta}{L^2} \tag{13-8}$$

Equations (13-7) and (13-8) represent stiffness relations for member translation.

As an illustration of the technique of analyzing frames containing joint translation, let us consider the frame of Fig. 13-16. This frame was previously analyzed in Example 12-2 by the slope-deflection method. The relative moments of inertia of each member are as shown. E is constant. We first temporarily restrain the structure against horizontal translation of points ② and ③. This is accomplished with the restraint at ③, as shown in Fig. 13-17. The unknown restraining force is denoted as X and is considered positive in the direction shown. The frame is then analyzed for the member moments by the usual moment-distribution process. For the given loads the fixed-end moments are found to be

$$\text{FEM}_{12} = -\frac{(8)(4.5)^2}{12} = -13.5 \text{ kN-m} \qquad \text{FEM}_{21} = 13.5 \text{ kN-m}$$

$$\text{FEM}_{23} = -\frac{(9.6)(4.5)(1.5)^2}{6^2} = -2.7 \text{ kN-m}$$

$$\text{FEM}_{32} = \frac{(9.6)(4.5)^2(1.5)}{6^2} = 8.1 \text{ kN-m}$$

The moment-distribution calculations are performed on a sketch of the frame, as shown in Fig. 13-18. As for beams, the calculations for each

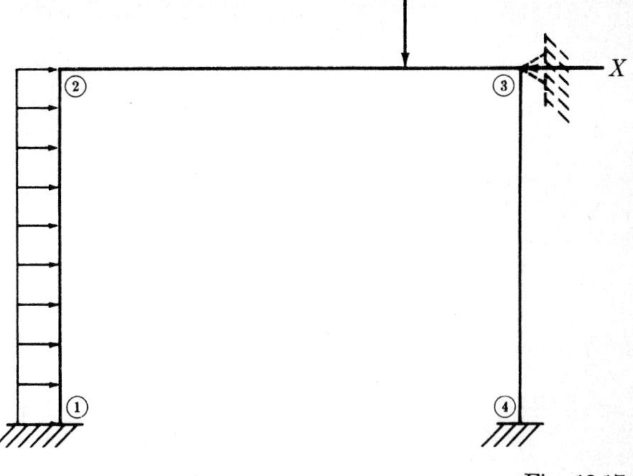

Fig. 13.17

member are laid out in a direction normal to the axis of the member.

The value of the temporary restraining force X in Fig. 13-17 is obtained by first evaluating the shear in the columns at points ① and ④. The values of shear corresponding to the member moments obtained in Fig. 13-18 are obtained from free-body considerations of the columns as shown in

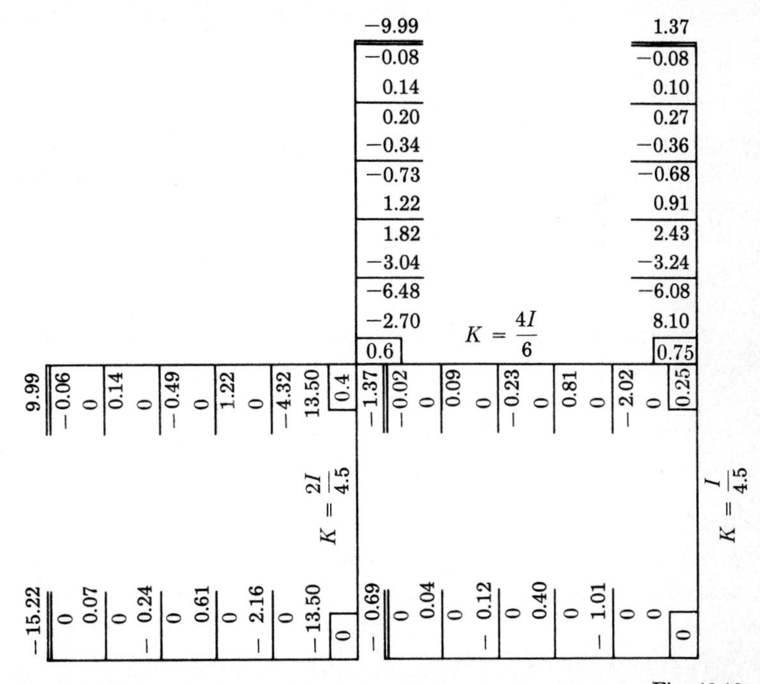

Fig. 13.18

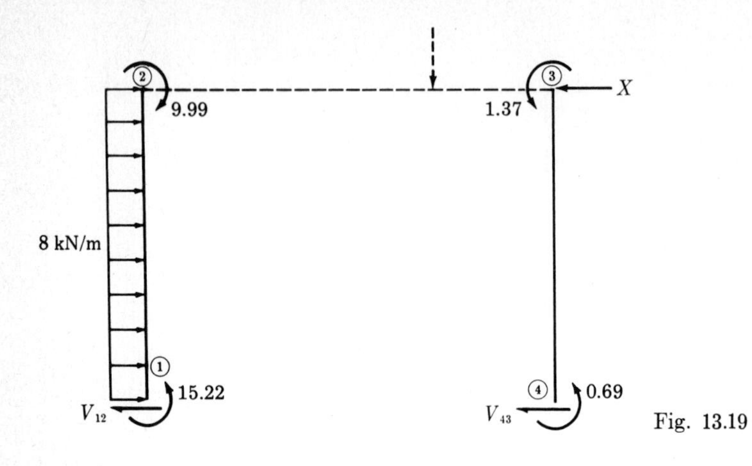

Fig. 13.19

Fig. 13-19. Thus

$$V_{12} = \frac{15.22 - 9.99 + (8)(4.5)(2.25)}{4.5} = 19.16 \text{ kN}$$

$$V_{43} = \frac{0.69 + 1.37}{4.5} = 0.46 \text{ kN}$$

For equilibrium of forces on the entire frame in the horizontal direction

$$(8)(4.5) - 19.16 - 0.46 - X = 0$$

or $\quad X = 16.38$ kN

The frame is therefore restrained from translating by a force of 16.38 kN acting to the left. To achieve the true condition of the structure we apply a force equal and opposite to X to the frame at ③. The member moments resulting from this condition of loading are combined with the moments obtained from the restrained condition to obtain the actual values of moment in the frame.

The moments in the frame due to the application of $-X$ are obtained in a rather indirect manner. The frame is considered to be loaded with an unknown force X', as shown in Fig. 13-20. With the assumption that the member ends are temporarily restrained against rotation, the resulting deflected shape of the frame is as shown in Fig. 13-20. The moments in the vertical members corresponding to a deflection Δ are given by Eq. (13-7). We can arbitrarily assume a value of moment in one vertical member, and because Δ is the same for each member, we can determine from Eq. (13-7) the corresponding moment in the other vertical member. Thus if we assume a value of -10.0 kN-m for the fixed-end moments in the left column, we see from Eq. (13-7) that for the values of moment of inertia in Fig. 13-16 the corresponding fixed-end moments in the right

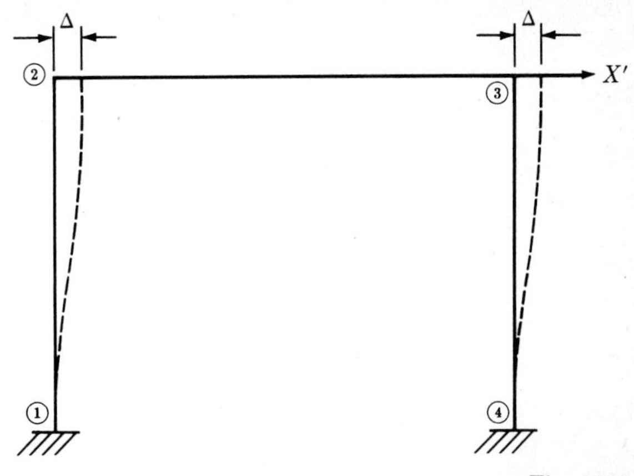

Fig. 13.20

member are

$$M_{43} = M_{34} = \frac{6EL}{L^2}\frac{-10.0L^2}{(6E)(2I)} = -5.0 \text{ kN-m}$$

There is a certain value of translation Δ that corresponds to these values of fixed-end moments. To obtain the value of lateral force X' that is holding the frame in this translated position we perform a moment-distribution analysis for these fixed-end moments as shown in Fig. 13-21. From free-body considerations of the columns the shears at ① and ④ corresponding to the member moments obtained in Fig. 13-21 are found to be

$$V_{12} = \frac{8.20 + 6.33}{4.5} = 3.23 \text{ kN} \qquad V_{43} = \frac{4.73 + 4.42}{4.5} = 2.03 \text{ kN}$$

The value of the force X' is then found from the equation for equilibrium of forces in the horizontal direction to be

$$X' = 3.23 + 2.03 = 5.26 \text{ kN}$$

The true value of horizontal force to be applied to the frame is 16.38 kN. Therefore the actual values of moment in the frame are obtained by multiplying the values of moment obtained in Fig. 13-21 by the factor 16.38/5.26 = 3.11 and adding these values to the values of moment in Fig. 13-18. Thus the actual moments in the frame are found to be

$$M_{12} = -15.22 + (3.11)(-8.20) = -40.72 \text{ kN-m}$$
$$M_{21} = -M_{23} = 9.99 + (3.11)(-6.33) = -9.70 \text{ kN-m}$$
$$M_{32} = -M_{34} = 1.37 + (3.11)(4.42) = 15.12 \text{ kN-m}$$
$$M_{43} = -0.69 + (3.11)(-4.73) = -15.40 \text{ kN-m}$$

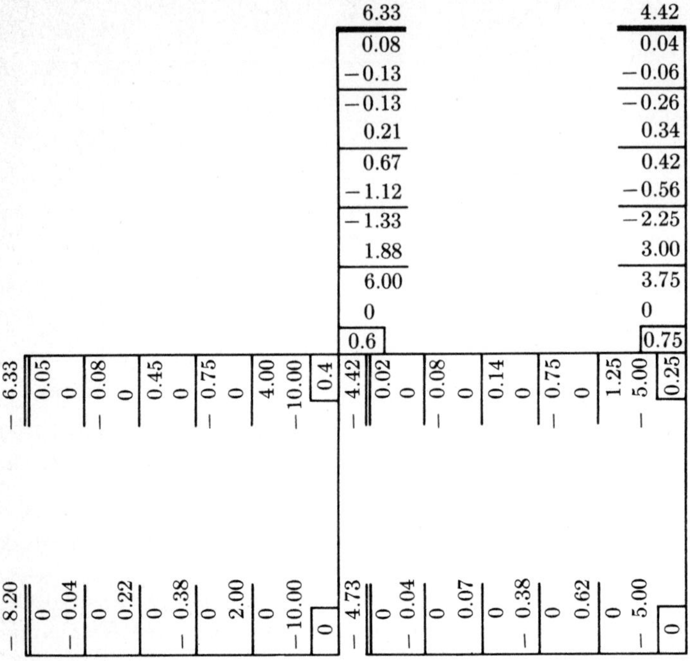

6.33
0.08
−0.13
−0.13
0.21
0.67
−1.12
−1.33
1.88
6.00
0
0.6

4.42
0.04
−0.06
−0.26
0.34
0.42
−0.56
−2.25
3.00
3.75
0
0.75

− 6.33 | 0.05 0 | − 0.08 0 | 0.45 0 | − 0.75 0 | 4.00 −10.00 | 0.4 | − 4.42 | 0.02 0 | − 0.08 0 | 0.14 0 | − 0.75 0 | 1.25 − 5.00 | 0.25

− 8.20 | 0 − 0.04 | 0 0.22 | 0 − 0.38 | 0 2.00 | 0 −10.00 | 0 | − 4.73 | 0 − 0.04 | 0 0.07 | 0 − 0.38 | 0 0.62 | 0 − 5.00 | 0

Fig. 13.21

If a frame contains a pinned support, then Eq. (13-8) can be used instead of Eq. (13-7) to determine the initial moments in the pinned member for the translated condition. The stiffness of the pinned member for purposes of evaluating the distribution factors is the reduced stiffness of Eq. (13-4). If, for example, the support at ④ in the frame just considered were pinned, then for the translated-moment-distribution operation of Fig. 13-21 we would begin with $FEM_{43} = 0$ and $FEM_{34} = -2.50$, according to Eq. (13-8). No moment is carried back to ④ in the remaining cycles of distribution.

The moment-distribution technique can be extended to include the analysis of multistory frames, although manual computations for such structures can become quite cumbersome if the number of stories is large.

As an illustration of the moment-distribution approach to multistory frames, consider the two-story frame of Fig. 13-22. The actual moments in the frame are found by superimposing the results of the three separate steps, as shown. In step 1 translation of both floors is prevented. The moments in the structure corresponding to this condition of restraint are evaluated, and the values of the restraining forces X_1 and X_2 are determined from equilibrium equations. In step 2 the top floor is permitted to displace. On the basis of assumed values of fixed-end moments in the upper columns, the values of X_1' and X_2' are obtained from a moment-distribution analysis. Similarly, X_1'' and X_2'' are determined in step 3, where the lower floor is permitted to translate. The values of the constants A and B by which the values of moments obtained in steps 2 and 3 are to be multiplied are determined from equilibrium considerations of the points of temporary restraint. Thus, considering the forces shown in Fig. 13-22 to be positive, we obtain

$$-X_1 + AX_1' - BX_1'' = 0 \qquad -X_2 - AX_2' + BX_2'' = 0$$

$$\text{or} \quad \mathrm{A}\mathrm{X}_1' - \mathrm{B}\mathrm{X}_1'' = \mathrm{X}_1 \qquad -\mathrm{A}\mathrm{X}_2' + \mathrm{B}\mathrm{X}_2'' = \mathrm{X}_2$$

These simultaneous equations are solved for the constants A and B. The actual moments in the structure are then obtained by the superimposing

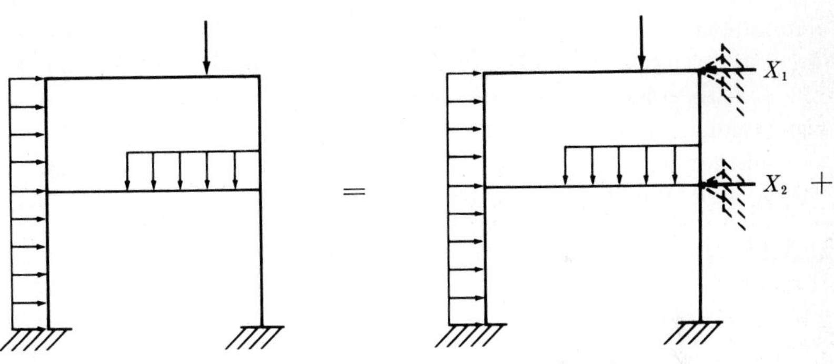

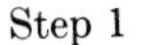

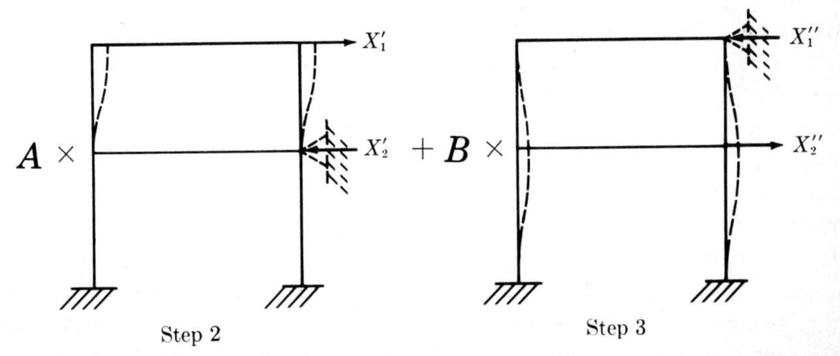

Fig. 13.22

operation of Fig. 13-22. Thus for the two-story frame of Fig. 13-22 three separate sets of moment-distribution computations are required. The number of sets obviously increases with the number of stories, as does the number of simultaneous equations to be solved.

The moment-distribution method can also be used to a limited extent for the analysis of three-dimensional frames. The procedure is illustrated in the following example. In addition to showing how a three-dimensional frame can be analyzed, the example affords deeper consideration of the meanings of the various terms used in moment distribution due to the fact that some of the expressions change for use in three-dimensional analysis.

EXAMPLE 13-2 The bending moments and the twisting moments in the members of the three-dimensional frame of Fig. 13-23 are to be evaluated by moment distribution. The polar moment of inertia J for each member is twice the value of I for the member. $G = 0.4E =$ constant. The coordinate system shown is to be used when force quantities are referred to. Note first that no joint translations are permitted. This fact considerably simplifies the computations and makes the moment-distribution method an attractive one for analyzing the frame. To avoid confusion in recording the calculations, the analysis is performed in two steps, with a two-dimensional view of the frame in each step.

Consider first the view of the frame as seen toward the origin along the X axis. We see that for a rotation about the X axis at joint ②, not

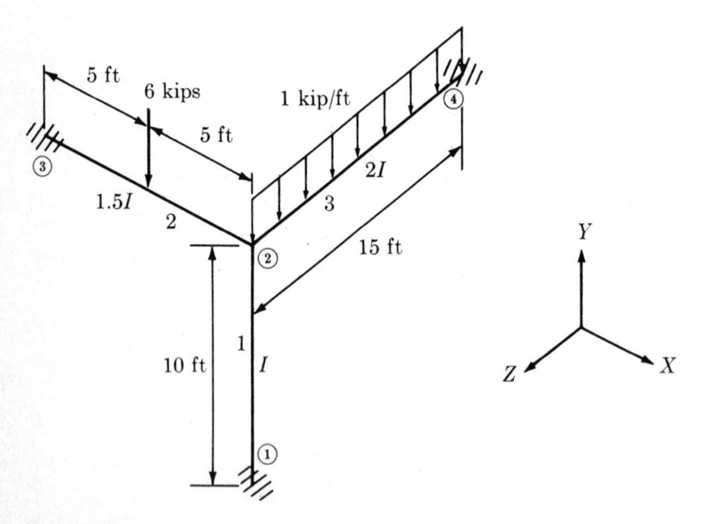

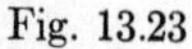
Fig. 13.23

only are the bending stiffnesses of members 1 and 3 to be considered, but also the twisting stiffness of member 2. The stiffness of a member due to twist is given in Eq. (9-33) as

$$k = \frac{JG}{L}$$

which for the given general relations between J and I and between G and E becomes

$$k = \frac{(2I)(0.4E)}{L} = \frac{0.8EI}{L}$$

This expression for twisting stiffness is in the same form as the expression for bending stiffness of Eq. (13-2),

$$k_{11} = \frac{4EI}{L}$$

From the basic definition of a carryover factor, its value is seen to be 1.0 for twist. Twisting moment in member 2 is considered to be positive when acting opposite to the direction shown in Fig. 9-19. Thus its action on joint ② will agree with the moment-distribution sign convention of Fig. 13-2.

Only the load on member 3 is considered in the first step of the analysis. The fixed-end moments for the uniform load on the member are

$$\text{FEM}_{24} = -\frac{(1)(15)^2}{12} = -18.75 \text{ ft-kips}$$
$$\text{FEM}_{42} = 18.75 \text{ ft-kips}$$

The distribution factors at joint ② are found from the above expressions for twisting and bending stiffness to be

$$\text{DF}_{24} = \frac{(4E)(2I)/15}{(4E)(2I)/15 + 4EI/10 + (0.8E)(1.5I)/10} = \frac{8EI/15}{31.6EI/30} = 0.506$$
$$\text{DF}_{21} = 0.380 \qquad \text{DF}_{23} = 0.114$$

The moment-distribution calculations are shown in Fig. 13-24. For convenience, the calculations for member 2 have been recorded in the same plane as the other two members.

For the second step of the analysis consider the view of the frame as seen toward the origin of the coordinate system in Fig. 13-23 along the Z axis. For consistent sign conventions for moment at joint ②, twist in member 3 is considered to be positive when acting in the direction oppo-

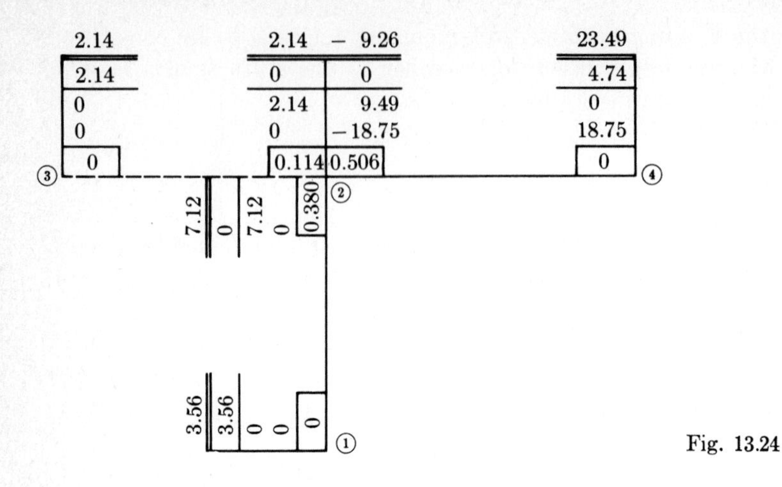

Fig. 13.24

site to that shown in Fig. 9-19. The fixed-end moments for member 2 are found to be

$$\mathrm{FEM}_{23} = 7.5 \text{ ft-kips} \qquad \mathrm{FEM}_{32} = -7.5 \text{ ft-kips}$$

The distribution factors at joint ② are

$$\mathrm{DF}_{23} = 0.542 \qquad \mathrm{DF}_{21} = 0.362 \qquad \mathrm{DF}_{24} = 0.096$$

The moment-distribution calculations for this step of the analysis are shown in Fig. 13-25. The actual twisting moments and bending moments in the frame are obtained by superimposing the results of the two analyses in Figs. 13-24 and 13-25.

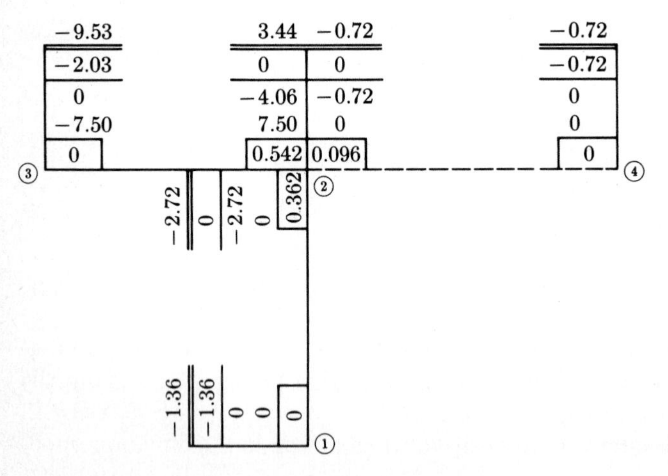

Fig. 13.25

13-4 MEMBERS WITH VARIABLE MOMENT OF INERTIA

The moment-distribution terms developed in Sec. 13-1 change in value when a member has a variable moment of inertia. The carryover factor is no longer ½, nor is it necessarily equal in both directions on the member. Also, the absolute stiffness factor as defined in Eq. (13-2) is not equal to $4EI/L$.

To evaluate the moment-distribution terms for a member with variable moment of inertia, let us consider the member shown in Fig. 13-26*a*. If, for a unit value of moment applied at ①, we can determine the resulting moment at ②, then the carryover factor from ① to ② is equal to the value of M_2. Also, if the corresponding rotation at ① can be determined, then the absolute stiffness at ① will be equal to $1/\phi_1$.

The principles of the moment-area method can be used to determine the desired results. The moment diagram for the end moments of Fig. 13-26*a*, constructed according to the beam sign convention, is shown in Fig. 13-26*b*. We should note that the moment M_x at any position x on the member can be expressed as

$$M_x = 1 - (1 + M_2)\frac{x}{L} \tag{13-9}$$

where the origin of x is at the left end of the member. A required deflection condition of the member in Fig. 13-26*a* in terms of the second moment-area theorem in Eq. (7-6) is

$$t_{12} = \int_0^L \frac{M_x}{EI_x} x\,dx = 0 \tag{13-10}$$

The subscript x is used with the moment-of-inertia term to indicate that it varies over the length of the member. It is assumed that E is constant for the member.

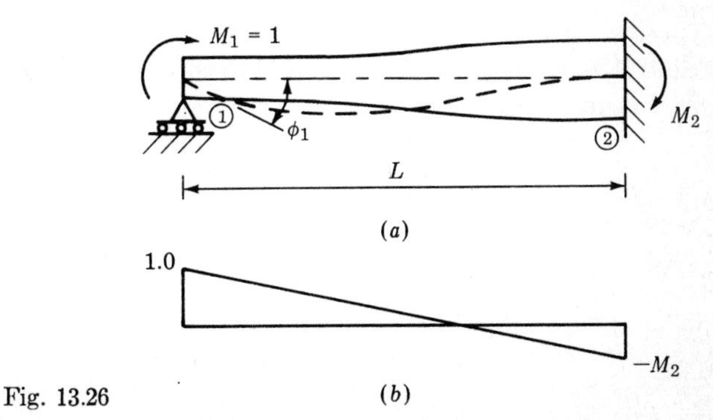

Fig. 13.26

We see that if the expression for M_x of Eq. (13-9) is substituted into Eq. (13-10) we obtain one equation with the desired quantity M_2 as the unknown. The difficulty that arises is in the evaluation of the integral of Eq. (13-10), particularly when I_x is a complicated function. Various approximate techniques are available for the evaluation of Eq. (13-10). One approach is to use a semigraphical or summation process whereby the integral of Eq. (13-10) is replaced by the summation

$$\sum_{i=1}^{N} \frac{M_i}{EI_i} x_i \, \Delta x = 0$$

wherein the length of the member is divided into N equal segments of length Δx and the moment and moment of inertia are evaluated at the midpoint of each segment, giving values of M_i and I_i. The value of x_i is the distance from the left end of the member to the midpoint of the ith segment.

Another approach is to use numerical integration. Two frequently used methods of numerical integration are the *trapezoidal rule* and *Simpson's one-third rule.* Recall from calculus that the value of the integral of some function of x, y_x, between the limits a and b, evaluated according to the trapezoidal rule is

$$\int_a^b y_x \, dx = \frac{h}{2} (y_0 + 2y_1 + 2y_2 + \cdots + 2y_{n-1} + y_n) \tag{13-11}$$

where $h = (b - a)/n$. The values of $y_0, y_1, y_2, \ldots, y_n$ are the values of the function y_x evaluated at $x = x_0$, $x = x_1$, and so on. In deriving the trapezoidal rule, the original function y_x is approximated by a series of straight-line segments. If the original function is approximated by segments of second-order polynomials, Simpson's one-third rule is obtained:

$$\int_a^b y_x \, dx = \frac{h}{3} (y_0 + 4y_1 + 2y_2 + 4y_3 + \cdots + 4y_{n-1} + y_n) \tag{13-12}$$

In this expression n must be an even number. We see that in each of Eqs. (13-11) and (13-12) y_i is equal to the value of the quantity $(M_i/EI_i)x_i$ or, because E is constant it can be factored out, so that $y_i = (M_i/I_i)x_i$.

After the value of M_2 is obtained, and thus the carryover factor from ① to ②, the value of ϕ_1 can be obtained. Using the first moment-area theorem as stated in Eq. (7-3), the value of ϕ_1 is obtained from the expression

$$\phi_1 = \int_0^L \frac{M_x}{EI_x} \, dx \tag{13-13}$$

The techniques suggested above can be used to evaluate the integral of Eq. (13-13).

The carryover factor from ② to ① and the absolute stiffness at ② can be obtained in a similar manner, by reversing the conditions shown in Fig. 13-26. The fixed-end moments for a member with variable moment of inertia can be obtained by using a numerical approach with one of the previously discussed methods for statically indeterminate analysis. Numerical integration of virtual work combined with the method of consistent displacements is one convenient method for evaluating fixed-end moments. The carryover factors, stiffnesses, and fixed-end moments for typical haunched members are also available in tabular and chart forms.[1]

13-5 MATRIX FORMULATION OF THE MOMENT-DISTRIBUTION METHOD

The moment-distribution method can be formulated in terms of matrices, a form which lends itself well to programming for digital computer solutions. Because moment distribution is a cyclic process, the computer can readily perform the repetitive matrix operations.

The matrices as used for moment distribution are thought of as *operators*. The operative nature of matrices can, for example, be compared to the operative nature of an integral. When some function is operated on by an integral, the result is another function. Similarly, when a matrix is operated on by another matrix, the result is another matrix. As an example of matrix formulation, let us consider the moment-distribution calculations for the frame of Fig. 13-16. We shall concern ourselves with the condition that the frame is temporarily restrained against translation. The calculations for this condition were shown in Fig. 13-18 and are repeated in Fig. 13-27 for convenient reference.

The process of moment distribution begins with values of fixed-end moments. These can be represented by a column matrix (vector) written as $\mathbf{X}^1$, where the superscript 1 denotes the first value of unbalanced moments. The unbalanced moments are then distributed by operating on $\mathbf{X}^1$ with a distribution operator $\mathbf{A}$ to produce values denoted as $\mathbf{Y}$. Thus

$$\mathbf{Y}^1 = \mathbf{A}\mathbf{X}^1 \tag{13-14}$$

where the matrix $\mathbf{A}$ is defined as the *distribution operator*. From Fig. 13-27

[1] "Handbook of Frame Constants," Portland Cement Association, Skokie, Ill., 1958.

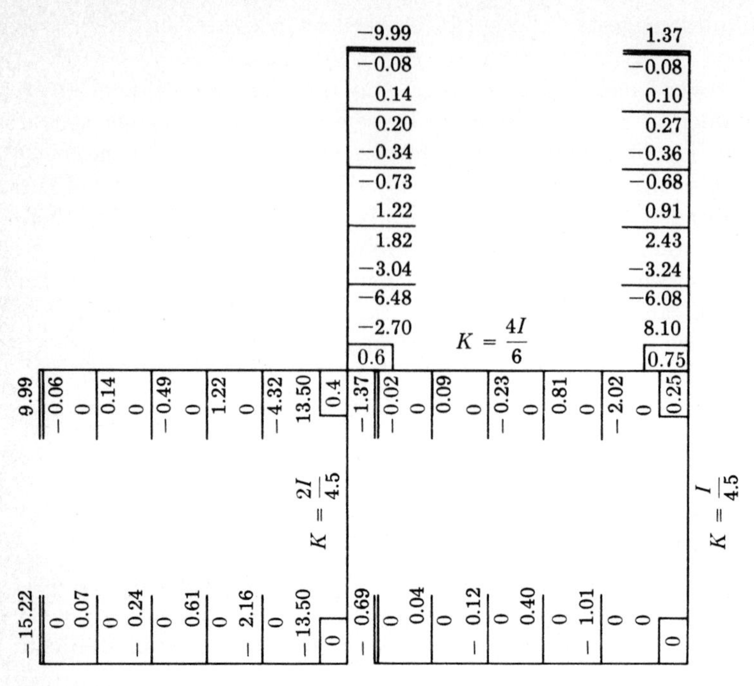

Fig. 13.27

we see that

$$\mathbf{X}^1 = \begin{Bmatrix} -13.5 \\ 13.5 \\ -2.7 \\ 8.1 \\ 0 \\ 0 \end{Bmatrix} \qquad \mathbf{A} = \begin{bmatrix} 0 & 0 & 0 & 0 & 0 & 0 \\ 0 & -0.4 & -0.4 & 0 & 0 & 0 \\ 0 & -0.6 & -0.6 & 0 & 0 & 0 \\ 0 & 0 & 0 & -0.75 & -0.75 & 0 \\ 0 & 0 & 0 & -0.25 & -0.25 & 0 \\ 0 & 0 & 0 & 0 & 0 & 0 \end{bmatrix}$$

The fixed-end moments of $\mathbf{X}^1$ are shown in the order in which the moments are encountered in going clockwise around the frame. Other orders could be used. The nonzero elements of **A** are the distribution factors placed in the appropriate positions and with negative signs such that when the matrix operation of Eq. (13-14) is performed, the distributed moments will be obtained.

The values of **Y** resulting from the distribution process are then carried over to the opposite ends of the members, resulting in new values of unbalanced moment. The process of carrying over moments can be represented by an operator **B**, which relates $\mathbf{X}^2$ to $\mathbf{Y}^1$ in the form

$$\mathbf{X}^2 = \mathbf{B}\mathbf{Y}^1 \tag{13-15}$$

The matrix **B** is defined as the *carryover operator*. For the frame under consideration

$$\mathbf{B} = \begin{bmatrix} 0 & 0.5 & 0 & 0 & 0 & 0 \\ 0.5 & 0 & 0 & 0 & 0 & 0 \\ 0 & 0 & 0 & 0.5 & 0 & 0 \\ 0 & 0 & 0.5 & 0 & 0 & 0 \\ 0 & 0 & 0 & 0 & 0 & 0.5 \\ 0 & 0 & 0 & 0 & 0.5 & 0 \end{bmatrix}$$

The nonzero elements of **B** are the carryover factors located such that when the matrix operation of Eq. (13-15) is performed, the moments will be carried over to the proper member end.

This cycle of operations is repeated successively and can therefore be represented by the general expression

$$\mathbf{X}^i = \mathbf{C}\mathbf{X}^{i-1} \tag{13-16}$$

where $\mathbf{C} = \mathbf{BA}$. The moments in the frame, denoted as **M**, can be represented at the end of any cycle by the expression

$$\mathbf{M}^i = \mathbf{M}^{i-1} + \mathbf{D}\mathbf{X}^i \tag{13-17}$$

where $\mathbf{D} = \mathbf{I} + \mathbf{A}$ and **I** is a unit matrix. The process is complete when the elements of a vector $\Delta\mathbf{M} = \mathbf{M}^i - \mathbf{M}^{i-1}$, which from Eq. (13-17) is equal to $\mathbf{D}\mathbf{X}^i$, are within the desired degree of accuracy.

For the above values of **A** and **B** the values of **C** and **D** are

$$\mathbf{C} = \mathbf{BA} = \begin{bmatrix} 0 & -0.2 & -0.2 & 0 & 0 & 0 \\ 0 & 0 & 0 & 0 & 0 & 0 \\ 0 & 0 & 0 & -0.375 & -0.375 & 0 \\ 0 & -0.3 & -0.3 & 0 & 0 & 0 \\ 0 & 0 & 0 & 0 & 0 & 0 \\ 0 & 0 & 0 & -0.125 & -0.125 & 0 \end{bmatrix}$$

$$\mathbf{D} = \mathbf{I} + \mathbf{A} = \begin{bmatrix} 1 & 0 & 0 & 0 & 0 & 0 \\ 0 & 0.6 & -0.4 & 0 & 0 & 0 \\ 0 & -0.6 & 0.4 & 0 & 0 & 0 \\ 0 & 0 & 0 & 0.25 & -0.75 & 0 \\ 0 & 0 & 0 & -0.25 & 0.75 & 0 \\ 0 & 0 & 0 & 0 & 0 & 1 \end{bmatrix}$$

Performing the operations of Eq. (13-17) with a beginning value of

$\mathbf{M}^0 = \mathbf{0}$, we obtain the following values of moment at the end of each cycle:

$$\mathbf{M}^1 = \begin{Bmatrix} -13.50 \\ 9.18 \\ -9.18 \\ 2.02 \\ -2.02 \\ 0 \end{Bmatrix} \quad \mathbf{M}^2 = \begin{Bmatrix} -15.66 \\ 10.40 \\ -10.40 \\ 1.22 \\ -1.22 \\ -1.01 \end{Bmatrix} \quad \mathbf{M}^3 = \begin{Bmatrix} -15.05 \\ 9.91 \\ -9.91 \\ 1.44 \\ -1.44 \\ -0.61 \end{Bmatrix} \quad \mathbf{M}^4 = \begin{Bmatrix} -15.30 \\ 10.05 \\ -10.05 \\ 1.35 \\ -1.35 \\ -0.72 \end{Bmatrix} \quad \mathbf{M}^5 = \begin{Bmatrix} -15.23 \\ 9.99 \\ -9.99 \\ 1.38 \\ -1.38 \\ -0.68 \end{Bmatrix}$$

The value of $\mathbf{M}^5$ corresponds to the final values of moment obtained in Fig. 13-18.

There are only a few different types of matrix operations in the above procedure. The repeated operations can readily be handled by a rather short computer program. The input to a computer program is the nonzero values of the **A** and **B** matrices and the elements of the $\mathbf{X}^1$ matrix. Matrices **A** and **B** are characteristic of the structure, so for any value of **A** and **B** a variety of loadings can be investigated. The moments for the translated condition in the frame considered previously can be obtained simply by using the $\mathbf{X}^1$ vector of fixed-end moments in Fig. 13-21 with the same **A** and **B** matrices. A Fortran IV computer program for the moment-distribution operations presented above is listed in Appendix B.

With a digital computer moment distribution in matrix form can be used for the analysis of large structures. The matrix formulation also offers an advantage over hand calculations for structures whose members are oriented in such a way that it is difficult to record the calculations on a sketch of the structure.

PROBLEMS

13-1 through 13-4. For the beams shown, (*a*) determine the moments at the member ends; (*b*) construct shear and moment diagrams for the beams using the beam sign convention. E is constant.

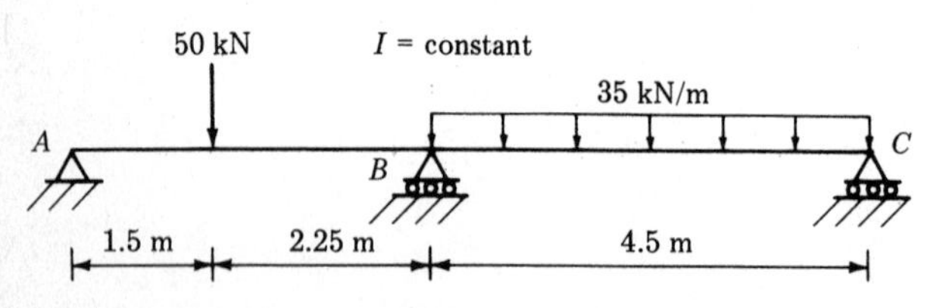

P13.1

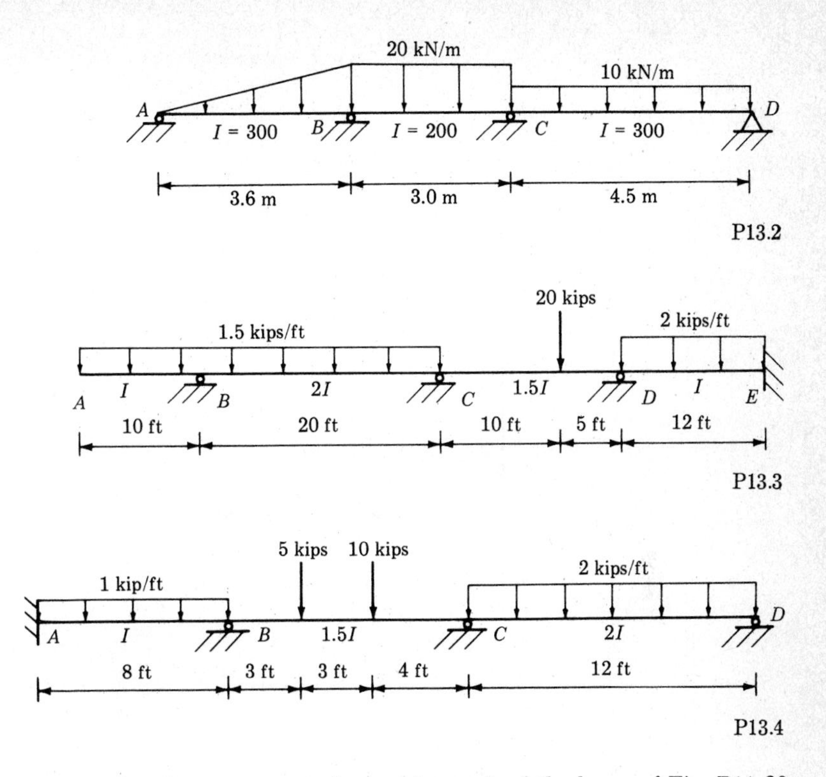

P13.2

P13.3

P13.4

13-5. Determine the moments at the member ends of the beam of Fig. P11-20 using the moment-distribution method.

13-6. Determine the moments at the member ends of the beam of Fig. P11-21 using the moment-distribution method.

13-7 through 13-9. For the frames shown, (*a*) determine the moments in the members; (*b*) construct the moment diagram for the frames on either (*i*) the compression side of the members or (*ii*) the tension side of the members. *E* is constant.

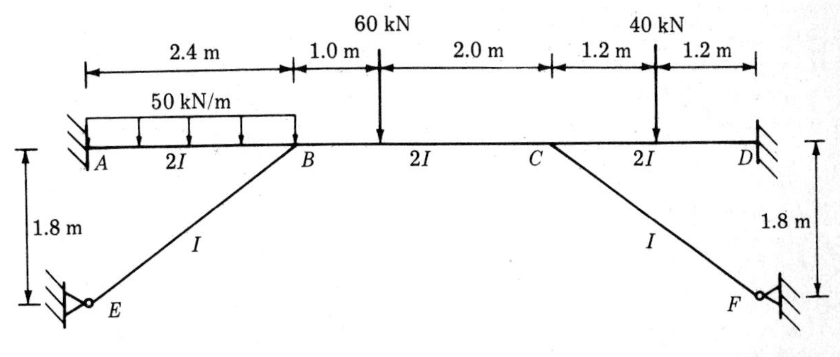

P13.7

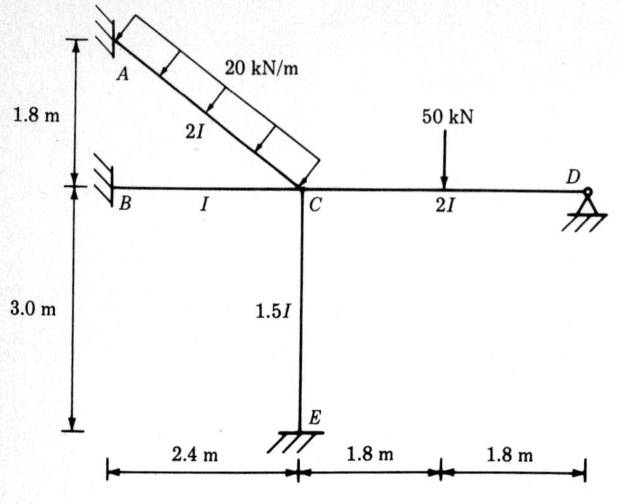
A
20 kN/m
1.8 m
2I
50 kN
D
B
I
C
2I
3.0 m
1.5I
E
2.4 m
1.8 m
1.8 m

P13.8

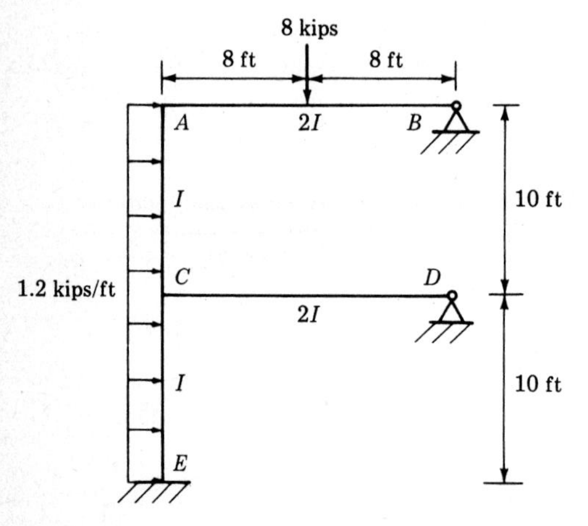
8 kips
8 ft
8 ft
A
2I
B
I
10 ft
C
D
1.2 kips/ft
2I
I
10 ft
E

P13.9

13-10. Use an appropriate method to determine the fixed-end moments in member BD, and determine the moments in the members of the frame shown. EI is constant.

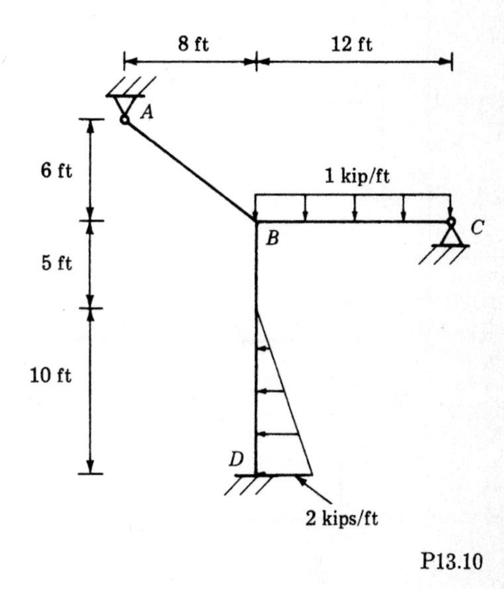

P13.10

13-11 through 13-14. For the frames shown, (*a*) identify the type of joint translation that is possible; (*b*) determine the moments in the members of the frames. E is constant.

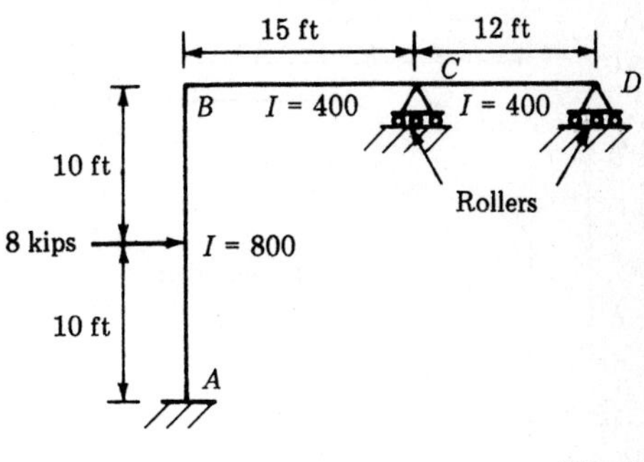

P13.11

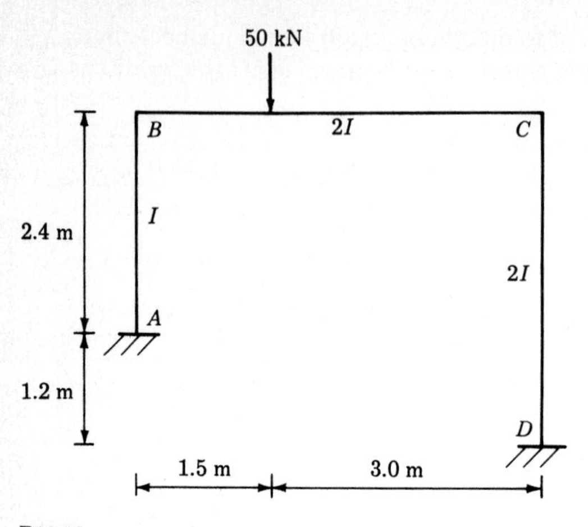

P13.12

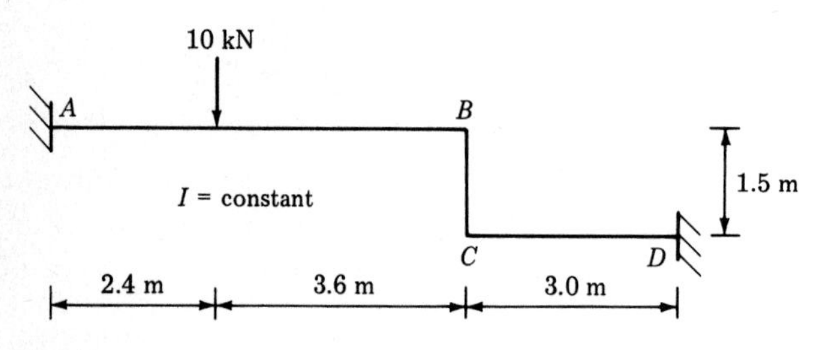

P13.13

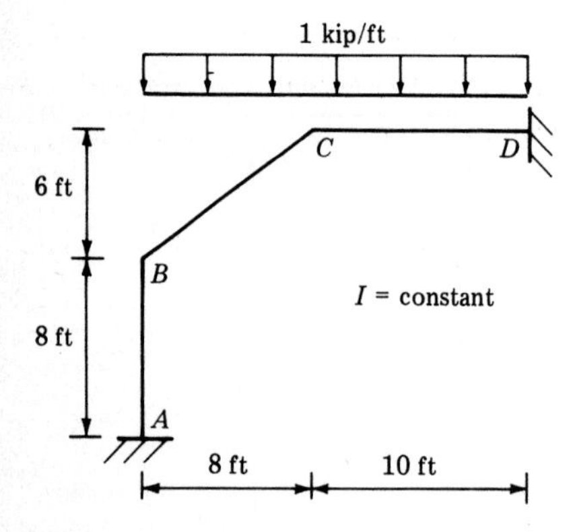

P13.14

13-15. Determine the moments in the members of the frame of Fig. P11-11 using the moment-distribution method.

13-16. Determine the moments in the members of the frame of Fig. P12-8 using the moment-distribution method.

13-17 and 13-18. The members shown have rectangular cross-sections that vary in depth as shown and have unit width. Divide the length of the members into an appropriate number of segments, and use one of the suggested techniques of Sec. 13-4 to determine (*a*) the value of the carryover factor from A to B; (*b*) the value of the carryover factor from B to A; (*c*) the expression for the absolute stiffness at A. E is constant.

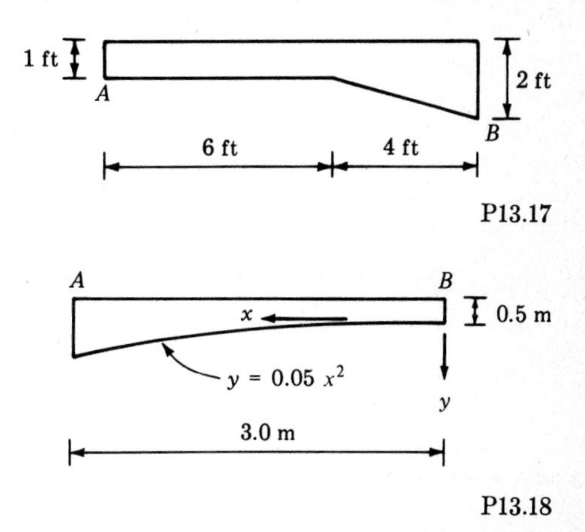

13-19. For the beam of Fig. P7-23, determine (*a*) the value of the carryover factor from A to C; (*b*) the value of the carryover factor from C to A.

CHAPTER

FOURTEEN

MATRIX DISPLACEMENT (STIFFNESS) METHOD

14-1 INTRODUCTION

The *matrix displacement method*, which is also known as the *stiffness method*, is a method of analysis that uses the stiffness properties of the elements of a structure to form a set of simultaneous equations relating displacements of the structure to loads acting on the structure. The governing matrix equation is of the form

$$\mathbf{Q} = \mathbf{K}\mathbf{D} \tag{14-1}$$

where **Q** is a column matrix of external loads, **K** is the stiffness matrix of the structure, and **D** is a column matrix of external displacements. For a given set of external loads, the corresponding displacements are obtained by solving the set of simultaneous equations. Forces in the elements of the structure are obtained through the use of the calculated displacements and the element stiffness properties. The matrix displacement method can be used for the analysis of various types of structures and, with the use of a digital computer, for very large and complex structural systems.

In analyzing a structure by the matrix displacement method, it is advantageous to consider the structure in a broad fashion. A structure is considered to be an assembly of structural elements connected at a finite number of points referred to as *nodal points*. The plane truss of

Fig. 14-1*a*, for example, can be considered an assembly of elements connected at nodal points by pins. Resistance to loading and displacements is provided by the stiffness of the pin-connected members. The plane frame of Fig. 14-1*b* can likewise be considered an assembly of elements. The elements in this type of structure provide stiffness in the form of axial stiffness and bending stiffness. Nodal points are selected at points connecting the members and at points of support. Additional nodal points could, however, be introduced at other points on the members of the structure.

A more advanced application of the matrix displacement method can be illustrated by the frame and shear-wall structure of Fig. 14-1*c*. For purposes of analysis, the shear wall can be represented by a series of triangular finite elements as shown in Fig. 14-1*d*. The finite elements are pin-connected to each other at their corners. If a sufficient number of elements is used and if the stiffness of each element is properly formulated, the discretized model closely represents the behavior of the given shear wall. The representation of continuous structural elements such as plates, walls, and shells by a finite number of elements is known as the *finite-element method* of analysis. The finite-element method will be considered in the next chapter.

Although reference has been made to planar structures in Fig. 14-1, the matrix displacement method can be applied as well to the analysis of three-dimensional structures.

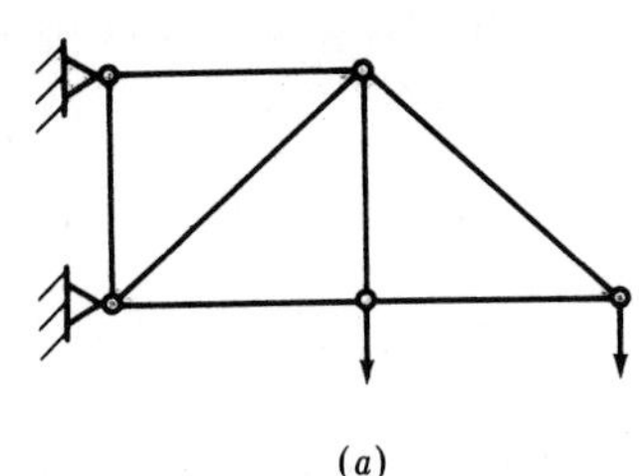

(*a*)

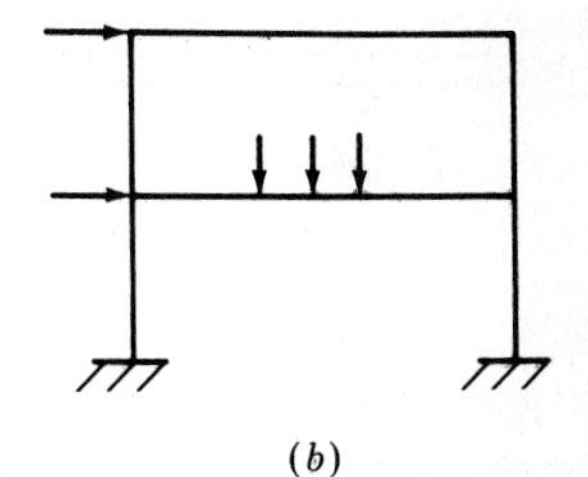

(*b*)

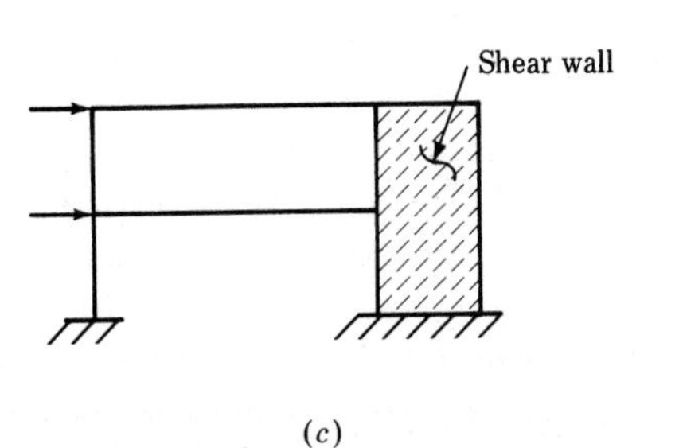

(*c*)

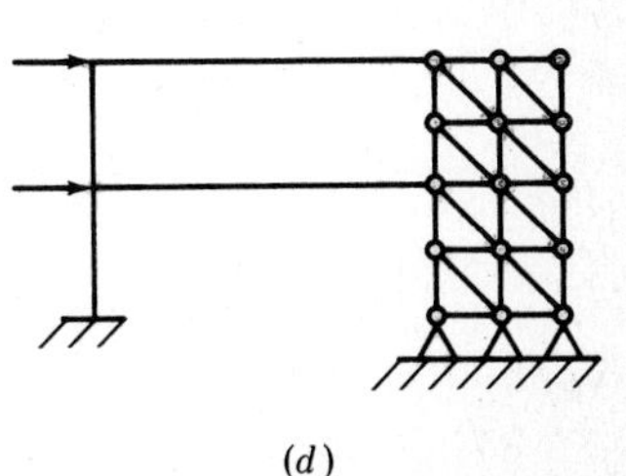

(*d*)

Fig. 14.1

The development of the matrix displacement method begins in the next section with a consideration of the stiffness matrices of types of elements that are commonly encountered in structural analysis. These stiffness matrices are then used to develop the stiffness matrix for the entire structure.

14-2 ELEMENT-STIFFNESS MATRICES

We begin by considering element stiffnesses in terms of their *local coordinate system*, i.e., in terms of a coordinate system coincident with the orientation of the member. Because elements of a complete structure are often oriented in various directions, the element stiffnesses will be transformed in a later section into stiffnesses described with reference to a common coordinate system for the structure. This common coordinate system is referred to as the *global* or *framework* coordinate system. With all element stiffnesses described in terms of a common coordinate system, the stiffness matrix for the entire structure can be developed more effectively.

Element stiffnesses have been developed previously in Sec. 9-4 in terms of their local coordinate systems. The two-dimensional stiffnesses for axial force and bending are summarized in Fig. 14-2. Force and dis-

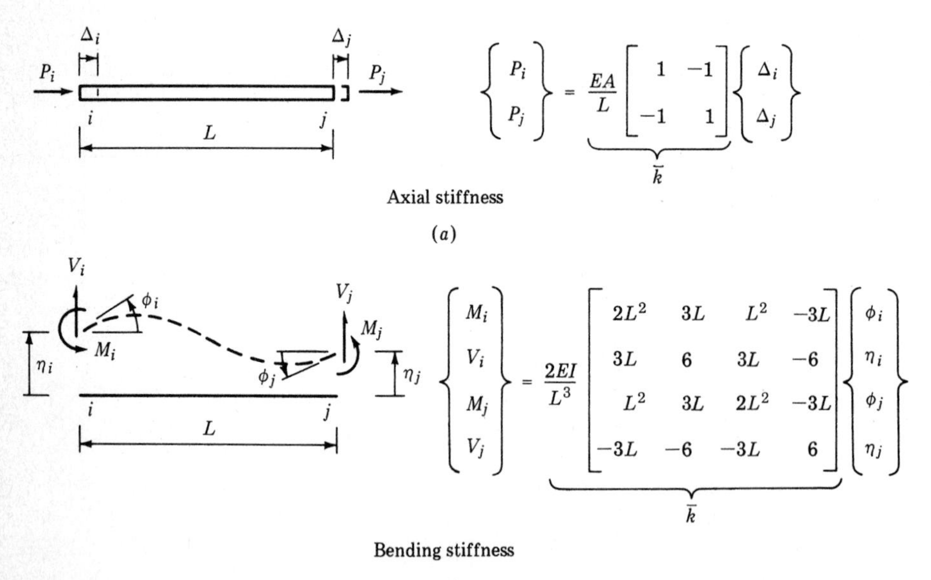

$$\begin{Bmatrix} P_i \\ P_j \end{Bmatrix} = \underbrace{\frac{EA}{L}\begin{bmatrix} 1 & -1 \\ -1 & 1 \end{bmatrix}}_{\bar{k}} \begin{Bmatrix} \Delta_i \\ \Delta_j \end{Bmatrix}$$

Axial stiffness

(*a*)

$$\begin{Bmatrix} M_i \\ V_i \\ M_j \\ V_j \end{Bmatrix} = \underbrace{\frac{2EI}{L^3}\begin{bmatrix} 2L^2 & 3L & L^2 & -3L \\ 3L & 6 & 3L & -6 \\ L^2 & 3L & 2L^2 & -3L \\ -3L & -6 & -3L & 6 \end{bmatrix}}_{\bar{k}} \begin{Bmatrix} \phi_i \\ \eta_i \\ \phi_j \\ \eta_j \end{Bmatrix}$$

Bending stiffness

(*b*)

Fig. 14.2

placement quantities as shown are considered positive. It should be noted that member end rotations and moments will be considered positive when they act counterclockwise in the development of the matrix displacement method. They thus correspond to previously defined positive rotation about the Z axis for a three-dimensional coordinate system. For clarity in the development that follows, the stiffness matrix of an element described in its *local coordinate system* will be denoted with a bar, for example, $\bar{\mathbf{k}}$.

14-3 THE CONCEPT OF ASSEMBLING ELEMENT STIFFNESSES

The displacements that can occur at the nodal points of a structure are referred to as *external displacements* and are denoted by D. Forces that act on the structure at the nodal points are referred to as *external forces* and are denoted by Q. The external displacements D and the external forces Q are expressed in terms of global components. For the analysis of a frame such as that in Fig. 14-3*a*, we establish a global coordinate system such as that shown in Fig. 14-3*b*. If we consider both axial and bending deformations in the members of the frame, there are three possible components of displacement at both nodal points ② and ③. For the supports shown, there are no possible displacements at nodal points ① and ④. The six possible external displacements constitute the column matrix of external displacements $\mathbf{D}$. The elements of $\mathbf{D}$ are denoted by $D_2{}^1$, $D_2{}^2$, . . . , $D_3{}^3$, where the subscript denotes the nodal point at which the displacement occurs. The external forces are likewise expressed in terms of global components at the nodal points and constitute the column matrix of external forces $\mathbf{Q}$. The elements of $\mathbf{Q}$ must correspond to the order of the elements used in $\mathbf{D}$, and the sign of the force components must be in accordance with positive quantities as indicated by

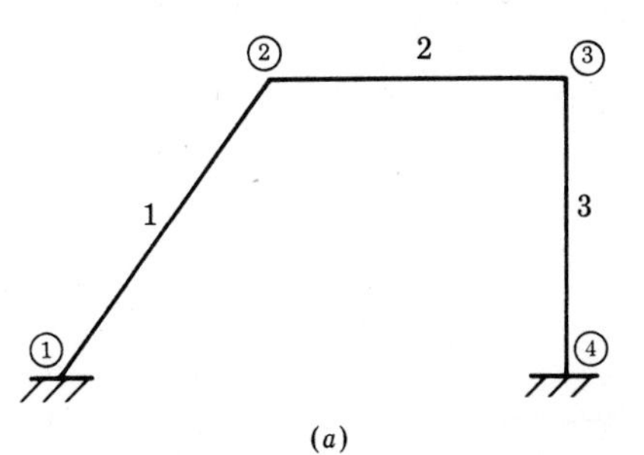

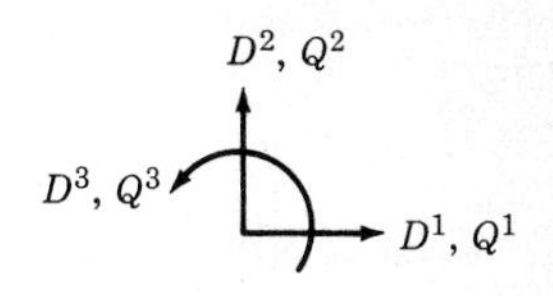

External (global) coordinate system for deflections and forces

(*a*) (*b*)

Fig. 14.3

the external (global) coordinate system. As indicated previously in Eq. (14-1), the external displacements are related to the external forces by the structure-stiffness matrix $\mathbf{K}$ in the form

$$\mathbf{Q} = \mathbf{KD}$$

The elements of $\mathbf{K}$ are stiffness influence coefficients relating external displacement components to external force components. The values of the elements in $\mathbf{K}$ reflect the cumulative stiffnesses of the elements meeting at a nodal point.

To develop a feeling for the manner in which individual elements can be assembled into the structure-stiffness matrix $\mathbf{K}$, let us consider an elementary combination of elements. Consider the two springs arranged as shown in Fig. 14-4*a*. The springs have stiffnesses of k_1 and k_2. The springs provide stiffness only for horizontal forces and displacements at the nodal points; therefore, it is sufficient to consider a single component of external displacement D and a single component of external force Q at each nodal point. The positive direction for D and Q is indicated in Fig. 14-4*a*. The possible external forces on the structure are shown in Fig. 14-4*b*. The desired stiffness relationship for the structure is

$$\begin{Bmatrix} Q_1 \\ Q_2 \\ Q_3 \end{Bmatrix} = \begin{bmatrix} k_{11} & k_{12} & k_{13} \\ k_{21} & k_{22} & k_{23} \\ k_{31} & k_{32} & k_{33} \end{bmatrix} \begin{Bmatrix} D_1 \\ D_2 \\ D_3 \end{Bmatrix} \tag{14-2}$$

Let us consider the first equation in Eq. (14-2):

$$Q_1 = k_{11}D_1 + k_{12}D_2 + k_{13}D_3 \tag{14-3}$$

We shall see that this and the other two equations represent the condition for equilibrium at the nodal points in terms of displacements. To demonstrate this, we shall use the stiffness relationship for an individual spring as shown in Fig. 14-5*a*. Quantities as shown are considered positive. The stiffness relationship is obtained from that developed for an axially loaded member in Sec. 9-4 with EA/L replaced by the spring stiffness k. The forces acting *on the nodal points* of the assembly in Fig. 14-4 are as

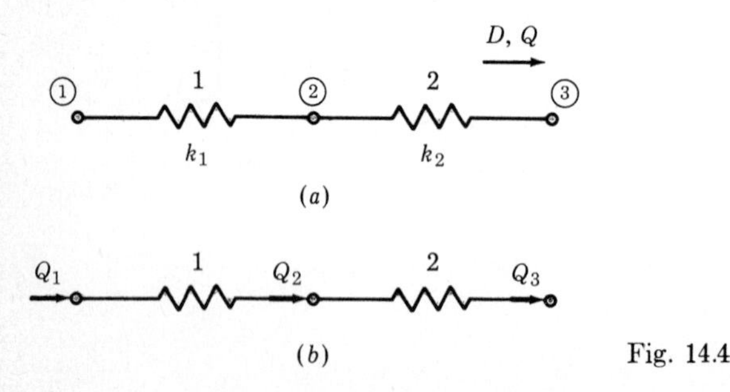

Fig. 14.4

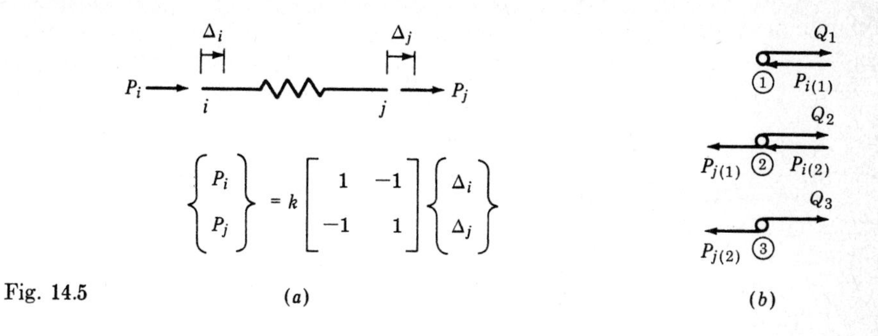

Fig. 14.5 (a) (b)

shown in Fig. 14-5*b*. Note that the spring forces acting on the nodal points are equal and opposite to the forces acting on the springs. Subscripted numbers in parentheses in Fig. 14-5*b* denote the member to which the force pertains. For equilibrium of nodal point ① in Fig. 14-5*b* we can write

$$Q_1 - P_{i(1)} = 0 \qquad \text{or} \qquad Q_1 = P_{i(1)}$$

For spring 1, $\Delta_i = D_1$ and $\Delta_j = D_2$. Therefore, from the stiffness relationship in Fig. 14-5*a*, we obtain

$$Q_1 = P_{i(1)} = k_1 D_1 - k_1 D_2$$

We see, therefore, in Eq. (14-3), that $k_{11} = k_1$, $k_{12} = -k_1$, and $k_{13} = 0$. The second equation in Eq. (14-2) represents equilibrium of nodal point ②. From Fig. 14-5*b* we can write the equilibrium equation for nodal point ② as

$$Q_2 - P_{j(1)} - P_{i(2)} = 0 \qquad \text{or} \qquad Q_2 = P_{j(1)} + P_{i(2)}$$

Using the stiffness relationship of Fig. 14-5*a* for both springs 1 and 2, we obtain

$$\begin{aligned} Q_2 &= (-k_1 D_1 + k_1 D_2) + (k_2 D_2 - k_2 D_3) \\ &= -k_1 D_1 + (k_1 + k_2) D_2 - k_2 D_3 \end{aligned}$$

We see therefore in Eq. (14-2) that $k_{21} = -k_1$, $k_{22} = k_1 + k_2$, and $k_{23} = -k_2$. Proceeding in a similar manner with equilibrium at nodal point ③, we can obtain the elements of the third row in the stiffness matrix of Eq. (14-2). The resulting stiffness matrix for the spring structure is therefore

$$\mathbf{K} = \begin{array}{c} \begin{matrix} (D_1) & (D_2) & (D_3) \end{matrix} \\ \begin{bmatrix} k_1 & -k_1 & 0 \\ -k_1 & k_1 + k_2 & -k_2 \\ 0 & -k_2 & k_2 \end{bmatrix} \end{array}$$

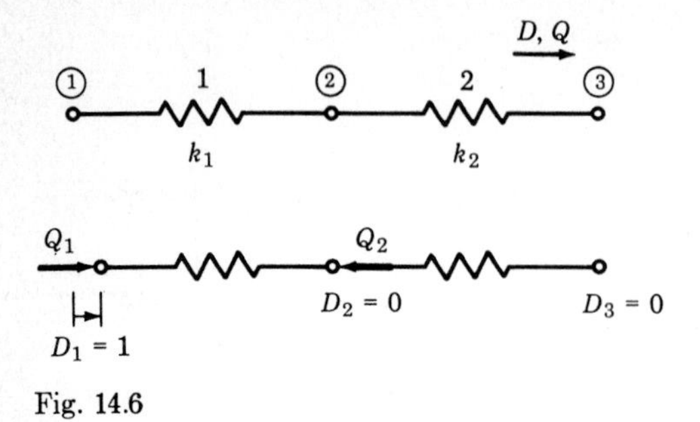

Fig. 14.6

In the above procedure the elements of **K** were developed row by row. It is also possible to develop the **K** matrix column by column by proceeding as we did in Sec. 9-4, where unit values of displacement are successively applied to the structure and the corresponding forces are determined. Thus, to determine the values of the first column of **K** for the spring structure, we let $D_1 = 1$ and $D_2 = D_3 = 0$ as shown in Fig. 14-6. For this condition we see that

$$Q_1 = k_1 D_1 \qquad Q_2 = -k_1 D_1 \qquad Q_3 = 0$$

These results correspond to the values obtained previously for the first column of the **K** matrix. The values of the second and third columns can likewise be determined by successively applying unit values of D_2 and D_3.

14-4 TRANSFORMATION OF ELEMENT-STIFFNESS MATRICES

The fact that the spring stiffness in Fig. 14-5*a* was expressed in terms of forces P and displacements Δ that were coincident with external components of force and displacement for the spring structure of Fig. 14-4 contributed to the ease with which the structure-stiffness matrix was developed. For conditions in which members are oriented in various directions, it is convenient to have member stiffnesses described in terms of a common coordinate system for the entire structure. Using appropriate transformations, element stiffnesses described in local coordinates can be expressed in terms of a global coordinate system. To demonstrate the procedure for obtaining a transformed member-stiffness matrix, let us consider the pinned-pinned member of Fig. 14-7. The member lies in the xy plane and is oriented at an angle ω to the horizontal (x) direction. It is desired to express the member stiffness in terms of the global components of displacement d and the global components of force q at the i

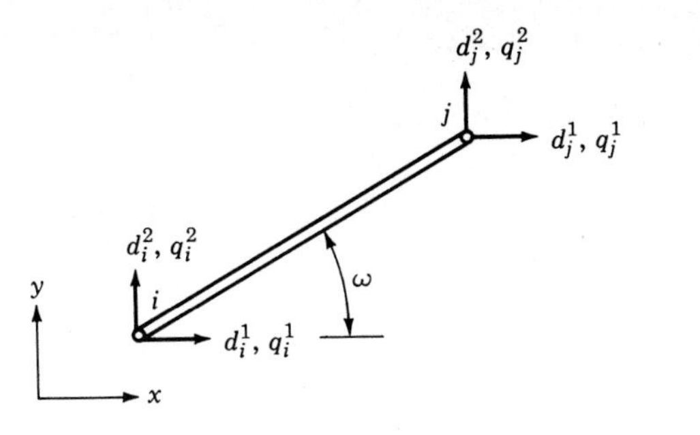

Fig. 14.7

and j ends of the member. The global system is shown to the right in Fig. 14-7. Horizontal components of displacement and force are denoted with a superscript 1, and vertical components of displacement and force are denoted with a superscript 2.

The member-stiffness relationship to be developed is of the form

$$\begin{Bmatrix} q_i^1 \\ q_i^2 \\ q_j^1 \\ q_j^2 \end{Bmatrix} = \begin{bmatrix} k_{11} & k_{12} & k_{13} & k_{14} \\ k_{21} & k_{22} & k_{23} & k_{24} \\ k_{31} & k_{32} & k_{33} & k_{34} \\ k_{41} & k_{42} & k_{43} & k_{44} \end{bmatrix} \begin{Bmatrix} d_i^1 \\ d_i^2 \\ d_j^1 \\ d_j^2 \end{Bmatrix} \tag{14-4}$$

The stiffness influence coefficients k_{ij} can be determined by the procedure used in Sec. 9-4, where unit values of member-end displacements are applied successively to the member and the corresponding member-end forces are determined. The values of the first column of the stiffness matrix of Eq. (14-4) are obtained from the condition shown in Fig. 14-8*a*, where $d_i^1 = 1$ and $d_i^2 = d_j^1 = d_j^2 = 0$. The stiffness relationship between axial loading P and axial displacement Δ at the member ends as shown in

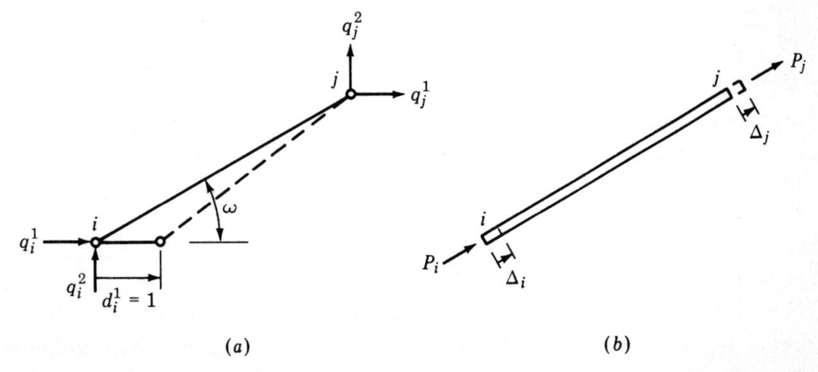

Fig. 14.8

Fig. 14-8*b* has been expressed previously in Eq. (9-15) as

$$\begin{Bmatrix} P_i \\ P_j \end{Bmatrix} = \frac{EA}{L}\begin{bmatrix} 1 & -1 \\ -1 & 1 \end{bmatrix}\begin{Bmatrix} \Delta_i \\ \Delta_j \end{Bmatrix}$$

For the conditions of Fig. 14-8, $\Delta_i = (1)\cos\omega$ and $\Delta_j = 0$. Therefore

$$P_i = \frac{EA}{L}\cos\omega \tag{14-5}$$

$$P_j = -\frac{EA}{L}\cos\omega \tag{14-6}$$

The global components of force q at the i end of the member can be expressed in terms of the axial force P_i at the i end as

$$q_i^1 = P_i\cos\omega \qquad q_i^2 = P_i\sin\omega$$

Substituting the value of P_i from Eq. (14-5), we obtain

$$q_i^1 = \frac{EA}{L}\cos^2\omega \qquad q_i^2 = \frac{EA}{L}\sin\omega\cos\omega$$

By proceeding in the same manner at the j end and using the expression for P_j in Eq. (14-6), the values of k_{i1} are found to be

$$k_{11} = \frac{EA}{L}\cos^2\omega \qquad k_{21} = \frac{EA}{L}\sin\omega\cos\omega$$

$$k_{31} = -\frac{EA}{L}\cos^2\omega \qquad k_{41} = -\frac{EA}{L}\sin\omega\cos\omega$$

The remaining columns of the stiffness matrix are obtained by applying successive unit values of d_i^2, d_j^1, and d_j^2. The desired stiffness relationship is found to be

$$\begin{Bmatrix} q_i^1 \\ q_i^2 \\ q_j^1 \\ q_j^2 \end{Bmatrix} = \frac{EA}{L}\begin{bmatrix} c^2 & sc & -c^2 & -sc \\ sc & s^2 & -sc & -s^2 \\ -c^2 & -sc & c^2 & sc \\ -sc & -s^2 & sc & s^2 \end{bmatrix}\begin{Bmatrix} d_i^1 \\ d_i^2 \\ d_j^1 \\ d_j^2 \end{Bmatrix} \tag{14-7}$$

where $s \equiv \sin\omega$ and $c \equiv \cos\omega$.

EXAMPLE 14-1 To demonstrate the use of transformed element stiffnesses, let us consider the spring structure shown in Fig. 14-9. Two external displacements are possible, and the desired stiffness relationship for the structure is

$$\begin{Bmatrix} Q_2^2 \\ Q_3^1 \end{Bmatrix} = \begin{bmatrix} k_{11} & k_{12} \\ k_{21} & k_{22} \end{bmatrix}\begin{Bmatrix} D_2^2 \\ D_3^1 \end{Bmatrix} \tag{14-8}$$

The transformed stiffness relationship developed for a pin-connected member in Eq. (14-7) can be used for each of the springs by replacing

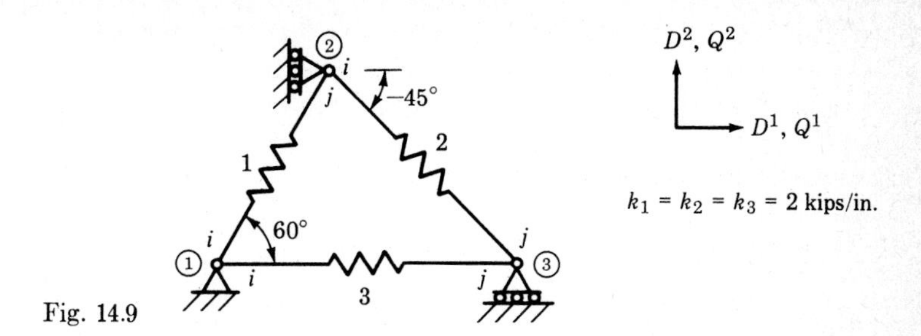

Fig. 14.9

EA/L with the spring stiffness k. The internal displacements and forces in global components act on a typical spring as shown in Fig. 14-10. The ends of the springs in Fig. 14-9 are arbitrarily denoted i and j.

The first equation in Eq. (14-8) represents the equilibrium of forces in the vertical direction at nodal point ② in terms of the two external displacements:

$$Q_2^{\,2} = k_{11}D_2^{\,2} + k_{12}D_3^{\,1} \tag{14-9}$$

Proceeding as we did in the previous spring problem, we can write the equilibrium equation for nodal point ② in terms of the force components of springs 1 and 2 at ②:

$$Q_2^{\,2} = q_{j(1)}^{\;2} + q_{i(2)}^{\;2} \tag{14-10}$$

As before, the parenthesized subscripts denote the spring number. We recognize for spring 1 that $d_{j(1)}^{\;2} = D_2^{\,2}$ and for spring 2 that $d_{i(2)}^{\;2} = D_2^{\,2}$ and $d_{j(2)}^{\;1} = D_3^{\,1}$. Using the appropriate elements from the stiffness matrix in Eq. (14-7) (element k_{44} for $q_{j(1)}^{\;2}$ and elements k_{22} and k_{23} for $q_{i(2)}^{\;2}$), Eq. (14-10) becomes

$$Q_2^{\,2} = (ks^2D_2^{\,2})_{(1)} + [ks^2D_2^{\,2} + k(-sc)D_3^{\,1}]_{(2)}$$

For the given values of ω and k in Fig. 14-9, this becomes

$$Q_2^{\,2} = (2)(\sin 60°)^2D_2^{\,2} + (2)(\sin -45°)^2D_2^{\,2} + (2)[-(\sin -45°)(\cos -45°)]D_3^{\,1}$$

$$= 2.5D_2^{\,2} + 1.0D_3^{\,1}$$

We see therefore in Eq. (14-9) that $k_{11} = 2.5$ and $k_{12} = 1.0$.

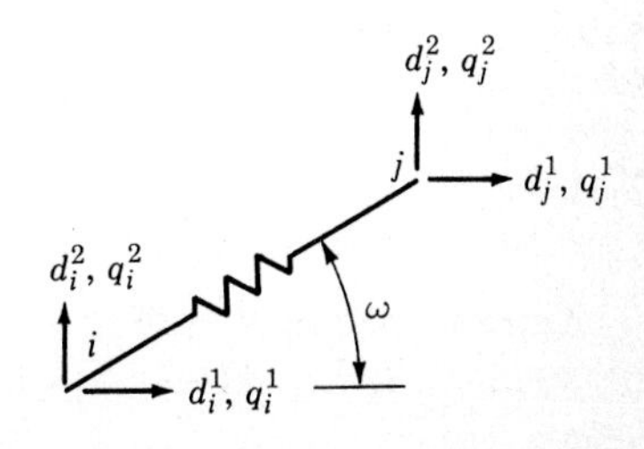

Fig. 14.10

The same procedure can be followed for the second equation in Eq. (14-8), using the equilibrium of horizontal forces at nodal point ③. The resulting stiffness matrix for the structure is

$$\mathbf{K} = \begin{bmatrix} 2.5 & 1.0 \\ 1.0 & 3.0 \end{bmatrix} \quad \begin{matrix} (D_2^2) & (D_3^1) \end{matrix}$$

The **K** matrix can also be developed by successively applying unit values of displacements D_2^2 and D_3^1 and thereby obtaining the columns of **K**. A displacement of $D_2^2 = 1$, with $D_3^1 = 0$, results in $d_{j(1)}{}^2 = 1$ and $d_{i(2)}{}^2 = 1$. For these element displacements we see from Eq. (14-7) that we obtain for the element forces $q_{j(1)}{}^2 = (ks^2)_{(1)}$ and $q_{i(2)}{}^2 = (ks^2)_{(2)}$. The sum of these two forces represents the value of k_{11}. When it is evaluated, we obtain the same value for k_{11} as above. The value of k_{21} is obtained by noting that for this displacement condition, a horizontal force is generated in spring 2 at nodal point ③ and is $q_{j(2)}{}^1 = (-ksc)_{(2)}$. The values in the second column can be obtained in a similar manner by letting $D_3^1 = 1$ with $D_2^2 = 0$.

Transformations of element-stiffness matrices can also be obtained by the use of virtual work, or other energy concepts, and the use of *displacement transformations*. To demonstrate this technique, let us again consider the pin-connected member of Fig. 14-7. For the member to be in equilibrium under the end forces q and the internal axial force P, the external virtual work must be equal to the internal virtual work; i.e.,

$$\delta W_E = \delta W_I$$

Denoting quantities in the local coordinate system with bars, let

$$\text{Internal displacements} \equiv \bar{\mathbf{d}} = \begin{Bmatrix} \Delta_i \\ \Delta_j \end{Bmatrix}$$

$$\text{Internal forces} \equiv \bar{\mathbf{q}} = \begin{Bmatrix} P_i \\ P_j \end{Bmatrix}$$

$$\text{Member stiffness} \equiv \bar{\mathbf{k}} = \frac{EA}{L}\begin{bmatrix} 1 & -1 \\ -1 & 1 \end{bmatrix}$$

$$\text{External displacements} \equiv \mathbf{d} = \begin{Bmatrix} d_i^1 \\ d_i^2 \\ d_j^1 \\ d_j^2 \end{Bmatrix}$$

$$\text{External forces} \equiv \mathbf{q} = \begin{Bmatrix} q_i^1 \\ q_i^2 \\ q_j^1 \\ q_j^2 \end{Bmatrix}$$

In terms of external and internal forces $\mathbf{q}$ and $\bar{\mathbf{q}}$, and external and internal virtual displacements $\delta\mathbf{d}$ and $\delta\bar{\mathbf{d}}$, the expression for equality of external and internal virtual work as developed in Sec. 8-5 can be written as

$$\mathbf{q}'\delta\mathbf{d} = \bar{\mathbf{q}}'\delta\bar{\mathbf{d}} \tag{14-11}$$

or $$\delta\mathbf{d}'\mathbf{q} = \delta\bar{\mathbf{d}}'\bar{\mathbf{q}} \tag{14-12}$$

From Fig. 14-2*a*,

$$\bar{\mathbf{q}} = \bar{\mathbf{k}}\bar{\mathbf{d}}$$

Therefore, the internal virtual work can be expressed as

$$\delta W_I = \delta\bar{\mathbf{d}}'\bar{\mathbf{k}}\bar{\mathbf{d}} \tag{14-13}$$

It is next necessary to express the internal displacements in terms of the external displacements. This can be done by introducing a *displacement transformation matrix* **A** which results in the relationship

$$\bar{\mathbf{d}} = \mathbf{A}\mathbf{d} \tag{14-14}$$

For the pinned-pinned member, this relationship is

$$\begin{Bmatrix} \Delta_i \\ \Delta_j \end{Bmatrix} = \underbrace{\begin{bmatrix} \cos\omega & \sin\omega & 0 & 0 \\ 0 & 0 & \cos\omega & \sin\omega \end{bmatrix}}_{\mathbf{A}} \begin{Bmatrix} d_i^1 \\ d_i^2 \\ d_j^1 \\ d_j^2 \end{Bmatrix} \tag{14-15}$$

The elements of **A** can be obtained as shown in Fig. 14-11.

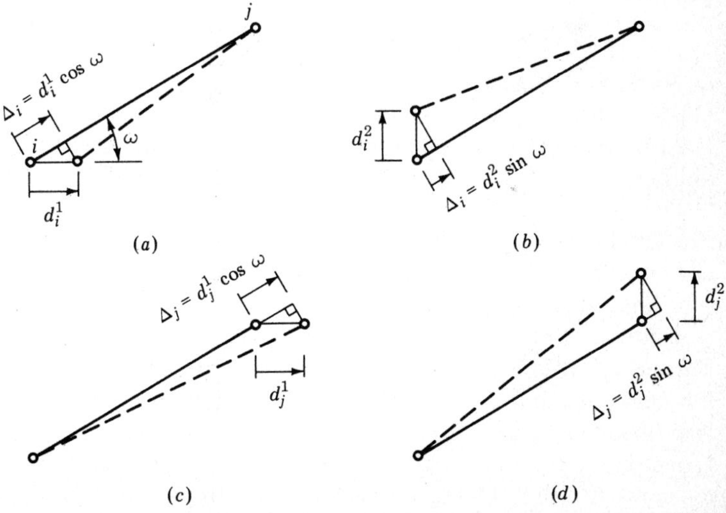

Fig. 14.11

In terms of the displacement transformation of Eq. (14-15), the internal work can be expressed as

$$\delta W_I = \delta\bar{\mathbf{d}}'\bar{\mathbf{k}}\mathbf{A}\mathbf{d}$$

Internal virtual displacements $\delta\bar{\mathbf{d}}$ are related to the external virtual displacements $\delta\mathbf{d}$ by the same $\mathbf{A}$ matrix; i.e.,

$$\delta\bar{\mathbf{d}} = \mathbf{A}\delta\mathbf{d}$$

or $\qquad \delta\bar{\mathbf{d}}' = \delta\mathbf{d}'\mathbf{A}'$

Therefore

$$\delta W_I = \delta\mathbf{d}'\mathbf{A}'\bar{\mathbf{k}}\mathbf{A}\mathbf{d}$$

and the equality of external and internal virtual work can be written as

$$\delta\mathbf{d}'\mathbf{q} = \delta\mathbf{d}'\mathbf{A}'\bar{\mathbf{k}}\mathbf{A}\mathbf{d} \tag{14-16}$$

The values of the virtual displacements $\delta\mathbf{d}$ are arbitrary. We can apply unit values of virtual displacements in the form of the identity matrix; i.e.,

$$\delta\mathbf{d} = \delta\mathbf{d}' = \mathbf{I} = \begin{bmatrix} 1 & 0 & 0 & 0 \\ 0 & 1 & 0 & 0 \\ 0 & 0 & 1 & 0 \\ 0 & 0 & 0 & 1 \end{bmatrix}$$

Equation (14-16) thereby becomes

$$\mathbf{q} = \mathbf{A}'\bar{\mathbf{k}}\mathbf{A}\mathbf{d} = \mathbf{k}\mathbf{d} \tag{14-17}$$

where $\qquad \mathbf{k} = \mathbf{A}'\bar{\mathbf{k}}\mathbf{A} \qquad$ (14-18)

The matrix $\mathbf{k}$ is the stiffness of the member in terms of global components of force and displacement. Performing the matrix operations of Eq. (14-18) with $\bar{\mathbf{k}}$ as given in Fig. 14-2*a* and the $\mathbf{A}$ matrix as given in Eq. (14-15), we obtain the same result for the pin-connected member as we did before in Eq. (14-7).

The expression for the transformed stiffness of a member as given in Eq. (14-18) is a general expression that can be applied to various types of members. It is necessary to know the stiffness of the member in local coordinates, $\bar{\mathbf{k}}$, and the displacement transformation matrix $\mathbf{A}$ which transforms global components of displacement into local displacements. Various forms of $\bar{\mathbf{k}}$ were developed earlier in the text in Sec. 9-4, and those for axial force and bending were summarized in Fig. 14-2. The determination of the elements of the displacement transformation matrix $\mathbf{A}$ can be generalized by considering the basic concept of coordinate transformations as shown in Fig. 14-12. For general purposes, the components in the global coordinate system are denoted x and y, and the components in the

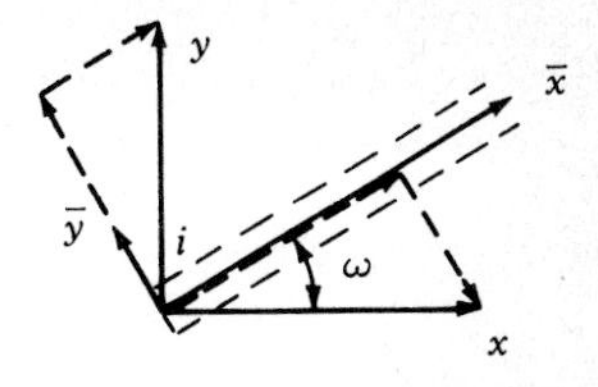

Fig. 14.12

local coordinate system are denoted $\bar{x}$ and $\bar{y}$. For the subject matter under consideration these represent displacements at the i end of the member. For a member inclined at an angle ω, we see from Fig. 14-12 that

$$\bar{x} = x \cos \omega + y \sin \omega$$
$$\bar{y} = -x \sin \omega + y \cos \omega$$

or, in matrix form,

$$\begin{Bmatrix} \bar{x} \\ \bar{y} \end{Bmatrix} = \begin{bmatrix} \cos \omega & \sin \omega \\ -\sin \omega & \cos \omega \end{bmatrix} \begin{Bmatrix} x \\ y \end{Bmatrix} \tag{14-19}$$

Rotation at the member end can be included directly as it is independent of ω. If we denote rotation by ϕ, the general transformation for both the i and j ends of the member can be written as

$$\begin{Bmatrix} \bar{x}_i \\ \bar{y}_i \\ \bar{\phi}_i \\ \bar{x}_j \\ \bar{y}_j \\ \bar{\phi}_j \end{Bmatrix} = \begin{bmatrix} \cos \omega & \sin \omega & 0 & 0 & 0 & 0 \\ -\sin \omega & \cos \omega & 0 & 0 & 0 & 0 \\ 0 & 0 & 1 & 0 & 0 & 0 \\ 0 & 0 & 0 & \cos \omega & \sin \omega & 0 \\ 0 & 0 & 0 & -\sin \omega & \cos \omega & 0 \\ 0 & 0 & 0 & 0 & 0 & 1 \end{bmatrix} \begin{Bmatrix} x_i \\ y_i \\ \phi_i \\ x_j \\ y_j \\ \phi_j \end{Bmatrix} \tag{14-20}$$

We see that the displacement transformation for a pin-connected member as given in Eq. (14-15) is a limited case of the transformation in Eq. (14-20), with the rows associated with $\bar{y}$ and $\bar{\phi}$ and the columns associated with ϕ eliminated.

The transformed stiffness matrix **k** for a two-dimensional member considering both axial and bending effects can be obtained using Eq. (14-18). The global components of displacement that are to be considered are those shown in Fig. 14-13. The $\bar{\mathbf{k}}$ matrix is given in Eq. (9-36), and the **A** matrix is the full transformation of Eq. (14-20). The transformed stiffness of a member in three dimensions can be obtained by use of the three-dimensional member stiffness in Eq. (9-35) and a modified three-dimensional form of the transformation of Eq. (14-20).

Before proceeding further, we should take note of certain details in evaluating the transformed element stiffnesses, particularly when developing a computer program. It is convenient, prior to the beginning of

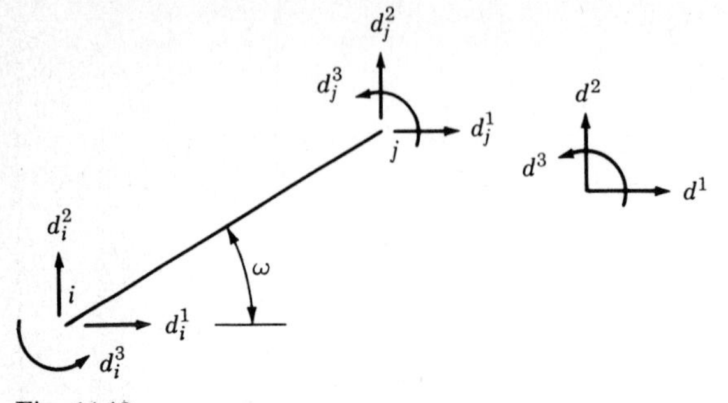

Fig. 14.13

the analysis, to express the location of the nodal points in terms of x and y coordinates that have a common origin. The length of the member can conveniently be obtained by using the values of the x and y coordinates of the member ends and the expression

$$L = [(x_j - x_i)^2 + (y_j - y_i)^2]^{1/2} \tag{14-21}$$

The orientation of the member is measured from the horizontal by the angle ω. The value of the member length can be used with the values of the x and y coordinates of the member ends to evaluate the sine and cosine terms; i.e.,

$$\sin \omega = \frac{y_j - y_i}{L} \qquad \cos \omega = \frac{x_j - x_i}{L} \tag{14-22}$$

The use of these expressions for evaluating $\sin \omega$ and $\cos \omega$ maintains the proper sign for the terms for various orientations of a member.

It is feasible in some cases to express the transformed member stiffness in a general form, as was done in Eq. (14-7) for a pin-connected member. For more involved situations the transformed stiffness can be obtained by the matrix operations of Eq. (14-18), using the values of $\bar{\mathbf{k}}$ and $\mathbf{A}$ as presented in the previous paragraphs.

14-5 GENERAL PROCEDURES FOR DEVELOPING THE K MATRIX

The procedure for developing the structure-stiffness matrix $\mathbf{K}$ by accumulating the individual element stiffnesses as demonstrated in the previous sections is generally referred to as the *direct method*. The method can be summarized by considering the equilibrium of some nodal point ⓘ. The nodal point is in equilibrium due to external forces Q and internal element forces q. Considering, for example, the X components of these forces, we can write the equilibrium equation

$$Q_i^X = q_{i(a)}^X + q_{i(b)}^X + \cdots \tag{14-23}$$

for all elements a, b, . . . meeting at ⓘ. For an element such as b, the element forces are related to the displacements at the nodal points of the element by the element-stiffness coefficients, i.e.,

$$q_{i(b)}^X = k_{iXj(b)}D_j + k_{iXk(b)}D_k + \cdots \tag{14-24}$$

for all j, k, . . . deflections D that influence the element force in the X direction at ⓘ. Thus it is seen that an appropriate summation of the element stiffnesses can be used to generate the **K** matrix.

The direct method of developing the **K** matrix can be used for structures containing different types of elements. Care must be taken to recognize whether or not all members contribute to all components of stiffness at a nodal point, even though the member element is connected to the point. The following example illustrates these points.

EXAMPLE 14-2 The structure stiffness matrix **K** is to be developed for the structure shown in Fig. 14-14. The structure is composed of a continuous member from nodal point ① to ③ and a pin-connected member from ② to ④. Axial and bending deformations are to be considered in members 1 and 2; only axial deformation is considered in member 3. We shall assume that the given values of the member properties have units that are consistent for the analysis.

There are three components of displacement, D_2^1, D_2^2, and D_2^3. They will be considered in the order stated. The axial stiffness and the bending stiffness of Fig. 14-2 can be used to obtain the structure stiffness. Considering first the equilibrium of nodal point ② in the horizontal direction and following the approach expressed in Eqs. (14-23) and (14-24), we

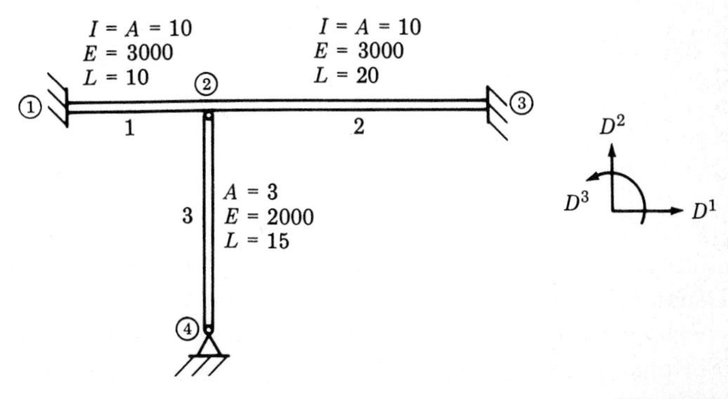

Fig. 14.14

obtain

$$Q_2{}^1 = q_{2(1)}{}^1 + q_{2(2)}{}^1$$

We note that there is no contribution for member 3, as it provides no stiffness for a horizontal movement of ②. In terms of member stiffnesses we obtain

$$\begin{aligned} Q_2{}^1 &= \left(\frac{EA}{L}\right)_{(1)} D_2{}^1 + \left(\frac{EA}{L}\right)_{(2)} D_2{}^1 \\ &= \frac{(3000)(10)}{10} D_2{}^1 + \frac{(3000)(10)}{20} D_2{}^1 \\ &= 4500 D_2{}^1 \end{aligned}$$

For equilibrium of nodal point ② in the vertical direction we obtain

$$Q_2{}^2 = q_{2(1)}{}^2 + q_{2(2)}{}^2 + q_{2(3)}{}^2$$

A term such as $q_{2(1)}{}^2$ is represented by V_j in Fig. 14-2b with $\eta_j = D_2{}^2$ and $\phi_j = D_2{}^3$. The axial stiffness of member 3 is considered for this displacement component. Therefore

$$\begin{aligned} Q_2{}^2 &= \left(\frac{12EI}{L^3} D_2{}^2 - \frac{6EI}{L^2} D_2{}^3\right)_{(1)} \\ &\qquad + \left(\frac{12EI}{L^3} D_2{}^2 + \frac{6EI}{L^2} D_2{}^3\right)_{(2)} + \left(\frac{EA}{L} D_2{}^2\right)_{(3)} \\ &= \frac{(12)(3000)(10)}{10^3} D_2{}^2 - \frac{(6)(3000)(10)}{10^2} D_2{}^3 + \frac{(12)(3000)(10)}{20^3} D_2{}^2 \\ &\qquad + \frac{(6)(3000)(10)}{20^2} D_2{}^3 + \frac{(2000)(3)}{15} D_2{}^2 \\ &= 805 D_2{}^2 - 1350 D_2{}^3 \end{aligned}$$

Proceeding in a similar manner with the equilibrium of moment about nodal point ②, we obtain

$$Q_2{}^3 = -1350 D_2{}^2 + 18000 D_2{}^3$$

The desired structure-stiffness matrix is therefore

$$\mathbf{K} = \begin{bmatrix} 4500 & 0 & 0 \\ 0 & 805 & -1350 \\ 0 & -1350 & 18000 \end{bmatrix} \quad \text{columns: } (D_2{}^1)\ (D_2{}^2)\ (D_2{}^3)$$

We see from the above example and by consideration of Eq. (14-24) that the process of forming the **K** matrix by the direct method lends itself well to being programmed for computer calculation. The process is one of placing element stiffnesses in their proper location in the **K** matrix. A convenient algorithm for formulating the **K** matrix is presented in the

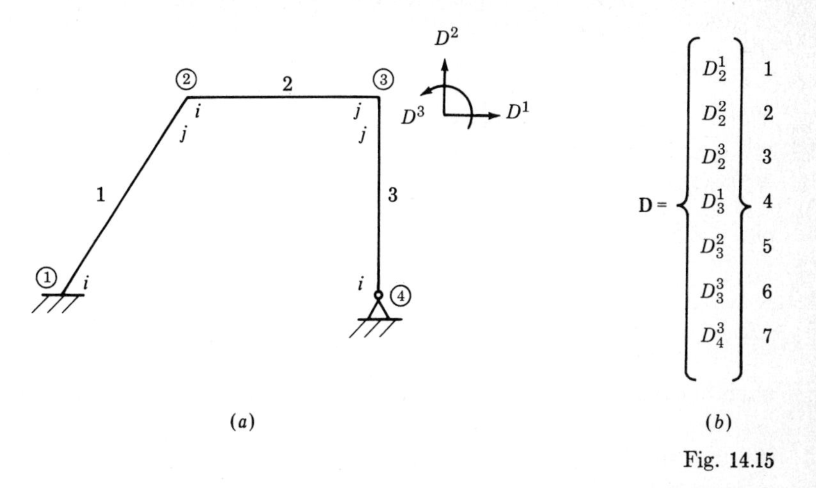

Fig. 14.15

following paragraphs. The technique is referred to as the *code-number technique.*[1] A Fortran program for the analysis of two-dimensional frames using the code-number technique is listed in Appendix B.

To demonstrate the technique, let us consider the frame of Fig. 14-15*a*. If we consider both axial and bending deformations in the members of the frame, the frame has seven components of external displacement. The external displacements, and the order in which they will be considered, are shown in Fig. 14-15*b*. The order of the external displacements is arbitrary but, as noted earlier, the elements of **Q** must be considered in the same order as that used for **D**. For purposes of using the code-number technique, the external displacements are numbered as they appear in the **D** matrix. We recall from previous developments that there is a correspondence between the external-displacement components and the member displacements if the member stiffnesses have been transformed into global components. The elements of the member displacement matrix for the type of member under consideration are shown in Fig. 14-16. These member displacements are numbered as shown. A code number is developed for each member of the structure. The number of positions in each code number is equal to the number of displacement components for the member. For a two-dimensional member, considering axial and bending deformation, there are six positions. By comparison, for a two-dimensional pin-connected member, there would be four positions. The code numbers for the frame of Fig. 14-15 are shown in Fig. 14-17. A zero in a code number indicates no displacement.

The **K** matrix is developed from the member code numbers. Each

[1] S. S. Tezcan, discussion of P. M. Wright, Simplified Formation of Stiffness Matrices, *J. Structural Div., ASCE,* vol. 89, no. ST6, December, 1963.

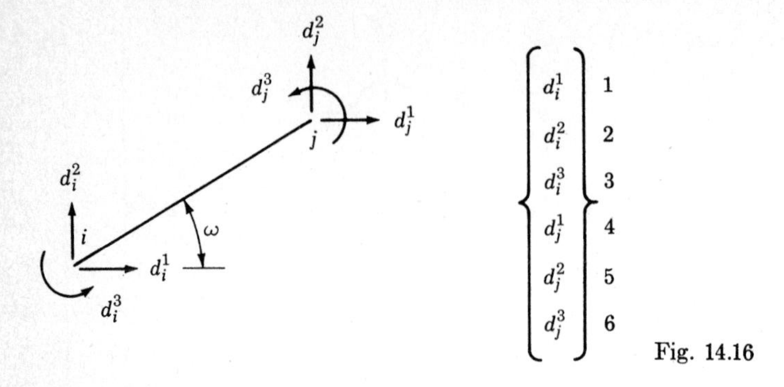

Fig. 14.16

member displacement number is coupled first with itself and then with all member displacement numbers to the right to provide the row-and-column subscript for each element to be taken from the member-stiffness matrix **k**. Each position value of the member code number is then coupled first with itself and then with all position values to the right to provide the row-and-column subscript indicating the position of the element in the **K** matrix. Zero values in the code numbers need not be considered. From the code number of Fig. 14-17 we see that the following operations would be performed for member 1:

Element of $\mathbf{k}_1$	Position in **K**
44	11
45	12
46	13
55	22
56	23
66	33

The same procedure is followed for members 2 and 3. The values in **K**

	Member displacement number					
	1	2	3	4	5	6
Member	d_i^1	d_i^2	d_i^3	d_j^1	d_j^2	d_j^3
1	0	0	0	1	2	3
2	1	2	3	4	5	6
3	0	0	7	4	5	6

Member code numbers Fig. 14.17

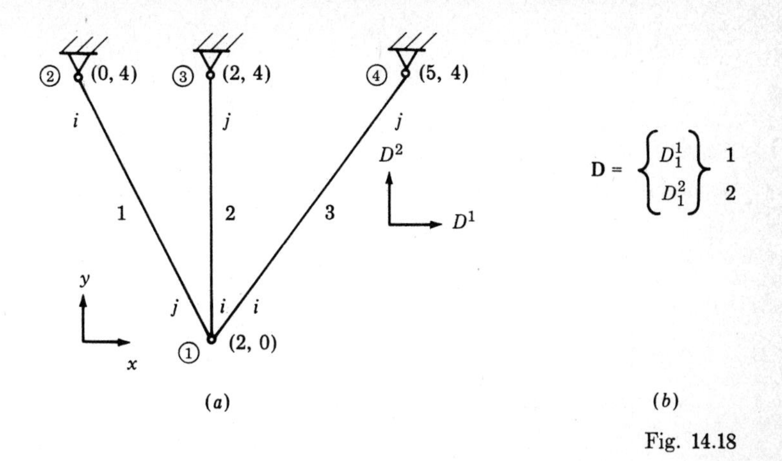

Fig. 14.18

are superimposed when necessary. The final form of the stiffness matrix **K** is obtained using its property of symmetry.

EXAMPLE 14-3 The structure stiffness matrix **K** is to be developed for the pin-connected truss shown in Fig. 14-18*a* using the code-number technique. E and A are the same for each member. The truss configuration is expressed in terms of x and y values of the nodal points as shown.

There are two components of external displacement; a horizontal and a vertical component at ①. The external-displacement matrix **D** is indicated in Fig. 14-18*b* with the elements numbered as shown.

As mentioned earlier, there are four positions in the code numbers for a two-dimensional, pin-connected member. The displacements and the transformed member stiffness have been indicated previously in Fig. 14-7 and Eq. (14-7), respectively. The code numbers for the truss of Fig. 14-18 are shown in Fig. 14-19.

	Member displacement number			
	1	2	3	4
Member	d_i^1	d_i^2	d_j^1	d_j^2
1	0	0	1	2
2	1	2	0	0
3	1	2	0	0

Member code numbers

Fig. 14.19

Using the code numbers of Fig. 14-19 and the general form of k_i in Eq. (14-7), we obtain the following:

Member	Element of $\mathbf{k}_i$	Value of element, $\times EA$	Position in $\mathbf{K}$
1	33	c_1^2/L_1	11
	34	s_1c_1/L_1	12
	44	s_1^2/L_1	22
2	11	c_2^2/L_2	11
	12	s_2c_2/L_2	12
	22	s_2^2/L_2	22
3	11	c_3^2/L_3	11
	12	s_3c_3/L_3	12
	22	s_3^2/L_3	22

With the expressions for L, sin ω, and cos ω of Eqs. (14-21) and (14-22), and the nodal-point coordinate values as given in Fig. 14-18, these operations result in the following values of $\mathbf{K}$:

$$\mathbf{K} = EA \begin{bmatrix} 0.1167 & 0.0066 \\ & 0.5569 \end{bmatrix}$$

The $\mathbf{K}$ matrix is completed by using its symmetric property; that is, $k_{21} = k_{12}$.

Once we have obtained the structure stiffness matrix $\mathbf{K}$, the displacements $\mathbf{D}$ corresponding to given external loads $\mathbf{Q}$ are obtained by solving the equations $\mathbf{Q} = \mathbf{KD}$. Techniques for solving sets of simultaneous equations are presented in Appendix A, and Fortran programs are listed in Appendix B.

The member forces, which are often the quantities of most interest, can be obtained by recalling that member forces are related to member displacements by the member stiffness; i.e.,

$$\mathbf{q}_i = \mathbf{k}_i\mathbf{d}_i$$

where i denotes the member number. The member displacements d correspond to particular elements of the calculated displacements $\mathbf{D}$. In the code-number technique, the member displacements d can conveniently be identified with the elements of $\mathbf{D}$ by the member code numbers.

In using the matrix displacement method to solve problems, attention must be given to the units that are used for the various terms. The most convenient approach is to use the same units throughout. For example, in the truss of Example 14-3 we could select units of inches and kips.

Thus, A would be in square inches, L in inches, E in kips per square inch, and Q in kips. The resulting displacements **D** would be in inches, and the axial forces in kips. In SI units we could select units of meters and newtons. Thus, A would be in square meters, L in meters, E in Pascals (N/m^2), and Q in newtons. The resulting displacements would be in meters, and the axial forces in newtons.

14-6 DISTRIBUTED AND INTERMEDIATE LOADS

In previous developments of the matrix displacement method, we have assumed that external loads were concentrated at the nodal points. Distributed loads and, if desired, intermediate concentrated loads on a member can be accounted for by the procedure shown in Fig. 14-20. The given distributed load condition is equivalent to the superposition of the conditions shown in Figs. 14-20*b* and *c*.[1] In Fig. 14-20*b*, no nodal-point displacements are permitted, and forces are evaluated that correspond to fixed-end conditions for the loaded member. The negatives of the fixed-end quantities are the nodal-point forces used in the analysis as shown in Fig. 14-20*c*. If adjacent members have fixed-end quantities, the net moments and forces are applied at the common nodal point. To obtain the true state of force in the member having distributed loading, the member values obtained from the analysis in Fig. 14-20*c* are superimposed with the fixed-end forces in Fig. 14-20*b*.

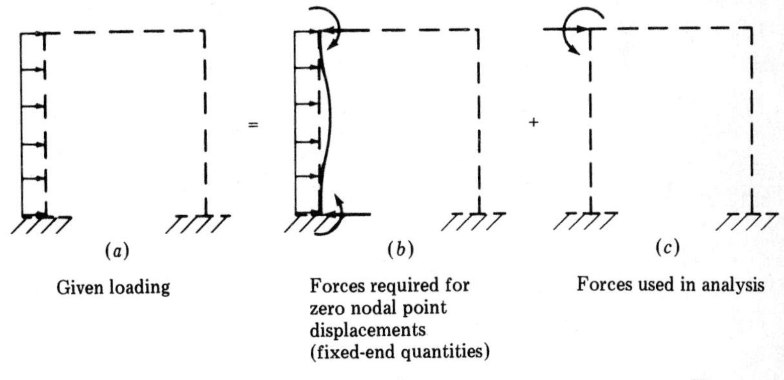

Fig. 14.20

[1] For a general proof of the use of equivalent concentrated forces, see J. S. Przemieniecki, 'Theory of Matrix Structural Analysis," McGraw-Hill Book Company, New York, 1968.

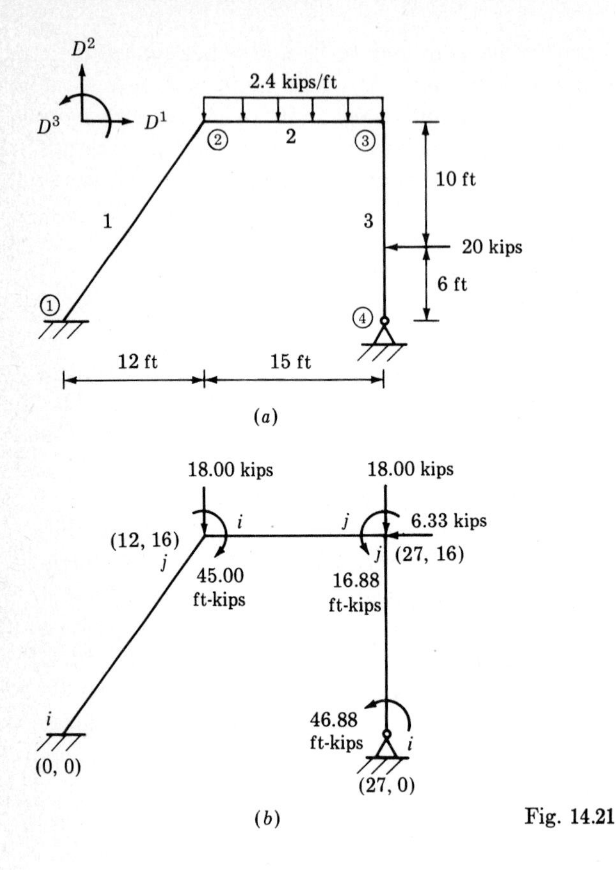

Fig. 14.21

EXAMPLE 14-4 The frame shown in Fig. 14-21*a* is to be analyzed using the matrix displacement method and the code-number technique. The computer program of Appendix B will be used to perform the calculations. For all members, $A = 20$ in.², $I = 100$ in.⁴, and $E = 30 \times 10^3$ ksi.

A nodal point could be introduced at the location of the 20-kip load; however, to demonstrate the use of fixed-end moments, we shall consider the load as an intermediate load on member 3. The magnitudes of the fixed-end moments for members 2 and 3 are:

$$\text{FEM}_{23} = \text{FEM}_{32} = \frac{(2.4)(15)^2}{12} = 45.00 \text{ ft-kips}$$

$$\text{FEM}_{34} = \frac{(20)(6)^2(10)}{16^2} = 28.12 \text{ ft-kips}$$

$$\text{FEM}_{43} = \frac{(20)(6)(10)^2}{16^2} = 46.88 \text{ ft-kips}$$

The loading for analysis is shown in Fig. 14-21*b*. The external displacements **D**, the external loads **Q**, and the code numbers for the members are shown in Fig. 14-22. Using the computer program of Appendix B, the

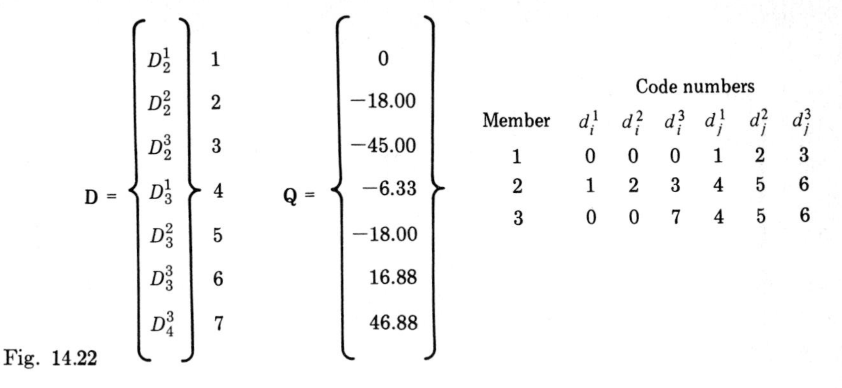

$$\mathbf{D} = \begin{Bmatrix} D_2^1 \\ D_2^2 \\ D_2^3 \\ D_3^1 \\ D_3^2 \\ D_3^3 \\ D_4^3 \end{Bmatrix} \begin{matrix} 1 \\ 2 \\ 3 \\ 4 \\ 5 \\ 6 \\ 7 \end{matrix} \qquad \mathbf{Q} = \begin{Bmatrix} 0 \\ -18.00 \\ -45.00 \\ -6.33 \\ -18.00 \\ 16.88 \\ 46.88 \end{Bmatrix}$$

	Code numbers					
Member	d_i^1	d_i^2	d_i^3	d_j^1	d_j^2	d_j^3
1	0	0	0	1	2	3
2	1	2	3	4	5	6
3	0	0	7	4	5	6

Fig. 14.22

following results are obtained:

$$\begin{Bmatrix} D_2{}^1 \\ D_2{}^2 \\ D_2{}^3 \\ D_3{}^1 \\ D_3{}^2 \\ D_3{}^3 \\ D_4{}^3 \end{Bmatrix} = \begin{Bmatrix} 0.02685 \\ -0.02091 \\ -0.00492 \\ 0.02656 \\ -0.00057 \\ 0.00126 \\ 0.00588 \end{Bmatrix} \text{ft and radian} \qquad \begin{Bmatrix} M_i \\ M_j \\ P_{j\ (1)} \\ \hdashline M_i \\ M_j \\ P_{j\ (2)} \\ \hdashline M_i \\ M_j \\ P_{j\ (3)} \end{Bmatrix} = \begin{Bmatrix} 0.38 \\ -9.87 \\ -18.43 \\ -35.13 \\ -17.97 \\ -11.44 \\ 46.88 \\ 34.85 \\ -21.54 \end{Bmatrix} \text{ft-kips and kips}$$

Positive values of member moments and axial force are as indicated in Fig. 14-2. Units of kips and feet were used in the solution. The given values of A (in.^{2}), I (in.^{4}), and E (ksi) were converted into values with units of square feet, feet^4, and kips per square feet, respectively, prior to the computer solution.

The true values of moment at the ends of members 2 and 3 are obtained by superimposing the fixed-end values with those obtained from the computer solution. These operations can be visualized as shown in Fig. 14-23. The member moments can be summarized in equation form as

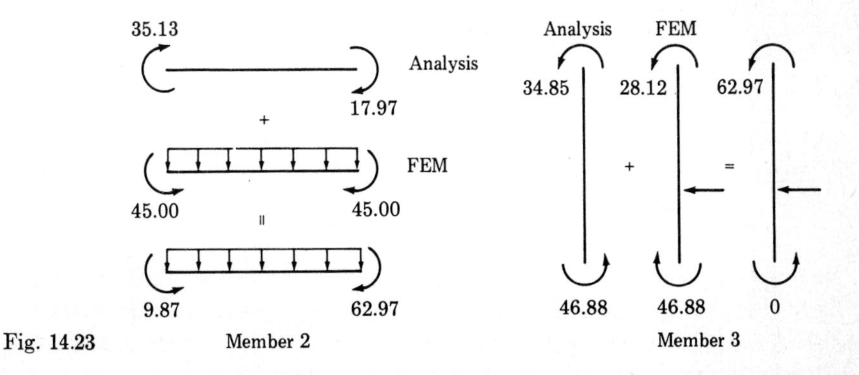

Fig. 14.23

follows:

$$\begin{Bmatrix} M_i \\ M_{j\ (1)} \\ M_i \\ M_{j\ (2)} \\ M_i \\ M_{j\ (3)} \end{Bmatrix} = \begin{Bmatrix} 0.38 \\ -9.87 \\ -35.13 \\ -17.97 \\ 46.88 \\ 34.85 \end{Bmatrix} \begin{matrix} \\ \\ +45.00 = 9.87 \\ -45.00 = -62.97 \\ -46.88 = 0 \\ +28.12 = 62.97 \end{matrix}$$

The axial forces in the members are given in the computer results above in terms of the force at the j end, P_j.

14-7 SUPPORT (BOUNDARY) CONDITIONS

In the use of the code-number technique in the previous section, the support, or boundary, conditions were satisfied by setting certain values of the code numbers equal to zero. The resulting structure-stiffness matrix **K** represents the displacements that are permissible for the structure. Such a procedure works well if possible displacements at support points coincide with the global components of displacement. For example, if a structure is supported by a roller that is permitted to displace in the horizontal direction such as point B in Fig. 14-24a, the horizontal displacement is treated as another degree of freedom when formulating the analysis.

For an inclined support condition such as that shown at point A in Fig. 14-24a, the possible displacement along the incline does not coincide with a global component of displacement. Such a condition can be represented in analysis by introducing an additional member as shown in Fig. 14-24b. The added member is given a very large axial stiffness and is oriented perpendicular to the incline. It thereby provides a reaction to

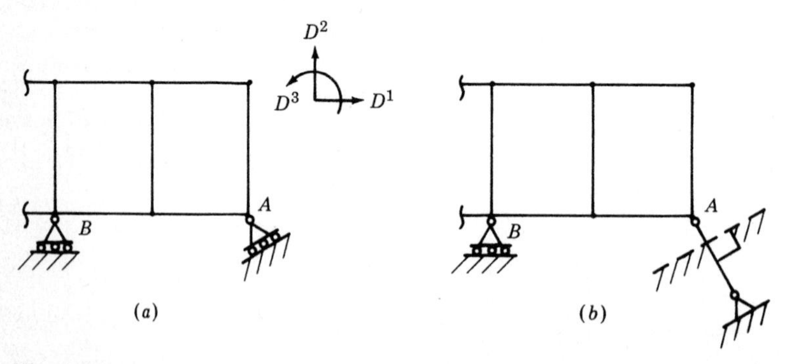

Fig. 14.24

the structure that is equivalent to the roller. Rather than use a pinned-pinned member as shown, it is possible to use an axial and bending type member and stipulate a very small value for its moment of inertia and a very large value for its axial stiffness.

In the development of the previous section, it is noted that the values of the reaction components were not obtained in the solution. If the values of reaction are desired, members such as that in Fig. 14-24*b* can be added to the structure and oriented in the direction of the desired reaction components. In more complex analyses, particularly in three dimensions, linear and rotational spring elements can be used to model support conditions.

An alternative approach to the development of the **K** matrix is to consider all elements without regard to support conditions; i.e., the **K** matrix is assembled in terms of all displacements that can occur at the unrestrained nodal points. Such a procedure was followed in the spring assembly of Fig. 14-4. The **K** matrix obtained in this manner represents the structure as a free body, and the **K** matrix is singular; i.e., it does not possess an inverse. A solution can be obtained by specifying that appropriate displacements be zero. The rows and columns of **K** corresponding to these displacements are eliminated, resulting in a **K** matrix that represents the structure in equilibrium.

14-8 BANDING OF THE K MATRIX

The **K** matrix for many types of structures can be developed such that nonzero values appear only in a band along the main diagonal. Such a matrix is referred to as a *banded matrix*. There are distinct advantages to having a banded **K** matrix. Less computer space is required to store only the elements of the band instead of the entire **K** matrix. Various efficient techniques have been developed to solve the governing simultaneous equations using only the elements of **K** that are in the band. Because of the symmetry of **K**, only the main-diagonal elements and the nonzero elements above or below the diagonal need be considered.

To see how a banded **K** matrix is obtained, let us consider the frame of Fig. 14-25*a*. Considering axial and bending deformations in the members, we would have three external displacements at each nodal point. The elements of **D** are placed in the order in which the nodal points are numbered. For the nodal points numbered as shown, we would obtain a stiffness relationship for the structure as shown in Fig. 14-25*b*. Nonzero values appear only in the shaded zone (band) of the **K** matrix. The reason for banding can be seen by considering a single external displace-

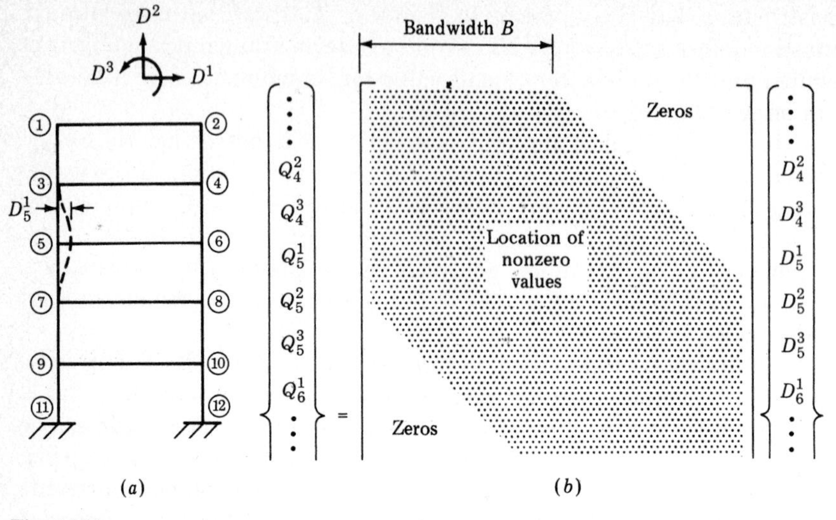

Fig. 14.25

ment imposed on the structure, such as $D_5{}^1$ in Fig. 14-25*a*. From the basic definition of a stiffness influence coefficient, the elements in the column of **K** corresponding to $D_5{}^1$ represent the forces required to maintain the configuration of Fig. 14-25*a*. We see that external forces are required only in the vicinity of the imposed displacement, i.e., at nodal points ③, ⑥, and ⑦. Thus, to obtain good banding, it is necessary to keep the nodal-point numbers as "near" each other as possible.

The value of the bandwidth B for a structure can be determined by considering the code numbers for the members. The value of B is equal to the largest difference between code numbers for a single member plus 1. Zero values in a code number are not considered. Thus, for the numbering system shown in Fig. 14-25*a*, $B = 9$, which can be obtained by considering any vertical member above the lower level. The value of B can also be obtained in some cases from the expression

$$B = (\Delta + 1)N \tag{14-25}$$

where Δ is the largest difference between any two external nodal-point numbers for a single element, and N is the number of degrees of freedom at each node. This expression is valid only if all nodal points in the sequence being checked have all N degrees of freedom.

A general guideline for obtaining good banding for a structure is to number the nodal points across the dimension of the structure that has the least nodal points. The value of $B = 9$ above should be compared to that if the nodal points for the structure of Fig. 14-25a were inappropriately numbered from top to bottom on the left side and then top to bottom on the right side. It would be found that $B = 18$ for such a numbering pattern.

14-9 DEVELOPMENT OF THE DISPLACEMENT METHOD USING VIRTUAL WORK

It is well to understand that the governing equations of the matrix displacement method can also be developed using energy methods. The energy techniques are particularly useful in developing concepts for complex finite-element analyses. The use of the virtual-work method will be discussed here.

As in previous developments of the displacement method, we define the external displacements at the nodal points by the column matrix $\mathbf{D}$, and the corresponding external loads by the column matrix $\mathbf{Q}$. The element stiffnesses $\mathbf{k}_i$ are assumed to be expressed in terms of global components. The subscript i denotes the element number. We next define a matrix $\mathbf{k}$ that is made up of all the $\mathbf{k}_i$ matrices in the structure. For a structure with N members, the $\mathbf{k}$ matrix would be

$$\mathbf{k} = \begin{bmatrix} \mathbf{k}_1 & \mathbf{0} & \mathbf{0} & \cdots & \mathbf{0} \\ \mathbf{0} & \mathbf{k}_2 & \mathbf{0} & \cdots & \mathbf{0} \\ \mathbf{0} & \mathbf{0} & \mathbf{k}_3 & \cdots & \mathbf{0} \\ \cdot & \cdot & \cdot & \cdots & \cdot \\ \mathbf{0} & \mathbf{0} & \mathbf{0} & \cdots & \mathbf{k}_N \end{bmatrix} \tag{14-26}$$

If we denote all member displacements by the column matrix $\mathbf{d}$, and the corresponding member forces by the column matrix $\mathbf{q}$, the member displacements are related to the member forces by the $\mathbf{k}$ matrix in the form

$$\mathbf{q} = \mathbf{kd} \tag{14-27}$$

The external displacements $\mathbf{D}$ can be related to the member displacements $\mathbf{d}$ by a displacement transformation matrix $\mathbf{A}$; that is,

$$\mathbf{d} = \mathbf{AD} \tag{14-28}$$

Because the member displacements $\mathbf{d}$ and the external displacements $\mathbf{D}$ are both expressed in global components, the $\mathbf{A}$ matrix contains only ones and zeros. To demonstrate this point, consider the truss of Fig. 14-18.

The transformation of Eq. (14-28) for the truss is

$$\begin{Bmatrix} d_i^1 \\ d_i^2 \\ d_j^1 \\ d_j^2 \;(1) \\ \hdashline d_i^1 \\ d_i^2 \\ d_j^1 \\ d_j^2 \;(2) \\ \hdashline d_i^1 \\ d_i^2 \\ d_j^1 \\ d_j^2 \;(3) \end{Bmatrix} = \begin{bmatrix} 0 & 0 \\ 0 & 0 \\ 1 & 0 \\ 0 & 1 \\ 1 & 0 \\ 0 & 1 \\ 0 & 0 \\ 0 & 0 \\ 1 & 0 \\ 0 & 1 \\ 0 & 0 \\ 0 & 0 \end{bmatrix} \begin{Bmatrix} D_1^1 \\ D_1^2 \end{Bmatrix}$$

The first column of **A** represents member displacements due to an application of D_1^1 only; the second column represents member displacements due to an application of D_1^2 only.

Using the principal of virtual work, the expression for equality of external and internal virtual work due to the application of external virtual displacements $\boldsymbol{\delta}\mathbf{D}$ and the corresponding internal virtual displacements $\boldsymbol{\delta}\mathbf{d}$ can be written as

$$\boldsymbol{\delta}\mathbf{D}'\mathbf{Q} = \boldsymbol{\delta}\mathbf{d}'\mathbf{q}$$

In terms of **q** as given in Eq. (14-27), this becomes

$$\boldsymbol{\delta}\mathbf{D}'\mathbf{Q} = \boldsymbol{\delta}\mathbf{d}'\mathbf{k}\mathbf{d}$$

Using the displacement transformation **A** as defined in Eq. (14-28) for both real and virtual displacements, we obtain

$$\boldsymbol{\delta}\mathbf{D}'\mathbf{Q} = \boldsymbol{\delta}\mathbf{D}'\mathbf{A}'\mathbf{k}\mathbf{A}\mathbf{D}$$

A set of virtual displacements in the form of an identity matrix **I** can be used; this results in

$$\mathbf{Q} = \mathbf{A}'\mathbf{k}\mathbf{A}\mathbf{D} = \mathbf{K}\mathbf{D}$$

$$\text{where} \qquad \mathbf{K} = \mathbf{A}'\mathbf{k}\mathbf{A} \tag{14-29}$$

K, as before, is the structure-stiffness matrix. The internal member forces **q** can be expressed in terms of the calculated external displacements **D** as

$$\mathbf{q} = \mathbf{k}\mathbf{A}\mathbf{D}$$

We see that the evaluation of **K** by the matrix operations of Eq. (14-29) would be inefficient because of the number of zeros in the **k** and **A** matrices.

The process of superimposing, or summing, the element stiffnesses as in the direct method of developing **K** is equivalent to the operations of Eq. (14-29). The code-number technique presented earlier provides an efficient method for performing the necessary operations.

14-10 TEMPERATURE AND LACK-OF-FIT ANALYSES

The effects of temperature change in elements and lack of fit in connecting elements can be considered through slight modifications of previously developed concepts. Let us consider a truss for illustrative purposes, and denote the change in temperature in a member by ΔT. The change in length e of a member, if the member is free to deflect, is

$$e = \alpha L \, \Delta T$$

where α is the coefficient of thermal expansion for the member material, and L is the length of the member. The axial force developed at the ends of the member, when displacement at the ends is prevented, is

$$P = \frac{EA}{L} e = EA\alpha \, \Delta T$$

In terms of the end forces as shown in Fig. 14-2a, this becomes

$$P_i = EA\alpha \, \Delta T$$
$$P_j = -EA\alpha \, \Delta T$$

These axial forces can be transformed into global components of force q at the member ends by using the same procedure as in Sec. 14-4 to develop Eq. (14-17). We would find for temperature considerations that the transformed member forces are given by

$$\mathbf{q} = \mathbf{A}'\bar{\mathbf{q}}$$

where $\bar{\mathbf{q}}$ represents the axial forces P_1 and P_2 given above. **A** for a pinned-pinned member is given in Eq. (14-15). The end forces q can be assembled into a column matrix $\mathbf{Q}_T$ that represents all the forces at the nodal points due to temperature changes in the members. The governing matrix equation for the structure thereby becomes:

$$\mathbf{Q} = \mathbf{KD} + \mathbf{Q}_T$$

With no external loads **Q**, the set of simultaneous equations is solved with known values of $\mathbf{Q}_T$. If external loads **Q** and temperature forces $\mathbf{Q}_T$ are both present, the matrices **Q** and $\mathbf{Q}_T$ are combined and the simultaneous equations are solved for **D**.

Lack-of-fit problems can be considered in a similar manner by setting e in the above expression equal to the amount of misfit. The resulting axial forces at the ends are transformed into global components and combined into nodal-point forces as was done for $\mathbf{Q}_T$ above.

Consideration of temperature and lack of fit can be applied to other types of elements by extending the procedures developed above for a pinned-pinned member.

14-11 EVALUATION OF INFLUENCE LINES

The matrix displacement method can be used to evaluate influence lines for complex structural systems. Recall the basic definition of an influence line from Chap. 10; the influence line for the moment at A for a unit load moving from B to C in Fig. 14-26 could be obtained by applying a unit load successively to the points from B to C. The external load vector $\mathbf{Q}$ for each position would be all zeros except for a value of one in the appropriate position. The procedure could be carried out in one operation by using a load matrix in the form

$$\mathbf{Q} = \begin{bmatrix} 1 & 0 & \cdots \\ 0 & 1 & \cdots \\ 0 & 0 & \cdots \\ \cdots & \cdots & \cdots \end{bmatrix}$$

Depending on the flexibility of the computer program being used and the size of the structure, this may not be a very efficient approach.

Another approach is to use the Müller-Breslau principle for statically indeterminate structures. To illustrate this approach, consider the continuous beam of Fig. 14-27. The influence line for the reaction at ①, R_1, is to be evaluated for a unit load moving from ② to ③. The position of the unit load on the beam is denoted by nodal point ⓧ. If the effects of R_1 are removed and then applied as a unit force, the requirement for

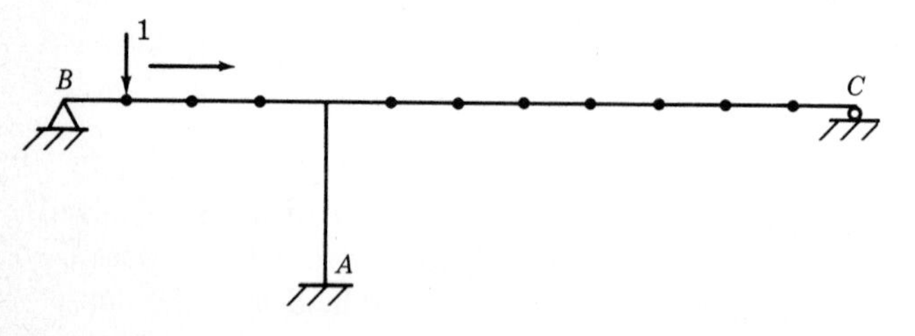

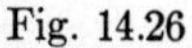
Fig. 14.26

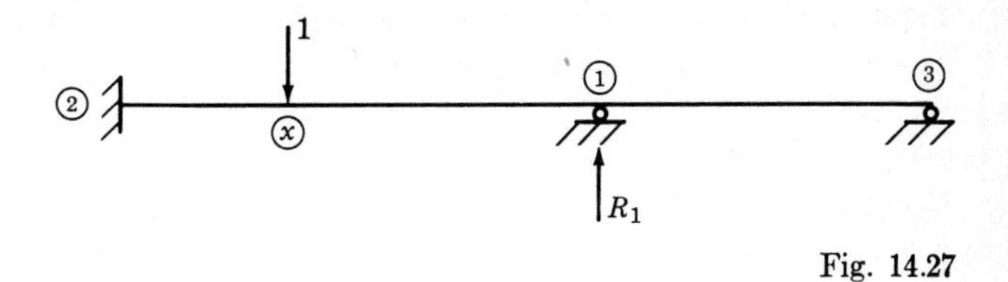

Fig. 14.27

consistent displacement at ① is

$$D_1^Q + f_{11}R_1 = 0$$

Forces and displacements are considered positive in the upward direction. Therefore, in terms of the unit load at some position ⓧ, we can write

$$f_{1x}(-1) + f_{11}R_1 = 0$$

or $$R_1 = \frac{f_{1x}}{f_{11}}(1)$$

However, from Maxwell's reciprocal theorem, $f_{1x} = f_{x1}$. Therefore,

$$R_1 = \frac{f_{x1}}{f_{11}}$$

f_{x1} is the deflection at ⓧ due to $R_1 = 1$; f_{11} is the deflection at ① due to $R_1 = 1$. The matrix displacement method can conveniently be used to evaluate these deflections. To do so, a unit value of R_1 is applied to the structure, and the vertical deflections at prescribed points on the beam are determined as well as the deflection corresponding to the unit value of R_1. The values of the influence line are therefore the values of the vertical displacement at the prescribed points divided by the displacement associated with $R_1 = 1$.

EXAMPLE 14-5 The influence line for the vertical reaction at D, for a unit load moving from A to B, is to be determined for the structure shown in Fig. 14-28a. $I = 100$ in.^{4}, $A = 20$ in.^{2}, and $E = 30 \times 10^3$ ksi.

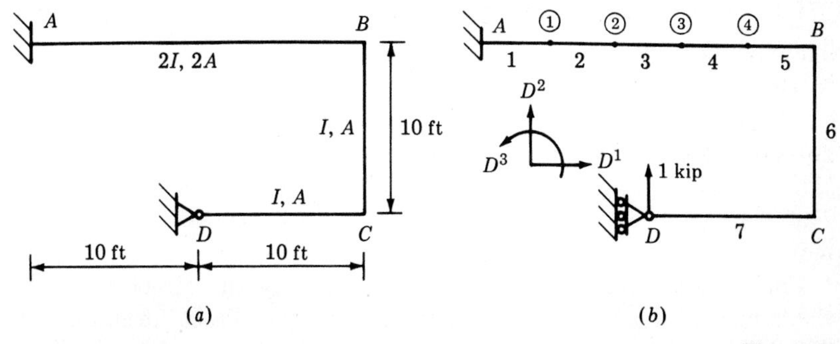

Fig. 14.28

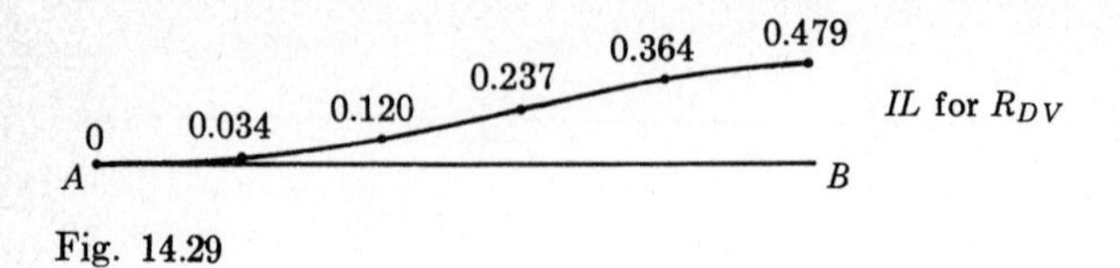

Fig. 14.29

The effects of the vertical reaction at D are removed, and a unit value of force is applied as shown in Fig. 14-28b. Length AB is divided into five equal segments. The following values of displacement are obtained from a computer solution of the structure in Fig. 14-28b:

$$\begin{Bmatrix} D_D{}^2 \\ D_1{}^2 \\ D_2{}^2 \\ D_3{}^2 \\ D_4{}^2 \\ D_B{}^2 \end{Bmatrix} = \begin{Bmatrix} 7.102 \\ 0.238 \\ 0.851 \\ 1.684 \\ 2.585 \\ 3.399 \end{Bmatrix} \times 10^{-2} \text{ ft}$$

Other nodal-point displacements are not listed here. Using these displacement values and the expression

$$R_{DV} = \frac{f_{xD}}{f_{DD}}$$

we obtain the influence line for R_{DV} as shown in Fig. 14-29.

When this technique is used for the construction of influence lines for quantities such as moment, care must be taken to use the total relative rotation of the member ends at the pin for the term in the denominator if this point is an interior point on the structure.

14-12 AN ANALOGOUS APPROACH— THE MATRIX FORCE METHOD

The matrix force method is a method of analysis that is similar in form to the matrix displacement method although the approach is different. Of the two methods, the matrix displacement method has become the most often used, particularly in finite-element analyses. There are, however, occasions where the matrix force method can be used effectively. The basic concepts of the matrix force method will be presented in this section.

The matrix force method is characterized by the direct formation of the structure-flexibility matrix rather than the direct formation of the structure-stiffness matrix as in the matrix displacement method. Use is made of element-flexibility matrices, as discussed in Chap. 9, rather than element stiffnesses.

The method will be developed here using virtual work and, for illustrative purposes, we shall consider the pin-connected truss shown in Fig. 14-30. The flexibility relationship for an axially loaded member as shown previously in Fig. 9-12 can be obtained from Eq. (9-14) and written as

$$e = \frac{L}{EA} P \tag{14-30}$$

P is the constant axial force in the member. For equilibrium of the member, $P_2 = -P_1 = P$. The net change in the length of the member is denoted by e, which is equal to $\Delta_2 - \Delta_1$.

We define a matrix **f**, similar to the **k** matrix defined in Eq. (14-26), that relates all member forces to their corresponding displacements in one expression. Thus, for the truss of Fig. 14-30, the **f** matrix is written as

$$\mathbf{f} = \begin{bmatrix} \mathbf{f}_1 & \mathbf{0} & \mathbf{0} & \mathbf{0} & \mathbf{0} \\ \mathbf{0} & \mathbf{f}_2 & \mathbf{0} & \mathbf{0} & \mathbf{0} \\ \mathbf{0} & \mathbf{0} & \mathbf{f}_3 & \mathbf{0} & \mathbf{0} \\ \mathbf{0} & \mathbf{0} & \mathbf{0} & \mathbf{f}_4 & \mathbf{0} \\ \mathbf{0} & \mathbf{0} & \mathbf{0} & \mathbf{0} & \mathbf{f}_5 \end{bmatrix}$$

The submatrices $\mathbf{f}_i$ are 1×1 matrices representing the flexibility in Eq. (14-30) of a pinned-pinned member. The complete relation between member forces **q** and member displacements **d** can therefore be written as

$$\mathbf{d} = \mathbf{f}\mathbf{q} \tag{14-31}$$

If, as discussed in Sec. 8-5, the external and internal forces are considered to be virtual forces, and if the displacements given to the structure are the real displacements of the structure for some condition of loading, then the statement of equality of external and internal virtual work can

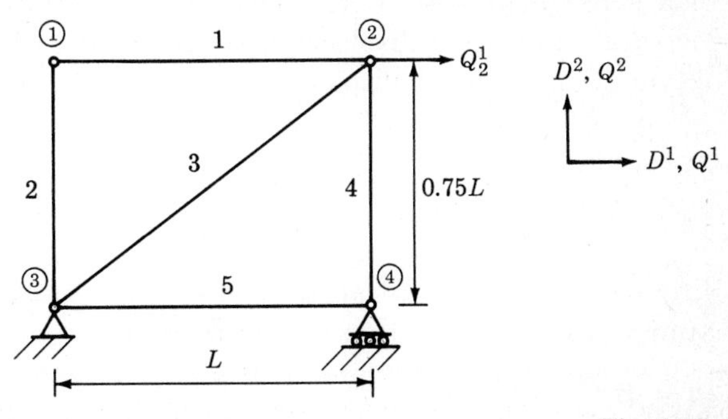

Fig. 14.30

be written as

$$\delta\mathbf{Q}'\mathbf{D} = \delta\mathbf{q}'\mathbf{d} \tag{14-32}$$

where $\delta\mathbf{Q}$ and $\delta\mathbf{q}$ represent external and internal virtual forces, respectively. In this form the expression for internal virtual work is

$$\delta W_I = \delta\mathbf{q}'\mathbf{d}$$

which, with the expression for **d** of Eq. (14-31), becomes

$$\delta W_I = \delta\mathbf{q}'\mathbf{f}\mathbf{q} \tag{14-33}$$

The internal work can be described in terms of external forces by introducing a force transformation matrix **B**, which relates external forces to internal forces in the form

$$\mathbf{q} = \mathbf{B}\mathbf{Q} \tag{14-34}$$

The elements of **B** can be thought of as influence coefficients in that the *i*th column of **B** determines the member forces **q** resulting from a unit value of the *i*th element of **Q**, all other external forces being zero. For the truss under consideration there is one nonzero external load Q_2^1. From statics, the force transformation is found to be

$$\begin{Bmatrix} P_1 \\ P_2 \\ P_3 \\ P_4 \\ P_5 \end{Bmatrix} = \begin{bmatrix} 0 \\ 0 \\ 1.25 \\ -0.75 \\ 0 \end{bmatrix} \{Q_2^1\}$$

It should be noted at this point that a **B** matrix could be developed for all possible external components of loading, which would result in a 5×5 **B** matrix for the truss rather than the 5×1 matrix above. We shall subsequently see that to use the **B** matrix above we obtain only the value of external displacement associated with Q_2^1, whereas we could obtain all nodal-point displacements if we used the 5×5 **B** matrix.

From Eq. (14-34) we find that the expression for internal virtual work of Eq. (14-33) is

$$\delta W_I = \delta\mathbf{q}'\mathbf{f}\mathbf{B}\mathbf{Q} \tag{14-35}$$

Internal virtual forces can be related to external virtual forces by the same force transformation matrix **B**; that is,

$$\delta\mathbf{q} = \mathbf{B}\delta\mathbf{Q}$$

or $\quad \delta\mathbf{q}' = \delta\mathbf{Q}'\mathbf{B}'$

Equation (14-35) can therefore be written as

$$\delta W_I = \delta \mathbf{Q}'\mathbf{B}'\mathbf{f}\mathbf{B}\mathbf{Q}$$

The condition for equality of external and internal virtual work of Eq. (14-32) becomes

$$\delta \mathbf{Q}'\mathbf{D} = \delta \mathbf{Q}'\mathbf{B}'\mathbf{f}\mathbf{B}\mathbf{Q} \tag{14-36}$$

The external virtual forces $\delta\mathbf{Q}$ can be applied in the form of successive unit values, as was done with the external virtual displacements in the displacement method. This entails the stipulation that

$$\delta \mathbf{Q} = \delta \mathbf{Q}' = \mathbf{I}$$

For this value of $\delta\mathbf{Q}'$, we find that Eq. (14-36) and the relation between external forces $\mathbf{Q}$ and external displacements $\mathbf{D}$ become

$$\mathbf{D} = \mathbf{B}'\mathbf{f}\mathbf{B}\mathbf{Q} \tag{14-37}$$

or $$\mathbf{D} = \mathbf{F}\mathbf{Q} \tag{14-38}$$

where $$\mathbf{F} = \mathbf{B}'\mathbf{f}\mathbf{B} \tag{14-39}$$

Matrix $\mathbf{F}$ is the flexibility matrix of the structure. $\mathbf{F}$ was determined previously in the displacement method by inverting the structure-stiffness matrix $\mathbf{K}$. The internal member displacements can be determined in terms of $\mathbf{f}$ and $\mathbf{B}$ for any value of $\mathbf{Q}$ from the relationship

$$\mathbf{d} = \mathbf{f}\mathbf{B}\mathbf{Q} \tag{14-40}$$

To demonstrate the application of the matrix force method to the analysis of statically indeterminate structures, let us consider the truss of Fig. 14-31*a*. The truss is statically indeterminate to the first degree. If we select the axial force in member 6 as the redundant, we obtain the primary structure as shown in Fig. 14-31*b*. The primary structure is loaded with the given external load Q_2^1 and the redundant force X_6. A $\mathbf{B}$ matrix can be developed that relates the external load Q_2^1 and the

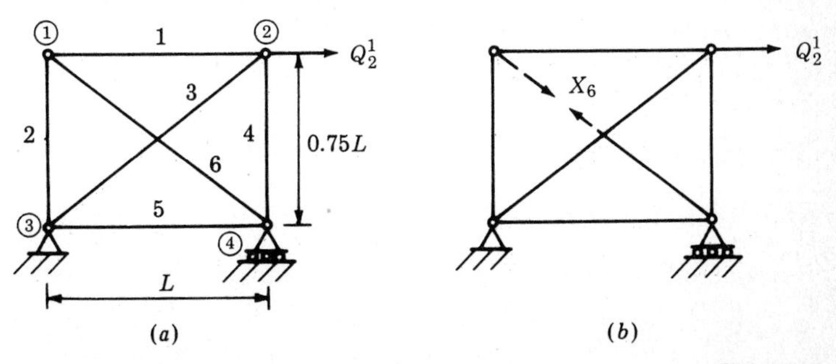

Fig. 14.31

redundant force X_6 to the member axial forces in the partitioned form:

$$\begin{Bmatrix} P_1 \\ P_2 \\ P_3 \\ P_4 \\ P_5 \\ P_6 \end{Bmatrix} = \left[\begin{array}{c|c} 0 & -0.8 \\ 0 & -0.6 \\ 1.25 & 1.0 \\ -0.75 & -0.6 \\ 0 & -0.8 \\ 0 & 1.0 \end{array}\right] \begin{Bmatrix} Q_2{}^1 \\ \hline X_6 \end{Bmatrix}$$

The force transformation can be expressed in general form as

$$\mathbf{q} = [\mathbf{B}_1 \mid \mathbf{B}_2] \begin{Bmatrix} \mathbf{Q} \\ \hline \mathbf{X} \end{Bmatrix} \tag{14-41}$$

where $\mathbf{B}_1$ is associated with the external loads, and $\mathbf{B}_2$ is associated with the redundants $\mathbf{X}$. Using Eq. (14-37), the general relation between external loads and external displacements can be written as

$$\begin{Bmatrix} \mathbf{D} \\ \hline \mathbf{0} \end{Bmatrix} = \begin{bmatrix} \mathbf{B}_1' \\ \hline \mathbf{B}_2' \end{bmatrix} [\mathbf{f}][\mathbf{B}_1 \mid \mathbf{B}_2] \begin{Bmatrix} \mathbf{Q} \\ \hline \mathbf{X} \end{Bmatrix} \tag{14-42}$$

The condition that there be zero displacements associated with the redundant forces in the given structure is satisfied by setting these displacements equal to zero in Eq. (14-42). A more convenient form of Eq. (14-42) is

$$\begin{Bmatrix} \mathbf{D} \\ \hline \mathbf{0} \end{Bmatrix} = \left[\begin{array}{c|c} \mathbf{E}_{11} & \mathbf{E}_{12} \\ \hline \mathbf{E}_{21} & \mathbf{E}_{22} \end{array}\right] \begin{Bmatrix} \mathbf{Q} \\ \hline \mathbf{X} \end{Bmatrix} \tag{14-43}$$

where

$$\mathbf{E}_{11} = \mathbf{B}_1'\mathbf{f}\mathbf{B}_1 \tag{14-44}$$
$$\mathbf{E}_{12} = \mathbf{B}_1'\mathbf{f}\mathbf{B}_2 \tag{14-45}$$
$$\mathbf{E}_{21} = \mathbf{B}_2'\mathbf{f}\mathbf{B}_1 \tag{14-46}$$
$$\mathbf{E}_{22} = \mathbf{B}_2'\mathbf{f}\mathbf{B}_2 \tag{14-47}$$

The values of the redundants can be found in terms of the given loads from Eq. (14-43) and the partitioned matrix relation associated with zero displacements. Thus,

$$\mathbf{X} = -\mathbf{E}_{22}{}^{-1}\mathbf{E}_{21}\mathbf{Q} \tag{14-48}$$

The expression for $\mathbf{D}$ as given in Eq. (14-43) can therefore be written as

$$\mathbf{D} = (\mathbf{E}_{11} - \mathbf{E}_{12}\mathbf{E}_{22}{}^{-1}\mathbf{E}_{21})\mathbf{Q} \tag{14-49}$$

The flexibility matrix for the statically indeterminate structure is therefore

$$\mathbf{F} = \mathbf{E}_{11} - \mathbf{E}_{12}\mathbf{E}_{22}{}^{-1}\mathbf{E}_{21} \tag{14-50}$$

The expression for member forces is found from Eqs. (14-41) and (14-48), which result in

$$\mathbf{q} = (\mathbf{B}_1 - \mathbf{B}_2\mathbf{E}_{22}^{-1}\mathbf{E}_{21})\mathbf{Q} \tag{14-51}$$

The force method as discussed above with reference to a truss can be applied in a similar manner to other types of structures. The **f** matrix would be modified to reflect the type of element flexibilities that are to be considered.

A review of the equations developed for the matrix displacement method in Sec. 14-9 and for the matrix force method in this section, particularly Eqs. (14-29) and (14-39), will verify that there is a similarity in the types of expressions in the two methods of analysis. There are, however, distinct differences in the two approaches. One distinguishing difference is the use of a displacement transformation matrix **A** in the displacement method and the use of a force transformation matrix **B** in the force method. The ease with which these transformations can be developed, or the ease with which **K** and **F** can be generated, is a major factor in the choice of methods. Another distinguishing difference in the two methods is the need to solve the simultaneous stiffness equations in the governing equation of the displacement method. The development of efficient techniques for generating the **K** matrix as presented earlier in this chapter and the availability of high-speed computers and improved techniques for solving simultaneous equations have resulted in popular use of the matrix displacement method.

PROBLEMS

14-1. For the spring assembly shown, (*a*) remove each spring from the assembly, and sketch its deflected shape corresponding to a positive value of external displacement $D_1{}^X$, with all other displacements equal to zero; (*b*) for the displacement $D_1{}^X = 1$ in (*a*), define the values of the internal displacements d as shown in Fig. 14-10; (*c*) use the displacements defined in (*b*) and the stiffness relationship of Eq. (14-7) with EA/L replaced by the spring stiffnesses k to determine the element forces q acting on each spring; (*d*) indicate these element forces q on the sketches of the deflected springs; (*e*) repeat steps (*a*) through (*d*) for a positive unit value of external displacement $D_1{}^Y$; (*f*) indicate how each of the applicable element forces q acts on nodal point ①; (*g*) write the equilibrium equations for forces acting in the X and Y directions on nodal point ① in terms of the element forces q of (*f*) and possible X and Y components of external forces Q_1^X and Q_1^Y acting at ①; (*h*) assemble the results in the form of the stiffness matrix **K**.

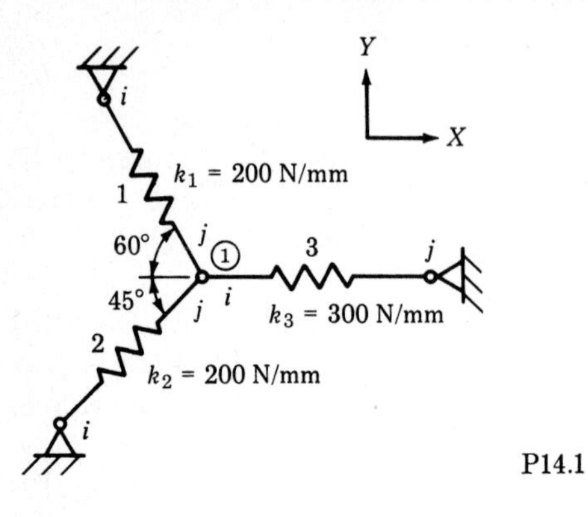

P14.1

14-2. Repeat Prob. 14-1 for the pin-connected truss shown. The external displacements to be considered are D_2^2 and D_3^1. For each member, $A = 2$ in.² and $E = 10 \times 10^3$ ksi. Express the results in units of kips per inch.

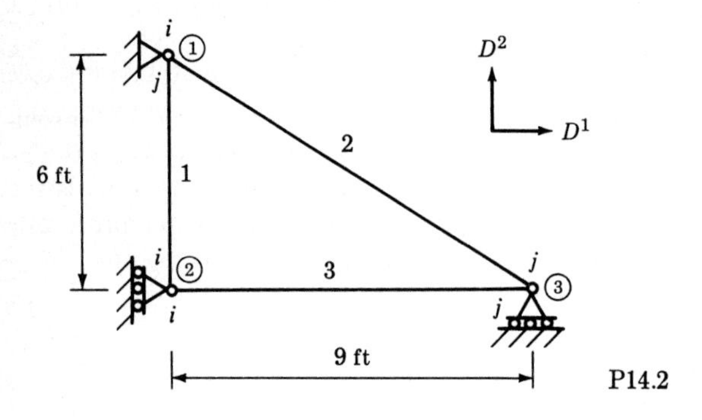

P14.2

14-3. Axial and bending deformations are to be considered in member 3 of the structure shown; only axial deformation is to be considered in members 1 and 2. Determine the stiffness matrix **K** for the structure. For members 1 and 2, $A = 1.5$ in.² and $E = 30 \times 10^3$ ksi. For member 3, $I = 200$ in.⁴, $A = 15$ in.², and $E = 1.5 \times 10^3$ ksi.

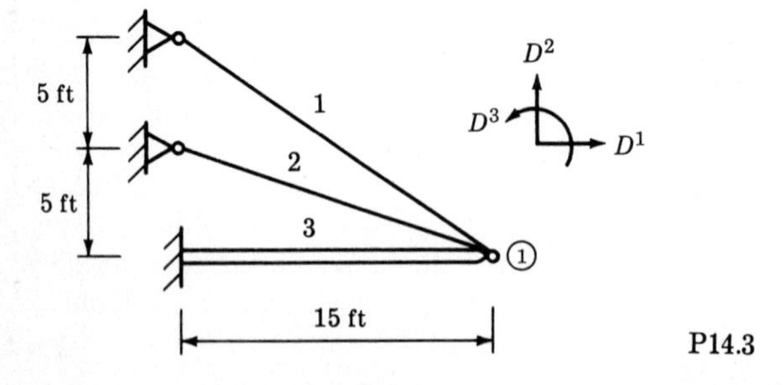

P14.3

14-4. Determine the stiffness matrix **K** for the structure shown. $I = 100 \times 10^6$ mm⁴, $A = 15 \times 10^3$ mm², and $E = 200$ GPa.

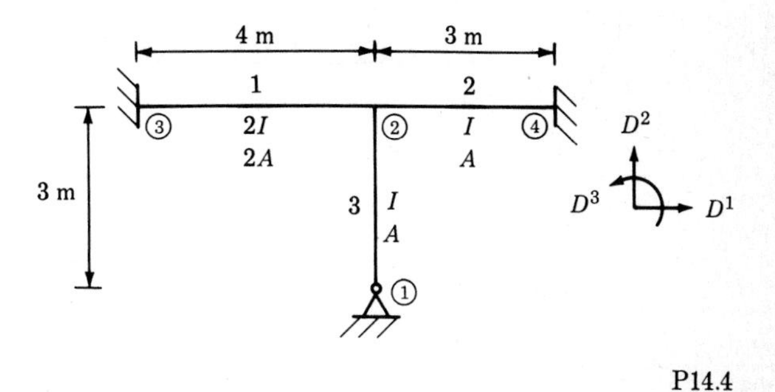

P14.4

14-5. For a matrix displacement analysis of the frame shown, (*a*) determine the value of the external load vector **Q**; (*b*) develop the code numbers for the members; (*c*) determine the value of the bandwidth B; (*d*) if computer facilities are available, analyze the frame using the computer program of Appendix B or another program that is available; (*e*) where necessary, adjust the member moments obtained from the computer solution for the fixed-end moment conditions used in establishing the external loads **Q**. Axial and bending deformations are to be considered in the members. Consider the external displacements in the order shown. For each member, $I = 200 \times 10^6$ mm⁴, $A = 25 \times 10^3$ mm², and $E = 200$ GPa.

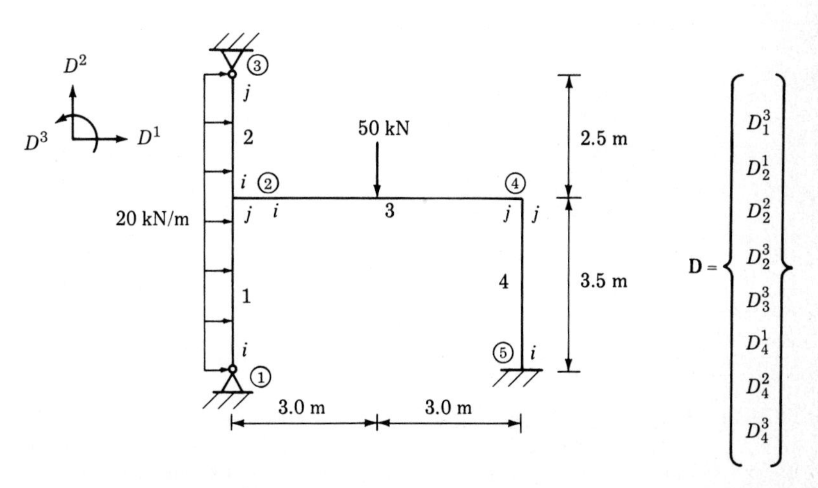

P14.5

14-6. Repeat Prob. 14-5 for the frame shown. For each member, $I = 200$ in.^{4}, $A = 20$ in.^{2}, and $E = 30 \times 10^3$ ksi.

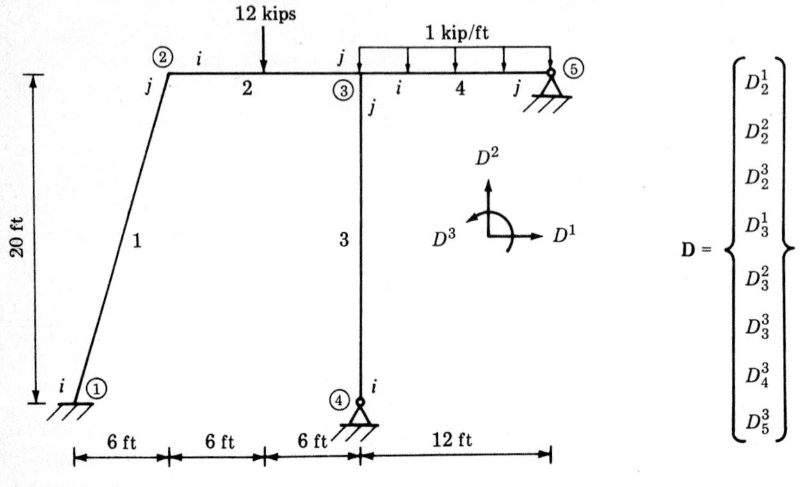

P14.6

14-7. Repeat Prob. 14-5 for the frame shown. $I = 100$ in.^{4}, $A = 10$ in.^{2}, and $E = 30 \times 10^3$ ksi. Consider the displacements in the order in which the nodal points are numbered.

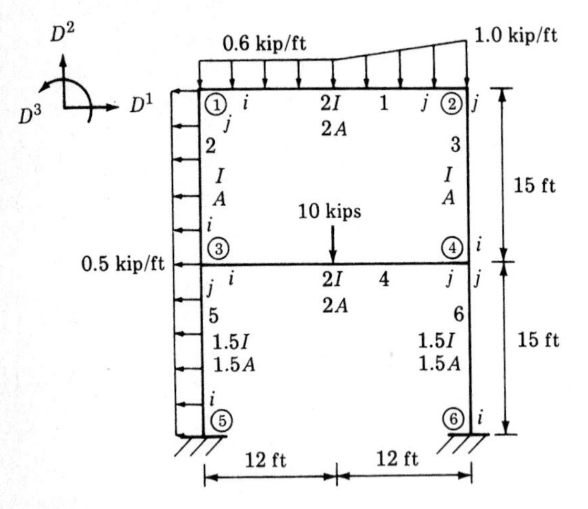

P14.7

14-8. Analyze the frame of Fig. P12-7 using the matrix displacement method and the computer program of Appendix B or another program that is available. Let $I = 500$ in.^{4}, $A = 0.1$ times the numerical value of I, in inches squared, and $E = 10 \times 10^3$ ksi. Construct the moment diagram for the frame on the tension side of the members.

14-9. Analyze the frame of Fig. 13-12 using the matrix displacement method and the computer program of Appendix B or another program that is available. Let $I = 100 \times 10^6$ mm^4, $A = 0.0001$ times the numerical value of I, in millimeters squared, and $E = 200$ GPa. Construct the moment diagram for the frame on the tension side of the members.

14-10. Repeat Prob. 14-5 for the frame shown. $I = 100$ in.^{4}, $A = 10$ in.^{2}, and $E = 30 \times 10^3$ ksi. Note that the ends of member 3 are pinned to continuous members at nodal points ② and ④. The pinned conditions can be represented in the analysis by establishing two separate external rotations at ② and ④, where one rotation represents the rotation of the continuous member at the nodal point, and the other rotation represents the possible rotation of the end of member 3. Or, because there are no transverse loads on member 3, a small value of I can be assumed for member 3 and the rotations at the ends considered to be the same as those for the continuous members. Consider the displacements in the order in which the nodal points are numbered.

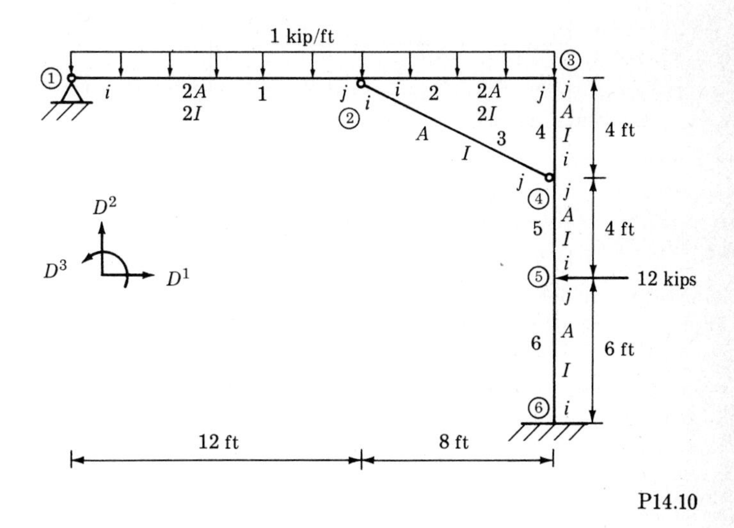

P14.10

14-11. The overpass structure shown is to be analyzed for a standard truck loading using the matrix displacement method. Axial and bending deformations are to be considered in the members. In the analysis, (*a*) number the nodal points such that the minimum bandwidth is obtained for **K**; (*b*) analyze the structure using the computer program of Appendix B or another program that is available; (*c*) construct the moment diagram for the structure on the tension side of the members. For members 1 through 8, $I = 5000$ in.^{4} and $A = 50$ in.^{2}. For members 9 through 12, $I = 2000$ in.^{4} and $A = 20$ in.^{2}. $E = 30 \times 10^3$ ksi for all members.

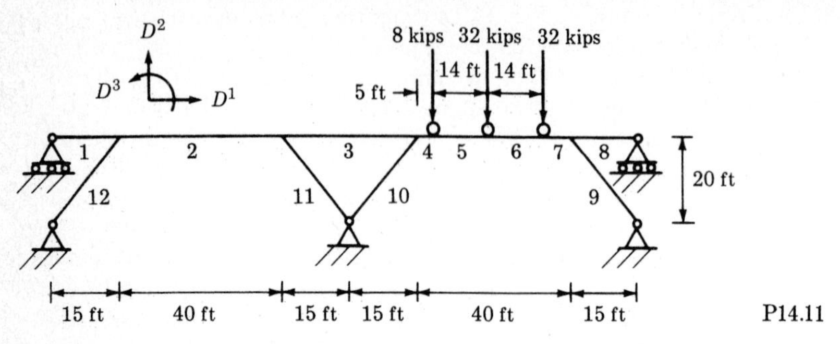

P14.11

14-12. Repeat Prob. 14-11 for the structure shown. Construct the moment diagram for members 1, 4, and 7 on the tension side of the members. Assume for each member that $I = 200$ in.^{4}, $A = 10$ in.^{2}, and $E = 30 \times 10^3$ ksi.

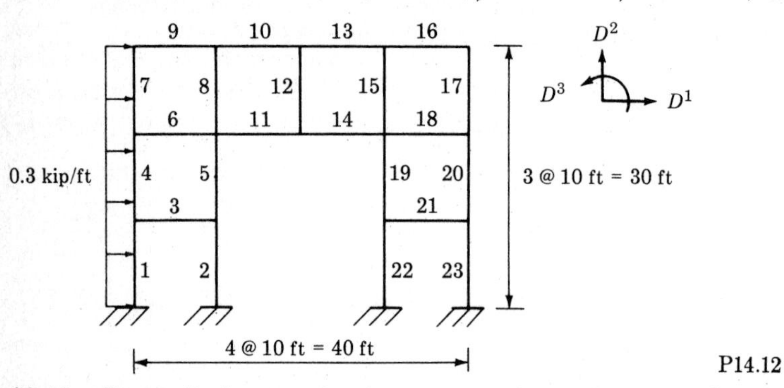

P14.12

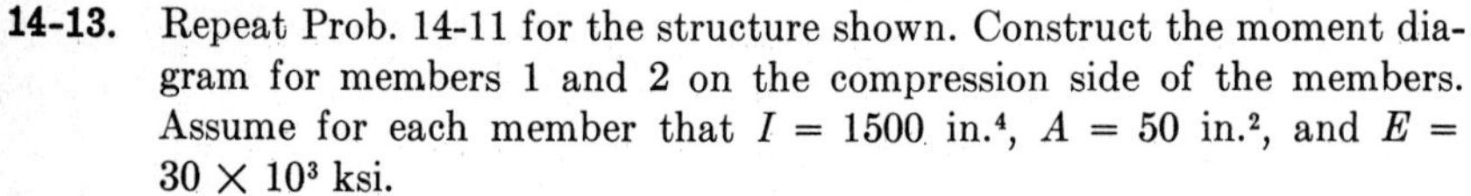

14-13. Repeat Prob. 14-11 for the structure shown. Construct the moment diagram for members 1 and 2 on the compression side of the members. Assume for each member that $I = 1500$ in.^{4}, $A = 50$ in.^{2}, and $E = 30 \times 10^3$ ksi.

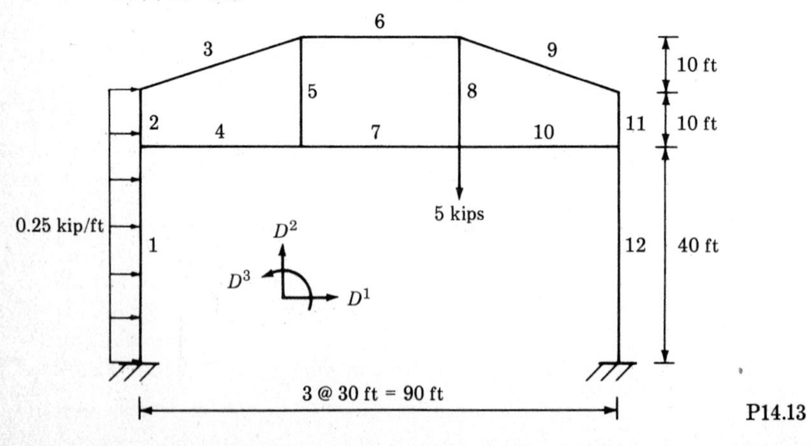

P14.13

14-14. Analyze the frame shown using the matrix displacement method. The support condition at ⑥ can be modeled according to the suggestions of Sec. 14-6. For each member, $I = 600$ in.^{4}, $A = 20$ in.^{2}, and $E = 30 \times 10^3$ ksi.

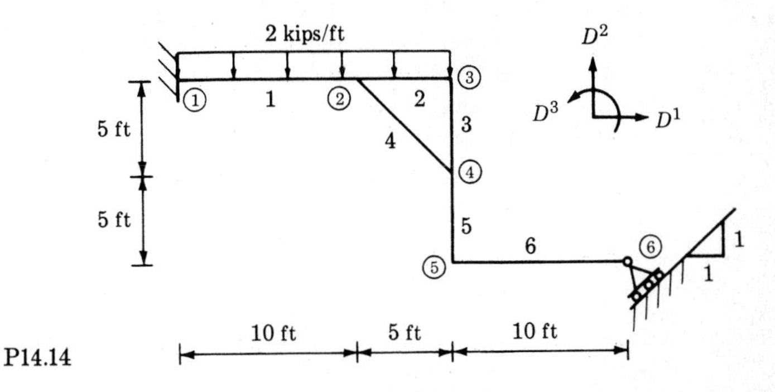

P14.14

14-15. For a unit load moving from ① to ⑤ on the beam shown, develop the influence line for the reaction at ⑤. Determine the value of the influence line at the points indicated, and plot the results. EI is constant. Note that for such a beam the deflection calculations for use with the Müller-Breslau principle can be performed manually using a method such as moment-area.

①　②　③　④　⑤

4 @ 3 ft = 12 ft

P14.15

14-16. For a unit load moving from A to C on the structure shown, develop the influence lines for (*a*) the horizontal component of reaction at D; (*b*) the moment in member BD just below point B. The computer program of Appendix B and the Müller-Breslau principle can be used to develop the influence lines. Determine the value of the influence lines every 5 ft along AC, and plot your results. E is constant. Assume an appropriate value for I, and assign A a large value corresponding to the condition of neglecting axial deformations.

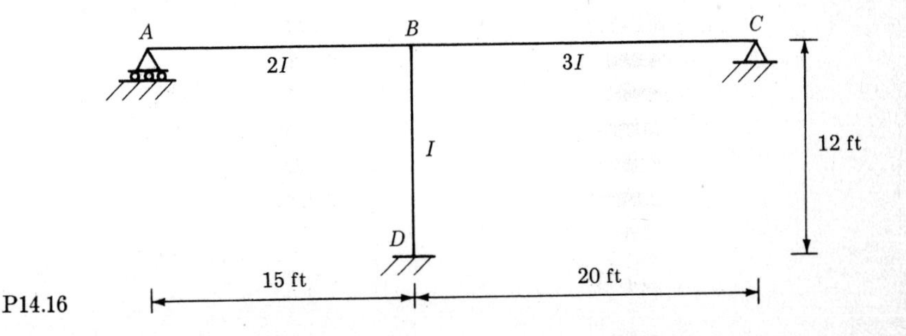

P14.16

14-17. For the beam of Prob. 10-15, develop the influence lines for (*a*) the reaction at C; (*b*) the moment at A. Let $L = 8$ ft, and assume appropriate values for E, I, and A. Divide the length L into four equal segments for the analysis.

14-18. For the beam of Prob. 10-16, develop the influence lines for (*a*) the reaction at B; (*b*) the moment at C. Let $L = 4$ m, and assume appropriate values for E, I, and A. Divide the length L into four equal segments for the analysis.

14-19 and 14-20. Analyze the frames of Probs. 6-7 and 6-8 using the matrix displacement method, and compare the results to those obtained by the approximate methods of analysis. Assume that I is the same for each member.

CHAPTER

FIFTEEN

AN INTRODUCTION TO THE FINITE-ELEMENT METHOD

15-1 THE BASIC CONCEPT

The determination of stresses and displacements for continuous structural systems such as shells, walls, and footings can be accomplished by representing the given structural system by a finite number of elements and performing an analysis on the finite-element model using the matrix techniques presented in the previous chapter. For example, the vertical plate structure of Fig. 15-1*a* can be analyzed for the given loading by representing the given structure by a series of quadrilateral finite elements, as shown in Fig. 15-1*b*. As shown in the inset of Fig. 15-1*b*, the elements are connected to each other at their corner points. The connection points, as before, are referred to as *nodal points* of the structure. The structure is thus made up of individual elements, just as the structures considered in the previous chapter were made up of individual members. This similarity has led to the interpretation by some that the wording "finite-element method" can include both conventional structures and the process described here. The wording "finite-element method" as used here signifies the process of dividing a continuous structural system into finite elements for purposes of analysis.

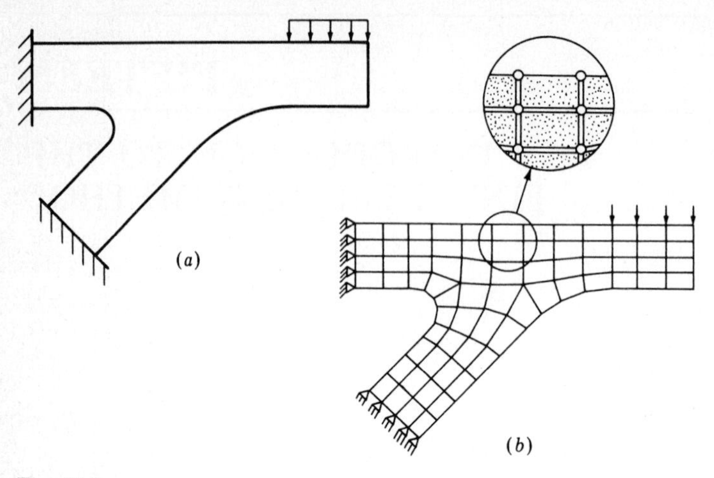

Fig. 15.1

The finite-element method can be applied to a wide variety of problems. Two possible uses are shown in Fig. 15-2. In Fig. 15-2*a*, a flat plate is represented for analysis by a series of quadrilateral finite elements. The plate elements are formulated such that they represent the bending properties of a plate. The elements are connected to each other at nodal points as indicated by the dots in Fig. 15-2*a*. The problem of analyzing for stresses in a buried pipe and in the surrounding soil, as shown in Fig. 15-2*b*, lends itself well to a solution by the finite-element method. The pipe wall and the soil in the vicinity of the pipe are modeled by a series of finite elements. The difference in material properties of the soil and the pipe is readily accounted for by establishing nodal points on the

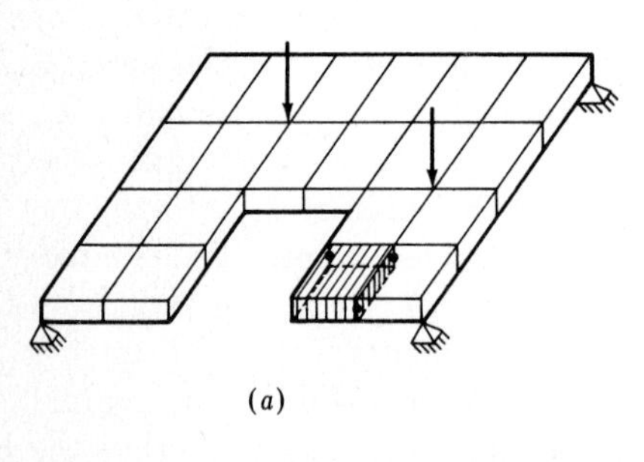

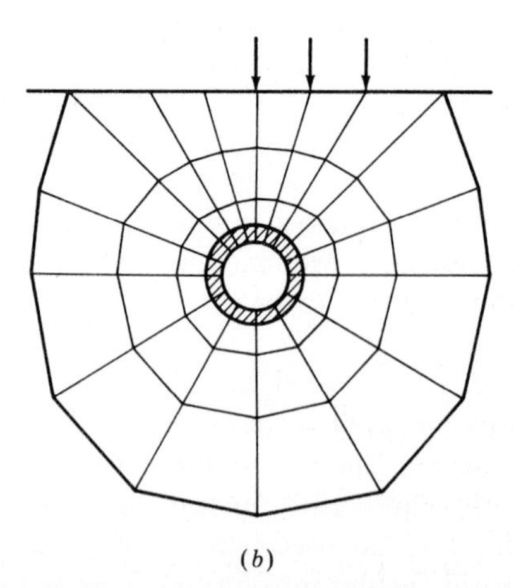

Fig. 15.2

surface between the pipe and the soil and using two different types of elements, one representing the material properties of the pipe and the other representing the material properties of the soil.

Various shapes of elements are used. The use of a particular shape is dependent on the type of problem that is being considered and the degree of refinement to which the element has been developed.

The finite-element technique can also be used in other areas such as seepage flow, heat conduction, electrical networks, and fluid dynamics. The presentation here will pertain only to structural systems, and it is intended to be only an introduction to the method. For comprehensive treatments of the finite-element method, the reader is referred to texts devoted entirely to the subject.[1]

15-2 DEVELOPMENT OF A PLANE-STRESS ELEMENT

An essential part of the finite-element method is the development of the element stiffnesses. To demonstrate the basic concepts of developing an element stiffness, we shall consider a plane-stress triangular element. The developments presented here are based on the original presentation of the finite-element method.[2] More sophisticated elements have been developed in recent years, but the triangular element presented here will serve to demonstrate the basic procedure for developing an element stiffness.

Consider the triangular element shown in Fig. 15-3. The nodal points of the element are denoted i, j, and k, and their locations are denoted by the coordinates x and y. The x component of displacement at any point on the element is denoted by u, and the y component of displacement is denoted by v. The stiffness of the element is to be developed in terms of the nodal-point displacements u_i, u_j, u_k, v_i, v_j, and v_k. These nodal-point displacements will be denoted by the column matrix **d**.

[1] K. J. Bathe and E. L. Wilson, "Numerical Methods in Finite Element Analysis," Prentice-Hall, Inc., Englewood Cliffs, N.J., 1976; R. D. Cook, "Concepts and Applications of Finite Element Analysis," John Wiley & Sons, Inc., New York, 1974; C. S. Desai and J. F. Abel, "Introduction to the Finite Element Method," Van Nostrand Reinhold Co., New York, 1972; R. H. Gallagher, "Finite Element Analysis Fundamentals," Prentice-Hall, Inc., Englewood Cliffs, N.J., 1975; O. C. Zienkiewicz, "The Finite Element Method in Engineering Science," McGraw-Hill Publishing Company Limited, London, 1971. See also appropriate sections in J. S. Przemieniecki, "Theory of Matrix Structural Analysis," McGraw-Hill Book Company, New York, 1968.

[2] M. J. Turner, R. W. Clough, H. C. Martin, and L. J. Topp, Stiffness and Deflection Analysis of Complex Structures, *J. Aero. Sci.*, vol. 23, no. 9, pp. 805–823, 1956.

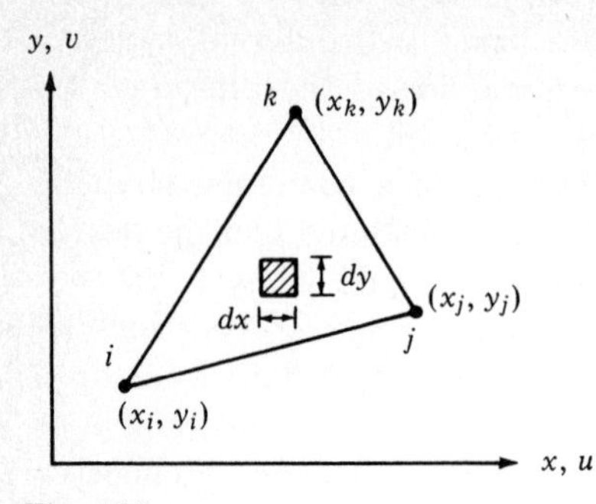

Fig. 15.3

For a differential element of the triangle as shown in Fig. 15-3, we can write the relationship between stress and strain from Eq. (8-22) as

$$\begin{Bmatrix} \epsilon_x \\ \epsilon_y \\ \gamma_{xy} \end{Bmatrix} = \frac{1}{E} \begin{bmatrix} 1 & -\nu & 0 \\ -\nu & 1 & 0 \\ 0 & 0 & 2(1+\nu) \end{bmatrix} \begin{Bmatrix} \sigma_x \\ \sigma_y \\ \tau_{xy} \end{Bmatrix} \tag{15-1}$$

We note in Eq. (8-22) that for plane stress, $\sigma_z = \tau_{yz} = \tau_{zx} = 0$. Equation (15-1) can be inverted to obtain the stiffness relationship for the differential element,

$$\begin{Bmatrix} \sigma_x \\ \sigma_y \\ \tau_{xy} \end{Bmatrix} = \frac{E}{1-\nu^2} \begin{bmatrix} 1 & \nu & 0 \\ \nu & 1 & 0 \\ 0 & 0 & \dfrac{(1-\nu)}{2} \end{bmatrix} \begin{Bmatrix} \epsilon_x \\ \epsilon_y \\ \gamma_{xy} \end{Bmatrix} \tag{15-2}$$

which can be written as

$$\boldsymbol{\sigma} = \bar{\mathbf{k}}\boldsymbol{\epsilon} \tag{15-3}$$

where
$$\bar{\mathbf{k}} = \frac{E}{1-\nu^2} \begin{bmatrix} 1 & \nu & 0 \\ \nu & 1 & 0 \\ 0 & 0 & \dfrac{(1-\nu)}{2} \end{bmatrix} \tag{15-4}$$

It is necessary to express the strains $\boldsymbol{\epsilon}$ of Eq. (15-2) in terms of the displacements at the nodal points $\mathbf{d}$ by means of a displacement transformation $\mathbf{A}$; that is,

$$\boldsymbol{\epsilon} = \mathbf{A}\mathbf{d} \tag{15-5}$$

To accomplish this, we assume displacement functions of the form

$$u = C_1 + C_2x + C_3y \tag{15-6a}$$
$$v = C_4 + C_5x + C_6y \tag{15-6b}$$

The selection of appropriate displacement functions is a major aspect of developing finite elements. To fully appreciate the concept of displacement functions, the reader is encouraged to study the presentations found in the finite-element texts cited earlier. It should be noted that for the functions assumed in Eqs. (15-6) there are six unknown constants C, which is equal to the number of nodal-point displacements. The displacement functions can therefore be expressed in terms of the displacements at nodal points i, j, and k. From Eq. (15-6a) we obtain

$$\begin{aligned} u_i &= C_1 + C_2 x_i + C_3 y_i \\ u_j &= C_1 + C_2 x_j + C_3 y_j \\ u_k &= C_1 + C_2 x_k + C_3 y_k \end{aligned}$$

or
$$\begin{Bmatrix} u_i \\ u_j \\ u_k \end{Bmatrix} = \begin{bmatrix} 1 & x_i & y_i \\ 1 & x_j & y_j \\ 1 & x_k & y_k \end{bmatrix} \begin{Bmatrix} C_1 \\ C_2 \\ C_3 \end{Bmatrix} \tag{15-7}$$

Solving for C_1, C_2, and C_3 by using inversion, we obtain

$$\begin{Bmatrix} C_1 \\ C_2 \\ C_3 \end{Bmatrix} = \frac{1}{2A} \begin{bmatrix} x_j y_k - x_k y_j & x_k y_i - x_i y_k & x_i y_j - x_j y_i \\ y_j - y_k & y_k - y_i & y_i - y_j \\ x_k - x_j & x_i - x_k & x_j - x_i \end{bmatrix} \begin{Bmatrix} u_i \\ u_j \\ u_k \end{Bmatrix} \tag{15-8}$$

where A is the area of the triangle ijk. The area is equal to one-half the value of the determinant of the 3×3 matrix in Eq. (15-7). It should be noted that the sign of the area is positive when the nodal points of the triangle are labeled counterclockwise around the element as shown in Fig. 15-3. The sign of the area would be negative if labeled clockwise.

By substituting the values of C of Eq. (15-8) into Eq. (15-6a), the expression for the displacement u is found to be

$$\begin{aligned} u = \frac{1}{2A} \{ & [(x_j y_k - x_k y_j) + (y_j - y_k)x + (x_k - x_j)y]u_i \\ & + [(x_k y_i - x_i y_k) + (y_k - y_i)x + (x_i - x_k)y]u_j \\ & + [(x_i y_j - x_j y_i) + (y_i - y_j)x + (x_j - x_i)y]u_k \} \end{aligned} \tag{15-9}$$

Proceeding in the same manner with the evaluation of the constants C_4, C_5, and C_6 in Eq. (15-6b), we obtain

$$\begin{aligned} v = \frac{1}{2A} \{ & [(x_j x_k - x_k y_j) + (y_j - y_k)x + (x_k - x_j)y]v_i \\ & + [(x_k y_i - x_i y_k) + (y_k - y_i)x + (x_i - x_k)y]v_j \\ & + [(x_i y_j - x_j y_i) + (y_i - y_j)x + (x_j - x_i)y]v_k \} \end{aligned} \tag{15-10}$$

To obtain the desired form of the displacement transformation as stated in Eq. (15-5), we make use of the fact that strains can be expressed as the partial derivatives of displacements,[1] i.e.,

$$\epsilon_x = \frac{\partial u}{\partial x} \qquad \epsilon_y = \frac{\partial v}{\partial y} \qquad \gamma_{xy} = \frac{\partial u}{\partial y} + \frac{\partial v}{\partial x} \tag{15-11}$$

After the necessary differentiation of the expressions for u and v in Eqs. (15-9) and (15-10), the desired displacement transformation between nodal-point displacements and the differential strains becomes

$$\begin{Bmatrix} \epsilon_x \\ \epsilon_y \\ \gamma_{xy} \end{Bmatrix} = \underbrace{\frac{1}{2A}\begin{bmatrix} y_j - y_k & y_k - y_i & y_i - y_j & 0 & 0 & 0 \\ 0 & 0 & 0 & x_k - x_j & x_i - x_k & x_j - x_i \\ x_k - x_j & x_i - x_k & x_j - x_i & y_j - y_k & y_k - y_i & y_i - y_j \end{bmatrix}}_{\mathbf{A}} \begin{Bmatrix} u_i \\ u_j \\ u_k \\ v_i \\ v_j \\ v_k \end{Bmatrix} \tag{15-12}$$

The final expression for the stiffness matrix **k** for the triangular element can be obtained using virtual work. As indicated in Sec. 8-6, we can write the expression for internal virtual work on a three-dimensional differential element as

$$\delta W_I = \delta\boldsymbol{\epsilon}'\boldsymbol{\sigma}\, dV \tag{15-13}$$

For the plane triangular element $dV = t\,dA$, where t denotes the constant thickness of the element. In Eq. (15-13) $\delta\boldsymbol{\epsilon}$ represents a set of virtual strains, and $\boldsymbol{\sigma}$ for the triangle under consideration is the set of stresses in Eq. (15-2). Using the relationship of Eq. (15-3), Eq. (15-13) can be written as

$$\delta W_I = \delta\boldsymbol{\epsilon}'\bar{\mathbf{k}}\boldsymbol{\epsilon}t\, dA \tag{15-14}$$

The real and virtual strains can be expressed in terms of nodal-point displacements by the **A** matrix of Eq. (15-12). Equation (15-14) thereby becomes

$$\delta W_I = \delta\mathbf{d}'\mathbf{A}'\bar{\mathbf{k}}\mathbf{A}\mathbf{d}t\, dA \tag{15-15}$$

The total internal virtual work for the triangle is obtained by integrating Eq. (15-15) over the area A of the triangle. Thus, for the triangle, we can write

$$\delta W_I = \delta\mathbf{d}'\left[\int_A \mathbf{A}'\bar{\mathbf{k}}\mathbf{A}t\, dA\right]\mathbf{d} \tag{15-16}$$

Equating external virtual work to internal virtual work and applying virtual displacements in the form of an identity matrix as was done in the previous chapter, we obtain the following stiffness relationship be-

[1] J. T. Oden, "Mechanics of Elastic Structures," McGraw-Hill Book Company, New York, 1967.

tween nodal-point forces $\mathbf{q}$ and nodal-point displacements $\mathbf{d}$:

$$\mathbf{q} = \left[\int_A \mathbf{A}'\bar{\mathbf{k}}\mathbf{A} t \, dA \right] \mathbf{d} \tag{15-17}$$

The quantity inside the brackets represents the stiffness matrix of the element $\mathbf{k}$. We note that because t is constant, and because $\mathbf{A}$ as given in Eq. (15-12) and $\bar{\mathbf{k}}$ as given in Eq. (15-4) are not functions of x and y, the expression for the stiffness matrix of the triangular element becomes

$$\mathbf{k} = At\mathbf{A}'\bar{\mathbf{k}}\mathbf{A} \tag{15-18}$$

In the above development it is seen that we obtain constant values for strain when we substitute the assumed displacement functions of Eqs. (15-6) into the expressions for strain of Eq. (15-11). The element is therefore often referred to as the *constant-strain triangle.*

With the stiffness of the individual elements available, the stiffness matrix $\mathbf{K}$ of the finite-element assembly can be formed by techniques such as those presented in the previous chapter. External loading can be represented by equivalent concentrated loads at the nodal points. The solution of the resulting governing equations of Eq. (14-1) yields the external nodal-point displacements. The external displacements correspond to displacements at the nodes of the elements, and we see that the stresses in the elements can be obtained by means of Eqs. (15-12) and (15-2).

15-3 APPLICATION OF THE METHOD

As noted earlier, various elements have been developed in recent years and are used for a wide variety of problems. The use of the finite-element method to obtain a solution to a particular problem requires that consideration be given to several aspects, some of which will be mentioned in the following paragraphs. Details of the application of the method can be found in the finite-element texts cited earlier.

The representation of a given problem by finite elements is much a matter of good judgment. The proper element should be selected to represent the type of problem being solved. Because the finite element approximates the true solution, it is necessary to use smaller elements in areas where large changes in stress and strain occur. The number of elements that can be used may be limited by the size of the computer facility that is available. Proper numbering of the nodal points should be used to obtain good banding of the $\mathbf{K}$ matrix. Boundary conditions of the given problem must be satisfied by specifying proper restraint, or movement, of the nodal points in the finite-element mesh.

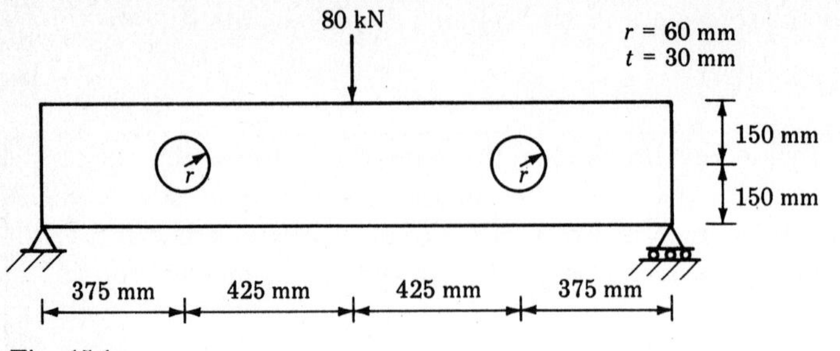

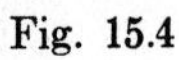
Fig. 15.4

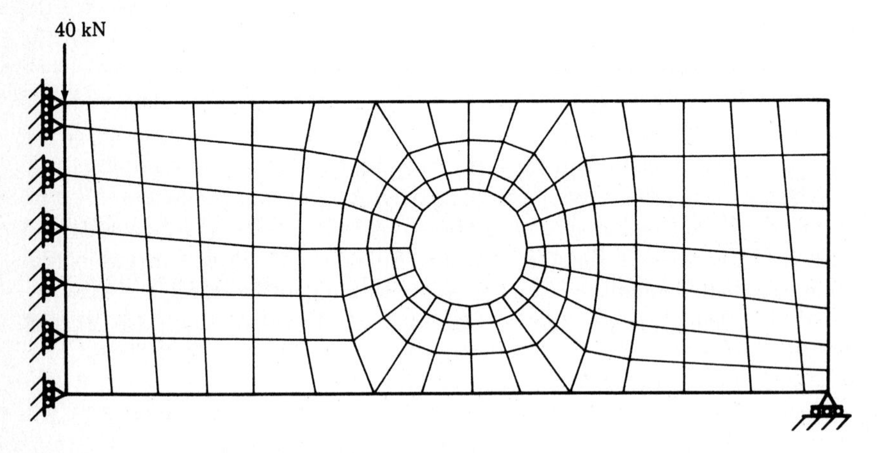

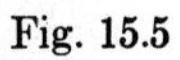
Fig. 15.5

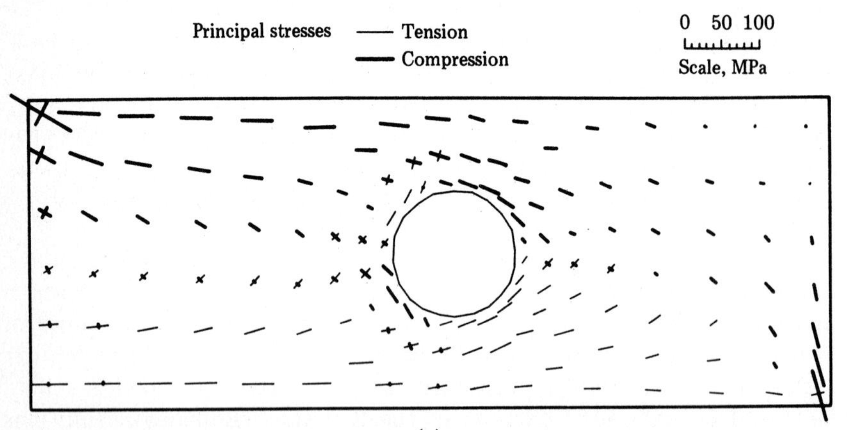

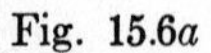
Fig. 15.6a

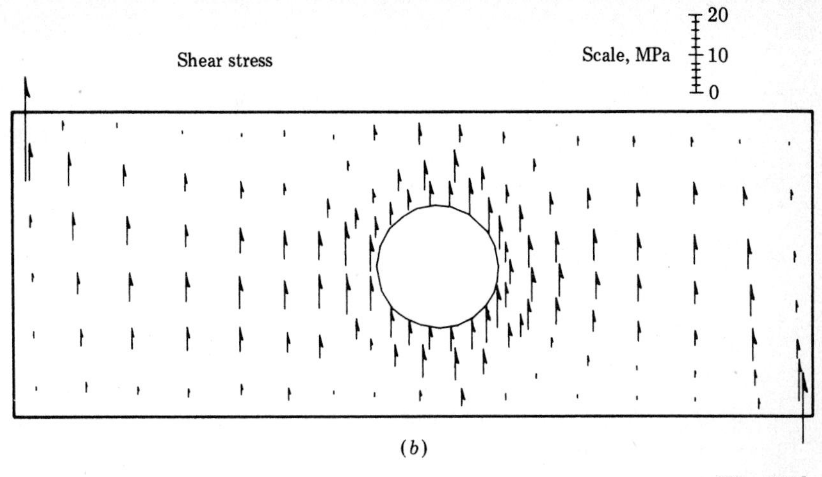

Fig. 15.6b

To demonstrate the use of the finite-element method, the beam of Fig. 15-4 is analyzed for stresses resulting from the 80-kN concentrated load. There is a hole with a radius of 60 mm near each end of the beam. The beam has a constant thickness of 30 mm. Plane-stress finite elements are used to determine the values of principal stresses and shear stresses in the beam. E is constant.

Because of symmetry, only one-half of the beam is considered in the analysis. The finite-element model for the solution is shown in Fig. 15-5. Quadrilateral elements are used. Nodal points on the centerline of the beam span are restrained in the horizontal direction and are permitted to displace in the vertical direction. An effort is made to use smaller elements in areas of the beam where large stress changes are expected to occur—at the point of loading, at the support point, and near the hole.

The results of a finite-element solution using the computer program SAP IV[1] are shown in Fig. 15-6. Principal stresses are plotted to scale in Fig. 15-6a, and vertical shear stresses are plotted to scale in Fig. 15-6b. The stresses shown are the stresses at the center of the elements.

[1] K. J. Bathe, E. L. Wilson, and F. E. Peterson, "SAP IV—A Structural Analysis Program for Static and Dynamic Response of Linear Systems," EERC Report No. 73-11, College of Engineering, University of California, Berkeley, 1974.

APPENDIX

A

MATRIX ALGEBRA

A-1 MATRIX DEFINITION

A matrix is defined as an $m \times n$ rectangular array of elements arranged in m rows and n columns. The general representation of an $m \times n$ matrix **A** is

$$\mathbf{A} = \begin{bmatrix} a_{11} & a_{12} & \cdots & a_{1n} \\ a_{21} & a_{22} & \cdots & a_{2n} \\ \cdots & \cdots & \cdots & \cdots \\ a_{m1} & a_{m2} & \cdots & a_{mn} \end{bmatrix}$$

An element of matrix **A** in the ith row and jth column is represented by the notation a_{ij}. An example of matrix representation are the coefficients in a set of simultaneous equations. In the set of equations

$$\begin{aligned} 2x_1 - 3x_2 + x_3 &= 4 \\ -x_1 + 2x_2 + x_3 &= 1 \\ 3x_1 + x_2 + 2x_3 &= -3 \end{aligned} \tag{A-1}$$

the coefficients can be represented by a matrix **A** as

$$\mathbf{A} = \begin{bmatrix} 2 & -3 & 1 \\ -1 & 2 & 1 \\ 3 & 1 & 2 \end{bmatrix} \tag{A-2}$$

where $a_{11} = 2$, $a_{12} = -3$, and so on.

Equations (A-1) can also be written in the form

$$\sum_{j=1}^{3} a_{ij}x_j = b_i \qquad i = 1, 2, 3$$

For example, when $i = 1$, the summation of j from 1 to 3 yields

$$a_{11}x_1 + a_{12}x_2 + a_{13}x_3 = b_1$$

which is, according to the values given in the first of Eqs. (A-1),

$$2x_1 - 3x_2 + x_3 = 4$$

The second and third of Eqs. (A-1) are represented by $i = 2$ and $i = 3$, respectively; i is often referred to as the *free index* and j the *dummy index.*

A-2 TYPES OF MATRICES

Square matrix

In a square matrix the number of rows is equal to the number of columns. The **A** matrix of Eq. (A-2) is a square matrix, where $m = n = 3$.

Column and row matrices (vectors)

A column matrix is a matrix with dimensions $m \times 1$, that is, m rows and one column. Similarly, a row matrix is a matrix with dimensions $1 \times n$. Column and row matrices are often referred to as *vectors*. Column matrices are denoted here by braces { }. The constant values on the right side of Eqs. (A-1) can be represented by a column matrix **B** as

$$\mathbf{B} = \begin{Bmatrix} b_1 \\ b_2 \\ b_3 \end{Bmatrix} = \begin{Bmatrix} 4 \\ 1 \\ -3 \end{Bmatrix} \tag{A-3}$$

where the notation b_1 really implies b_{11}. Similarly, $b_2 = b_{21}$ and $b_3 = b_{31}$. The second index is commonly omitted in denoting the elements of column matrices.

Diagonal matrix

A diagonal matrix is a square matrix which has zero elements everywhere except on the main diagonal. A diagonal matrix **A** requires that $a_{ij} = 0$

if $i \neq j$ and has the form

$$\mathbf{A} = \begin{bmatrix} a_{11} & 0 & \cdots & 0 \\ 0 & a_{22} & \cdots & 0 \\ \cdot & \cdot & \cdot & \cdot \\ 0 & 0 & \cdots & a_{nn} \end{bmatrix}$$

It is convenient to omit the zero elements in writing a diagonal matrix.

Unit, or identity, matrix

A diagonal matrix which has unit-valued elements on the diagonal is a unit, or identity, matrix. The unit matrix is commonly represented by the symbol **I** and is shown as

$$\mathbf{I} = \begin{bmatrix} 1 & & & & \\ & 1 & & & \\ & & \cdot & & \\ & & & \cdot & \\ & & & & \cdot & \\ & & & & & 1 \end{bmatrix}$$

Null matrix

A null matrix is one whose elements are all equal to zero. The common notation for the null matrix is **0**.

Symmetric matrix

A square matrix whose elements are symmetric about the main diagonal is defined as a symmetric matrix. If **A** is a symmetric matrix, then $a_{ij} = a_{ji}$.

Antisymmetric or skew-symmetric matrix

A matrix **A** in which $a_{ij} = -a_{ji}$ and $a_{ii} = 0$ is an antisymmetric or skew-symmetric matrix.

A-3 MATRIX ALGEBRA

Equality of matrices

Two matrices are equal if the corresponding elements of each are equal. It is necessary that the matrices be of the same size. Thus, if **A** and **B** are two matrices of the same size, the condition for equality can be written as

$$a_{ij} = b_{ij}$$

for each i and j.

Addition and subtraction of matrices

Matrices can be added or subtracted only if they are of the same size. The addition of two matrices **A** and **B** results in another matrix **C**, the elements of which are equal to the sum of the corresponding elements of **A** and **B**. The summation can be written as

$$\mathbf{A} + \mathbf{B} = \mathbf{C}$$
$$\text{or} \qquad a_{ij} + b_{ij} = c_{ij}$$

for each i and j. Similarly, the difference of two matrices **A** and **B** is another matrix **D**,

$$\mathbf{A} - \mathbf{B} = \mathbf{D}$$
$$\text{or} \qquad a_{ij} - b_{ij} = d_{ij}$$

for each i and j. It is obvious that the matrix resulting from the process of addition or subtraction is of the same size as the original matrices. Also, the process of addition and subtraction is applicable to more than two matrices.

EXAMPLE A-1 We are given two 2×3 matrices **A** and **B**,

$$\mathbf{A} = \begin{bmatrix} 7 & 3 & -1 \\ 2 & -5 & 6 \end{bmatrix} \qquad \mathbf{B} = \begin{bmatrix} 1 & 5 & 6 \\ -4 & -2 & 3 \end{bmatrix}$$

The sum of **A** and **B** is

$$\mathbf{C} = \mathbf{A} + \mathbf{B} = \begin{bmatrix} 7+1 & 3+5 & -1+6 \\ 2-4 & -5-2 & 6+3 \end{bmatrix} = \begin{bmatrix} 8 & 8 & 5 \\ -2 & -7 & 9 \end{bmatrix}$$

Similarly, the difference of **A** and **B** is

$$\mathbf{D} = \mathbf{A} - \mathbf{B} = \begin{bmatrix} 7-1 & 3-5 & -1-6 \\ 2+4 & -5+2 & 6-3 \end{bmatrix} = \begin{bmatrix} 6 & -2 & -7 \\ 6 & -3 & 3 \end{bmatrix}$$

Scalar multiplication of a matrix

The product of a matrix **A** and a scalar k is the process of multiplying each element of **A** by the scalar k. The result is another matrix **B**. Thus

$$k\mathbf{A} = ka_{ij} = \mathbf{B}$$

EXAMPLE A-2 The multiplication of the matrix **A** by 1.5, where

$$\mathbf{A} = \begin{bmatrix} 3 & 5 \\ 4 & 2 \end{bmatrix}$$

is $$1.5\begin{bmatrix} 3 & 5 \\ 4 & 2 \end{bmatrix} = \begin{bmatrix} (1.5)(3) & (1.5)(5) \\ (1.5)(4) & (1.5)(2) \end{bmatrix} = \begin{bmatrix} 4.5 & 7.5 \\ 6.0 & 3.0 \end{bmatrix}$$

Matrix multiplication

The product of two matrices **A** and **B** is written as

$$\mathbf{AB} = \mathbf{C}$$

where the elements of the resulting matrix **C** are, by definition,

$$\sum_{k=1}^{n} a_{ik}b_{kj} = c_{ij} \tag{A-4}$$

In the expression **AB**, **B** is said to be *premultiplied* by **A**. **A** is denoted as the *premultiplier* and **B** the *postmultiplier*. In an expression of the form **BA**, **B** is said to be *postmultiplied* by **A**. In general, $\mathbf{AB} \neq \mathbf{BA}$.

Two matrices can be multiplied only if the number of columns in the premultiplier is equal to the number of rows in the postmultiplier. The resulting matrix will have the same number of rows as the premultiplier and the same number of columns as the postmultiplier.

Now that we have defined matrix multiplication, we can write Eqs. (A-1) in the concise form

$$\mathbf{AX} = \mathbf{B}$$

where **A** is expressed by Eq. (A-2) and **B** is expressed by Eq. (A-3). From the definition of matrix multiplication, this can be written as

$$\sum_{k=1}^{3} a_{ik}x_{kj} = b_{ij}$$

With $i = 1$ and $j = 1$, the first of Eqs. (A-1) is obtained as

$$a_{11}x_{11} + a_{12}x_{21} + a_{13}x_{31} = b_{11}$$

or $$2x_1 - 3x_2 + x_3 = 4$$

As suggested before, the second index of the column matrices can be omitted.

EXAMPLE A-3 Given two matrices **A** and **B**,

$$\mathbf{A} = \begin{bmatrix} 5 & 3 & 8 \\ -1 & 4 & 7 \\ 0 & 1 & 1 \end{bmatrix} \qquad \mathbf{B} = \begin{bmatrix} 6 & 7 \\ 10 & 9 \\ 2 & -3 \end{bmatrix}$$

we are to determine the product **AB**. It is sometimes desirable to write the two matrices with their dimensions (number of rows times number of columns) under them. This makes it immediately apparent whether or not the matrices satisfy the conditions for multiplication. For this example

$$\mathbf{C} = \underset{3\times 3}{\mathbf{A}} \quad \underset{3\times 2}{\mathbf{B}}$$

The dimensions connected by solid lines verify that the number of columns in the premultiplier is equal to the number of rows in the postmultiplier. The dimensions connected by the dashed lines indicate that the size of the resulting matrix **C** will have dimensions of 3×2. The multiplication operation yields

$$\mathbf{C} = \begin{bmatrix} 5 & 3 & 8 \\ -1 & 4 & 7 \\ 0 & 1 & 1 \end{bmatrix} \begin{bmatrix} 6 & 7 \\ 10 & 9 \\ 2 & -3 \end{bmatrix}$$

$$= \begin{bmatrix} (5)(6) + (3)(10) + (8)(2) & (5)(7) + (3)(9) + (8)(-3) \\ (-1)(6) + (4)(10) + (7)(2) & (-1)(7) + (4)(9) + (7)(-3) \\ (0)(6) + (1)(10) + (1)(2) & (0)(7) + (1)(9) + (1)(-3) \end{bmatrix}$$

$$= \begin{bmatrix} 76 & 38 \\ 48 & 8 \\ 12 & 6 \end{bmatrix}$$

According to the rules for matrix multiplication and the previous definition of the unit matrix **I**, the product of **I** and any square matrix **A** is

$$\mathbf{AI} = \mathbf{IA} = \mathbf{A}$$

Transpose of matrices

The transpose of a matrix **A**, denoted as **A**′, is obtained by interchanging the rows and columns in the matrix **A**. Therefore, if the original matrix is written as

$$\mathbf{A} = \begin{bmatrix} a_{11} & a_{12} & \cdots & a_{1n} \\ a_{21} & a_{22} & \cdots & a_{2n} \\ \cdot & \cdot & \cdot & \cdot \\ a_{m1} & a_{m2} & \cdots & a_{mn} \end{bmatrix}$$

the transpose of **A** is

$$\mathbf{A}' = \begin{bmatrix} a_{11} & a_{21} & \cdots & a_{m1} \\ a_{12} & a_{22} & \cdots & a_{m2} \\ \cdots & \cdots & \cdots & \cdots \\ a_{1n} & a_{2n} & \cdots & a_{mn} \end{bmatrix}$$

By the laws of addition and multiplication of matrices, for two matrices **A** and **B** we have

$$(\mathbf{A} + \mathbf{B})' = \mathbf{A}' + \mathbf{B}'$$
$$(k\mathbf{A})' = k\mathbf{A}'$$

Also, if

$$\mathbf{C} = \mathbf{AB}$$

then $\quad \mathbf{C}' = \mathbf{B}'\mathbf{A}'$

Note the reversed order of **A** and **B** in transposing **C**.

EXAMPLE A-4 The transpose of matrix

$$\mathbf{A} = \begin{bmatrix} 4 & 0 & 7 \\ 5 & 1 & 2 \end{bmatrix} \quad \text{is} \quad \mathbf{A}' = \begin{bmatrix} 4 & 5 \\ 0 & 1 \\ 7 & 2 \end{bmatrix}$$

Partitioning of matrices

It is often desirable to partition a matrix into a number of smaller submatrices. If **A** is a 3×4 matrix, we can partition it and define its submatrices as

$$\mathbf{A} = \left[\begin{array}{c|ccc} a_{11} & a_{12} & a_{13} & a_{14} \\ a_{21} & a_{22} & a_{23} & a_{24} \\ \hline a_{31} & a_{32} & a_{33} & a_{34} \end{array}\right] = \begin{bmatrix} \mathbf{A}_{11} & \mathbf{A}_{12} \\ \mathbf{A}_{21} & \mathbf{A}_{22} \end{bmatrix}$$

where $\quad \mathbf{A}_{11} = \begin{bmatrix} a_{11} \\ a_{21} \end{bmatrix} \qquad \mathbf{A}_{12} = \begin{bmatrix} a_{12} & a_{13} & a_{14} \\ a_{22} & a_{23} & a_{24} \end{bmatrix}$

$$\mathbf{A}_{21} = [a_{31}] \qquad \mathbf{A}_{22} = [a_{32} \quad a_{33} \quad a_{34}]$$

Two matrices **A** and **B** which are of the same size can be added or subtracted by adding or subtracting their submatrices, provided they are partitioned into the same number and size of submatrices. Thus, if **A** and **B** are partitioned into four submatrices, their sum **C** can be written as

$$\mathbf{C} = \mathbf{A} + \mathbf{B} = \begin{bmatrix} \mathbf{A}_{11} + \mathbf{B}_{11} & \mathbf{A}_{12} + \mathbf{B}_{12} \\ \mathbf{A}_{21} + \mathbf{B}_{21} & \mathbf{A}_{22} + \mathbf{B}_{22} \end{bmatrix}$$

Multiplication of two matrices can be performed in terms of their submatrices, provided the columns and rows of the original matrices and the columns and rows of the submatrices are such that they conform to the requirements for the multiplication of matrices.

EXAMPLE A-5 Matrices **A** and **B** of Example A-3 are partitioned in the following manner:

$$\mathbf{A} = \left[\begin{array}{cc:c} 5 & 3 & 8 \\ -1 & 4 & 7 \\ \hdashline 0 & 1 & 1 \end{array}\right] \qquad \mathbf{B} = \left[\begin{array}{cc} 6 & 7 \\ 10 & 9 \\ \hdashline 2 & -3 \end{array}\right]$$

The product of **A** and **B** in terms of their submatrices is

$$\mathbf{C} = \mathbf{AB} = \begin{bmatrix} \mathbf{A}_{11} & \mathbf{A}_{12} \\ \mathbf{A}_{21} & \mathbf{A}_{22} \end{bmatrix} \begin{bmatrix} \mathbf{B}_{11} \\ \mathbf{B}_{21} \end{bmatrix} = \begin{bmatrix} \mathbf{A}_{11}\mathbf{B}_{11} + \mathbf{A}_{12}\mathbf{B}_{21} \\ \mathbf{A}_{21}\mathbf{B}_{11} + \mathbf{A}_{22}\mathbf{B}_{21} \end{bmatrix}$$

where $\mathbf{A}_{11} = \begin{bmatrix} 5 & 3 \\ -1 & 4 \end{bmatrix} \qquad \mathbf{A}_{12} = \begin{bmatrix} 8 \\ 7 \end{bmatrix} \qquad \mathbf{B}_{11} = \begin{bmatrix} 6 & 7 \\ 10 & 9 \end{bmatrix}$

$\mathbf{A}_{21} = [0 \quad 1] \qquad \mathbf{A}_{22} = [1] \qquad \mathbf{B}_{21} = [2 \quad -3]$

Performing the required multiplications, we obtain

$$\mathbf{A}_{11}\mathbf{B}_{11} = \begin{bmatrix} 5 & 3 \\ -1 & 4 \end{bmatrix} \begin{bmatrix} 6 & 7 \\ 10 & 9 \end{bmatrix} = \begin{bmatrix} 60 & 62 \\ 34 & 29 \end{bmatrix}$$

$$\mathbf{A}_{12}\mathbf{B}_{21} = \begin{bmatrix} 8 \\ 7 \end{bmatrix} [2 \quad -3] = \begin{bmatrix} 16 & -24 \\ 14 & -21 \end{bmatrix}$$

$$\mathbf{A}_{21}\mathbf{B}_{11} = [0 \quad 1] \begin{bmatrix} 6 & 7 \\ 10 & 9 \end{bmatrix} = [10 \quad 9]$$

$$\mathbf{A}_{22}\mathbf{B}_{21} = [1][2 \quad -3] = [2 \quad -3]$$

$$\mathbf{C} = \begin{bmatrix} \begin{bmatrix} 60 & 62 \\ 34 & 29 \end{bmatrix} + \begin{bmatrix} 16 & -24 \\ 14 & -21 \end{bmatrix} \\ [10 \quad 9] \; + \; [2 \quad -3] \end{bmatrix} = \begin{bmatrix} 76 & 38 \\ 48 & 8 \\ 12 & 6 \end{bmatrix}$$

Additional laws of matrix algebra

By applying the previously defined matrix operations, we can verify the following rules:

1. The addition of two matrices is commutative:

   $\mathbf{A} + \mathbf{B} = \mathbf{B} + \mathbf{A}$

2. The product of two matrices, in general, is not commutative:

   $\mathbf{AB} \neq \mathbf{BA}$

   An exception is the case where both **A** and **B** are diagonal matrices.

3. The distributive law applies:

   $\mathbf{A}(\mathbf{B} + \mathbf{C}) = \mathbf{AB} + \mathbf{AC}$

4. The associative law applies:

   $\mathbf{A}(\mathbf{BC}) = (\mathbf{AB})\mathbf{C}$

   and therefore can be written as **ABC**.

A-4 DETERMINANTS

Since determinants are often involved in matrix algebra, we shall discuss the fundamentals of determinant theory. Consider the set of equations

$$\begin{aligned} a_{11}x_1 + a_{12}x_2 + a_{13}x_3 &= b_1 \\ a_{21}x_1 + a_{22}x_2 + a_{23}x_3 &= b_2 \\ a_{31}x_1 + a_{32}x_2 + a_{33}x_3 &= b_3 \end{aligned} \tag{A-5}$$

The determinant of the coefficients a_{ij} is written as

$$|\mathbf{A}| = \begin{vmatrix} a_{11} & a_{12} & a_{13} \\ a_{21} & a_{22} & a_{23} \\ a_{31} & a_{32} & a_{33} \end{vmatrix} \tag{A-6}$$

It is essential to note the basic difference between a matrix and a determinant. A determinant, when expanded, has a numerical value, while matrices cannot be reduced to a single value. Matrices are in fact operators, and it is helpful to think of them as such. Further, a determinant is always composed of a square array of elements.

According to Cramer's rule, in Eqs. (A-5)

$$x_1 = \frac{\begin{vmatrix} b_1 & a_{12} & a_{13} \\ b_2 & a_{22} & a_{23} \\ b_3 & a_{32} & a_{33} \end{vmatrix}}{|\mathbf{A}|} \qquad x_2 = \frac{\begin{vmatrix} a_{11} & b_1 & a_{13} \\ a_{21} & b_2 & a_{23} \\ a_{31} & b_3 & a_{33} \end{vmatrix}}{|\mathbf{A}|} \qquad x_3 = \frac{\begin{vmatrix} a_{11} & a_{12} & b_1 \\ a_{21} & a_{22} & b_2 \\ a_{31} & a_{32} & b_3 \end{vmatrix}}{|\mathbf{A}|} \tag{A-7}$$

where $|\mathbf{A}|$ is expressed in Eq. (A-6). The solution for the unknown x terms is contingent upon the fact that $|\mathbf{A}| \neq 0$.

If $b_1 = b_2 = b_3 = 0$, then Eqs. (A-5) are described as being *homogeneous*. For a set of homogeneous equations where $x_1, x_2, \ldots, x_n$ are not zero (that is, where a nontrivial solution is possible), the determinant of the coefficients of the unknowns is equal to zero.

Minor of an element

The minor of the element a_{ij} of a determinant of size n is the determinant of size $n - 1$ obtained by deleting the ith row and the jth column of the original determinant. In Eq. (A-6) the minor M_{ij} of element a_{23} is

$$M_{23} = \begin{vmatrix} a_{11} & a_{12} \\ a_{31} & a_{32} \end{vmatrix}$$

Cofactor of an element

The cofactor of an element a_{ij} of a determinant is the minor of the element and is either a positive or a negative quantity. The sign of the minor is determined from the relationship -1^{i+j}. Thus the cofactor C_{ij} of element a_{23} in Eq. (A-6) is

$$C_{23} = -1^{2+3}M_{23} = -M_{23}$$

Laplace expansion of determinants

The numerical value of an $n \times n$ determinant $|\mathbf{A}|$ can be found from either

$$|\mathbf{A}| = \sum_{j=1}^{n} a_{ij}C_{ij} \tag{A-8}$$

for any value of i, or

$$|\mathbf{A}| = \sum_{i=1}^{n} a_{ij}C_{ij} \tag{A-9}$$

for any value of j. Equation (A-8) describes the expansion of the determinant about a row, and Eq. (A-9) describes the expansion of the determinant about a column.

EXAMPLE A-6 We are to use the Laplace expansion to find the value of the determinant

$$|\mathbf{A}| = \begin{vmatrix} 1 & 1 & 4 & 2 \\ 0 & 0 & 3 & 0 \\ -1 & 4 & 0 & 2 \\ 2 & 3 & 2 & 3 \end{vmatrix}$$

The determinant will be expanded about the second row to reduce the number of computations, since there are three values of zero in the second row. Equation (A-8), with $i = 2$ and $j = 1, 2, 3, 4$, is applicable. It is sometimes more convenient to determine the sign of the minor by visualizing the sign pattern in the determinant from the relation -1^{i+j}:

$$\begin{vmatrix} + & - & + & - \\ - & + & - & + \\ + & - & + & - \\ - & + & - & + \end{vmatrix}$$

From Eq. (A-8) we obtain

$$
\begin{aligned}
|\mathbf{A}| &= a_{21}C_{21} + a_{22}C_{22} + a_{23}C_{23} + a_{24}C_{24} \\
&= 0(-M_{21}) + 0M_{22} + 3(-M_{23}) + 0M_{24} \\
&= -3M_{23}
\end{aligned}
$$

$$
\text{where} \qquad M_{23} = \begin{vmatrix} 1 & 1 & 2 \\ -1 & 4 & 2 \\ 2 & 3 & 3 \end{vmatrix}
$$

The computations can be further simplified by expanding M_{23} about the first column, which leaves only 2×2 determinants to evaluate. Expansion of M_{23} in this manner requires the use of Eq. (A-9), with $j = 1$ and $i = 1, 2, 3$. The value of the determinant becomes

$$
\begin{aligned}
|\mathbf{A}| &= -3\left\{1\begin{vmatrix} 4 & 2 \\ 3 & 3 \end{vmatrix} - (-1)\begin{vmatrix} 1 & 2 \\ 3 & 3 \end{vmatrix} + 2\begin{vmatrix} 1 & 2 \\ 4 & 2 \end{vmatrix}\right\} \\
&= -3[(1)(6) + (1)(-3) + (2)(-6)] \\
&= (-3)(-9) = 27
\end{aligned}
$$

Properties of determinants

The following rules are essential in working with determinants:

1. The value of a determinant is unchanged if rows and columns are interchanged.
2. If each element in a row or column of a determinant is zero, the value of the determinant is zero.
3. If two rows or two columns of a determinant are interchanged, the value of the determinant changes sign.
4. If two rows or two columns of a determinant are proportional, the value of the determinant is zero.
5. If each of the elements of a row or column of a determinant is multiplied by a factor k, the value of the determinant is multiplied by k.

A-5 SOLUTION OF LINEAR EQUATIONS

The solution of n linear equations by Cramer's rule, as in Eq. (A-7), becomes impractical when n is large, because of the need to evaluate numerous and large determinants. Methods entailing elimination processes serve well to obtain the solutions of large systems of equations. Two of the simplest and most practical of the elimination methods are discussed in the following paragraphs. For a more elaborate and comprehensive treatment of solution methods the student is referred to texts devoted more to this topic.[1]

Gauss elimination method

This method can best be described by an example.

EXAMPLE A-7 The following set of equations will be solved using the Gauss method:

$$\begin{aligned} 2x_1 + x_2 - 3x_3 &= 11 \\ 4x_1 - 2x_2 + 3x_3 &= 8 \\ -2x_1 + 2x_2 - x_3 &= -6 \end{aligned} \tag{A-10}$$

The first of the equations is divided by a_{11}, which in this example is 2. The result is the first of Eqs. (A-11), which is then multiplied by $-a_{21}$ or -4 and added to the second of Eqs. (A-10). The result is the second of Eqs. (A-11). The first of Eqs. (A-11) is then multiplied by $-a_{31}$ or 2 and added to the third of Eqs. (A-10). The result is the third of Eqs. (A-11):

$$\begin{aligned} x_1 + 0.5x_2 - 1.5x_3 &= 5.5 \\ -4x_2 + 9x_3 &= -14 \\ 3x_2 - 4x_3 &= 5 \end{aligned} \tag{A-11}$$

This procedure is repeated for the second and third of Eqs. (A-11). The second of Eqs. (A-11) is divided by -4, and the resulting equation is then multiplied by -3 and added to the third equation. The result is the set of equations

$$\begin{aligned} x_1 + 0.5x_2 - 1.50x_3 &= 5.5 \\ x_2 - 2.25x_3 &= 3.5 \\ 2.75x_3 &= -5.5 \end{aligned} \tag{A-12}$$

Equations (A-12) can then be solved by back substitution. From the third of Eqs. (A-12), we have $x_3 = -2$. Substituting this value of x_3 into the second equation gives

$$x_2 = (2.25)(-2) + 3.5 = -1$$

Similarly, from the first equation we obtain

$$x_1 = -(0.5)(-1) + (1.5)(-2) + 5.5 = 3$$

[1] V. N. Faddeeva, "Computational Methods of Linear Algebra," Dover Publications, Inc., New York, 1959; E. Bodewig, "Matrix Calculus," 2d ed., Interscience Publishers, Inc., New York, 1959.

We can express the operations used to proceed from Eqs. (A-10) to Eqs. (A-11) as

$$x_1 + \frac{a_{12}}{a_{11}} x_2 + \frac{a_{13}}{a_{11}} x_3 = \frac{b_1}{a_{11}}$$

$$(-a_{21} + a_{21})x_1 + \left(\frac{-a_{21}a_{12}}{a_{11}} + a_{22}\right) x_2 + \left(\frac{-a_{21}a_{13}}{a_{11}} + a_{23}\right) x_3 = \frac{-a_{21}b_1}{a_{11}} + b_2$$

$$(-a_{31} + a_{31})x_1 + \left(\frac{-a_{31}a_{12}}{a_{11}} + a_{32}\right) x_2 + \left(\frac{-a_{31}a_{13}}{a_{11}} + a_{33}\right) x_3 = \frac{-a_{31}b_1}{a_{11}} + b_3$$

In general, the nonzero values of a_{ij} and b_i in the second and third equations for this first set of operations can be expressed as

$$a_{ij}^{(1)} = a_{ij} - a_{i1}\left(\frac{a_{1i}}{a_{11}}\right) \qquad i, j = 2, 3$$

$$b_i^{(1)} = b_i - a_{i1}\left(\frac{b_1}{a_{11}}\right) \qquad i = 2, 3$$

Also $$x_1 = \frac{b_1}{a_{11}} - \sum_{i=2}^{3} \left(\frac{a_{1i}}{a_{11}}\right) x_i$$

For the general case of n simultaneous equations, we can express the nonzero values of a_{ij} in the kth set of operations as

$$a_{ij}^{(k)} = a_{ij}^{(k-1)} - a_{ik}^{(k-1)}\left(\frac{a_{kj}^{(k-1)}}{a_{kk}^{(k-1)}}\right) \qquad i, j = k + 1, \ldots, n \qquad \text{(A-12}a\text{)}$$

where $(k - 1)$ with $k = 1$ implies the given quantity. The values of b_i are

$$b_i^{(k)} = b_i^{(k-1)} - a_{ik}^{(k-1)}\left(\frac{b_i^{(k-1)}}{a_{kk}^{(k-1)}}\right) \qquad i = k + 1, \ldots, n \qquad \text{(A-12}b\text{)}$$

The corresponding expression for x_k is

$$x_k = \frac{b_k^{(k-1)}}{a_{kk}^{(k-1)}} - \sum_{i=k+1}^{n} \left(\frac{a_{ki}^{(k-1)}}{a_{kk}^{(k-1)}}\right) x_i \qquad \text{(A-12}c\text{)}$$

The above set of operations is performed $n - 1$ times; that is, $k = 1, \ldots, n - 1$. In the final step we obtain the expression

$$a_{nn}^{(n-1)} x_n = b_n^{(n-1)}$$

This expression is solved for x_n. By using back substitution and the expression for x_k in Eq. (A-12c), the remaining values of x are obtained.

For a symmetric banded matrix with a bandwidth of m, the upper limits for Eqs. (A-12a), (A-12b), and (A-12c) become $k + m - 1$, and Eq. (A-12a) can be simplified to

$$a_{ij}^{(k)} = a_{ij}^{(k-1)} - a_{ik}^{(k-1)} \left(\frac{a_{kj}^{(k-1)}}{a_{kk}^{(k-1)}} \right) \qquad \begin{array}{l} i = k+1, \ldots, k+m-1 \\ j = i, \ldots, k+m-1 \end{array}$$

Gauss-Jordan method

A modification of the above elimination method is the complete elimination, or Gauss-Jordan, method. As the name implies, a complete elimination is performed, and there is therefore no need for back substitution. The method can best be described by application to Eqs. (A-10). Instead of proceeding from Eqs. (A-11) to Eqs. (A-12) as before, we include an additional step by also eliminating the x_2 term in the first of Eqs. (A-11). This is done in the same manner as with the third of Eqs. (A-11). The result is

$$\begin{aligned} x_1 \qquad - 0.375x_3 &= 3.75 \\ x_2 - 2.25x_3 &= 3.5 \\ 2.75x_3 &= -5.5 \end{aligned} \tag{A-13}$$

The elimination process is then performed to reduce the coefficients of x_3. The third of Eqs. (A-13) is divided by 2.75. The resulting equation is then multiplied by 2.25 and added to the second of Eqs. (A-13), and then multiplied by 0.375 and added to the first of Eqs. (A-13). The result of these operations is

$$\begin{aligned} x_1 &= 3 \\ x_2 &= -1 \\ x_3 &= -2 \end{aligned}$$

An obvious difficulty in using this elimination method is that if any of the diagonal elements of the coefficient matrix of the unknowns are equal to zero, the equations must first be reordered so that the zero-valued elements are not on the diagonal.

A-6 MATRIX INVERSION

The operation of an inverse matrix can be compared to the operation of a reciprocal quantity in elementary analysis; that is, given the relationship $3y = 7$, we can evaluate y by operating on 7 with the reciprocal of 3. Thus

$$y = (1/3)(7)$$

For similar operations with the inverse matrix, consider a set of n linear equations with n unknown values of x,

$$\sum_{j=1}^{n} a_{ij}x_j = b_i$$

In matrix form this becomes

$$\mathbf{AX} = \mathbf{B} \tag{A-14}$$

A general expression for determining the values of x_i by Cramer's rule, given in Eqs. (A-7), is

$$x_i = \frac{1}{|\mathbf{A}|} \sum_{j=1}^{n} C_{ji}b_j \tag{A-15}$$

This can be written in matrix form as

$$\mathbf{X} = \mathbf{A}^{-1}\mathbf{B} \tag{A-16}$$

where $\mathbf{A}^{-1}$ is defined as the inverse of matrix $\mathbf{A}$ of Eq. (A-14). The value of $\mathbf{A}^{-1}$, according to Eq. (A-15), is

$$\mathbf{A}^{-1} = \frac{1}{|\mathbf{A}|} \begin{bmatrix} C_{11} & C_{21} & \cdots & C_{n1} \\ C_{12} & C_{22} & \cdots & C_{n2} \\ \cdots & \cdots & \cdots & \cdots \\ C_{1n} & C_{2n} & \cdots & C_{nn} \end{bmatrix} \tag{A-17}$$

The square matrix made up of the cofactors C_{ji} in Eq. (A-17) is often referred to as the *adjoint matrix* of the matrix **A**. Note that the adjoint matrix is the transpose of the matrix whose elements are C_{ij}.

It is apparent that inverses can be obtained only for square matrices and for matrices that are nonsingular (matrices whose determinants are not equal to zero).

Although matrix **B** of Eq. (A-16) was considered to be a column matrix, this need not be the case. **B** could have any number of columns. The result of premultiplying **B** by $\mathbf{A}^{-1}$ would be a matrix **X** having a size equal to the number of rows in $\mathbf{A}^{-1}$ and the number of columns in **B**. For example, where the solutions of linear equations are to be determined for various values of the column vector **B**, the various values of the vector **B** would make up a matrix **B**. In premultiplying this matrix **B** by $\mathbf{A}^{-1}$ the entire set of solutions is obtained directly and is represented as the columns of **X**.

The following properties of matrices are often used:

$$\mathbf{A}^{-1}\mathbf{A} = \mathbf{I} = \mathbf{A}\mathbf{A}^{-1} \qquad (\mathbf{A}^{-1})' = (\mathbf{A}')^{-1} \qquad (\mathbf{AB})^{-1} = \mathbf{B}^{-1}\mathbf{A}^{-1}$$

Note the order in the last expression.

When the size of a matrix is greater than 4×4, its inversion by Eq. (A-17) becomes very cumbersome. Many additional methods have been developed for inverting matrices.[1] One of the most frequent methods is one involving the Gauss-Jordan, or complete elimination method. As an illustration, consider the matrix of coefficients in Eqs. (A-10),

$$\mathbf{A} = \begin{bmatrix} 2 & 1 & -3 \\ 4 & -2 & 3 \\ -2 & 2 & -1 \end{bmatrix} \tag{A-18}$$

The inverse of **A** is obtained by forming a rectangular matrix, which is **A** followed by the identity matrix **I**:

$$\begin{bmatrix} 2 & 1 & -3 & 1 & 0 & 0 \\ 4 & -2 & 3 & 0 & 1 & 0 \\ -2 & 2 & -1 & 0 & 0 & 1 \end{bmatrix}$$

This process is referred to as *augmenting* the matrix **A**. The Gauss-Jordan elimination process is applied to the rectangular matrix, reducing the left part of the matrix to an identity matrix, with the right part obtaining elements denoted as b_{ij}. The resulting rectangular matrix will be of the form

$$\begin{bmatrix} 1 & 0 & 0 & b_{11} & b_{12} & b_{13} \\ 0 & 1 & 0 & b_{21} & b_{22} & b_{23} \\ 0 & 0 & 1 & b_{31} & b_{32} & b_{33} \end{bmatrix}$$

The inverse of **A** is

$$\mathbf{A}^{-1} = \mathbf{B} = \begin{bmatrix} b_{11} & b_{12} & b_{13} \\ b_{21} & b_{22} & b_{23} \\ b_{31} & b_{32} & b_{33} \end{bmatrix}$$

Let us apply the procedure to matrix **A** of Eq. (A-18). After elimination of the first column, the augmented matrix is

$$\begin{bmatrix} 1 & 0.5 & -1.5 & 0.5 & 0 & 0 \\ 0 & -4 & 9 & -2 & 1 & 0 \\ 0 & 3 & -4 & 1 & 0 & 1 \end{bmatrix}$$

After the second column is reduced, the matrix becomes

$$\begin{bmatrix} 1 & 0 & -0.375 & 0.25 & 0.125 & 0 \\ 0 & 1 & -2.25 & 0.5 & -0.25 & 0 \\ 0 & 0 & 2.75 & -0.5 & 0.75 & 1 \end{bmatrix}$$

[1] *Ibid.*

Reducing the third column yields

$$\begin{bmatrix} 1 & 0 & 0 & 0.1818 & 0.2273 & 0.1364 \\ 0 & 1 & 0 & 0.0909 & 0.3636 & 0.8182 \\ 0 & 0 & 1 & -0.1818 & 0.2727 & 0.3636 \end{bmatrix}$$

Therefore

$$\mathbf{A}^{-1} = \begin{bmatrix} 0.1818 & 0.2273 & 0.1364 \\ 0.0909 & 0.3636 & 0.8182 \\ -0.1818 & 0.2727 & 0.3636 \end{bmatrix} \tag{A-19}$$

The result can be verified by the relationship

$$\mathbf{A}^{-1}\mathbf{A} = \mathbf{I}$$

$$\begin{bmatrix} 0.1818 & 0.2273 & 0.1364 \\ 0.0909 & 0.3636 & 0.8182 \\ -0.1818 & 0.2727 & 0.3636 \end{bmatrix} \begin{bmatrix} 2 & 1 & -3 \\ 4 & -2 & 3 \\ -2 & 2 & -1 \end{bmatrix} = \begin{bmatrix} 1 & 0 & 0 \\ 0 & 1 & 0 \\ 0 & 0 & 1 \end{bmatrix}$$

Having obtained the value of $\mathbf{A}^{-1}$ in Eq. (A-19), we can use this to determine the values of $\mathbf{X}$ in Eq. (A-10) from the expression

$$\mathbf{X} = \mathbf{A}^{-1}\mathbf{B}$$

or

$$\mathbf{X} = \begin{bmatrix} 0.1818 & 0.2273 & 0.1364 \\ 0.0909 & 0.3636 & 0.8182 \\ -0.1818 & 0.2727 & 0.3636 \end{bmatrix} \begin{Bmatrix} 11 \\ 8 \\ -6 \end{Bmatrix} = \begin{Bmatrix} 3 \\ -1 \\ -2 \end{Bmatrix}$$

A-7 MATRIX INTEGRATION AND DIFFERENTIATION

If the elements of a matrix $\mathbf{A}$ are functions of a parameter x,

$$\mathbf{A}(x) = \begin{bmatrix} a_{11}(x) & a_{12}(x) & \cdots & a_{1n}(x) \\ a_{21}(x) & a_{22}(x) & \cdots & a_{2n}(x) \\ \cdots & \cdots & \cdots & \cdots \\ a_{m1}(x) & a_{m2}(x) & \cdots & a_{mn}(x) \end{bmatrix}$$

the integral of $\mathbf{A}$ with respect to x is

$$\int \mathbf{A}\,dx = \begin{bmatrix} \int a_{11}\,dx & \int a_{12}\,dx & \cdots & \int a_{1n}\,dx \\ \int a_{21}\,dx & \int a_{22}\,dx & \cdots & \int a_{2n}\,dx \\ \cdots & \cdots & \cdots & \cdots \\ \int a_{m1}\,dx & \int a_{m2}\,dx & \cdots & \int a_{mn}\,dx \end{bmatrix} \tag{A-20}$$

Similarly, the derivative of **A** with respect to x is

$$\frac{d\mathbf{A}}{dx} = \begin{bmatrix} \frac{da_{11}}{dx} & \frac{da_{12}}{dx} & \cdots & \frac{da_{1n}}{dx} \\ \frac{da_{21}}{dx} & \frac{da_{22}}{dx} & \cdots & \frac{da_{2n}}{dx} \\ \cdots & \cdots & \cdots & \cdots \\ \frac{da_{m1}}{dx} & \frac{da_{m2}}{dx} & \cdots & \frac{da_{mn}}{dx} \end{bmatrix}$$

EXAMPLE A-8 We are to determine the value of the integral of the matrix **A** between the limits 0 and L, where

$$\mathbf{A} = \begin{bmatrix} 1 - \frac{2x}{L} + \frac{x^2}{L^2} & \frac{x}{L} - \frac{x^2}{L^2} \\ \frac{x}{L} - \frac{x^2}{L^2} & \frac{x^2}{L^2} \end{bmatrix}$$

According to Eq. (A-20),

$$\int_0^L \mathbf{A}\,dx = \begin{bmatrix} \int_0^L \left(1 - \frac{2x}{L} + \frac{x^2}{L^2}\right) dx & \int_0^L \left(\frac{x}{L} - \frac{x^2}{L^2}\right) dx \\ \int_0^L \left(\frac{x}{L} - \frac{x^2}{L^2}\right) dx & \int_0^L \frac{x^2}{L^2}\,dx \end{bmatrix}$$

$$= \begin{bmatrix} \left(x - \frac{x^2}{L} + \frac{x^3}{3L^2}\right)\Big|_0^L & \left(\frac{x^2}{2L} - \frac{x^3}{3L^2}\right)\Big|_0^L \\ \left(\frac{x^2}{2L} - \frac{x^3}{3L^2}\right)\Big|_0^L & \left(\frac{x^3}{3L^2}\right)\Big|_0^L \end{bmatrix}$$

$$= \begin{bmatrix} \frac{L}{3} & \frac{L}{6} \\ \frac{L}{6} & \frac{L}{3} \end{bmatrix} = \frac{L}{6}\begin{bmatrix} 2 & 1 \\ 1 & 2 \end{bmatrix}$$

PROBLEMS

The following matrices are to be used in the solution of Probs. A-1 through A-8:

$$\mathbf{A} = \begin{bmatrix} 1 & 0 \\ -1 & 4 \end{bmatrix} \qquad \mathbf{B} = \begin{bmatrix} 3 & 1 & 0 & 1 \\ -1 & 0 & 2 & 3 \\ 4 & -2 & 0 & -1 \end{bmatrix} \qquad \mathbf{C} = \begin{Bmatrix} 2 \\ 0 \\ -4 \\ 1 \end{Bmatrix}$$

$$\mathbf{D} = [-1 \quad -1 \quad 3] \qquad \mathbf{E} = \begin{bmatrix} -\frac{1}{2} & 0 & 1 \\ 0 & \frac{1}{2} & -1 \end{bmatrix} \qquad \mathbf{F} = \begin{bmatrix} -1 & 2 & -1 \\ 0 & 3 & -1 \\ 3 & 1 & 2 \end{bmatrix}$$

$$\mathbf{G} = \begin{bmatrix} 1.25 & 0 \\ 0 & 3.50 \end{bmatrix}$$

A-1. What are the dimensions of matrices **A**, **B**, **C**, **D**, and **F**?

A-2. Evaluate

(*a*) $\mathbf{A} + \mathbf{G}$
(*b*) $\mathbf{D}(\mathbf{I} - \mathbf{F})$
(*c*) $\mathbf{E}'\mathbf{A}$
(*d*) $\mathbf{DB}$
(*e*) $\mathbf{BCD}$
(*f*) $\mathbf{B}'\mathbf{FB}$
(*g*) $\mathbf{B}'\mathbf{D}' - \mathbf{C}$
(*h*) $\mathbf{DF} + (\mathbf{BC})'$

A-3. Demonstrate whether or not **AG** is equal to **GA**.

A-4. If $\mathbf{H} = \mathbf{D}'\mathbf{D}$, demonstrate whether or not **HF** is equal to **FH**.

A-5. Determine the inverse of matrices **G** and **F** by the method of cofactors.

A-6. Determine the inverse of **F** by the method associated with complete elimination.

A-7. Determine the value of **X** in the expression $\mathbf{AGX} = \mathbf{EB}$.

A-8. If matrices **B** and **F** are partitioned in the form

$$\mathbf{B} = \left[\begin{array}{cc:cc} 3 & 1 & 0 & 1 \\ -1 & 0 & 2 & 3 \\ \hdashline 4 & -2 & 0 & -1 \end{array}\right] \qquad \mathbf{F} = \left[\begin{array}{cc:c} -1 & 2 & -1 \\ \hdashline 0 & 3 & -1 \\ 3 & 1 & 2 \end{array}\right]$$

use the submatrices to determine the product **FB**. Verify your answer by direct multiplication.

A-9. For the matrix

$$\mathbf{A} = \begin{bmatrix} \cos\omega & \sin\omega \\ -\sin\omega & \cos\omega \end{bmatrix}$$

show that $\mathbf{A}^{-1} = \mathbf{A}'$.

A-10. Evaluate the determinant

$$\begin{vmatrix} 1 & 1 & -3 & -1 \\ 0 & -5 & 1 & -2 \\ -2 & 0 & 4 & 0 \\ 1 & 0 & 3 & 2 \end{vmatrix}$$

A-11. The area A of a triangle described by coordinates x_i and y_i as shown in Fig. PA-11 can be obtained by evaluating the determinant

$$A = \tfrac{1}{2}\begin{vmatrix} 1 & x_1 & y_1 \\ 1 & x_2 & y_2 \\ 1 & x_3 & y_3 \end{vmatrix}$$

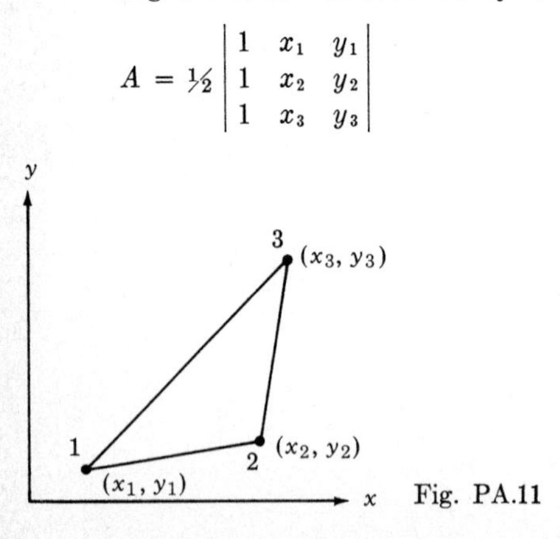

Fig. PA.11

For the values $x_1 = 2$, $y_1 = 1$, $x_2 = 8$, $y_2 = 2$, $x_3 = 9$, and $y_3 = 8$, (*a*) determine the value of the area of the triangle; (*b*) show that a negative value of area is obtained if the vertices of the triangle are numbered clockwise around the triangle rather than counterclockwise as shown in Fig. PA-11.

A-12. Given the set of equations

$$\begin{aligned} 12x_1 - 8x_2 + 8x_3 &= -15 \\ -6x_1 \qquad\quad - 10x_3 &= -4 \\ 7x_1 - 2x_2 - 3x_3 &= 10 \end{aligned}$$

(*a*) write the equations in the matrix form $\mathbf{AX} = \mathbf{B}$; (*b*) determine the value of $\mathbf{A}^{-1}$; (*c*) determine the values of x_1, x_2, and x_3 using either the Gauss elimination method or the Gauss-Jordan complete elimination method.

A-13. Determine the values of x_1, x_2, and x_3 in the equations

$$\begin{aligned} 3.2x_1 - 1.5x_2 + 0.8x_3 &= -4.6 \\ -2.2x_1 - 3.6x_2 + 1.8x_3 &= 6.5 \\ 0.6x_1 + 2.4x_2 - 1.2x_3 &= -2.8 \end{aligned}$$

Verify your results using the computer program of Sec. B-1 or a similar computer program.

APPENDIX

B

COMPUTER PROGRAMS

Following are three computer programs, written in Fortran IV, which can be used for solving various problems discussed in the text. Included with each program listing is an example computer solution. The programs are written so that they can be used with either a remote terminal or with punched cards. The input statements are written for free-form input which can readily be modified if not applicable on the user's computer system. The description of the required input is given in comment statements and can be typed out when using a remote terminal.

B-1 SOLUTION OF SIMULTANEOUS EQUATIONS AND MATRIX INVERSION

The following program solves the matrix equation **AX** = **B** and can be used to determine the inverse of an **A** matrix. The determinant of the **A** matrix is also determined. The program uses a modified version of a subroutine included in an earlier IBM Share Library.[1] Because of the wide use of these operations, the user may find that a similar program is already available in permanent storage at the computer facility being used.

[1] B. S. Garbow, Argonne National Laboratory, Lemont, Ill., February, 1959.

```
      PROGRAM EQSOL(INPUT,OUTPUT)
C ********
C PROGRAM FOR SOLVING SIMULTANEOUS EQUATIONS OF THE FORM AX=B AND
C  OBTAINING THE INVERSE OF AN A MATRIX.
C IF PROGRAM IS USED WITH A TERMINAL, INCLUDE THE STATEMENTS BEGINNING
C  WITH THE LETTERS 'CT' BY REPLACING THE LETTERS 'CT' WITH BLANKS.
C INPUT IS FREE FORM.
C ********
      DIMENSION A(50,50),B(50,1)
      PRINT 4
    4 FORMAT(///" *** SOLUTION OF SIMULTANEOUS EQUATIONS AND MATRIX ",
     ."INVERSION ***"//" INPUT IS IN FREE FORM."/)
CT    PRINT 8
CT  8 FORMAT(" TYPE THE NUMBER OF ROWS IN A AND INDICATE THE DESIRED"
CT   ./"  OPERATION - TYPE 1 IF SOLVING SIMULTANEOUS EQUATIONS, TYPE 0"
CT   ./"  IF OBTAINING ONLY AN INVERSE OF A.")
      READ*,N,IND
CT    PRINT 12
CT 12 FORMAT(" TYPE THE ELEMENTS OF A - 1ST ROW, 2ND ROW, ETC.")
      DO 16 I=1,N
   16 READ*,(A(I,J),J=1,N)
      IF(IND.EQ.0)GO TO 24
CT    PRINT 20
CT 20 FORMAT(" TYPE THE ELEMENTS OF B.")
      READ*,(B(I,1),I=1,N)
   24 PRINT 26
   26 FORMAT(//" INPUT VALUES:"//" THE ELEMENTS OF A ARE:"/)
      DO 32 I=1,N
      PRINT 28,I
   28 FORMAT(" ROW",I3,":")
   32 PRINT 34,(A(I,J),J=1,N)
   34 FORMAT(1X,4E14.6)
      IF(IND.EQ.0)GO TO 44
      PRINT 36
   36 FORMAT(/" THE ELEMENTS OF B ARE:"/)
      DO 40 I=1,N
   40 PRINT 42,B(I,1)
   42 FORMAT(2X,E13.6)
   44 INDX=1
      IF(IND.EQ.0)INDX=-1
      CALL MATINV(A,N,B,INDX,DET)
      IF(IND.EQ.1)GO TO 56
      PRINT 46
   46 FORMAT(//" THE ELEMENTS OF A INVERSE ARE:"/)
      DO 52 I=1,N
      PRINT 48,I
   48 FORMAT(" ROW",I3)
   52 PRINT 34,(A(I,J),J=1,N)
      GO TO 64
   56 PRINT 58
   58 FORMAT(/" THE SOLUTIONS ARE:"/)
      DO 60 I=1,N
   60 PRINT 62,I,B(I,1)
   62 FORMAT("  X(",I2,") = ",E12.6)
   64 PRINT 68,DET
   68 FORMAT(/" THE DETERMINANT OF A = ",E12.6)
      END
      SUBROUTINE MATINV(A,N,B,M,DETERM)
      DIMENSION IPIVOT(50),A(50,50),B(50,1),INDEX(50,2),
     1PIVOT(50)
      DETERM=1.0
      DO 20 J=1,N
   20 IPIVOT(J)=0
      DO 550 I=1,N
```

```
C     SEARCH FOR PIVOT ELEMENT
      AMAX=0.0
      DO 105 J=1,N
      IF (IPIVOT(J)-1) 60, 105, 60
   60 DO 100 K=1,N
      IF (IPIVOT(K)-1) 80, 100, 740
   80 IF (ABS(AMAX)-ABS(A(J,K))) 85, 100, 100
   85 IROW=J
      ICOLUM=K
      AMAX=A(J,K)
  100 CONTINUE
  105 CONTINUE
      IPIVOT(ICOLUM)=IPIVOT(ICOLUM)+1
C     INTERCHANGE ROWS TO PUT PIVOT ELEMENT ON DIAGONAL
      IF (IROW-ICOLUM) 140, 260, 140
  140 DETERM=-DETERM
      DO 200 L=1,N
      SWAP=A(IROW,L)
      A(IROW,L)=A(ICOLUM,L)
  200 A(ICOLUM,L)=SWAP
      IF(M) 260, 260, 210
  210 DO 250 L=1, M
      SWAP=B(IROW,L)
      B(IROW,L)=B(ICOLUM,L)
  250 B(ICOLUM,L)=SWAP
  260 INDEX(I,1)=IROW
      INDEX(I,2)=ICOLUM
      PIVOT(I)=A(ICOLUM,ICOLUM)
      DETERM=DETERM*PIVOT(I)
C     DIVIDE PIVOT ROW BY PIVOT ELEMENT
      A(ICOLUM,ICOLUM)=1.0
      DO 350 L=1,N
  350 A(ICOLUM,L)=A(ICOLUM,L)/PIVOT(I)
      IF(M) 380, 380, 360
  360 DO 370 L=1,M
  370 B(ICOLUM,L)=B(ICOLUM,L)/PIVOT(I)
C     REDUCE NON-PIVOT ROWS
  380 DO 550 L1=1,N
      IF(L1-ICOLUM) 400, 550, 400
  400 T=A(L1,ICOLUM)
      A(L1,ICOLUM)=0.0
      DO 450 L=1,N
  450 A(L1,L)=A(L1,L)-A(ICOLUM,L)*T
      IF(M) 550, 550, 460
  460 DO 500 L=1,M
  500 B(L1,L)=B(L1,L)-B(ICOLUM,L)*T
  550 CONTINUE
C     INTERCHANGE COLUMNS
      DO 710 I=1,N
      L=N+1-I
      IF (INDEX(L,1)-INDEX(L,2)) 630, 710, 630
  630 JROW=INDEX(L,1)
      JCOLUM=INDEX(L,2)
      DO 705 K=1,N
      SWAP=A(K,JROW)
      A(K,JROW)=A(K,JCOLUM)
      A(K,JCOLUM)=SWAP
  705 CONTINUE
  710 CONTINUE
  740 RETURN
      END
```

The following is a computer solution of the simultaneous equations on page 64 using the above program with a remote terminal.

```
*** SOLUTION OF SIMULTANEOUS EQUATIONS AND MATRIX INVERSION ***


INPUT IS IN FREE FORM.


TYPE THE NUMBER OF ROWS IN A AND INDICATE THE DESIRED
 OPERATION - TYPE 1 IF SOLVING SIMULTANEOUS EQUATIONS, TYPE 0
 IF OBTAINING ONLY AN INVERSE OF A.
? 6 1
TYPE THE ELEMENTS OF A - 1ST ROW, 2ND ROW, ETC.
? -0.6 0.6 0 0 0 0
? -0.8 -0.8 0 0 0 0
? 0.6 0 1 1 0 0
? 0.8 0 0 0 1 0
? 0 -0.6 -1 0 0 0
? 0 0.8 0 0 0 1
TYPE THE ELEMENTS OF B.
? -20 0 0 0 0 0


 INPUT VALUES:

 THE ELEMENTS OF A ARE:

 ROW  1:
   -.600000E+00   .600000E+00  0.            0.
   0.            0.
 ROW  2:
   -.800000E+00  -.800000E+00  0.            0.
   0.            0.
 ROW  3:
    .600000E+00  0.             .100000E+01   .100000E+01
   0.            0.
 ROW  4:
    .800000E+00  0.            0.            0.
    .100000E+01  0.
 ROW  5:
   0.            -.600000E+00  -.100000E+01  0.
   0.            0.
 ROW  6:
   0.             .800000E+00  0.            0.
   0.             .100000E+01


 THE ELEMENTS OF B ARE:

   -.200000E+02
   0.
   0.
   0.
   0.
   0.


 THE SOLUTIONS ARE:

  X( 1) =  .166667E+02
  X( 2) = -.166667E+02
  X( 3) =  .100000E+02
  X( 4) = -.200000E+02
  X( 5) = -.133333E+02
  X( 6) =  .133333E+02

 THE DETERMINANT OF A = -.960000E+00
```

B-2 MOMENT-DISTRIBUTION METHOD

The following is a program for the moment-distribution method as formulated in Sec. 13-5.

```
      PROGRAM STRUCM(INPUT,OUTPUT)
C ********
C PROGRAM FOR MOMENT DISTRIBUTION CALCULATIONS.
C IF PROGRAM IS USED WITH A TERMINAL, INCLUDE THE STATEMENTS BEGINNING
C   WITH THE LETTERS 'CT' BY REPLACING THE LETTERS 'CT' WITH BLANKS.
C INPUT IS FREE FORM.
C ********
      DIMENSION A(20,20),B(20,20),C(20,20),X1(20),X2(20),XIN(10),
     .IDPIN(10),IDCON(10,7),STIF(20),XL(10),TEMP(6),XM(20,20),DF(20)
      PRINT 4
    4 FORMAT(///" *** MOMENT DISTRIBUTION ANALYSIS ***"/,
     ./" INPUT IS IN FREE FORM."/)
CT    PRINT 8
CT  8 FORMAT(" TYPE THE NO. OF MEMBERS, THE NUMBER OF CYCLES DESIRED,"
CT   ./"  THE NO. OF SETS OF LOADS, AND INDICATE IF YOU WANT THE "/
CT   ."  RESULTS AT THE END OF EACH CYCLE TYPED OUT - TYPE 1 IF YES,"
CT   ./"  TYPE 0 IF NO.")
      READ*,NM,NCYC,NSETS,IND
CT    PRINT 20
CT 20 FORMAT(" TYPE THE MOMENTS OF INERTIA - I(1), I(2), ETC.")
      READ*,(XIN(I),I=1,NM)
CT    PRINT 24
CT 24 FORMAT(" TYPE THE MEMBER LENGTHS - L(1), L(2), ETC.")
      READ*,(XL(I),I=1,NM)
CT    PRINT 28
CT 28 FORMAT(" TYPE THE NUMBER OF JOINTS THAT ARE PINNED.")
      READ*,NPIN
      IF(NPIN.EQ.0)GO TO 34
CT    PRINT 32
CT 32 FORMAT(" TYPE THE IDENTIFICATION NUMBERS AT THE PINNED JOINTS.")
      READ*,(IDPIN(I),I=1,NPIN)
   34 CONTINUE
CT    PRINT 36
CT 36 FORMAT(" TYPE THE NO. OF JOINTS THAT HAVE MORE THAN ONE MEMBER "
CT   ./" MEETING AT THE JOINT.")
      READ*,NJTS
CT    PRINT 40
CT 40 FORMAT(" FOR EACH OF THESE JOINTS, TYPE THE NUMBER OF MEMBERS "/
CT   ." THAT MEET AT THE JOINT AND THE MEMBER IDENTIFICATION "/
CT   ." NUMBERS AT THE JOINT.")
      DO 44 I=1,NJTS
      READ*,J,(TEMP(K),K=1,J)
      L=J+1
      DO 42 K=2,L
   42 IDCON(I,K)=TEMP(K-1)
   44 IDCON(I,1)=J
C EVALUATE MEMBER STIFFNESSES.
      NMOM=2*NM
      DO 46 I=1,NMOM
   46 DF(I)=0.0
      DO 48 I=2,NMOM,2
      J=I/2
      STIF(I-1)=XIN(J)/XL(J)
   48 STIF(I)=STIF(I-1)
      IF(NPIN.EQ.0)GO TO 53
```

```
      DO 52 I=1,NPIN
      IF(IDPIN(I)/2.0.GT.IDPIN(I)/2)GO TO 50
      J=IDPIN(I)-1
      GO TO 51
   50 J=IDPIN(I)+1
   51 STIF(J)=0.75*STIF(J)
   52 DF(IDPIN(I))=1.0
C DEVELOPMENT OF A MATRIX.
   53 DO 54 I=1,NMOM
      DO 54 J=1,NMOM
   54 A(I,J)=0.0
      DO 58 I=1,NJTS
      J=IDCON(I,1)
      SUM=0.0
      DO 56 K=1,J
   56 SUM=SUM+STIF(IDCON(I,K+1))
      DO 58 K=1,J
      DF(IDCON(I,K+1))=STIF(IDCON(I,K+1))/SUM
      DO 58 L=1,J
   58 A(IDCON(I,K+1),IDCON(I,L+1))=-DF(IDCON(I,K+1))
C DEVELOPMENT OF B MATRIX
      DO 60 I=1,NMOM
      DO 60 J=1,NMOM
   60 B(I,J)=0.0
      DO 68 I=1,NMOM,2
      J1=I+1
      IF(NPIN.EQ.0)GO TO 63
      DO 62 J=1,NPIN
      IF(IDPIN(J).EQ.I)GO TO 64
   62 CONTINUE
   63 B(I,J1)=0.5
      IF(NPIN.EQ.0)GO TO 67
   64 DO 66 J=1,NPIN
      IF(IDPIN(J).EQ.J1)GO TO 68
   66 CONTINUE
   67 B(J1,I)=0.5
   68 CONTINUE
      DO 136 II=1,NSETS
CT    PRINT 70
CT 70 FORMAT(" TYPE THE ELEMENTS OF X1.")
      READ*,(X1(I),I=1,NMOM)
      IF(II.GT.1)GO TO 77
      PRINT 71
   71 FORMAT(//"INPUT VALUES:"//" MEMBER",18X,"COMPUTED"/
     ." NUMBER  LENGTH     I     DIST. FACTORS"/)
      DO 72 I=2,NMOM,2
      J=I/2
      K=I-1
   72 PRINT 74,J,XL(J),XIN(J),DF(K),DF(I)
   74 FORMAT(2X,I4,2F8.2,F10.3/22X,F10.3)
   77 PRINT 78
   78 FORMAT(//," THE FIXED-END MOMENTS (X1) ARE:")
      DO 80 I=1,NMOM
   80 PRINT 82, X1(I)
   82 FORMAT(2X,E13.5)
C BALANCE PINNED ENDS.
      IF(NPIN.EQ.0)GO TO 88
      DO 86 I=1,NPIN
      J=IDPIN(I)
      IF(J/2.0.GT.J/2)GO TO 84
      X1(J-1)=X1(J-1)-0.5*X1(J)
      GO TO 86
   84 X1(J+1)=X1(J+1)-0.5*X1(J)
   86 X1(J)=0.0
   88 IF(II-1)90,90,96
```

```
C DEVELOPMENT OF THE C MATRIX.
   90 DO 92 I=1,NMOM
      DO 92 J=1,NMOM
      C(I,J)=0.0
      DO 92 K=1,NMOM
   92 C(I,J)=C(I,J)+B(I,K)*A(K,J)
C DEVELOPMENT OF D MATRIX BY MODIFYING THE A MATRIX.
      DO 94 I=1,NMOM
   94 A(I,I)=1.0+A(I,I)
   96 DO 118 I=1,NCYC
      JJ=I+1
      DO 104 J=1,NMOM
      X2(J)=0.0
      DO 104 K=1,NMOM
  104 X2(J)=X2(J)+C(J,K)*X1(K)
      DO 108 J=1,NMOM
      XM(J,JJ)=0.0
      DO 108 K=1,NMOM
  108 XM(J,JJ)=XM(J,JJ)+A(J,K)*X1(K)
      DO 112 J=1,NMOM
  112 XM(J,JJ)=XM(J,JJ)+XM(J,I)
      DO 116 J=1,NMOM
  116 X1(J)=X2(J)
  118 CONTINUE
      IF(IND.EQ.1)GO TO 128
      PRINT 120,NCYC
  120 FORMAT(/," THE MOMENTS AT THE END OF ",I2," CYCLES ARE:")
      DO 124 I=1,NMOM
  124 PRINT 82,XM(I,JJ)
      GO TO 136
  128 DO 132 I=1,NCYC
      JJ=I+1
      PRINT 120,I
      DO 132 J=1,NMOM
  132 PRINT 82,XM(J,JJ)
  136 CONTINUE
      PRINT 140
  140 FORMAT(/" ANALYSIS COMPLETE."///)
      END
```

The following is a computer solution of Prob. 13-4. To establish the format for calculations using this program, the members of the structure are numbered from 1 through the number of members, and member identification numbers are assigned to each end of the members. Thus, for member 1, the member identification numbers are 1 and 2; for member 2, the member identification numbers are 3 and 4; etc. The order of the numbers for an individual member is arbitrary. In the following example (the beam in Fig. P13-4), the members are numbered from left to right, and the member identification numbers are assigned left end first. The computer program includes a selection for the number of loading conditions which can be used conveniently, for example, in joint translation analyses. Only the fixed-end moments need to be input after the first set is analyzed.

```
*** MOMENT DISTRIBUTION ANALYSIS ***

 INPUT IS IN FREE FORM.

 TYPE THE NO. OF MEMBERS, THE NUMBER OF CYCLES DESIRED,
  THE NO. OF SETS OF LOADS, AND INDICATE IF YOU WANT THE
  RESULTS AT THE END OF EACH CYCLE TYPED OUT - TYPE 1 IF YES,
  TYPE 0 IF NO.
? 3 6 1 1
 TYPE THE MOMENTS OF INERTIA - I(1), I(2), ETC.
? 10 15 20
 TYPE THE MEMBER LENGTHS - L(1), L(2), ETC.
? 8 10 12
 TYPE THE NUMBER OF JOINTS THAT ARE PINNED.
? 1
 TYPE THE IDENTIFICATION NUMBERS AT THE PINNED JOINTS.
? 6
 TYPE THE NO. OF JOINTS THAT HAVE MORE THAN ONE MEMBER
 MEETING AT THE JOINT.
? 2
 FOR EACH OF THESE JOINTS, TYPE THE NUMBER OF MEMBERS
 THAT MEET AT THE JOINT AND THE MEMBER IDENTIFICATION
 NUMBERS AT THE JOINT.
? 2 2 3
? 2 4 5
 TYPE THE ELEMENTS OF X1.
? -5.33 5.33 -16.95 17.55 -24 24


INPUT VALUES:

 MEMBER                     COMPUTED
 NUMBER   LENGTH     I     DIST. FACTORS

     1      8.00   10.00      0.000
                               .455
     2     10.00   15.00       .545
                               .545
     3     12.00   20.00       .455
                              1.000


 THE FIXED-END MOMENTS (X1) ARE:
    -.53300E+01
     .53300E+01
    -.16950E+02
     .17550E+02
    -.24000E+02
     .24000E+02


 THE MOMENTS AT THE END OF  1 CYCLES ARE:
    -.53300E+01
     .10612E+02
    -.10612E+02
     .27614E+02
    -.27614E+02
    0.
       .
       .
       .
       .
```

```
THE MOMENTS AT THE END OF  5 CYCLES ARE:
    -.37213E+01
     .85766E+01
    -.85766E+01
     .28491E+02
    -.28491E+02
    0.

THE MOMENTS AT THE END OF  6 CYCLES ARE:
    -.37067E+01
     .85639E+01
    -.85639E+01
     .28499E+02
    -.28499E+02
    0.

ANALYSIS COMPLETE.
```

B-3 FRAME ANALYSIS USING THE CODE-NUMBER TECHNIQUE

The following is a program for the analysis of two-dimensional frames using the matrix displacement (stiffness) method and the code-number technique as developed in Sec. 14-5. The program includes a subroutine for the solution of a banded set of equations.

```
      PROGRAM STRUCF(INPUT,OUTPUT)
C ********
C PROGRAM FOR 2-D FRAMES USING CODE NUMBERS.
C IF PROGRAM IS USED WITH A TERMINAL, INCLUDE THE STATEMENTS BEGINNING
C   WITH THE LETTERS 'CT' BY REPLACING THE LETTERS 'CT' WITH BLANKS.
C INPUT IS IN FREE FORM.
C ********
      DIMENSION BGK(100,30),XL(60),SINW(60),COSW(60),AR(60),XIN(60),
     .NC(60,6),QL(100),T(3,3),SK(6,6),XK(3,3),FEV(3,6),SQ(60,3)
      PRINT 4
    4 FORMAT(///" *** FRAME ANALYSIS USING CODE NUMBERS ***"/,
     ./" INPUT IS IN FREE FORM."/)
CT    PRINT 8
CT  8 FORMAT(" TYPE THE NO. OF MEMBERS, THE NO. OF EXTERNAL ",
CT   ."DISPLACEMENTS,"/"  AND THE VALUE OF E.")
      READ*,NM,ND,EM
CT    PRINT 12
CT 12 FORMAT(" TYPE THE MEMBER END COORDINATES, MEMBER 1, 2, ETC.,"/
CT   ." IN THE ORDER:"/"  XI  YI  XJ  YJ")
      DO 16 I=1,NM
      READ*,XI,YI,XJ,YJ
      XL(I)=SQRT((YJ-YI)**2+(XJ-XI)**2)
      SINW(I)=(YJ-YI)/XL(I)
   16 COSW(I)=(XJ-XI)/XL(I)
CT    PRINT 20
CT 20 FORMAT(" TYPE THE MEMBER AREAS - A(1), A(2), ETC.")
      READ*,(AR(I),I=1,NM)
CT    PRINT 24
CT 24 FORMAT(" TYPE THE MOMENTS OF INERTIA - I(1), I(2), ETC.")
      READ*,(XIN(I),I=1,NM)
CT    PRINT 28
```

```
CT 28 FORMAT(" TYPE THE CODE NUMBERS; MEMBER 1, 2, ETC.")
      DO 32 I=1,NM
   32 READ*,(NC(I,J),J=1,6)
CT    PRINT 36
CT 36 FORMAT(" TYPE THE ELEMENTS OF Q.")
      READ*,(QL(I),I=1,ND)
      PRINT 40
   40 FORMAT(//," INPUT VALUES:"//" MEMBER",25X,
     ."   MOMENT OF          CODE"/" NUMBER    LENGTH        AREA ",
     ."        INERTIA         NUMBERS"/)
      DO 44 I=1,NM
   44 PRINT 48,I,XL(I),AR(I),XIN(I),(NC(I,J),J=1,6)
   48 FORMAT(2X,I4,3E13.5,1X,6I3)
      PRINT 52,EM
   52 FORMAT(/" THE VALUE OF E IS ",E11.5)
      PRINT 56
   56 FORMAT(1X/" ID NO.   LOADS APPLIED "/)
      DO 60 I=1,ND
   60 PRINT 64,I,QL(I)
   64 FORMAT(2X,I3,4X,E13.5)
      IBND=0
      DO 72 I=1,NM
      DO 72 J=1,5
      JJ=J+1
      DO 72 K=JJ,6
      IF(NC(I,J).EQ.0.OR.NC(I,K).EQ.0)GO TO 72
      M=IABS(NC(I,J)-NC(I,K))
      IF(M.GT.IBND)IBND=M
   72 CONTINUE
      IBND=IBND+1
      PRINT 76,IBND
   76 FORMAT(/" THE BANDWIDTH IS ",I2)
      IF(IBND.LE.30)GO TO 79
      PRINT 77
   77 FORMAT(/" BANDWIDTH EXCEEDS DIMENSION."/)
      GO TO 200
   79 DO 80 I=1,ND
      DO 80 J=1,IBND
   80 BGK(I,J)=0.0
      DO 98 N=1,NM
      DO 82 I=1,3
      DO 82 J=1,3
      XK(I,J)=0.0
   82 T(I,J)=0.0
      T(1,1)=COSW(N)
      T(2,2)=COSW(N)
      T(1,2)=SINW(N)
      T(2,1)=-SINW(N)
      T(3,3)=1.0
      S11=AR(N)*EM/XL(N)
      S22=12.0*EM*XIN(N)/XL(N)**3
      S32=6.0*EM*XIN(N)/XL(N)**2
      S33=4.0*EM*XIN(N)/XL(N)
      XK(1,1)=S11
      XK(2,2)=S22
      XK(3,2)=S32
      XK(2,3)=S32
      XK(3,3)=S33
      CALL TTKT(T,XK)
      DO 84 I=1,3
      DO 84 J=1,3
      SK(I,J)=XK(I,J)
   84 XK(I,J)=0.0
      XK(1,1)=-S11
```

```
      XK(2,2)=-S22
      XK(3,2)=S32
      XK(2,3)=-S32
      XK(3,3)=S33/2.0
      CALL TTKT(T,XK)
      DO 86 I=1,3
      DO 86 J=1,3
      IP=I+3
      SK(IP,J)=XK(I,J)
   86 XK(I,J)=0.0
      XK(1,1)=S11
      XK(2,2)=S22
      XK(3,2)=-S32
      XK(2,3)=-S32
      XK(3,3)=S33
      CALL TTKT(T,XK)
      DO 88 I=1,3
      DO 88 J=1,3
      IP=I+3
      JP=J+3
   88 SK(IP,JP)=XK(I,J)
      DO 90 I=1,6
      DO 90 J=I,6
   90 SK(I,J)=SK(J,I)
      DO 98 I=1,6
      DO 98 J=I,6
      K=NC(N,I)
      L=NC(N,J)
      IF(K.EQ.0.OR.L.EQ.0)GO TO 98
      IF(K.LE.L)GO TO 96
      IT=K
      K=L
      L=IT
   96 IPOS=L-K+1
      BGK(K,IPOS)=BGK(K,IPOS)+SK(I,J)
   98 CONTINUE
      CALL BNDSOL(BGK,QL,ND,IBND)
      DO 104 N=1,NM
      X=6.0*EM*XIN(N)/XL(N)**2
      FEV(1,1)=-SINW(N)*X
      FEV(1,2)=COSW(N)*X
      FEV(1,3)=4.0*EM*XIN(N)/XL(N)
      FEV(1,4)=-FEV(1,1)
      FEV(1,5)=-FEV(1,2)
      FEV(1,6)=FEV(1,3)/2.0
      FEV(2,1)=FEV(1,1)
      FEV(2,2)=FEV(1,2)
      FEV(2,3)=FEV(1,6)
      FEV(2,4)=FEV(1,4)
      FEV(2,5)=FEV(1,5)
      FEV(2,6)=FEV(1,3)
      X=AR(N)*EM/XL(N)
      FEV(3,1)=-COSW(N)*X
      FEV(3,2)=-SINW(N)*X
      FEV(3,3)=0.0
      FEV(3,4)=-FEV(3,1)
      FEV(3,5)=-FEV(3,2)
      FEV(3,6)=0.0
      DO 104 I=1,3
      SQ(N,I)=0.0
      DO 104 J=1,6
      K=NC(N,J)
      IF(K.EQ.0)GO TO 104
      SQ(N,I)=SQ(N,I)+FEV(I,J)*QL(K)
  104 CONTINUE
      PRINT 108
```

```
108 FORMAT(//" OUTPUT VALUES "//"  ID NO. DEFLECTIONS",13X,
   ."MEMBER FORCES"/22X,"MOMENT(I)     ",
   ."MOMENT(J)      AXIAL(J)"/)
    IF(NM-ND)112,122,122
112 DO 114 I=1,NM
114 PRINT 116,I,QL(I),(SQ(I,J),J=1,3)
116 FORMAT(3X,I3,4E13.5)
    J=NM+1
    DO 118 I=J,ND
118 PRINT 120,I,QL(I)
120 FORMAT(3X,I3,E13.5)
    GO TO 130
122 DO 124 I=1,ND
124 PRINT 116,I,QL(I),(SQ(I,J),J=1,3)
    IF(NM.EQ.ND)GO TO 130
    J=ND+1
    DO 126 I=J,NM
126 PRINT 128,I,(SQ(I,K),K=1,3)
128 FORMAT(3X,I3,13X,3E13.5)
130 PRINT 132
132 FORMAT(/" ANALYSIS COMPLETE."///)
200 STOP
    END
    SUBROUTINE TTKT(T,XK)
    DIMENSION TEM(3,3),T(3,3),XK(3,3)
    DO 4 I=1,3
    DO 4 J=1,3
    TEM(I,J)=0.0
    DO 4 K=1,3
  4 TEM(I,J)=TEM(I,J)+T(K,I)*XK(K,J)
    DO 6 I=1,3
    DO 6 J=1,3
    XK(I,J)=0.0
    DO 6 K=1,3
  6 XK(I,J)=XK(I,J)+TEM(I,K)*T(K,J)
    RETURN
    END
    SUBROUTINE BNDSOL(BGK,Q,NDIS,MB)
    DIMENSION BGK(100,30),Q(100),F(30)
    N=0
500 N=N+1
    Q(N)=Q(N)/BGK(N,1)
    IF(N-NDIS)550,700,550
550 DO 600 K=2,MB
    F(K)=BGK(N,K)
600 BGK(N,K)=BGK(N,K)/BGK(N,1)
    DO 660 L=2,MB
    I=N+L-1
    IF(NDIS-I)660,640,640
640 J=0
    DO 650 K=L,MB
    J=J+1
650 BGK(I,J)=BGK(I,J)-F(L)*BGK(N,K)
    Q(I)=Q(I)-F(L)*Q(N)
660 CONTINUE
    GO TO 500
700 N=N-1
    IF(N)750,900,750
750 DO 800 K=2,MB
    L=N+K-1
    IF(NDIS-L)800,770,770
770 Q(N)=Q(N)-BGK(N,K)*Q(L)
800 CONTINUE
    GO TO 700
900 RETURN
    END
```

The following is a computer solution of the frame in Example 14-4 using the above program with a remote terminal.

```
*** FRAME ANALYSIS USING CODE NUMBERS ***
INPUT IS IN FREE FORM.
 TYPE THE NO. OF MEMBERS, THE NO. OF EXTERNAL DISPLACEMENTS,
  AND THE VALUE OF E.
? 3 7 4320000
 TYPE THE MEMBER END COORDINATES, MEMBER 1, 2, ETC.,
 IN THE ORDER:
  XI  YI  XJ  YJ
? 0 0 12 16
? 12 16 27 16
? 27 0 27 16
 TYPE THE MEMBER AREAS - A(1), A(2), ETC.
? 0.1389 0.1389 0.1389
 TYPE THE MOMENTS OF INERTIA - I(1), I(2), ETC.
? 0.004823 0.004823 0.004823
 TYPE THE CODE NUMBERS; MEMBER 1, 2, ETC.
? 0 0 0 1 2 3
? 1 2 3 4 5 6
? 0 0 7 4 5 6
 TYPE THE ELEMENTS OF Q.
? 0 -18 -45 -6.33 -18 16.88 46.88

INPUT VALUES:

MEMBER                                    MOMENT OF           CODE
NUMBER    LENGTH        AREA              INERTIA            NUMBERS

    1    .20000E+02   .13890E+00   .48230E-02   0  0  0  1  2  3
    2    .15000E+02   .13890E+00   .48230E-02   1  2  3  4  5  6
    3    .16000E+02   .13890E+00   .48230E-02   0  0  7  4  5  6

THE VALUE OF E IS  .43200E+07

ID NO.   LOADS APPLIED

   1       0.
   2       -.18000E+02
   3       -.45000E+02
   4       -.63300E+01
   5       -.18000E+02
   6        .16880E+02
   7        .46880E+02

THE BANDWIDTH IS  6

OUTPUT VALUES

 ID NO. DEFLECTIONS                    MEMBER FORCES
                          MOMENT(I)     MOMENT(J)      AXIAL(J)

    1    .26850E-01    .38283E+00   -.98677E+01   -.18431E+02
    2   -.20906E-01   -.35132E+02   -.17966E+02   -.11438E+02
    3   -.49198E-02    .46880E+02    .34846E+02   -.21540E+02
    4    .26564E-01
    5   -.57435E-03
    6    .12594E-02
    7    .58800E-02

ANALYSIS COMPLETE.
```

APPENDIX

C

CONVERSIONS BETWEEN SI AND U.S. CUSTOMARY UNITS

The International System of Units, generally known as the SI system, consists of seven base units: meter, kilogram, second, ampere, kelvin, mole, and candela, and two supplementary units: radian and steradian Complete definitions of these units and detailed explanations of the use of the SI system can be found in guides for the use of the SI system.[1,2] Of particular interest in structural engineering are the quantities of length, mass, time, and temperature which in the SI system are described respectively in the base units of meter (m), kilogram (kg), second (s), and kelvin (K). The following definitions are presented as aids in converting quantities between the SI system and the U.S. Customary (USC) system.

Length

The basic unit of length in the SI system is the *meter* (m). The recommended submultiple of the meter is the *millimeter* (mm) which is equal

[1] "Metric Practice Guide," E 380-74, American Society for Testing and Materials (ASTM).

[2] "Unit of Weights and Measures (United States Customary and Metric): Definitions and Tables of Equivalents," NBS misc. publ. MP286, May, 1967.

to 10^{-3} m. The relationships between these SI units of length and generally used U.S. Customary units are:

$$1 \text{ ft} = 0.3048 \text{ m} \qquad 1 \text{ m} = 3.281 \text{ ft}$$
$$1 \text{ in.} = 25.40 \text{ mm} \qquad 1 \text{ mm} = 0.03937 \text{ in.}$$

Mass

The basic unit of mass in the SI system is the *kilogram* (kg). The kilogram is a nongravimetric unit. The U.S. Customary unit of *pound-mass avoirdupois* (lbm) is related to the kilogram as

$$1 \text{ lbm (avdp)} = 0.4536 \text{ kg} \qquad 1 \text{ kg} = 2.205 \text{ lbm (avdp)}$$

Time

The basic unit of time in the SI system is the *second* (s).

Temperature

The basic unit of temperature in the SI system is the *kelvin* (K). Wide use is also made of the Celsius (C) scale. One degree Celsius is equal to an interval of 1 K. The Celsius temperature 0°C is equal to 273.15 K. The relationships of temperature t between the kelvin, Celsius, and Fahrenheit scales are:

$$t_F^\circ = \tfrac{9}{5}t_C^\circ + 32 \qquad t_C^\circ = \tfrac{5}{9}(t_F^\circ - 32)$$
$$t_F^\circ = \tfrac{9}{5}t_K - 459.67 \qquad t_K = \tfrac{5}{9}(t_F^\circ + 459.67)$$
$$t_C^\circ = t_K - 273.15 \qquad t_K = t_C^\circ + 273.15$$

Force and Weight

The units of force and weight in the SI system are derived from the base units. Weight is defined as the force required to restrain a body against the acceleration of gravity. The unit of force in the SI system is the *newton* (N). A newton is defined as the force that gives an acceleration of 1 m/s² to a mass of 1 kg. Using the standard value for the acceleration of gravity g of 9.807 m/s², we obtain the expression for the weight W of a 1-kg mass:

$$W = mg = (1 \text{ kg})(9.807 \text{ m/s}^2) = 9.807 \text{ N}$$

A *pound-force* (lbf) is defined as the force equal to the weight of 1 lbm subjected to the standard value of the acceleration of gravity. Using the above relationship between the pound-mass and the kilogram, we obtain

$$1 \text{ lbf} = (0.4536 \text{ kg})(9.807 \text{ m/s}^2) = 4.448 \text{ N}$$

A commonly used unit of force in the U.S. Customary system is the *kilopound* (kip) which is equal to a force of 1000 lb. Therefore

$$1 \text{ kip} = 4.448 \times 10^3 \text{ N}$$

Stress

The unit of stress in the SI system is the *Pascal* (Pa) which is defined as 1 N/m².

It is recommended in using the SI system that prefixes be used to indicate orders of magnitude of the units. A partial listing of the prefixes and the symbols that are used with SI units and the corresponding multiples and submultiples is given in Table C-1.

Thus, an E value of 200×10^9 N/m² (Pa) can be expressed as 200 GPa. It is recommended that numbers having four or more digits be placed in groups of three separated by a space rather than a comma. This avoids confusion in countries where the comma is used for the decimal point. Thus, for example, 6,422.4 should be written as 6 422.4 or 6422.4.

To perform a detailed conversion of a quantity from one system to another, care should be taken to account for all units as well as the numerical values. Thus, to convert the U.S. Customary quantity 1 ksi to the SI units of Pascals (N/m²), we can proceed as follows:

$$1 \text{ ksi} = \left(\frac{1 \text{ kip}}{1 \text{ in.}^2}\right)\left(\frac{1000 \text{ lb}}{1 \text{ kip}}\right)\left(\frac{1 \text{ in.}}{0.0254 \text{ m}}\right)^2\left(\frac{1 \text{ N}}{0.2248 \text{ lb}}\right)$$

Canceling out units when they appear in both the numerator and the denominator and carrying out the numerical computations, we obtain

$$1 \text{ ksi} = 6.895 \times 10^6 \text{ N/m}^2 = 6.895 \text{ MPa}$$

TABLE C-1

Symbol	Prefix	Multiples and Submultiples
G	giga	10^9
M	mega	10^6
k	kilo	10^3
m	milli	10^{-3}
μ	micro	10^{-6}
n	nano	10^{-9}

Conversion factors for quantities commonly encountered in structural engineering are summarized in the following:

Dimensions

1 in.	= 25.40 mm	1 mm	= 39.37×10^{-3} in.
	= 2.540 cm	1 cm	= 0.3937 in.
	= 25.40×10^{-3} m	1 m	= 39.37 in.
1 ft	= 304.8 mm	1 mm	= 3.281×10^{-3} ft
	= 30.48 cm	1 cm	= 32.81×10^{-3} ft
	= 0.3048 m	1 m	= 3.281 ft

Area

1 in.^{2}	= 645.2 mm^2	1 mm^2	= 1.550×10^{-3} in.^{2}
	= 6.452 cm^2	1 cm^2	= 0.1550 in.^{2}
	= 645.2×10^{-6} m^2	1 m^2	= 1550 in.^{2}
1 ft^2	= 92.90×10^{3} mm^2	1 mm^2	= 10.76×10^{-6} ft^2
	= 929.0 cm^2	1 cm^2	= 1.076×10^{-3} ft^2
	= 92.90×10^{-3} m^2	1 m^2	= 10.76 ft^2

Section modulus

1 in.^{3}	= $16\ 39 \times 10^{3}$ mm^3	1 mm^3	= 61.02×10^{-6} in.^{3}
	= 16.39 cm^3	1 cm^3	= 61.02×10^{-3} in.^{3}
	= 16.39×10^{-6} m^3	1 m^3	= 61.02×10^{3} in.^{3}
1 ft^3	= 28.32×10^{-3} m^3	1 m^3	= 35.31 ft^3

Moment of inertia

1 in.^{4}	= 0.4162×10^{6} mm^4	1 mm^4	= 2.403×10^{-6} in.^{4}
	= 41.62 cm^4	1 cm^4	= 24.03×10^{-3} in.^{4}
	= 0.4162×10^{-6} m^4	1 m^4	= 2.403×10^{6} in.^{4}
1 ft^4	= 8.631×10^{-3} m^4	1 m^4	= 115.9 ft^4

Loading

1 lb	= 4.448 N	1 N	= 0.2248 lb
1 kip	= 4.448 kN	1 kN	= 0.2248 kip
1 lb/ft	= 14.59 N/m	1 N/m	= 0.6853×10^{-3} lb/ft
1 kip/ft	= 14.59 kN/m	1 kN/m	= 0.6853×10^{-3} kip/ft

Moment

1 in-lb = 0.1130 N-m   1 N-m = 8.850 in-lb
1 ft-lb = 1.356 N-m   1 N-m = 0.7376 ft-lb
1 ft-kip = 1.356 kN-m   1 kN-m = 0.7376 ft-kip

Stress, pressure, and modulus of elasticity

1 psi = 6.895 kPa (kN/m^2)   1 Pa = 0.1450×10^{-3} psi
1 ksi = 6.895 MPa   1 Pa = 0.1450×10^{-6} ksi
1 psf = 47.88 Pa   1 Pa = 20.88×10^{-3} psf

Unit weight

1 lb/ft^3 = 157.1 N/m^3 1 N/m^3 = 6.366×10^{-3} lb/ft^3

Expressions for converting temperature were given in a previous paragraph.

ANSWERS TO EVEN-NUMBERED PROBLEMS

Chapter 1

1-2. See Appendix C.

1-4. 8000 lb = 35.58 kN; 32,000 lb = 142.34 kN; 14 ft = 4.27 m; 6 ft = 1.83 m; 14 to 30 ft = 4.27 to 9.14 m.

1-6. $q = 0.0472V^2$.

1-8. 40 psf = 1.915 kN/m^2.

Chapter 2

2-2. $R_{BH} = 0$, $R_{BV} = 60$ kN$\uparrow$, $R_{DH} = 0$, $R_{DV} = 13.13$ kN$\uparrow$, $R_{EH} =$ 40 kN$\rightarrow$, $R_{EV} = 69.38$ kN$\uparrow$.

2-4. $R_{AH} = 36$ kips$\leftarrow$, $R_{AV} = 3$ kips$\downarrow$, $R_{DV} = 15$ kips$\uparrow$, $R_{DH} = 0$.

2-6. (*a*) $R_{AV} = 31.25$ kN$\uparrow$, $R_{BV} = 28.75$ kN$\uparrow$, $R_{AH} = R_{BH} = 0$; (*b*) V to the right of $A = +31.25$ kN, $V = 0$ 4.625 m right of A, V to the left of $B = -28.75$ kN, $M_{\max}$ at zero shear $= +84.45$ kN-m.

2-8. (*a*) $R_{AV} = 2.44$ kips$\downarrow$, $R_{BV} = 0.44$ kips$\uparrow$, $R_{AH} = R_{BH} = 0$; (*b*) V to the left of $A = 0$, V to the right of $A = -2.44$ kips, V to the left of $B = -0.44$ kip, M at $A = +4.0$ ft-kips, M at 4-kip load $= -3.33$ ft-kips.

2-10. $R_{AH} = 0$, $R_{AV} = 4$ kips$\uparrow$, $R_{BH} = 0$, $R_{BV} = 17$ kips$\uparrow$, V to the right of $A = +4$ kips, V to the right of 9-kip load $= -5$ kips, V to the left of $B = -9$ kips, V to the right of $B = +8$ kips, M at 9-kip load $= +16$ ft-kips, M at $B = -32$ ft-kips.

2-12. $R_{BH} = 0$, $R_{BV} = 3.75$ kN↑, $R_{CH} = 5$ kN→, $R_{CV} = 7.25$ kN↑, V to the right of $A = -2$ kN, V to the right of $B = +1.75$ kN, V to the left of $C = -1.25$ kN, V to the right of $C = +6$, $V_D = 0$, $V = 0$ at 0.42 m left of C, M at $B = -3.0$ kN-m, M at zero shear $= -0.74$ kN-m, M at $C = -1.0$ kN-m, M at $D = +5.0$ kN-m.

2-14. $R_{AH} = 1.52$ kips←, $R_{AV} = 10.03$ kips↑, $R_B = 2.54$ kips↘, $R_{BH} = 1.52$ kips→, $R_{BV} = 2.03$ kips↓, $V_C = 0$, V to the left of $E = -8$ kips, V to the right of $E = +2.03$ kips, V to the left of $D = +2.03$ kips, $M_C = 0$, M to the left of $E = -32.0$ ft-kips, M to the right of $E = -9.2$ ft-kips, $M_D = +11.14$ ft-kips.

2-16. With x measured in meters from the left end: $V = 0$ at $x = 0$, $V = 0.436$ MN at $x = 2$, $V = -0.545$ MN at $x = 3$, $V = 0.545$ MN at $x = 8$, $V = -0.436$ MN at $x = 9$, $V = 0$ at $x = 11$. Moment is symmetrical about centerline, $M = 0$ at $x = 0$, $M = 0.436$ MN-m at $x = 2$, $M = 0.533$ MN-m at $x = 2.44$, $M = 0.382$ MN-m at $x = 3$, $M = -0.3$ MN-m at $x = 5.5$.

2-18. Moment diagram on bottom side from B to C, on top side from C to D. $M_B = 0$, M at 10-kip load $= 80$ ft-kips, M to the left of $C = 120$ ft-kips, M to the right of $C = 42$ ft-kips, $M_D = 0$.

2-20. (*a*) $R_{AV} = 37.5$ kN↓, $R_{AH} = 250$ kN←, $R_{EV} = 237.5$ kN↑, $R_{EH} = 0$; (*b*) moment diagram on right side of AB, on bottom side of BD, zero for DE, $M_B = 625.0$ kN-m, M at 200 kN load $=$ 475 kN-m.

2-22. (*a*) $R_{AH} = 18$ kips←, $R_{AV} = 1.67$ kips↓, $R_{CH} = 0$, $R_{CV} = 9.67$ kips↑; (*b*) moment diagram on outside faces of members, $M_A = 0$, $M_B = 72$ ft-kips, M at 5-kip load $= 65.33$ ft-kips, M at 3-kip load $= 38.67$ ft-kips, $M_C = 0$.

2-24. (*a*) $R_{AH} = 9$ kN←, $R_{AV} = 7.625$ kN↓, $R_{DH} = 0$, $R_{DV} = 12.625$ kN↑; (*b*) moment diagram on right side of AB, on left side of BC, on bottom side of BD, $M_A = 0$, M at B on $BA = 32.5$ kN-m, M at B on $BC = 8.0$ kN-m, M at B on $BD = 40.5$ kN-m, M at 5-kN load $= 25.25$ kN-m, $M_C = M_D = 0$.

2-26. See answers for Prob. 3-6.

Chapter 3

3-2. $R_{9V} = 44$ kN↑, $R_{9H} = 15$ kN→, $R_{10V} = 16$ kN↑, partial results: $P_1 = -32.0$ kN, $P_8 = +25.0$ kN, $P_{10} = -45.0$ kN.

3-4. $R_{AV} = 10.5$ kips↑, $R_{BV} = 4.5$ kips↑, $R_{AH} = R_{DH} = 0$, partial results: $P_5 = -9.56$ kips, $P_{12} = 0$, $P_{16} = +14.06$ kips.

3-6. $R_{1H} = 23.81$ kips←, $R_{1V} = 20.0$ kips↑, $R_{3H} = 23.81$ kips→, $R_{3V} = 0$, $P_1 = +23.81$ kips, $P_2 = +20.0$ kips, $P_3 = -10.36$ kips,

$P_4 = -20.83$ kips, $P_5 = -2.5$ kips, $P_6 = +19.43$ kips, $P_7 = -16.67$ kips.

3-8. $R_{1H} = 65.21\rightarrow$, $R_{1V} = 112.9\uparrow$, $R_{2H} = 0$, $R_{2V} = 138.1\downarrow$, $R_5 =$ $P_1 = 92.23$, $P_2 = -118.5$, $P_3 = 71.52$, $P_4 = 101.1$, $P_5 = -130.4$ (kN).

3-10. $R_{1H} = R_{15H} = 0$, $R_{1V} = R_{15V} = 57.44\uparrow$, partial results: $P_1 = -112.38$, $P_6 = +86.16$, $P_{11} = +57.44$ (kN).

3-12. $P_1 = -50$, $P_2 = -41.23$, $P_3 = +74.54$, $P_4 = +106.7$ (kN).

3-14. $R_{HV} = 19.33\uparrow$, $R_{EV} = 10.67\uparrow$, $R_{EH} = 15.0\leftarrow$, $R_{HH} = 0$, $P_1 = +37.57$, $P_2 = +32.22$, $P_3 = -27.85$, $P_4 = +10.36$ (kN).

3-16. $P_1 = -9.33$, $P_2 = +21.33$, $P_3 = -16.67$, $P_4 = -13.20$ (k).

3-18. See answers for Prob. 3-8.

3-20. (*a*) Stable, indeterminate, third degree; (*c*) unstable; (*e*) unstable.

3-22. See answers for Prob. 3-6.

Chapter 4

4-2. $P_1 = 155.59$, $P_2 = P_3 = -94.34$, $P_4 = P_6 = -68.75$, $P_5 = 71.25$, $R_{2Y} = -110$, $R_{2Z} = 0$, $R_{3X} = R_{4X} = -15$, $R_{3Y} = R_{4Y} = 80$ (kN).

4-4. $R_{2Y} = -50.00$, $R_{2Z} = -40.00$, $R_{3X} = -58.33$, $R_{3Y} = -13.33$, $R_{4X} = +48.33$, $R_{4Y} = +93.33$, $P_1 = +70.72$, $P_2 = +15.72$, $P_3 = -110.06$, $P_4 = -64.58$, $P_5 = +33.75$, $P_6 = +2.08$ (kN).

4-6. (*a*) Members 1, 2, 3, 7, 8, 9, 10, 14, and 15; (*b*) $P_4 = -4$, $P_6 = -5.92$; (*c*) $R_{7Y} = 0$, $R_{6Z} = 1.0$, $R_{8X} = 1.33$ (kN).

4-8. (*a*) Members 1, 2, 3, 4, 5, 6, and 9; (*b*) $R_{5Z} = +33.33$, $R_{6Y} = -31.25$; (*c*) $P_4 = 0$, $P_7 = +39.16$ (kips).

4-10. $P_1 = 18.78$, $P_2 = 3.75$, $P_3 = -14.22$, $P_4 = -14.22$, $P_5 = -13.77$, $P_6 = -2.50$, $P_7 = -2.75$, $P_8 = 2.50$, $R_{2Y} = -12.53$, $R_{3X} = -2.00$, $R_{3Y} = -2.50$, $R_{4Y} = 12.51$, $R_{4Z} = 3.51$, $R_{5Y} = 12.51$, $R_{5Z} = -7.51$ (kN).

Chapter 5

5-2. (*a*) $R_{AV} = R_{CV} = 400.0$ kN$\uparrow$, $R_{AH} = -R_{CH} = 333.3$ kN$\rightarrow$; (*b*) thrust at $D = 388.7$ compression, $M_D = V_D = 0$.

5-4. (*a*) $R_{AH} = 7.44\rightarrow$, $R_{AV} = 18.00\uparrow$, $R_{EH} = 0.56\rightarrow$, $R_{EV} = 2.00\uparrow$ (kips); (*b*) $M_A = M_D = 0$, $M_B = 35.33$, $M_C = 9.04$ (ft-kips) on outer face.

5-6. (*a*) $R_{AH} = 18.33\leftarrow$, $R_{AV} = 8.75\downarrow$, $R_{CH} = 11.67\leftarrow$, $R_{CV} = 8.75\uparrow$; (*b*) $P_{DE} = -35.00$, $P_{BF} = +24.92$, $P_{GH} = -27.67$ (kips).

5-8. (*a*) $d_E = 8.52$ m; (*b*) T_{max} in $EF = 331.58$ kN; (*c*) $R_{BV} = 306.12$ kN$\uparrow$, $R_{BH} = 153.06$ kN$\leftarrow$.

5-10. (*a*) $R_{AH} = 15.23\leftarrow$, $R_{AV} = 4.48\uparrow$; (*b*) $T_{\max}$ at $C = 16.20$; (*c*) $P_{DE} = -41.04$ (kips).

5-12. (*a*) $d_F = 7.93$ m; (*b*) $T_{\max}$ at $B = 252.65$ kN; (*c*) $T_{AB} = 437.82$ kN; (*d*) $T_{GI} = 404.77$ kN.

Chapter 6

6-2. Partial results: (*a*) $P_{10} = 0$, $P_{11} = +10.67$, $P_{17} = 0$, $P_{16} = +24.00$; (*b*) $P_{11} = -P_{10} = +5.33$, $P_{16} = -P_{17} = +12.00$ (kips).

6-4. (*a*) $R_{AH} = 12\rightarrow$, $R_{AV} = 9.33\uparrow$, $R_{GH} = 12\rightarrow$, $R_{GV} = 9.33\downarrow$; (*b*) moment diagram on left side, $M_B = 72$, $M_A = M_C = 0$; (*c*) $P_{BD} = -36$, $P_{CE} = +10.0$, $P_{DE} = -9.33$, $P_{EF} = +16.82$ (kN and kN-m).

6-6. (*a*) $A_H = 49.46\leftarrow$, $A_V = 36.41\downarrow$, $M_A = 89.08\circlearrowleft$, $H_H = 43.54\leftarrow$, $H_V = 30.39\downarrow$, $M_H = 83.15\circlearrowleft$; (*b*) Moment-diagram values at intervals of 1.5 m starting with A: 89.08 (right side), 23.77 (right side), 23.76 (left side), 53.52 (left side), $M_C = 0$; (*c*) $P_{BD} = 49.50$, $P_{DE} = -29.51$, $P_{FG} = 23.44$ (kN and kN-m).

6-8. Partial results: $P_{BC} = -3.75$, $P_{DG} = 12.5$, $M_{\max}$ in $BE = 11.25$, $M_{\max}$ in $IL = 33.75$ (kN and kN-m).

6-10. Partial results: $V_{BF} = 6.67$, $P_{DH} = 2.00$, $M_{\max}$ in $AE = 5.00$, $V_{FJ} = 20.00$, $P_{HL} = 12.00$, $M_{\max}$ in $EI = 20.00$ (kN and kN-m).

6-12. Partial results: $P_{AC} = 1.12$, $V_{AC} = 1.50$, $M_{\max}$ in $AC = 9.00$, $P_{CF} = 2.64$, $V_{EH} = 4.04$, $M_{\max}$ in $EH = 28.28$ (kips and ft-kips).

Chapter 7

7-2. $\theta_D = 1.79 \times 10^{-3}$ rad$\circlearrowright$; $\Delta_D = 3.57$ mm$\downarrow$.

7-4. (*a*) $\theta_B = 1.38 \times 10^{-2}$ rad$\circlearrowright$; (*b*) $\Delta_B = 43.56$ mm$\downarrow$.

7-6. $\Delta_D = 0.192$ in.$\uparrow$.

7-8. (*a*) $\theta_A = 8.06 \times 10^{-3}$ rad$\circlearrowright$; (*b*) $\Delta_A = 13.89$ mm$\downarrow$.

7-10. (*a*) $\Delta_A = 9.72$ mm$\uparrow$; (*b*) $\theta_B = 4.05 \times 10^{-3}$ rad$\circlearrowright$.

7-12. (*a*) $\Delta_C = 12.2$ mm$\uparrow$; (*b*) $\Delta_{\max}$, 2.58 m from A, $= 12.9$ mm$\downarrow$.

7-14. $\Delta_{DH} = 14.5$ mm$\rightarrow$.

7-16. See answers for Prob. 7-2.

7-18. $\Delta_{\max}$, 3.366 m from A, $= 10.6$ mm$\downarrow$.

7-20. See answers for Prob. 7-8.

7-22. From left to right: 0, 0.10, 0.19, 0.27, 0.32, 0.34, 0.29, 0.16, 0 (down, in.).

7-24. Deflection of ③ is to the right and up.

7-26. See answers for Prob. 8-8.

Chapter 8

8-2. See answers for Prob. 8-8.

8-4. $D_D{}^H = 0.382$ in.←.

8-6. $\Delta_B = 11.89$ mm↓.

8-8. (*a*) $D_3{}^H = 0.150$ in.←; (*b*) $D_5{}^V = 0.648$ in.↓, $D_5{}^H = 0.227$ in.←.

8-10. (*a*) $D_{5H} = 0.014$ in.←; (*b*) $D_{5V} = 0.226$ in.↓.

8-12. See answers for Prob. 7-4.

8-14. See answers for Prob. 7-10.

8-16. (*a*) $D_A{}^V = 0.619$ in.↓; (*b*) $D_A{}^H = 0.263$ in.→; (*c*) $D_B{}^R = 5.91 \times 10^{-3}$ rad↻.

8-18. (*a*) $D_A{}^V = 8.67$ mm↓; (*b*) $D_C{}^V = 21.00$ mm↓.

8-20. $\Delta_{DH} = 2.02$ in.→, $\theta_D = 0.00496$ rad↺.

8-22. $\Delta_D = 0.641$ in. down the incline.

8-24. (*a*) $D_C{}^V = 0.289$ in.↓; (*b*) $D_C{}^V = 0.039$ in.↓.

8-26. See answer for Prob. 8-4.

8-28. See answers for Prob. 8-18.

Chapter 9

9-2. $f_{3X2Y} = -0.562L/EA$.

9-4. (*a*) $\begin{Bmatrix} D_{2Y} \\ D_{3Y} \end{Bmatrix} = \dfrac{L^3}{EI} \begin{bmatrix} 0.0117 & 0.0143 \\ 0.0143 & 0.0208 \end{bmatrix} \begin{Bmatrix} Q_{2Y} \\ Q_{3Y} \end{Bmatrix}$

(*b*) $\begin{Bmatrix} Q_{2Y} \\ Q_{3Y} \end{Bmatrix} = \dfrac{EI}{L^3} \begin{bmatrix} 535 & -368 \\ -368 & 301 \end{bmatrix} \begin{Bmatrix} D_{2Y} \\ D_{3Y} \end{Bmatrix}$

9-6. (*a*) $e = D \cos \omega$, $P = (EA/L)e = (EA/L)D \cos \omega$; (*b*) $\Sigma F_H = 0$ at ①: $Q = P \cos \omega = (EA/L) \cos^2 \omega D$.

Chapter 10

10-2. (*a*) $A = +1.5$, $B = +1.0$, $D = 0$, $E = -0.33$; (*b*) $A = +0.5$, $B = 0$, left of $D = -1.0$, right of $D = 0$, $E = -0.33$; (*c*) $A = +0.5$, $B = 0$, left of $C = -0.58$, right of $C = +0.42$, $D = 0$, $E = -0.33$; (*d*) $A = -2.50$, $B = 0$, $C = +2.92$, $D = 0$, $E = -2.33$; (*e*) $A = -6.0$, $B = C = D = E = 0$.

10-4. (*a*) $A = +1.0$, $B = C = D = 0$; (*b*) $B = +1.25$, $A = D = 0$; (*c*) $B = +0.25$, to the right of $C = +1.0$, $A = D =$ to the left of $C = 0$; (*d*) $B = -5.0$, $A = C = D = 0$.

10-6. P_1: $G = -1.5$, $C = D = E = F = 0$; P_2: $D = +0.707$, $G = +1.414$, $C = E = 0$; P_3: $D = -0.5$, $G = +1.0$, $C = E = 0$; P_4: $D = +0.5$, $G = -1.0$, $C = E = 0$.

10-8. From ④ to ⑧; P_1: 0, +1.250, +0.833, +0.417, 0; P_2: 0, −0.729, +0.486, +0.243, 0; P_3: 0, −0.637, −1.275, −0.637, 0; P_4: 0, 0, 0, +1.0, 0.

10-10. (*a*) $A = B = C = 0$, $D = +0.375$, $E = +1.0$; (*b*) $A = B = C = E = 0$, $D = 0.625$; (*c*) $A = D = E = 0$, $B = +5.83$, $C = +1.67$.

10-12. (*a*) From A to B, $R_D = 10.0$ kN↓; (*b*) from B to D; $M_C = 30.0$ kN-m.

10-14. (*a*) From A to C, $R_D = 48.0$ kN↓; (*b*) from A to 2 m right of I; $P_2 = -42.4$ kN.

10-16. (*a*) Positive from A to D, negative from D to E; (*b*) positive from A to B and B to D, negative from D to E; (*c*) positive from A to B, negative from B to D and D to E; (*d*) negative from A to B and D to E, positive from B to D.

10-18. (*a*) Left and center bay of both the floor and roof; (*b*) the left and right bay of the floor and the center bay of the roof; (*c*) the center bay of the floor and the left and right bay of the roof.

10-20. With the 80-kN load at B, $M_C = -513.3$ kN-m.

Chapter 11

11-2. $P_1 = +45.68$, $P_2 = -6.43$, $P_3 = -37.65$, $P_4 = P_5 = +22.59$ (kN).

11-4. $P_1 = -0.87$, $P_2 = +0.90$, $P_3 = +1.26$, $P_4 = -2.14$, $P_5 = -2.13$, $P_6 = +1.29$, $R_{3H} = 2.0$←, $R_{3V} = 2.0$↓, $R_{4H} = 0$, $R_{4V} = 4.0$↑ (kips).

11-6. $R_A = 4.28$, $R_B = 20.07$, $R_C = 10.65$ (kips).

11-8. $R_A = 2.65$, $R_B = 4.25$, $R_C = 0.10$ (kips).

11-10. (*a*) $R_{AH} = 45.0$ kN←, $R_{AV} = 5.62$ kN↓, $M_A = 50.64$ kN-m↻, $R_{CV} = 5.62$ kN↑; (*b*) at A: 50.64 on the left side of the member, at B on BA: 16.86 on the right side of the member, at B on BC: 16.86 on the bottom side of the member, $M_C = 0$.

11-12. $R_{AV} = 40.74$↑, $R_{AH} = 5.03$→, $R_{DV} = 39.26$↑, $R_{DH} = 55.03$← (kN).

11-14. $R_{AH} = 8.00$←, $R_{AV} = 1.19$↓, $R_{CV} = 1.51$↑, $R_{DV} = 0.32$↓ (kips).

11-16. $M_A = -43.8$ ft-kips, $P_{BD} = 12.7$ kips.

11-18. $P_1 = -1.07$, $P_2 = 0.96$, $P_3 = -3.41$, $P_4 = -1.96$ (kips).

11-20. (*a*) $M_A = -4.27$ ft-kips, $M_B = -7.47$ ft-kips, $M_C = -14.13$ ft-kips, $M_D = 0$; (*b*) $R_A = 3.60$ kips↑, $R_B = 10.33$ kips↑, $R_C = 9.25$ kips↑, $R_D = 6.82$ kips↑, partial results: $V = 0$ 3.60 ft right of A and $M = +2.21$ ft-kips, $V = 0$ 4.55 ft left of D and $M = +15.52$ ft-kips.

11-22. (*a*) $FEM_{AB} = -13.8$ ft-kips, $FEM_{BA} = 4.2$ ft-kips; (*b*) $FEM_{AB} = -5wL^2/192$, $FEM_{BA} = 11wL^2/192$; (*c*) $FEM_{AB} = -10.94$ kN-m, $FEM_{BA} = 12.81$ kN-m; (*d*) $FEM_{BA} = FEM_{AB} = M/4$ (clockwise on member positive).

Chapter 12

12-2. $M_{AB} = -10.24$, $M_{BA} = -M_{BC} = 4.52$, $M_{CB} = -M_{CD} = 10.00$, $M_{DC} = 0$ (ft-kips).

12-4. $M_{AB} = 0$, $M_{BA} = -M_{BC} = 17.26$, $M_{CB} = -M_{CD} = 16.87$, $M_{DC} = 21.57$ (kN-m).

12-6. $M_{12} = -11.35$, $M_{21} = +20.51$, $M_{23} = -22.35$, $M_{32} = +15.45$, $M_{34} = -11.88$, $M_{43} = -5.94$, $M_{25} = +1.84$, $M_{36} = -3.57$, $M_{52} = M_{63} = 0$ (kN-m).

12-8. $M_{AB} = -23.86$, $M_{BA} = -M_{BC} = -8.71$, $M_{CB} = -M_{CD} = 45.29$, $M_{DC} = -42.14$ (ft-kips).

12-10. $M_{AB} = -37.82$, $M_{BC} = -M_{BA} = 34.27$ (ft-kips).

Chapter 13

13-2. $M_A = M_D = 0$, $M_{BA} = -M_{BC} = 14.55$, $M_{CB} = -M_{CD} = 20.27$ (kN-m).

13-4. $M_{AB} = -3.71$, $M_{BA} = -M_{BC} = 8.56$, $M_{CB} = -M_{CD} = 28.50$, $M_D = 0$ (ft-kips).

13-6. See answers for Prob. 12-4.

13-8. $M_{AC} = -11.88$, $M_{CA} = 21.25$, $M_{BC} = 1.95$, $M_{CB} = 3.91$, $M_{EC} = 2.34$, $M_{CE} = 4.69$, $M_{CD} = -29.84$, $M_D = 0$ (kN-m).

13-10. $FEM_{BD} = -6.67$, $FEM_{DB} = 15.56$, $M_{DB} = 19.59$, $M_{BD} = 1.40$, $M_{BC} = -10.45$, $M_{BA} = 9.05$, $M_A = 0$ (ft-kips).

13-12. (*a*) Horizontal translation of BC; (*b*) $M_{AB} = 4.58$, $M_{BA} = -M_{BC} = 15.36$, $M_{CB} = -M_{CD} = 18.10$, $M_{DC} = -11.81$ (kN-m).

13-14. (*a*) One degree of translation described by horizontal movement of B and vertical movement of C; (*b*) $M_{AB} = 15.99$, $M_{BA} = -M_{BC} = 17.42$, $M_{CD} = -M_{CB} = 6.62$, $M_{DC} = 22.01$ (ft-kips).

13-16. See answers for Prob. 12-8.

13-18. (*a*) $C_{AB} = 0.344$; (*b*) $C_{BA} = 0.872$; (*c*) $M_A/\theta_A = 13.33EI_0/L$, where I_0 is the moment of inertia at B.

Chapter 14

14-2.

$$\begin{matrix} & D_2{}^2 & D_3{}^1 \end{matrix}$$

$$\mathbf{K} = \begin{bmatrix} 277.8 & 0 \\ 0 & 291.8 \end{bmatrix}$$

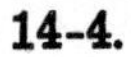

14-4.

$$\mathbf{K} = E \begin{bmatrix} 12.54 & 0 & 66.67 & 66.67 \\ 0 & 5.08 & -8.33 & 0 \\ 66.67 & -8.33 & 46.67 \times 10^4 & 6.67 \times 10^4 \\ 66.67 & 0 & 6.67 \times 10^4 & 13.33 \times 10^4 \end{bmatrix} \text{(mm)}.$$

(columns: D_2^1, D_2^2, D_2^3, D_1^3)

14-6. (*a*) $\mathbf{Q}' = [0 \ -6 \ -18 \ 0 \ -12 \ 6 \ 0 \ 12]$; (*b*) 1: 0 0 0 1 2 3; 2: 1 2 3 4 5 6; 3: 0 0 7 4 5 6; 4: 4 5 6 0 0 8; (*c*) $B = 6$; (*d*) and (*e*), partial results, $M_{i(1)} = -3.57$ ft-kips, $P_{j(1)} = -5.22$ kips, $M_{i(2)} = -10.78 + 18.00 = 7.22$ ft-kips, $M_{i(4)} = 7.71 + 12.00 = 19.71$ ft-kips, $M_{j(4)} = 12.00 - 12.00 = 0$.

14-8. Partial results: $D_{2V} = 0.30$ in.↓, $M_{21} = 311.1$ in.-kips, $M_{34} = -262.8$ in.-kips, $M_{43} = -998.6$ in.-kips, $P_{23} = -3.24$ kips.

14-10. Assuming a zero value for I_3:
(*a*) $\mathbf{Q}' = [-12 \ 0 \ -10 \ 6.667 \ 0 \ -4 \ 5.333 \ 0 \ 0 \ 0 \ -12 \ 0 \ 0]$; (*b*) 1: 0 0 1 2 3 4; 2: 2 3 4 5 6 7; 3: 2 3 4 8 9 10; 4: 8 9 10 5 6 7; 5: 11 12 13 8 9 10; 6: 0 0 0 11 12 13; (*c*) $B = 9$; (*d*) and (*e*) partial results, $M_{j(1)} = 19.90 - 12.00 = 7.90$ ft-kips, $M_{j(2)} = 4.95 - 5.33 = -0.38$ ft-kips, $M_{i(5)} = 11.55$ ft-kips, $P_{j(6)} = -13.34$ kips.

14-12. (*a*) Nodal point numbering, from left to right, roof first and proceeding downward: 7 8 10 11 13, 5 6 9 12 14, 3 4 15 16, 1 2 17 18; $B = 12$; (*b*) partial results, with i at the lower and left ends, $M_{i(1)} = 13.69 + 2.50 = 16.19$ ft-kips, $M_{j(1)} = 9.55 - 2.50 = 7.05$ ft-kips, $M_{i(4)} = 2.50 + 2.50 = 5.00$ ft-kips. $M_{j(4)} = 4.62 - 2.50 = 2.12$ ft-kips, $M_{i(7)} = 0.17 + 2.50 = 2.67$ ft-kips, $M_{j(7)} = 2.27 - 2.50 = -0.23$ ft-kips, $M_{i(3)} = -12.05$ ft-kips, $M_{j(3)} = -12.16$ ft-kips, $P_{j(1)} = 3.31$ kips, $D_7^1 = 0.167$ in.; (*c*) for member 4: $V = 0$ 2.63 ft below j, $M = 3.16$ ft-kips on the right side.

14-14. Partial results with the i at the lower and left ends, $M_{j(1)} = 27.41 - 16.67 = 10.74$ ft-kips, $M_{i(2)} = -13.15 + 4.17 = -8.98$ ft-kips, $M_{i(5)} = 71.48$ ft-kips, $P_{j(5)} = -7.15$ kips, $R_6 = 10.11$ kips↖.

14-16. Left to right on AC at 5 ft: (*a*) R_{DH} positive when acting to the right: 0, −0.0782, −0.0978, 0, 0.1155, 0.1320, 0.0825, 0; (*b*) M positive when acting counterclockwise on BD at B: 0, 0.626, 0.783, 0, −0.924, −1.056, −0.660, 0 (ft-kips).

14-18. Left to right, (*a*) R_B: 0 0.391 0.725 0.947 1.0 0.853 0.575 0.259 0 −0.131 −0.150 −0.094 0; (*b*) M_C: 0 −0.094 −0.150 −0.131 0 0.275 0.700, symmetric (kN-m).

14-20. Partial results, $M_{JG} = 54.01$, $M_{GJ} = 32.58$, $M_{GD} = 13.61$, $M_{DG} = 24.91$, $M_{DA} = 1.69$, $M_{AD} = 10.08$, $M_{KH} = 59.52$, M_{HK}

$= 43.99$, $M_{HE} = 29.98$, $M_{EH} = 36.46$, $M_{EB} = 8.35$, $M_{BE} = 16.30$, $M_{LI} = 51.54$, $M_{IL} = 28.36$, $M_{IF} = 8.73$, $M_{FI} = 21.30$, $M_{FC} = -0.04$, $M_{CF} = 8.62$ (kN-m).

Appendix A

A-2. (*b*) $[-11 \quad 1 \quad -5]$; (*d*) $[10 \quad -7 \quad -2 \quad -7]$;

(*f*) $\begin{bmatrix} 44 & -3 & 14 & 24 \\ -41 & 3 & 0 & -2 \\ -14 & 4 & 12 & 20 \\ -46 & 8 & 20 & 32 \end{bmatrix}$; (*h*) $[17 \quad -9 \quad 15]$.

A-4. **HF** $\neq$ **FH**.

A-6. $\mathbf{F}^{-1} = \begin{bmatrix} -1.75 & 1.25 & -0.25 \\ 0.75 & -0.25 & 0.25 \\ 2.25 & -1.75 & 0.75 \end{bmatrix}$

A-8. $\begin{bmatrix} -9 & 1 & 4 & 6 \\ -7 & 2 & 6 & 10 \\ 16 & -1 & 2 & 4 \end{bmatrix}$

A-10. -54.

A-12. $x_1 = 2.25$, $x_2 = 4.30$, $x_3 = -0.95$.

INDEX

Absolute maximum moment, 276–279
Absolute maximum shear, 276–279
Analysis:
  approximate, 113–126
  compared to design, 4
Approximate analysis, 113–126
Arch crown, 92, 93
Arch springings, 92, 93
Arches:
  hingeless, 92, 93
  three-hinged, 92–100
    analysis of, 94, 95
    graphical analysis of, 96–100
  two-hinged, 92, 93
Axial deformations:
  due to axial loading, 161
  due to temperature change, 161
Axial force:
  internal virtual work due to, 196
  internal work due to, 185
  member stiffness for, 236, 246
  sign convention for, 28, 29

Ball-and-socket support, 77, 78
Ball support, 78
Banded matrix, 387
Bandwidth, evaluation of, 388
Beam analogy for cables, 100–104
Bending:
  internal virtual work due to, 196
  internal work due to, 185
  (*See also* Moment)
Bending moment:
  relationship to load and shear, 36–38
  sign convention, 28, 29, 364, 365
Bending moment diagram, sign conventions for, 30–35
Bents, 16
Betti's theorem, 226, 227
Boundary conditions, representation of, 386, 387
Bow's notation, 66
Bridge live loads, 5, 7
Bridge trusses, types of, 2
Building live loads, 5, 6

Cable structures, 100–105
Cantilever method, 123–126
Cantilever parts, moment diagrams by, 141–145
Carryover factor, 330, 331
  for variable moment of inertia, 351–353
Castigliano's theorem, 210–216

Catenary, 104
Chord, member, 242
Code number, member, 380, 381
Code number technique, 379–382
Coefficient of thermal expansion, 161, 207, 391
Complete elimination method, 429
Composition of forces, 40
Conjugate-beam method, 147–152
Connection:
  pinned, 20
  rigid, 20, 111
Coordinate system:
  framework, 17, 174, 364
  global, 17, 364
  local, 17, 364
Cramer's rule, 424
Cross, Hardy, 330
Crown of an arch, 92, 93

Dead loads, 4, 5
Deflected shapes, 132–135
Deflections:
  by conjugate-beam method, 147–152
  by geometric methods, 132–167
  by moment-area method, 134–147
  by Newmark's method, 152–161
  by real work, 186–189
  due to temperature changes, 207–210
  by virtual work, 197–210
  by Williot-Mohr diagram, 161–167
Deformations, axial: due to axial loading, 161
    due to temperature change, 161
Design compared to analysis, 4
Determinacy, static: of plane trusses, 62
    of space trusses, 79
Determinants, 424–426
  cofactor of an element of, 425
  Laplace expansion of, 425
  minor of an element of, 424
  properties of, 426
Differential equation for bending, 135, 239
Direct stiffness method, 376
Displacement method, matrix, compared to matrix force method, 399
Displacement transformation matrix, 372–375, 389, 390, 412
Distribution factor, 333, 334
Dummy unit force, 197

Earthquake loads, 11–15
Energy:
  strain (*see* Internal work)
  virtual strain (*see* Internal virtual work)
Energy methods of computing deflections, 173–216
Equations, simultaneous, solution of, 426–429
Equilibrium equations, 26, 27
Equivalent concentrated forces, 383
Estimating shears, procedure for, 149–152

Finite element, plane stress, 409–413
Finite element method, definition of, 407
Finite elements, 363, 407
Fink truss, 2, 69, 73
Fixed-end moments:
  evaluation of, 304–308
  general expressions for, 308
Fixed support, 19
Flexibility influence coefficients, definition, 224
Flexibility matrix, 226
  member: due to axial force, 395
    due to bending, 243
  relationship to stiffness matrix, 233, 234
  of a structure, 394, 397
Floor system:
  influence lines for, 254–256
  typical, 254
Force method, matrix, compared to matrix displacement method, 399

Force polygon, 40, 41, 66–70, 97–100
Force transformation matrix, 396
Forces:
  composition of, 40
  redundant, 285
  resolution of, 40
Frame characteristics, 111–113
Frames:
  influence lines for, 252–254
  three-dimensional, analysis by matrix displacement method, 363
  typical, 3, 112
Framework coordinate system, 17, 174, 364
Free-body diagrams, 22–26
Funicular polygon, 42, 97–100

Gauss elimination method, 427
Gauss-Jordan method, 429
General method, 284
Geometric methods of computing deflections, 132–167
Girt, 16
Global coordinate system, 17, 364
Graphic analysis:
  of plane trusses, 65–70
  of three-hinged arches, 96–100
Graphic statics, 39–43
Graphical method for truss deflections, 161–167

Hingeless arch, 92, 93
Hooke's law, 39
Howe truss, 2

Impact factor, 11
Impact loads, 11
Inflection point, 115, 134
Influence coefficients:
  flexibility, definition of, 224
  force, 223, 224
  stiffness, definition of, 231
Influence lines:
  definition of, 249
  deflection, 259
  for floor systems, 254–256
  for frames, 252–254
  qualitative, for statically indeterminate structures, 263–266, 275, 276
  quantitative, for statically indeterminate structures, 392–394
  for simple beams, 250–252, 260–263
  for trusses, 255–258, 263
  use of: for concentrated loads, 266, 267
    for distributed loads, 267–269
    for moving loads, 269–275
Integration:
  numerical, 352
  visual, 201–203
Internal virtual work:
  due to axial force, 196
  due to bending, 196
  due to shear, 196
  due to temperature changes, 207, 208
  due to twist in a circular member, 196
Internal work:
  due to axial force, 185
  due to bending, 185
  due to shear, 185
  due to twist in a circular member, 185

Joints, method of, 54–58

Knee brace, 115

Lack of fit, included in matrix stiffness method, 391, 392
Live loads, 4–15
Load, relationship to shear and bending moment, 36–38

Loads:
  dead, 4, 5
  earthquake, 11–15
  impact, 11
  live, 4–15
    bridge, 5, 7
    building, 5, 6
  snow, 9–12
  wind, 7–10
Local coordinate system, 17–364

Matrices, types of, 417, 418
Matrix:
  adjoint, 430
  antisymmetric, 418
  augmented, 431
  banded, 387
  column, 417
  definition of, 416
  diagonal, 417
  differentiation of, 433
  displacement transformation, 372–375
  flexibility (*see* Flexibility matrix)
  force transformation, 396
  identity, 418
  integration of, 432
  inversion of, 429–432
  null, 418
  row, 417
  skew-symmetric, 418
  stiffness (*see* Stiffness matrix)
  symmetric, 418
  unit, 418
Matrix algebra, 418–424
Matrix displacement method compared to matrix force method, 399
Matrix force method compared to matrix displacement method, 399
Maxwell diagram, 67, 68
Maxwell's reciprocal theorem, 227, 228, 393
Mill bent, 114–117
Moment:
  absolute maximum, 276–279

Moment:
  of inertia, variable: carryover factor for, 351–353
    stiffness factor for, 351–353
  secondary, 53
Moment-area method, 134–147
Moment diagrams, 30–35
  by cantilever parts, 141–145
Müller-Breslau principle, 259, 260, 392

Newmark's method, 152–161
Newton, definition of, 450
Nodal point, 362, 407
Numerical integration, 352, 353
  Simpson's one-third rule for, 352
  trapezoidal rule for, 352

Panel, 255
Pascal, definition of, 451
Pinned connection, 20
Pinned support, 19
Plane trusses (*see* Trusses, plane)
Pole, 41
Polygon:
  force, 40, 41, 66–70, 97–100
  funicular, 42, 97–100
Portal, 117, 118
Portal method, 119–123
Pratt truss, 2
Primary structure, 285
Prismatic member, 304
Purlins, 16

Rays, 41
Real work, 174–189
  deflections by, 186–189
  internal, forms of (*see* Internal work)
Reciprocal theorem, Maxwell's, 227, 228, 393
Redundant force, 285
Resolution of a force, 40
Right-hand-screw rule, 28
Rigid connection, 20, 111

Roller support, 19
  for space truss, 78
Roof trusses, types of, 2

Secondary moments, 53
Sections, method of, 58–61
Settlement, support, 289
Shear:
  absolute maximum, 276–279
  internal virtual work due to, 196
  internal work due to, 185
  relationship to load and bending moment, 36–38
Shear diagrams, 30–35
Shear sign convention, 28, 29
Shear wall, 16, 17, 363
Shell structures, 3, 4
SI units:
  conversion factors for, 452, 453
  definitions of, 449–451
Sign convention:
  for axial force, 28, 29
  for bending moment, 28, 29
  for bending moment diagrams, 30–35
  for matrix displacement method, 364, 365
  for moment-distribution method, 331
  for shear, 28, 29
  for shear diagrams, 30–35
  for slope-deflection method, 317
  for twisting moment, 28, 29
Simpson's one-third rule for numerical integration, 352
Simultaneous equations, solution of, 426–429
Slope-deflection equations, 316
  simplified equation, 317
Slope-deflection method, sign convention for, 317
Snow loads, 9–12
Space diagram, 40
Springings of an arch, 92, 93
Stability, geometric, of plane trusses, 61–65
Static determinacy:
  of plane trusses, 62
  of space trusses, 79
Statics, graphic, 39–43
Stations, 156
Stiffness factor, 331, 332, 341, 342
  absolute, 331
  modified, 332
  for variable moment of inertia, 351, 353
Stiffness influence coefficients, definition, 231
Stiffness matrix, 233
  member: due to axial force, 236, 246
    due to bending, 239, 243
    due to twist, 244
    general, 245, 246
    transformed, 368–376
  relationship to flexibility matrix, 233, 234
  of a structure, 362, 366, 376, 377, 390
Stiffness method, matrix, compared to matrix force method, 399
Strain energy (*see* Internal work)
  virtual (*see* Internal virtual work)
Stringer, 15, 16
Strings, 42
Structural components, names of, 15–17
Superposition, principle of, 38, 39
Support:
  ball, 78
  ball-and-socket, 73, 74
  fixed, 19
  pinned, 19
  roller, 19
    for space truss, 74
Support settlements, 289

Temperature changes:
  axial deformations due to, 161
  considered in matrix stiffness method, 391
  deflections due to, 207–210
  internal virtual work due to, 207, 208

Thermal expansion, coefficient of, 161, 207, 391
Three-dimensional trusses (*see* Trusses, space)
Three-hinged arch, 92–100
Three-moment equation, 301
Torsional constant, 246
Trapezoidal rule for numerical integration, 352
Truck loading, standard, 272, 273
Trusses:
  bridge, types of, 2
  deflections by graphical methods, 161–167
  Fink, 2, 69, 73
  Howe, 2
  influence lines for, 255–258, 263
  plane, 52–70
    geometric stability of, 61–65
    graphical analysis of, 65–70
    method of joints, 54–58
    method of sections, 58–61
    notation for, 54
    static determinacy of, 61–65
  Pratt, 2
  roof, types of, 2
  space, 77–78
    forms of support for, 77, 78
    special theorems for, 79
  Warren, 2
Twist in a circular member:
  internal virtual work due to, 196
  internal work due to, 185
Twisting moment sign convention, 28, 29
Two-hinged arch, 92, 93

U.S. Customary units, conversion factors for, 452, 453

Variable moment of inertia, moment-distribution properties for, 351–353
Vectors, use of:
  plane truss analysis, 56, 57, 63, 64
    space truss analysis, 82, 86, 87
Virtual work, 189–210
  deflections by, 197–210
  internal, forms of (*see* Internal virtual work)
  principle of, 190
Visual integration, 201–203

Warren truss, 2
Williot diagram, 163, 164
Williot-Mohr diagrams, 161–167
Wind loads, 7–10
Work:
  real (*see* Real work)
  virtual (*see* Virtual work)